PROBLEMS & SOLUTIONS IN THEORETICAL & MATHEMATICAL PHYSICS

Problems and Solutions
in
Theoretical and Mathematical Physics

Volume II:
Advanced Level

3nd extended and revised edition

by
Willi-Hans Steeb
International School for Scientific Computing

in collaboration with
Yorick Hardy
International School for Scientific Computing

PROBLEMS & SOLUTIONS IN THEORETICAL & MATHEMATICAL PHYSICS

Third Edition

Volume II: Advanced Level

Willi-Hans Steeb
University of Johannesburg
South Africa

 **World Scientific**

NEW JERSEY · LONDON · SINGAPORE · BEIJING · SHANGHAI · HONG KONG · TAIPEI · CHENNAI

Published by

World Scientific Publishing Co. Pte. Ltd.

5 Toh Tuck Link, Singapore 596224

USA office: 27 Warren Street, Suite 401-402, Hackensack, NJ 07601

UK office: 57 Shelton Street, Covent Garden, London WC2H 9HE

British Library Cataloguing-in-Publication Data
A catalogue record for this book is available from the British Library.

ISBN-13 978-981-4282-16-1
ISBN-10 981-4282-16-2
ISBN-13 978-981-4282-17-8 (pbk)
ISBN-10 981-4282-17-0 (pbk)

Printed in Singapore.

Preface

The purpose of this book is to supply a collection of problems together with their detailed solution which will prove to be valuable to students as well as to research workers in the fields of mathematics, physics, engineering and other sciences. The topics range in difficulty from elementary to advanced. Almost all problems are solved in detail and most of the problems are self-contained. All relevant definitions are given. Students can learn important principles and strategies required for problem solving. Teachers will also find this text useful as a supplement, since important concepts and techniques are developed in the problems. The material was tested in my lectures given around the world.

The book is divided into two volumes. Volume I presents the introductory problems for undergraduate and advanced undergraduate students. In volume II the more advanced problems together with their detailed solutions are collected, to meet the needs of graduate students and researchers. Problems included cover most of the new fields in theoretical and mathematical physics such as Lax representation, Bäcklund transformation, soliton equations, Hilbert space theory, Lie algebra valued differential forms, Hirota technique, Painlevé test, the Bethe ansatz, the Yang-Baxter relation, wavelets, gauge theory, string theory, chaos, fractals, complexity, etc.

In the reference section some other books are listed having useful problems for students in theoretical physics and mathematical physics. Some or related problems to those given in volume I can be found in these books. In volume II references are given to books and original articles where some of the advanced problems can be found.

I wish to express my gratitude to Dr. Yorick Hardy for a critical reading of the manuscripts.

Any useful suggestions and comments are welcome.

Email addresses of the author:
steebwilli@gmail.com
steeb_wh@yahoo.com

Home page of the author:
http://issc.uj.ac.za

Contents

Notation

$\emptyset$	empty set
$A \subset B$	subset A of set B
$A \cap B$	the intersection of the sets A and B
$A \cup B$	the union of the sets A and B
$\mathbb{N}$	natural numbers
$\mathbb{Z}$	integers
$\mathbb{Q}$	rational numbers
$\mathbb{R}$	real numbers
$\mathbb{R}^+$	nonnegative real numbers
$\mathbb{C}$	complex numbers
$\mathbb{R}^n$	n-dimensional Euclidian space
$\mathbb{C}^n$	n-dimensional complex linear space
i	$:= \sqrt{-1}$
$\Re z$	real part of the complex number z
$\Im z$	imaginary part of the complex number z
$\mathbf{x} \in \mathbb{R}^n$	element $\mathbf{x}$ of $\mathbf{R}^n$ (column vector)
$\mathbf{0}$	zero vector (column vector)
$f \circ g$	composition of two mappings $(f \circ g)(x) = f(g(x))$
u	dependent variable
t	independent variable (time variable)
x	independent variable (space variable)
$\mathbf{x}^T = (x_1, x_2, \ldots, x_n)$	vector of independent variables, ^{T} means transpose
$\mathbf{u}^T = (u_1, u_2, \ldots, u_n)$	vector of dependent variables, ^{T} means transpose
$\|\cdot\|$	norm
$\mathbf{x} \cdot \mathbf{y} \equiv \mathbf{x}^T \mathbf{y}$	scalar product (inner product) in vector space $\mathbb{R}^n$
$\mathbf{x} \times \mathbf{y}$	vector product in vector space $\mathbb{R}^3$

det	determinant of a square matrix
tr	trace of a square matrix
0_n	$n \times n$ zero matrix
I_n	$n \times n$ unit matrix (identity matrix)
I	identity operator
$[\,,\,]$	commutator
$[\,,\,]_+$	anticommutator
δ_{jk}	Kronecker delta with $\delta_{jk} = 1$ for $j = k$ and $\delta_{jk} = 0$ for $j \neq k$
$\epsilon_{jk\ell}$	total antisymmetric tensor $\epsilon_{123} = 1$
$\text{sgn}(x)$	the sign of x, 1 if $x > 0$, -1 if $x < 0$, 0 if $x = 0$
λ	eigenvalue
ϵ	real parameter
$\otimes$	Kronecker product, tensor product
$\wedge$	Grassmann product (exterior product, wedge product)
L	Lagrange function
H	Hamilton function
$\mathcal{L}$	Lagrange density
$\hat{H}$	Hamilton operator
$\mathcal{H}$	Hilbert space
$L_2(\Omega)$	Hilbert space of square integrable functions
$\langle\,,\,\rangle$	scalar product in Hilbert space
δ	delta function

Chapter 1

Groups

Let G be a finite group. A group element h is said to be *conujugate* to the group element k, $h \sim k$, if there exists a $g \in G$ such that $k = ghg^{-1}$. For G the number of conjugacy classes is equal to the number of irreducible matrix representations. In a *character table* the rows correspond to irreducible group representations and columns to classes of group elements. If the irreducible matrix representation is given by an $n \times n$ matrix $(n \geq 2)$ then the trace of this square matrix is the element in the table.

Problem 1. Consider the 4×4 matrix

$$A = \begin{pmatrix} 0 & 0 & 1 & 0 \\ 0 & 0 & 0 & 1 \\ -1 & 0 & 0 & 0 \\ 0 & -1 & 0 & 0 \end{pmatrix}.$$

Obviously $\mathrm{rank}(A) = 4$ and therefore the inverse exists.
(i) Find the inverse of A.
(ii) Does the set $\{\, A, A^{-1}, I_4 \,\}$ form a group under matrix multiplication? If not can we find a finite extension to the set to obtain a group?

Solution 1. (i) We find

$$A^{-1} = \begin{pmatrix} 0 & 0 & -1 & 0 \\ 0 & 0 & 0 & -1 \\ 1 & 0 & 0 & 0 \\ 0 & 1 & 0 & 0 \end{pmatrix} = -A.$$

1

(ii) Since
$$A^2 = (A^{-1})^2 = -I_4$$

and $A^{-1} = -A$ we find that the set

$$\{\, A,\, -A,\, I_4,\, -I_4 \,\}$$

forms a group under matrix multiplication.

Problem 2.   Let D_n be the *dihedral group*. This is the group of rigid motions of an n-gon ($n \geq 3$)

$$D_n := \{\, a, b \,|\, a^n = b^2 = 1,\, ba = a^{-1}b \,\} \tag{1}$$

where a is a rotation by $2\pi/n$, b is a flip and 1 is the identity element. It is a noncommutative group of order $2n$. Find the *centre* of D_n. The *centre* C is defined as

$$C := \{\, c \in D_n \,|\, cx = xc \text{ for all } x \in D_n \,\}.$$

Solution 2.   To determine the centre of the group D_n it suffices to find those elements which commute with the generators a and b. Since $n \geq 3$, we have

$$a^{-1} \neq a.$$

Therefore
$$a^{r+1}b = a(a^r b) = (a^r b)a = a^{r-1}b.$$

It follows that $a^2 = 1$ a contradiction. Thus, no element of the form $a^r b$ is in the centre. Analogously, if for $1 \leq s < n$

$$a^s b = ba^s = a^{-s}b$$

then $a^{2s} = 1$, which is only possible if $2s = n$. Hence, a^s commutes with b if and only if $n = 2s$. Thus, if $n = 2s$ the centre of D_n is $\{\, 1, a^s \,\}$. If n is odd the centre is $\{\, 1 \,\}$.

Problem 3.   Consider the three 2×2 matrices

$$A = \begin{pmatrix} 2 & 1 \\ 1 & 2 \end{pmatrix}, \qquad I_2 = \begin{pmatrix} 1 & 0 \\ 0 & 1 \end{pmatrix}, \qquad C = \begin{pmatrix} 0 & 1 \\ 1 & 0 \end{pmatrix}.$$

(i) Show that
$$[A, I_2] = 0, \qquad [A, C] = 0.$$

(ii) Show that the matrices I_2 and C form a group under the matrix multiplication. Find the conjugacy classes.

(iii) Find the irreducible representations of I_2 and C. Give the *character table*.

(iv) Find the projection matrices from the character table. Use the projection matrices to find the invariant subspaces. Give the eigenvalues and eigenvectors of A.

Solution 3. (i) Calculating the commutators provides

$$[A, I_2] = AI_2 - I_2A = A - A = 0_2$$
$$[A, C] = AC - CA = \begin{pmatrix} 1 & 2 \\ 2 & 1 \end{pmatrix} - \begin{pmatrix} 1 & 2 \\ 2 & 1 \end{pmatrix} = 0_2,$$

where 0_2 is the 2×2 zero matrix.

(ii) We find

$$I_2I_2 = I_2, \qquad I_2C = C, \qquad CI_2 = C, \qquad CC = I_2.$$

Obviously the group properties are satisfied with $C^{-1} = C$. Since

$$I_2I_2I_2 = I_2, \qquad CI_2C = I_2, \qquad I_2CI_2 = C, \qquad CCC = C$$

we obtain the two classes $\{I_2\}$ and $\{C\}$.

(iii) Obviously, $I_2 \to 1$, $C \to 1$ and $I_2 \to 1$ $C \to -1$ are representations. The second one is a faithful representation. Since the representations are one-dimensional they are irreducible. The number of irreducible representations is equal to the number of classes. Taking into account the result from (ii) we find that the character table is given by

	$\{I_2\}$	$\{C\}$
A_1	1	1
A_2	1	-1

(iv) For the representation A_1 we find the projection matrix

$$\Pi_1 = \frac{1}{2}\sum_{g \in G} \chi_1(g^{-1})g = \frac{1}{2}\sum_{g \in G} \chi_1(g)g = \frac{1}{2}(I_2 + C) = \frac{1}{2}\begin{pmatrix} 1 & 1 \\ 1 & 1 \end{pmatrix}.$$

Analogously

$$\Pi_2 = \frac{1}{2}(I_2 - C) = \frac{1}{2}\begin{pmatrix} 1 & -1 \\ -1 & 1 \end{pmatrix}.$$

Applying the projection matrices to the standard basis in $\mathbb{R}^2$ gives

$$\Pi_1\begin{pmatrix} 1 \\ 0 \end{pmatrix} = \frac{1}{2}\begin{pmatrix} 1 \\ 1 \end{pmatrix}, \qquad \Pi_1\begin{pmatrix} 0 \\ 1 \end{pmatrix} = \frac{1}{2}\begin{pmatrix} 1 \\ 1 \end{pmatrix},$$

$$\Pi_2\begin{pmatrix} 1 \\ 0 \end{pmatrix} = \frac{1}{2}\begin{pmatrix} 1 \\ -1 \end{pmatrix}, \qquad \Pi_2\begin{pmatrix} 0 \\ 1 \end{pmatrix} = \frac{1}{2}\begin{pmatrix} -1 \\ 1 \end{pmatrix}.$$

When we normalize the right hand side we obtain the new basis

$$\frac{1}{\sqrt{2}}\begin{pmatrix} 1 \\ 1 \end{pmatrix}, \quad \frac{1}{\sqrt{2}}\begin{pmatrix} 1 \\ -1 \end{pmatrix}$$

which is called the *Hadamard basis*. In this new basis the matrix A takes the form

$$\tilde{A} = \begin{pmatrix} 3 & 0 \\ 0 & 1 \end{pmatrix}.$$

It follows that the eigenvalues of the matrix A are given by 3 and 1 and the normalized eigenvectors of A are given by the Hadamard basis.

Problem 4. (i) Consider the three 4×4 permutation matrices

$$U_1 = \begin{pmatrix} 0 & 0 & 0 & 1 \\ 0 & 0 & 1 & 0 \\ 0 & 1 & 0 & 0 \\ 1 & 0 & 0 & 0 \end{pmatrix}, \quad U_2 = \begin{pmatrix} 0 & 1 & 0 & 0 \\ 1 & 0 & 0 & 0 \\ 0 & 0 & 0 & 1 \\ 0 & 0 & 1 & 0 \end{pmatrix}, \quad U_3 = \begin{pmatrix} 0 & 0 & 1 & 0 \\ 0 & 0 & 0 & 1 \\ 1 & 0 & 0 & 0 \\ 0 & 1 & 0 & 0 \end{pmatrix}.$$

Show that the set

$$\{ I_4, U_1, U_2, U_3 \}$$

forms a group under matrix multiplication. Is the group commutative? Find the classes. Find the irreducible representations. Give the character table.

(ii) Let

$$A_1 = \begin{pmatrix} 0 & 1 & 1 & 0 \\ 1 & 0 & 0 & 1 \\ 1 & 0 & 0 & 1 \\ 0 & 1 & 1 & 0 \end{pmatrix}, \quad A_2 = \begin{pmatrix} 0 & -1 & -1 & 0 \\ -1 & 0 & 0 & -1 \\ -1 & 0 & 0 & -1 \\ 0 & -1 & -1 & 0 \end{pmatrix}$$

$$A_3 = \begin{pmatrix} 0 & -1 & 1 & 0 \\ -1 & 0 & 0 & 1 \\ 1 & 0 & 0 & -1 \\ 0 & 1 & -1 & 0 \end{pmatrix}, \quad A_4 = \begin{pmatrix} 0 & 1 & -1 & 0 \\ 1 & 0 & 0 & -1 \\ -1 & 0 & 0 & 1 \\ 0 & -1 & 1 & 0 \end{pmatrix}.$$

Show that

$$U_j A_k U_j = A_k$$

for $j = 1, 2, 3$ and $k = 1, 2, 3, 4$. Find the eigenvalues and eigenvectors of A_k.

Solution 4. (i) We find

$$I_4 U_1 = U_1, \quad I_4 U_2 = U_2, \quad I_4 U_3 = U_3, \quad U_1 U_2 = U_3$$

$$U_2 U_1 = U_3, \quad U_1 U_3 = U_2, \quad U_3 U_1 = U_2, \quad U_2 U_3 = U_1$$

$$U_3 U_2 = U_1, \qquad U_1 U_1 = I_4, \qquad U_2 U_2 = I_4, \qquad U_3 U_3 = I_4.$$

Thus the group is commutative. Since

$$\begin{array}{llll}
I_4 I_4 I_4 = I_4, & U_1 I_4 U_1 = I_4, & U_2 I_4 U_2 = I_4, & U_3 I_4 U_3 = I_4 \\
I_4 U_1 I_4 = U_1, & U_1 U_1 U_1 = U_1, & U_2 U_1 U_2 = U_1, & U_3 U_1 U_3 = U_1 \\
I_4 U_2 I_4 = U_2, & U_1 U_2 U_1 = U_2, & U_2 U_2 U_2 = U_2, & U_3 U_2 U_3 = U_2 \\
I_4 U_3 I_4 = U_3, & U_1 U_3 U_1 = U_3, & U_2 U_3 U_2 = U_3, & U_3 U_3 U_3 = U_3
\end{array}$$

we obtain four classes, namely

$$\{\{I_4\}, \{U_1\}, \{U_2\}, \{U_3\}\}.$$

Thus we have four one-dimensional irreducible representations. We find

$$\begin{array}{lllll}
\Gamma_1 : & I_4 \mapsto 1, & U_1 \mapsto 1, & U_2 \mapsto 1, & U_3 \mapsto 1 \\
\Gamma_2 : & I_4 \mapsto 1, & U_1 \mapsto -1 & U_2 \mapsto -1, & U_3 \mapsto 1 \\
\Gamma_3 : & I_4 \mapsto 1, & U_1 \mapsto -1, & U_2 \mapsto 1 & U_3 \mapsto -1 \\
\Gamma_4 : & I_4 \mapsto 1, & U_1 \mapsto 1 & U_2 \mapsto -1 & U_3 \mapsto -1.
\end{array}$$

Consequently, the character table is given by

	$\{I_4\}$	$\{U_1\}$	$\{U_2\}$	$\{U_3\}$
Γ_1	1	1	1	1
Γ_2	1	-1	-1	1
Γ_3	1	-1	1	-1
Γ_4	1	1	-1	-1

(ii) From the representation Γ_1 we find the projection matrix

$$\Pi_1 = \frac{1}{4} \sum_{g \in G} \chi_1(g^{-1}) g = \frac{1}{4}(I_4 + U_1 + U_2 + U_3) = \frac{1}{4} \begin{pmatrix} 1 & 1 & 1 & 1 \\ 1 & 1 & 1 & 1 \\ 1 & 1 & 1 & 1 \\ 1 & 1 & 1 & 1 \end{pmatrix}$$

where we have used that $g^{-1} = g$ for all $g \in G$. Analogously

$$\Pi_2 = \frac{1}{4}(I_4 - U_1 - U_2 + U_3)$$

$$\Pi_3 = \frac{1}{4}(I_4 - U_1 + U_2 - U_3)$$

$$\Pi_4 = \frac{1}{4}(I_4 + U_1 - U_2 - U_3).$$

Applying the projection matrix Π_1 to the standard basis in $\mathbb{R}^4$ gives

$$\Pi_1 \begin{pmatrix} 1 \\ 0 \\ 0 \\ 0 \end{pmatrix} = \frac{1}{4}\begin{pmatrix} 1 \\ 1 \\ 1 \\ 1 \end{pmatrix}, \qquad \Pi_1 \begin{pmatrix} 0 \\ 1 \\ 0 \\ 0 \end{pmatrix} = \frac{1}{4}\begin{pmatrix} 1 \\ 1 \\ 1 \\ 1 \end{pmatrix}$$

$$\Pi_1 \begin{pmatrix} 0 \\ 0 \\ 1 \\ 0 \end{pmatrix} = \frac{1}{4}\begin{pmatrix} 1 \\ 1 \\ 1 \\ 1 \end{pmatrix}, \qquad \Pi_1 \begin{pmatrix} 0 \\ 0 \\ 0 \\ 1 \end{pmatrix} = \frac{1}{4}\begin{pmatrix} 1 \\ 1 \\ 1 \\ 1 \end{pmatrix}.$$

Analogously, we apply the other projection operators to the standard basis. Then the new orthonormal basis is given by

$$\frac{1}{2}\begin{pmatrix} 1 \\ 1 \\ 1 \\ 1 \end{pmatrix}, \qquad \frac{1}{2}\begin{pmatrix} 1 \\ -1 \\ 1 \\ -1 \end{pmatrix}, \qquad \frac{1}{2}\begin{pmatrix} 1 \\ 1 \\ -1 \\ -1 \end{pmatrix}, \qquad \frac{1}{2}\begin{pmatrix} 1 \\ -1 \\ -1 \\ 1 \end{pmatrix}.$$

In this new basis the matrices A_j take the diagonal form

$$\begin{pmatrix} 2 & 0 & 0 & 0 \\ 0 & 0 & 0 & 0 \\ 0 & 0 & 0 & 0 \\ 0 & 0 & 0 & -2 \end{pmatrix}, \qquad \begin{pmatrix} -2 & 0 & 0 & 0 \\ 0 & 0 & 0 & 0 \\ 0 & 0 & 0 & 0 \\ 0 & 0 & 0 & 2 \end{pmatrix}$$

$$\begin{pmatrix} 0 & 0 & 0 & 0 \\ 0 & 2 & 0 & 0 \\ 0 & 0 & -2 & 0 \\ 0 & 0 & 0 & 0 \end{pmatrix}, \qquad \begin{pmatrix} 0 & 0 & 0 & 0 \\ 0 & -2 & 0 & 0 \\ 0 & 0 & 2 & 0 \\ 0 & 0 & 0 & 0 \end{pmatrix}.$$

Consequently, the eigenvalues of the matrices A_k are given by 2, -2, 0, 0 and the normalized eigenvectors are given above. We have $A_1 = U_2 + U_3$, $A_2 = -A_1$, $A_3 = -U_2 + U_3$, $A_4 = -A_3$.

Problem 5.   (i) The mapping ($z \in \mathbb{C}$)

$$w(z) = \frac{az + b}{cz + d}, \qquad ad - bc \neq 0 \tag{1}$$

is called the *fractional linear transformation*, where a, b, c, $d \in \mathbb{R}$. Show that these transformations form a group under the composition of mappings.
(ii) The *Schwarzian derivative* of a function f is defined as

$$D(f)_z := \frac{2f'(z)f'''(z) - 3[f''(z)]^2}{2[f''(z)]^2} \tag{2}$$

where $f'(z) \equiv df/dz$. Calculate $D(w)_z$.

Solution 5. (i) First we have to show that the transformation is closed under the composition. Let

$$W(w) = \frac{Aw + B}{Cw + D}, \qquad AD - BC \neq 0. \tag{3}$$

Inserting (1) yields

$$W(w(z)) = \frac{A(az+b) + B(cz+d)}{C(az+b) + D(cz+d)} = \frac{(Aa + Bc)z + (Ab + Bd)}{(Ca + Dc)z + (Cb + Dd)}. \tag{4}$$

Since $ad - bc \neq 0$ and $AD - BC \neq 0$ we find that

$$(Aa + Bc)(Cb + Dd) - (Ab + Bd)(Ca + Dc) \equiv (AD - BC)(ad - bc) \neq 0.$$

Thus mapping (1) is closed under composition. For $a = 1$, $b = 0$, $c = 0$, $d = 1$ we find the unit element $w(z) = z$. The inverse of mapping (1) is given by

$$z(w) = \frac{-dw + b}{cw - a}$$

since

$$
\begin{aligned}
z(w(z)) &= \frac{(-d(az+b) + b(cz+d))(cz+d)}{(cz+d)(c(az+b) - a(cz+d))} \\
&= \frac{-adz - db + bd + bcz}{bc - ad + acz - acz} \\
&= \frac{(-ad + bc)z}{(-ad + bc)} \\
&= z.
\end{aligned}
$$

Finally, it can be proved that the associative law is satisfied.
(ii) We find

$$D(w)_z = 0.$$

This means $w(z)$ is the general solution of the ordinary differential equation

$$D(f)_z = 0$$

since the solution includes three constants of integration.

Problem 6. Let P be the *parity operator*, i.e.

$$P\mathbf{r} := -\mathbf{r}. \tag{1}$$

Obviously, $P = P^{-1}$. We define

$$\mathbf{O}_P u(\mathbf{r}) := u(P^{-1}\mathbf{r}) \equiv u(-\mathbf{r}). \tag{2}$$

Now $\mathbf{r}$ can be expressed in spherical coordinates

$$\mathbf{r} = r(\sin\theta\cos\phi, \sin\theta\sin\phi, \cos\theta) \tag{3}$$

where $0 \le \phi < 2\pi$, $0 \le \theta < \pi$ and $0 \le r < \infty$.
(i) Calculate $P(r, \theta, \phi)$.
(ii) Consider the *spherical harmonics* defined by

$$Y_{lm}(\theta, \phi) := \frac{(-1)^{l+m}}{2^l l!} \left(\frac{2l+1}{4\pi} \frac{(l-m)!}{(l+m)!} \right)^{1/2}$$

$$\times (\sin\theta)^m \frac{d^{l+m}}{d(\cos\theta)^{l+m}} (\sin\theta)^{2l} e^{im\phi}$$

where $l = 0, 1, 2, \ldots$ and $m = -l, -l+1, \ldots, +l$. Find $\mathbf{O}_P Y_{lm}$.
(iii) Calculate

$$(\mathbf{O}_P \hat{L}_z) Y_{lm}, \qquad (\hat{L}_z \mathbf{O}_P) Y_{lm}$$

where

$$\hat{L}_z := -i\hbar \frac{\partial}{\partial\phi}.$$

Use these results to find the commutator $[\hat{L}_z, \mathbf{O}_P] Y_{lm}$.

Solution 6.   (i) From (1) we find that

$$P(r, \theta, \phi) = (r, \pi - \theta, \pi + \phi)$$

because the identities $\sin(\pi - \theta) \equiv \sin\theta$, $\cos(\pi - \theta) \equiv -\cos\theta$, $\sin(\pi + \phi) \equiv -\sin\phi$, $\cos(\pi + \phi) \equiv -\cos\phi$.
(ii) We find

$$\mathbf{O}_P Y_{lm}(\theta, \phi) = Y_{lm}(\pi - \theta, \pi + \phi) = (-1)^l Y_{lm}(\theta, \phi).$$

(iii) Since

$$\hat{L}_z Y_{lm} = m\hbar Y_{lm}$$

we obtain

$$\mathbf{O}_P \hat{L}_z Y_{lm} = \mathbf{O}_P(\hat{L}_z Y_{lm}) = m\hbar \mathbf{O}_P Y_{lm} = m\hbar(-1)^l Y_{lm}$$

and

$$(\hat{L}_z \mathbf{O}_P) Y_{lm} = \hat{L}_z(\mathbf{O}_P Y_{lm}) = (-1)^l \hat{L}_z Y_{lm} = (-1)^l m\hbar Y_{lm}.$$

Consequently, the commutator is given by

$$[\hat{L}_z, \mathbf{O}_P] Y_{lm} = 0.$$

Chapter 2

Kronecker and Tensor Products

Let A be an $m \times n$ matrix and let B be a $p \times q$ matrix. The *Kronecker product* of A and B is defined as

$$A \otimes B := \begin{pmatrix} a_{11}B & a_{12}B & \cdots & a_{1n}B \\ a_{21}B & a_{22}B & \cdots & a_{2n}B \\ \vdots & \vdots & \ddots & \vdots \\ a_{m1}B & a_{m2}B & \cdots & a_{mn}B \end{pmatrix}.$$

Thus $A \otimes B$ is an $mp \times nq$ matrix.

Assume that the matrix A is of order $m \times n$, B of order $p \times q$, C of order $n \times r$ and D of order $q \times s$. Then

$$(A \otimes B)(C \otimes D) = (AC) \otimes (BD).$$

Problem 1. Let A be an $m \times m$ matrix and let B be an $n \times n$ matrix. Show that

$$\mathrm{tr}(A \otimes B) \equiv (\mathrm{tr}A)(\mathrm{tr}B). \tag{1}$$

Solution 1. From the definition of the Kronecker product we find

$$\mathrm{diag}(a_{jj}B) = (a_{jj}b_{11}, a_{jj}b_{22}, \ldots, a_{jj}b_{nn}).$$

9

Therefore

$$\text{tr}(A \otimes B) = \sum_{j=1}^{m} \sum_{k=1}^{n} a_{jj} b_{kk}.$$

Since

$$\text{tr} A = \sum_{j=1}^{m} a_{jj}, \qquad \text{tr} B = \sum_{k=1}^{n} b_{kk}$$

we find (1).

Problem 2.   Let P and Q be two $n \times n$ matrices and I_n the $n \times n$ unit matrix. Assume that $P^2 = P$ and $Q^2 = Q$.
(i) Show that

$$(P \otimes Q)^2 = P \otimes Q.$$

(ii) Show that

$$(P \otimes (I_n - P))^2 = P \otimes (I_n - P).$$

(iii) Assume that $QP = 0$. Calculate $(Q \otimes Q)(P \otimes P)$.

Solution 2.   (i) We have

$$(P \otimes Q)(P \otimes Q) = (P^2 \otimes Q^2) = P \otimes Q.$$

(ii) We obtain

$$(P \otimes (I_n - P))(P \otimes (I_n - P)) = (P^2 \otimes (I_n - P)^2) = P \otimes (I_n - P - P + P^2).$$

Thus

$$(P \otimes (I_n - P))(P \otimes (I_n - P)) = P \otimes (I_n - P - P + P) = P \otimes (I_n - P).$$

(iii) We find

$$(Q \otimes Q)(P \otimes P) = (QP) \otimes (QP) = 0.$$

The equation $(P \otimes Q)^2 = P \otimes Q$ can be extended to N-factors. Let $P_1, \ldots, P_N$ be matrices with $P_j^2 = P_j$, where $j = 1, 2, \ldots, N$. Then

$$(P_1 \otimes P_2 \otimes \cdots \otimes P_N)^2 = P_1 \otimes P_2 \otimes \cdots \otimes P_N.$$

Problem 3.   (i) Let A be an $n \times n$ matrix and B be an $m \times m$ matrix. I_n is the $n \times n$ unit matrix and I_m the $m \times m$ unit matrix. Calculate the commutator

$$[A \otimes I_m, I_n \otimes B]. \tag{1}$$

(ii) Let A, B, C, D be $n \times n$ matrices. Assume that

$$[A, B] = 0_n, \qquad [C, D] = 0_n.$$

Calculate the commutator

$$[A \otimes C, B \otimes D].$$

Solution 3. (i) Straightforward calculation yields

$$
\begin{aligned}
[A \otimes I_m, I_n \otimes B] &= (A \otimes I_m)(I_n \otimes B) - (I_n \otimes B)(A \otimes I_m) \\
&= A \otimes B - A \otimes B \\
&= 0_{n \cdot m}.
\end{aligned}
$$

Consequently, the matrices $A \otimes I_m$ and $I_n \otimes B$ commute.
(ii) We find

$$
\begin{aligned}
[A \otimes C, B \otimes D] &= (A \otimes C)(B \otimes D) - (B \otimes D)(A \otimes C) \\
&= (AB) \otimes (CD) - (BA) \otimes (DC) \\
&= (AB) \otimes (CD) - (AB) \otimes (CD) \\
&= 0_{n \cdot m}
\end{aligned}
$$

where we have used (1).

Problem 4. (i) Let A be an $m \times m$ matrix with eigenvalue λ and corresponding eigenvector $\mathbf{a}$. Let B be an $n \times n$ matrix with eigenvalue μ corresponding eigenvector $\mathbf{b}$. Show that the matrix $A \otimes B$ has the eigenvalue $\lambda\mu$ with the corresponding eigenvector $\mathbf{a} \otimes \mathbf{b}$.
(ii) Consider the *Pauli spin matrices*

$$
\sigma_y := \begin{pmatrix} 0 & -i \\ i & 0 \end{pmatrix}, \qquad \sigma_z := \begin{pmatrix} 1 & 0 \\ 0 & -1 \end{pmatrix}.
$$

Find the eigenvalues of $\sigma_y \otimes \sigma_z$.

Solution 4. (i) From the eigenvalue equations

$$A\mathbf{a} = \lambda\mathbf{a}, \qquad B\mathbf{b} = \mu\mathbf{b}$$

we find

$$(A\mathbf{a}) \otimes (B\mathbf{b}) = \lambda\mu(\mathbf{a} \otimes \mathbf{b}).$$

Since

$$(A\mathbf{a}) \otimes (B\mathbf{b}) \equiv (A \otimes B)(\mathbf{a} \otimes \mathbf{b})$$

we obtain

$$(A \otimes B)(\mathbf{a} \otimes \mathbf{b}) = \lambda\mu(\mathbf{a} \otimes \mathbf{b}).$$

This is an eigenvalue equation. Consequently, $\mathbf{a} \otimes \mathbf{b}$ is an eigenvector of $A \otimes B$ with eigenvalue $\lambda\mu$.

(ii) Both matrices are hermitian, unitary and $\mathrm{tr}\sigma_y = 0$, $\mathrm{tr}\sigma_z = 0$. Thus the eigenvalues of σ_y and σ_z are given by $\{1, -1\}$. Since the eigenvalues of σ_y and σ_z are given by $\{1, -1\}$, we find that the eigenvalues of $\sigma_y \otimes \sigma_z$ are given by $\{1, 1, -1, -1\}$.

Problem 5.   (i) Let A be an $m \times m$ matrix and B be an $n \times n$ matrix. Let

$$A\mathbf{a} = \lambda\mathbf{a}, \qquad B\mathbf{b} = \mu\mathbf{b}$$

be their respective eigenvalue equations. Show that the matrix

$$A \otimes I_n + I_m \otimes B$$

where I_n is the $n \times n$ unit matrix and I_m is the $m \times m$ unit matrix, has the eigenvalue $\lambda + \mu$ with the corresponding eigenvector $\mathbf{a} \otimes \mathbf{b}$.

(ii) Consider the Pauli spin matrices

$$\sigma_y := \begin{pmatrix} 0 & -i \\ i & 0 \end{pmatrix}, \qquad \sigma_z := \begin{pmatrix} 1 & 0 \\ 0 & -1 \end{pmatrix}.$$

Find the eigenvalues of

$$\sigma_y \otimes I_2 + I_2 \otimes \sigma_z.$$

Solution 5.   (i) We have

$$
\begin{aligned}
(A \otimes I_n + I_m \otimes B)(\mathbf{a} \otimes \mathbf{b}) &= (A \otimes I_n)(\mathbf{a} \otimes \mathbf{b}) + (I_m \otimes B)(\mathbf{a} \otimes \mathbf{b}) \\
&= (A\mathbf{a}) \otimes (I_n\mathbf{b}) + (I_m\mathbf{a}) \otimes (B\mathbf{b}) \\
&= (\lambda\mathbf{a}) \otimes \mathbf{b} + \mathbf{a} \otimes (\mu\mathbf{b}) \\
&= (\lambda + \mu)(\mathbf{a} \otimes \mathbf{b}).
\end{aligned}
$$

Consequently, the eigenvectors and eigenvalues of $A \otimes I_n + I_m \otimes B$ admits the eigenvalue $\lambda + \mu$ and the corresponding eigenvector $\mathbf{a} \otimes \mathbf{b}$.

(ii) Since the eigenvalues of σ_y and σ_z are given by $\{1, -1\}$, we find that the eigenvalues of $\sigma_y \otimes I_2 + I_2 \otimes \sigma_z$ take the form $\{2, 0, 0, -2\}$.

Problem 6.   Let A, B be two arbitrary $n \times n$ matrices. Let I_n be the $n \times n$ unit matrix. Prove that

$$\mathrm{tr}e^{A \otimes I_n + I_n \otimes B} \equiv (\mathrm{tr}e^A)(\mathrm{tr}e^B).$$

Solution 6.   Since

$$[A \otimes I_n, I_n \otimes B] = 0_{n \cdot n}$$

we have

$$\mathrm{tr}e^{(A\otimes I_n + I_n\otimes B)} = \mathrm{tr}\left(e^{A\otimes I}e^{I_n\otimes B}\right).$$

Now

$$e^{A\otimes I_n} = \sum_{k=0}^{\infty}\frac{(A\otimes I_n)^k}{k!}, \qquad e^{I_n\otimes B} = \sum_{k=0}^{\infty}\frac{(I_n\otimes B)^k}{k!}.$$

An arbitrary term in the expansion of $e^{A\otimes I_n}e^{I_n\otimes B}$ is given by

$$\frac{1}{j!}\frac{1}{k!}(A\otimes I_n)^j(I_n\otimes B)^k.$$

Now we have

$$\frac{1}{j!}\frac{1}{k!}(A\otimes I_n)^j(I_n\otimes B)^k \equiv \frac{1}{j!}\frac{1}{k!}(A^j\otimes I_n^j)(I_n^k\otimes B^k)$$

$$\equiv \frac{1}{j!}\frac{1}{k!}(A^j\otimes I_n)(I_n\otimes B^k)$$

$$\equiv \frac{1}{j!}\frac{1}{k!}(A^j\otimes B^k).$$

Therefore

$$\mathrm{tr}\left(\frac{1}{j!}\cdot\frac{1}{k!}(A\otimes I_n)^j(I_n\otimes B)^k\right) = \frac{1}{j!}\cdot\frac{1}{k!}(\mathrm{tr}A^j)(\mathrm{tr}B^k).$$

Since

$$(\mathrm{tr}e^A)(\mathrm{tr}e^B) \equiv \mathrm{tr}\left(\sum_{k=0}^{\infty}\frac{A^k}{k!}\right)\mathrm{tr}\left(\sum_{k=0}^{\infty}\frac{B^k}{k!}\right) = \left(\sum_{k=0}^{\infty}\frac{1}{k!}\mathrm{tr}A^k\right)\left(\sum_{k=0}^{\infty}\frac{1}{k!}\mathrm{tr}B^k\right)$$

we have proved the identity.

Problem 7. Let A be an $n\times n$ matrix and I_n be the $n\times n$ identity matrix. We define

$$A_j := \overbrace{I_n\otimes\cdots\otimes I_n\otimes A\otimes I_n\otimes\cdots\otimes I_n}^{N-factors}$$

where A is at the j-th place. Let

$$X := A_1 + A_2 + \cdots + A_N.$$

Calculate

$$\frac{\mathrm{tr}\left(A_j e^X\right)}{\mathrm{tr}e^X}.$$

Solution 7. Since $[A_j, A_k] = 0$ for $j = 1, 2, \ldots, N$ and $k = 1, 2, \ldots, N$ we have

$$e^X \equiv e^{A_1 + A_2 + \cdots + A_N} \equiv e^{A_1} e^{A_2} \cdots e^{A_N}.$$

Now

$$e^{A_j} = \overbrace{I_n \otimes \cdots \otimes I_n \otimes e^A \otimes I_n \otimes \cdots \otimes I_n}^{N-factors}$$

where e^A is at the j-th place. It follows that

$$e^{A_1} e^{A_2} \cdots e^{A_N} = e^A \otimes e^A \otimes \cdots \otimes e^A.$$

Therefore

$$\mathrm{tr} e^X \equiv (\mathrm{tr} e^A)^N.$$

Now

$$A_j e^X = (I_n \otimes \cdots \otimes I_n \otimes A \otimes I_n \otimes \cdots \otimes I_n)(e^A \otimes e^A \otimes \cdots \otimes e^A)$$

or

$$A_j e^X = (e^A \otimes \cdots \otimes e^A \otimes Ae^A \otimes e^A \otimes \cdots \otimes e^A).$$

Thus

$$\mathrm{tr}(A_j e^X) = (\mathrm{tr} e^A)^{N-1}(\mathrm{tr}(Ae^A)).$$

Consequently

$$\frac{\mathrm{tr}(A_j e^X)}{\mathrm{tr} e^X} = \frac{\mathrm{tr}(Ae^A)}{\mathrm{tr} e^A}.$$

Problem 8. Consider the four normalized vectors in the vector space $\mathbb{R}^2$

$$\mathbf{u}_1 = \begin{pmatrix} 1 \\ 0 \end{pmatrix}, \qquad \mathbf{u}_2 = \begin{pmatrix} 0 \\ 1 \end{pmatrix}, \qquad \mathbf{v}_1 = \frac{1}{\sqrt{2}} \begin{pmatrix} 1 \\ 1 \end{pmatrix}, \qquad \mathbf{v}_2 = \frac{1}{\sqrt{2}} \begin{pmatrix} 1 \\ -1 \end{pmatrix}.$$

Calculate the vectors in the vector space $\mathbb{R}^4$

$$\mathbf{u}_1 \otimes \mathbf{u}_1, \qquad \mathbf{u}_1 \otimes \mathbf{u}_2, \qquad \mathbf{u}_2 \otimes \mathbf{u}_1, \qquad \mathbf{u}_2 \otimes \mathbf{u}_2$$

$$\mathbf{v}_1 \otimes \mathbf{v}_1, \qquad \mathbf{v}_1 \otimes \mathbf{v}_2, \qquad \mathbf{v}_2 \otimes \mathbf{v}_1, \qquad \mathbf{v}_2 \otimes \mathbf{v}_2$$

and the 2×2 matrices

$$\mathbf{u}_1^T \otimes \mathbf{u}_1, \qquad \mathbf{u}_1^T \otimes \mathbf{u}_2, \qquad \mathbf{u}_2^T \otimes \mathbf{u}_1, \qquad \mathbf{u}_2^T \otimes \mathbf{u}_2$$

$$\mathbf{v}_1^T \otimes \mathbf{v}_1, \qquad \mathbf{v}_1^T \otimes \mathbf{v}_2, \qquad \mathbf{v}_2^T \otimes \mathbf{v}_1, \qquad \mathbf{v}_2^T \otimes \mathbf{v}_2.$$

Discuss!

Solution 8. Using the definition of the Kronecker product we find

$$\mathbf{u}_1 \otimes \mathbf{u}_1 = \begin{pmatrix} 1 \\ 0 \\ 0 \\ 0 \end{pmatrix}, \qquad \mathbf{u}_1 \otimes \mathbf{u}_2 = \begin{pmatrix} 0 \\ 1 \\ 0 \\ 0 \end{pmatrix}$$

$$\mathbf{u}_2 \otimes \mathbf{u}_1 = \begin{pmatrix} 0 \\ 0 \\ 1 \\ 0 \end{pmatrix}, \qquad \mathbf{u}_2 \otimes \mathbf{u}_2 = \begin{pmatrix} 0 \\ 0 \\ 0 \\ 1 \end{pmatrix}.$$

The set $\{\mathbf{u}_1, \mathbf{u}_2\}$ is the standard basis of $\mathbb{R}^2$. We find that the set

$$\{\, \mathbf{u}_1 \otimes \mathbf{u}_1, \quad \mathbf{u}_1 \otimes \mathbf{u}_2, \quad \mathbf{u}_2 \otimes \mathbf{u}_1, \quad \mathbf{u}_2 \otimes \mathbf{u}_2 \,\}$$

is the standard basis in $\mathbb{R}^4$. Using the definition of the Kronecker product we obtain

$$\mathbf{v}_1 \otimes \mathbf{v}_1 = \frac{1}{2}\begin{pmatrix} 1 \\ 1 \\ 1 \\ 1 \end{pmatrix}, \qquad \mathbf{v}_1 \otimes \mathbf{v}_2 = \frac{1}{2}\begin{pmatrix} 1 \\ -1 \\ 1 \\ -1 \end{pmatrix}$$

$$\mathbf{v}_2 \otimes \mathbf{v}_1 = \frac{1}{2}\begin{pmatrix} 1 \\ 1 \\ -1 \\ -1 \end{pmatrix}, \qquad \mathbf{v}_2 \otimes \mathbf{v}_2 = \frac{1}{2}\begin{pmatrix} 1 \\ -1 \\ -1 \\ 1 \end{pmatrix}.$$

The set $\{\mathbf{v}_1, \mathbf{v}_2\}$ is an orthonormal basis of $\mathbb{R}^2$. Thus we find that the set

$$\{\, \mathbf{v}_1 \otimes \mathbf{v}_1, \quad \mathbf{v}_1 \otimes \mathbf{v}_2, \quad \mathbf{v}_2 \otimes \mathbf{v}_1, \quad \mathbf{v}_2 \otimes \mathbf{v}_2 \,\}$$

is an orthonormal basis in the vector space $\mathbb{R}^4$.

We find

$$\mathbf{u}_1^T \otimes \mathbf{u}_1 = (1,0) \otimes \begin{pmatrix} 1 \\ 0 \end{pmatrix} = \begin{pmatrix} 1 & 0 \\ 0 & 0 \end{pmatrix}, \quad \mathbf{u}_1^T \otimes \mathbf{u}_2 = (1,0) \otimes \begin{pmatrix} 0 \\ 1 \end{pmatrix} = \begin{pmatrix} 0 & 0 \\ 1 & 0 \end{pmatrix}$$

$$\mathbf{u}_2^T \otimes \mathbf{u}_1 = (0,1) \otimes \begin{pmatrix} 1 \\ 0 \end{pmatrix} = \begin{pmatrix} 0 & 1 \\ 0 & 0 \end{pmatrix}, \qquad \mathbf{u}_2^T \otimes \mathbf{u}_2 = (0,1) \otimes \begin{pmatrix} 0 \\ 1 \end{pmatrix} = \begin{pmatrix} 0 & 0 \\ 0 & 1 \end{pmatrix}.$$

Consequently,

$$\mathbf{u}_1^T \otimes \mathbf{u}_1 + \mathbf{u}_2^T \otimes \mathbf{u}_2$$

is the 2×2 unit matrix. We obtain

$$\mathbf{v}_1^T \otimes \mathbf{v}_1 = \frac{1}{2}(1,1) \otimes \begin{pmatrix} 1 \\ 1 \end{pmatrix} = \frac{1}{2}\begin{pmatrix} 1 & 1 \\ 1 & 1 \end{pmatrix}$$

$$\mathbf{v}_1^T \otimes \mathbf{v}_2 = \frac{1}{2}(1,1) \otimes \begin{pmatrix} 1 \\ -1 \end{pmatrix} = \frac{1}{2} \begin{pmatrix} 1 & 1 \\ -1 & -1 \end{pmatrix}$$

$$\mathbf{v}_2^T \otimes \mathbf{v}_1 = \frac{1}{2}(1,-1) \otimes \begin{pmatrix} 1 \\ 1 \end{pmatrix} = \frac{1}{2} \begin{pmatrix} 1 & -1 \\ 1 & -1 \end{pmatrix}$$

$$\mathbf{v}_2^T \otimes \mathbf{v}_2 = \frac{1}{2}(1,-1) \otimes \begin{pmatrix} 1 \\ -1 \end{pmatrix} = \frac{1}{2} \begin{pmatrix} 1 & -1 \\ -1 & 1 \end{pmatrix}.$$

Consequently,

$$\mathbf{v}_1^T \otimes \mathbf{v}_1 + \mathbf{v}_2^T \otimes \mathbf{v}_2$$

is the 2×2 unit matrix.

Problem 9.   Let A be a symmetric 4×4 matrix over $\mathbb{R}$. Assume that the eigenvalues are given by $\lambda_1 = 0$, $\lambda_2 = 1$, $\lambda_3 = 2$ and $\lambda_4 = 3$ with the corresponding normalized eigenvectors

$$\mathbf{u}_1 = \frac{1}{\sqrt{2}} \begin{pmatrix} 1 \\ 0 \\ 0 \\ 1 \end{pmatrix}, \qquad \mathbf{u}_2 = \frac{1}{\sqrt{2}} \begin{pmatrix} 1 \\ 0 \\ 0 \\ -1 \end{pmatrix}$$

$$\mathbf{u}_3 = \frac{1}{\sqrt{2}} \begin{pmatrix} 0 \\ 1 \\ 1 \\ 0 \end{pmatrix}, \qquad \mathbf{u}_4 = \frac{1}{\sqrt{2}} \begin{pmatrix} 0 \\ 1 \\ -1 \\ 0 \end{pmatrix}.$$

Find the symmetric matrix A with the help of the *spectral theorem*. Using the spectral theorem the matrix A we can reconstructed from the eigenvalues and normalized eigenvectors.

Solution 9.   The eigenvectors to different eigenvalues are orthogonal. Since the normalized eigenvectors are pairwise orthogonal we find from the spectral theorem

$$A = \sum_{j=1}^{4} \lambda_j \mathbf{u}_j^T \otimes \mathbf{u}_j.$$

Inserting the eigenvalues yields

$$A = \mathbf{u}_2^T \otimes \mathbf{u}_2 + 2\mathbf{u}_3^T \otimes \mathbf{u}_3 + 3\mathbf{u}_4^T \otimes \mathbf{u}_4.$$

Using the definition of the Kronecker product we find that

$$A = \frac{1}{2} \begin{pmatrix} 1 & 0 & 0 & -1 \\ 0 & 5 & -1 & 0 \\ 0 & -1 & 5 & 0 \\ -1 & 0 & 0 & 1 \end{pmatrix}.$$

Problem 10. Let U be a 4×4 unitary matrix. Assume that U can be written as

$$U = V \otimes W \tag{1}$$

where V and W are 2×2 unitary matrices. Show that U can be written as

$$U = \exp(iA \otimes I_2 + iI_2 \otimes B) \tag{2}$$

where A and B are 2×2 hermitian matrices and I_2 is the 2×2 unit matrix.

Solution 10. Any unitary matrix Z can be written as $Z = \exp(iX)$, where X is a hermitian matrix. Thus

$$V = \exp(iA), \qquad W = \exp(iB)$$

where A and B are 2×2 hermitian matrices. Consequently

$$U = V \otimes W = \exp(iA) \otimes \exp(iB).$$

Using the identity

$$\exp(iA) \otimes \exp(iB) \equiv \exp(iA \otimes I_2) \exp(iI_2 \otimes B)$$

and

$$[A \otimes I_2, I_2 \otimes B] = 0_4$$

we find equation (2).

Problem 11. Let A_i and B_j be $n \times n$ matrices over $\mathbb{C}$, where $i, j = 1, 2, \ldots, p$. We define

$$r_{12} := \sum_{i=1}^{p} A_i \otimes B_i \otimes I_n, \quad r_{13} := \sum_{j=1}^{p} A_j \otimes I_n \otimes B_j, \quad r_{23} := \sum_{k=1}^{p} I_n \otimes A_k \otimes B_k$$

$$\tag{1}$$

where I_n is the $n \times n$ unit matrix. Find

$$[r_{12}, r_{13}], \qquad [r_{12}, r_{23}], \qquad [r_{13}, r_{23}] \tag{2}$$

where $[\,,\,]$ denotes the commutator.

Solution 11. Since

$$(A_i \otimes B_i \otimes I_n)(A_j \otimes I_n \otimes B_j) - (A_j \otimes I_n \otimes B_j)(A_i \otimes B_i \otimes I_n)$$
$$= (A_i A_j) \otimes B_i \otimes B_j - (A_j A_i) \otimes B_i \otimes B_j$$
$$= (A_i A_j - A_j A_i) \otimes B_i \otimes B_j$$
$$= [A_i, A_j] \otimes B_i \otimes B_j$$

we have

$$[r_{12}, r_{13}] = \sum_{i=1}^{p} \sum_{j=1}^{p} [A_i, A_j] \otimes B_i \otimes B_j \,.$$

Analogously

$$[r_{12}, r_{23}] = \sum_{i=1}^{p} \sum_{k=1}^{p} A_i \otimes [B_i, A_k] \otimes B_k$$

and

$$[r_{13}, r_{23}] = \sum_{j=1}^{p} \sum_{k=1}^{p} A_j \otimes A_k \otimes [B_j, B_k] \,.$$

Assume that A_i and B_j are elements of a Lie algebra. The *Classical Yang-Baxter Equation* is given by

$$[r_{12}, r_{13}] + [r_{12}, r_{23}] + [r_{13}, r_{23}] = 0 \,.$$

Problem 12.   Let $X = (x_{ij})$ be an $m \times n$ matrix. We define the vector (*vec operator*)

$$v(X) := (x_{11}, x_{12}, \ldots, x_{1n}, x_{21}, \ldots, x_{2n}, \ldots, x_{mn})^T \qquad (1)$$

where ^{T} denotes transpose. Thus $v(X)$ is a column vector. Let C be a $p \times m$ matrix and let D be a $q \times n$ matrix. Then we have

$$v(CXD^T) = (C \otimes D)v(X) \,. \qquad (2)$$

Show that

$$D \otimes C = P(C \otimes D)Q \qquad (3)$$

where P and Q are permutation matrices depending only on p, q and m, n respectively.

Solution 12.   Consider the equation

$$CXD^T = E \,. \qquad (4)$$

Using (2) we find

$$(C \otimes D)v(X) = v(E) \,. \qquad (5)$$

Similarly, application of (2) to the transpose of (4) yields

$$(D \otimes C)v(X^T) = v(E^T) \,. \qquad (6)$$

In view of (1) it is clear that

$$v(E^T) = Pv(E), \qquad v(X) = Qv(X^T) \qquad (7)$$

for suitable permutation matrices P and Q which depend only on the dimensions of E and X, respectively. Equation(6), together with that resulting from substitution of (7) into (5), verifies (3).

Problem 13. The starting point in the construction of the *Yang-Baxter equation* is the 2×2 matrix

$$T := \begin{pmatrix} a & b \\ c & d \end{pmatrix} \tag{1}$$

where a, b, c, and d are $n \times n$ linear operators of an algebra over the complex numbers. In other words, T is an operator-valued matrix. Let I_2 be the 2×2 identity matrix. We define the 4×4 matrices

$$T_1 = T \otimes I_2, \qquad T_2 = I_2 \otimes T \tag{2}$$

where $\otimes$ denotes the Kronecker product. Thus T_1 and T_2 are operator-valued 4×4 matrices. Applying the rules for the Kronecker product we find

$$T_1 = T \otimes I_2 = \begin{pmatrix} a & 0 & b & 0 \\ 0 & a & 0 & b \\ c & 0 & d & 0 \\ 0 & c & 0 & d \end{pmatrix}, \qquad T_2 = I_2 \otimes T = \begin{pmatrix} a & b & 0 & 0 \\ c & d & 0 & 0 \\ 0 & 0 & a & b \\ 0 & 0 & c & d \end{pmatrix}.$$

The Yang-Baxter equation is given by

$$R_q T_1 T_2 = T_2 T_1 R_q$$

where R_q is a 4×4 matrix and q a nonzero complex number. Let

$$R_q := \begin{pmatrix} 1 & 0 & 0 & 0 \\ 0 & -1 & 0 & 0 \\ 0 & 1+q & q & 0 \\ 0 & 0 & 0 & 1 \end{pmatrix}.$$

Find the condition on a, b, c and d such that condition (1) is satisfied.

Solution 13. Altogether we find 16 relations which have to be satisfied. Equation (1) gives rise to the relations of the algebra elements a, b, c and d

$$ab = q^{-1}ba, \quad dc = qcd, \quad bc = -qcb$$

$$bd = -db, \quad ac = -ca, \quad [a,d] = (1+q^{-1})bc$$

where all the commutative relations have been omitted and $[\,,\,]$ denotes the commutator.

Problem 14.   Consider the normalized vector

$$|\psi\rangle = \frac{1}{\sqrt{2}} \begin{pmatrix} e^{i\phi} \\ 0 \\ 0 \\ 1 \end{pmatrix}$$

in the Hilbert space $\mathbb{C}^4$. Can the normalized vector $|\psi\rangle$ be written as the Kronecker product of two vectors $|u\rangle$, $|v\rangle$ in the Hilbert space $\mathbb{C}^2$?

Solution 14.   Let

$$|u\rangle = \begin{pmatrix} u_1 \\ u_2 \end{pmatrix}, \qquad |v\rangle = \begin{pmatrix} v_1 \\ v_2 \end{pmatrix}.$$

Then from $|\psi\rangle = |u\rangle \otimes |v\rangle$ we obtain the four equations

$$u_1 v_1 = e^{i\phi}/\sqrt{2}, \quad u_1 v_2 = 0, \quad u_2 v_1 = 0, \quad u_2 v_2 = 1/\sqrt{2}.$$

These four equations cannot be satisfied simultaneously. Thus $|\psi\rangle$ cannot be written as a Kronecker product of two vectors in $\mathbb{C}^2$. We call $|\psi\rangle$ *entangled*.

Problem 15.   Let A be an $n \times n$ matrix over $\mathbb{C}$. Consider the map

$$\rho(A) \to A \otimes I_n + I_n \otimes A.$$

Calculate the commutator

$$[\rho(A), \rho(B)].$$

Solution 15.   We find

$$
\begin{aligned}
[\rho(A), \rho(B)] &= [A \otimes I_n + I_n \otimes A, B \otimes I_n + I_n \otimes B] \\
&= [A \otimes I_n, B \otimes I_n] + [A \otimes I_n, I_n \otimes B] \\
&\quad + [I_n \otimes A, B \otimes I_n] + [I_n \otimes A, I_n \otimes B] \\
&= (AB) \otimes I_n - (BA) \otimes + I_n \otimes (AB) - I_n \otimes (BA) \\
&= [A, B] \otimes I_n + I_n \otimes [A, B] \\
&= \rho([A, B]).
\end{aligned}
$$

Problem 16.   Let

$$B(1) = \begin{pmatrix} 1 & 1 \\ 0 & 1 \end{pmatrix}$$

and $B(k)$ the $2^k \times 2^k$ matrix recursively defined by

$$B(k) := \begin{pmatrix} B(k-1) & B(k-1) \\ 0 & B(k-1) \end{pmatrix} = B(1) \otimes B(k-1), \qquad k = 2, 3, \dots$$

(i) Find $B(1)^{-1}$.

(ii) Find $B(k)^{-1}$ for $k = 2, 3, \dots$.

(iii) Let $A(k)$ be the $2^k \times 2^k$ anti-diagonal matrix defined by $A_{ij}(k) = 1$ if $i = 2^k + 1 - j$ and 0 otherwise with $j = 1, 2, \dots, 2^k$. Let

$$C(k) = A(k)B(k)^{-1}A(k).$$

Show that the matrices $B(k)$ and $C(k)$ satisfy the *braid-like relation*

$$B(k)C(k)B(k) = C(k)B(k)C(k).$$

Solution 16.   (i) We have

$$B(1)^{-1} = \begin{pmatrix} 1 & -1 \\ 0 & 1 \end{pmatrix}.$$

(ii) We find

$$B(k)^{-1} = (B(1)^{-1})^{\otimes k}.$$

(iii) We find

$$C(1) = \begin{pmatrix} 1 & 0 \\ -1 & 1 \end{pmatrix}.$$

The relation is true for $k = 1$ by a direct calculation. Since

$$B(k) = B(1)^{\otimes k}, \qquad C(k) = C(1)^{\otimes k}$$

and using the properties of the Kronecker product we find that the relation is true for every k.

Problem 17.   Let σ_x, σ_y, σ_z be the Pauli spin matrices. Let

$$R := \sigma_x \otimes \sigma_x + \sigma_y \otimes \sigma_y + \sigma_z \otimes \sigma_z.$$

(i) Find $\mathrm{tr}R$. Using this result what can be said about the eigenvalues of R?

(ii) Find R^2. Using this result and the result from (i) derive the eigenvalues of the matrix R.

(iii) Find

$$\frac{1}{4}(I_4 + R)^2.$$

Solution 17.   (i) Since $\mathrm{tr}\sigma_x = \mathrm{tr}\sigma_y = \mathrm{tr}\sigma_z = 0$ and $\mathrm{tr}(A \otimes B) = \mathrm{tr}A\,\mathrm{tr}B$ for any $n \times n$ matrices A and B we find $\mathrm{tr}R = 0$. Thus for the four eigenvalues of R we have $\lambda_1 + \lambda_2 + \lambda_3 + \lambda_4 = 0$.
(ii) Since $\sigma_x^2 = \sigma_y^2 = \sigma_z^2 = I_2$ we find

$$R^2 = 3I_4 - 2R.$$

From the eigenvalue equation $Ru = \lambda u$ we obtain $R^2 u = \lambda R u = \lambda^2 u$. Using $R^2 = 3I_4 - 2R$ we have

$$(3I_4 - 2R)u = (3 - 2\lambda)u = \lambda^2 u$$

or $(\lambda^2 + 2\lambda - 3)u = \mathbf{0}$. Thus $\lambda_+ = 1$ and $\lambda_- = -3$. Together with $\lambda_1 + \lambda_2 + \lambda_3 + \lambda_4 = 0$ we obtain the four eigenvalues

$$\lambda_1 = 1, \quad \lambda_2 = 1, \quad \lambda_3 = 1, \quad \lambda_4 = -3.$$

(iii) Using the result from (ii) we find

$$\frac{1}{4}(I_4 + R)^2 = \frac{1}{4}(I_4 + 2R + R^2) = I_4.$$

Problem 18.   (i) Consider the 4×4 matrix

$$T = \begin{pmatrix} 1 & 0 & 0 & 1 \\ 1 & 0 & 0 & -1 \\ 0 & 1 & 1 & 0 \\ 0 & -i & i & 0 \end{pmatrix}.$$

Show that the inverse T^{-1} of T exists and find the inverse.
(ii) Let

$$F_\alpha = \begin{pmatrix} \cos\alpha & -\sin\alpha \\ \sin\alpha & \cos\alpha \end{pmatrix}.$$

Calculate

$$T(F_\alpha \otimes F_\beta^T)T^{-1}.$$

Solution 18.   (i) The row vectors (or column) vectors in the matrix T are linearly independent. Thus $\mathrm{rank}(T) = 4$. Thus the inverse exists and is given by

$$T^{-1} = \frac{1}{2} \begin{pmatrix} 1 & 1 & 0 & 0 \\ 0 & 0 & 1 & i \\ 0 & 0 & 1 & -i \\ 1 & -1 & 0 & 0 \end{pmatrix}.$$

(ii) We find

$$T(F_\alpha \otimes F_\beta^T)T^{-1} = \begin{pmatrix} \cos(\alpha+\beta) & 0 & 0 & i\sin(\alpha+\beta) \\ 0 & \cos(\alpha-\beta) & -\sin(\alpha-\beta) & 0 \\ 0 & \sin(\alpha-\beta) & \cos(\alpha-\beta) & 0 \\ i\sin(\alpha+\beta) & 0 & 0 & \cos(\alpha+\beta) \end{pmatrix}.$$

Problem 19. Let A, B be $n \times n$ matrices such that $A^2 = I_n$ and $B^2 = I_n$. Calculate $\exp(z(A \otimes B))$, where $z \in \mathbb{C}$.

Solution 19. Since $(A \otimes B)^2 = I_n \otimes I_n$ we obtain

$$\exp(z(A \otimes B)) = (I_n \otimes I_n)\cosh(z) + (A \otimes B)\sinh(z).$$

Problem 20. Let $\mathbf{x}, \mathbf{y} \in \mathbb{R}^n$. We define

$$\mathbf{x} \wedge \mathbf{y} := \mathbf{x} \otimes \mathbf{y} - \mathbf{y} \otimes \mathbf{x}.$$

Is

$$((\mathbf{x} \wedge \mathbf{y}) \wedge \mathbf{z}) + ((\mathbf{z} \wedge \mathbf{x}) \wedge \mathbf{y}) + ((\mathbf{y} \wedge \mathbf{z}) \wedge \mathbf{x}) = \mathbf{0}\,?$$

Solution 20. Yes, the assumption is true.

Problem 21. Let A be an $m \times n$ matrix over $\mathbb{R}$, B be an $s \times t$ matrix over $\mathbb{R}$ and M be a $ms \times nt$ matrix over $\mathbb{R}$. Find A and B which minimizes

$$\|M - A \otimes B\|$$

by using the fact that the $mn \times st$ matrix $\mathrm{vec}A(\mathrm{vec}B)^T$ has the same entries as $A \otimes B$. The norm $\|\ \|$ is the Hilbert-Schmidt norm over $\mathbb{R}$ given by

$$\|X\| = \mathrm{tr}(XX^T) = \sqrt{\sum_{i=1}^{ms}\sum_{j=1}^{nt}(X)_{ij}^2}.$$

The order of the entries in the sum does not change the result. Thus we minimize

$$\|R(M) - \mathrm{vec}A(\mathrm{vec}B)^T\|$$

where $R(M)$ is the $mn \times st$ matrix with the same entries as M rearranged by R where $R(A \otimes B) = \mathrm{vec}A(\mathrm{vec}B)^T$. As an example consider $m = n = s = t = 2$ and the matrix

$$M = \begin{pmatrix} 1 & 0 & 0 & 1 \\ 0 & 0 & 0 & 0 \\ 0 & 0 & 0 & 0 \\ 1 & 0 & 0 & 1 \end{pmatrix}.$$

Solution 21. Since $\text{vec}A(\text{vec}B)^T$ has rank 1, we find the nearest rank 1 approximation of $R(M)$. Let

$$R(M) = U\Sigma V^T$$

be the *singular value decomposition* of $R(M)$, where $\Sigma = \text{diag}(\sigma_1, \sigma_2, \ldots)$ are the decreasing singular values of $R(M)$. Then the nearest rank 1 approximation is $U\text{diag}(\sigma_1, 0, \ldots)V^T$. Consequently we solve

$$U^T \text{vec}A(\text{vec}B)^T V = \text{diag}(\sigma_1, 0, \ldots).$$

The solution is given by $\text{vec}A = U(t\sigma_1, 0, \ldots)^T$ and $\text{vec}B = V(1/t, 0, \ldots)^T$ for $t \in \mathbb{R} \setminus \{0\}$. For $m = n = s = t = 2$ and $R(A \otimes B) = \text{vec}A(\text{vec}B)^T$ we find

$$R\begin{pmatrix} a_{11}b_{11} & a_{11}b_{12} & a_{12}b_{11} & a_{12}b_{12} \\ a_{11}b_{21} & a_{11}b_{22} & a_{12}b_{21} & a_{12}b_{22} \\ a_{21}b_{11} & a_{21}b_{12} & a_{22}b_{11} & a_{22}b_{12} \\ a_{21}b_{21} & a_{21}b_{22} & a_{22}b_{21} & a_{22}b_{22} \end{pmatrix} = \begin{pmatrix} a_{11}b_{11} & a_{11}b_{12} & a_{11}b_{21} & a_{11}b_{22} \\ a_{12}b_{11} & a_{12}b_{12} & a_{12}b_{21} & a_{12}b_{22} \\ a_{21}b_{11} & a_{21}b_{12} & a_{21}b_{21} & a_{21}b_{22} \\ a_{22}b_{11} & a_{22}b_{12} & a_{22}b_{21} & a_{22}b_{22} \end{pmatrix}.$$

Thus $R(M) = I_4$ with a singular value decomposition given by $U = V = \Sigma = I_4$. It follows that $\text{vec}A = I_4(t, 0, 0, 0)^T$ and $\text{vec}B = I_4(1/t, 0, 0, 0)^T$, i.e.

$$A = \begin{pmatrix} t & 0 \\ 0 & 0 \end{pmatrix}, \qquad B = \begin{pmatrix} \frac{1}{t} & 0 \\ 0 & 0 \end{pmatrix}.$$

More generally $V = U$ for an arbitrary 4×4 orthogonal matrix U and $\text{vec}A = U(t, 0, 0, 0)^T$ and $\text{vec}B = U(1/t, 0, 0, 0)^T$.

Problem 22. Calculate the eigenvalues of the Hamilton operator

$$\hat{H} := \lambda(S_x \otimes \hat{L}_x + S_y \otimes \hat{L}_y + S_z \otimes \hat{L}_z) \tag{1}$$

where we consider a subspace G_1 of the Hilbert space $L_2(S^2)$ with

$$S^2 := \{(x, y, z) \ : \ x^2 + y^2 + z^2 = 1\}.$$

Here $\otimes$ denotes the *tensor product*. The linear operators $\hat{L}_x, \hat{L}_y, \hat{L}_z$ act in the subspace G_1. A basis of subspace G_1 is

$$Y_{1,0} = \sqrt{\frac{3}{4\pi}} \cos\theta, \quad Y_{1,1} = -\sqrt{\frac{3}{8\pi}} \sin\theta e^{i\phi}, \quad Y_{1,-1} = \sqrt{\frac{3}{8\pi}} \sin\theta e^{-i\phi}.$$

The operators (matrices) S_x, S_y, S_z act in the Hilbert space $\mathbb{C}^2$. The standard basis is given by

$$\begin{pmatrix} 1 \\ 0 \end{pmatrix}, \qquad \begin{pmatrix} 0 \\ 1 \end{pmatrix}.$$

The *spin matrices* S_z, S_+ and S_- are given by

$$S_z := \frac{1}{2}\hbar \begin{pmatrix} 1 & 0 \\ 0 & -1 \end{pmatrix}, \qquad S_+ := \hbar \begin{pmatrix} 0 & 1 \\ 0 & 0 \end{pmatrix}, \qquad S_- := \hbar \begin{pmatrix} 0 & 0 \\ 1 & 0 \end{pmatrix}$$

where $S_\pm := S_x \pm iS_y$. The operators $\hat{L}_z$, $\hat{L}_+$ and $\hat{L}_-$ take the form

$$\hat{L}_z := -i\hbar\frac{\partial}{\partial\phi}$$

$$\hat{L}_+ := \hbar e^{i\phi}\left(\frac{\partial}{\partial\theta} + i\cot\theta\frac{\partial}{\partial\phi}\right)$$

$$\hat{L}_- := \hbar e^{-i\phi}\left(-\frac{\partial}{\partial\theta} + i\cot\theta\frac{\partial}{\partial\phi}\right)$$

where

$$\hat{L}_\pm := \hat{L}_x \pm i\hat{L}_y.$$

Give an interpretation of the Hamilton operator $\hat{H}$ and of the subspace G_1. In some textbooks we find the notation $\hat{H} = \lambda\mathbf{S}\cdot\mathbf{L}$.

Solution 22. The Hamilton operator (1) can be written as

$$\hat{H} = \lambda(S_z \otimes \hat{L}_z) + \frac{\lambda}{2}(S_+ \otimes \hat{L}_- + S_- \otimes \hat{L}_+).$$

In the tensor product space $\mathbb{C}^2 \otimes G_1$ a basis is given by

$$|1\rangle = \begin{pmatrix} 1 \\ 0 \end{pmatrix} \otimes Y_{1,0}, \qquad |2\rangle = \begin{pmatrix} 1 \\ 0 \end{pmatrix} \otimes Y_{1,-1}, \qquad |3\rangle = \begin{pmatrix} 1 \\ 0 \end{pmatrix} \otimes Y_{1,1}$$

$$|4\rangle = \begin{pmatrix} 0 \\ 1 \end{pmatrix} \otimes Y_{1,0}, \qquad |5\rangle = \begin{pmatrix} 0 \\ 1 \end{pmatrix} \otimes Y_{1,-1}, \qquad |6\rangle = \begin{pmatrix} 0 \\ 1 \end{pmatrix} \otimes Y_{1,1}.$$

In the following we use

$$\hat{L}_+Y_{1,1} = 0, \qquad \hat{L}_+Y_{1,0} = \hbar\sqrt{2}Y_{1,1}, \qquad \hat{L}_+Y_{1,-1} = \hbar\sqrt{2}Y_{1,0}$$
$$\hat{L}_-Y_{1,1} = \hbar\sqrt{2}Y_{1,0}, \qquad \hat{L}_-Y_{1,0} = \hbar\sqrt{2}Y_{1,-1}, \qquad \hat{L}_-Y_{1,-1} = 0$$
$$\hat{L}_zY_{1,1} = \hbar Y_{1,1}, \qquad \hat{L}_zY_{1,0} = 0, \qquad \hat{L}_zY_{1,-1} = -\hbar Y_{1,-1}.$$

For the state $|1\rangle$ we find

$$\hat{H}|1\rangle = \left[\lambda(S_z \otimes \hat{L}_z) + \frac{\lambda}{2}(S_+ \otimes \hat{L}_- + S_- \otimes \hat{L}_+)\right]\begin{pmatrix} 1 \\ 0 \end{pmatrix} \otimes Y_{1,0}.$$

Thus

$$\hat{H}|1\rangle = \lambda\left[S_z\begin{pmatrix} 1 \\ 0 \end{pmatrix} \otimes \hat{L}_zY_{1,0}\right] + \frac{\lambda}{2}\left[S_+\begin{pmatrix} 1 \\ 0 \end{pmatrix} \otimes \hat{L}_-Y_{1,0} + S_-\begin{pmatrix} 1 \\ 0 \end{pmatrix} \otimes \hat{L}_+Y_{1,0}\right].$$

Finally

$$\hat{H}|1\rangle = \frac{\lambda}{2} S_- \begin{pmatrix} 1 \\ 0 \end{pmatrix} \otimes \hat{L}_+ Y_{1,0} = \frac{\lambda}{\sqrt{2}} \hbar^2 |6\rangle.$$

Analogously, we find

$$\hat{H}|2\rangle = -\frac{\lambda\hbar^2}{2}|2\rangle + \frac{\lambda\hbar^2}{\sqrt{2}}|4\rangle$$

$$\hat{H}|3\rangle = \frac{\lambda\hbar^2}{2}|3\rangle$$

$$\hat{H}|4\rangle = \frac{\lambda\hbar^2}{2}|2\rangle$$

$$\hat{H}|5\rangle = \frac{\lambda\hbar^2}{2}|5\rangle$$

$$\hat{H}|6\rangle = -\frac{\lambda\hbar^2}{2}|6\rangle + \frac{\lambda\hbar^2}{\sqrt{2}}|1\rangle.$$

Hence the states $|3\rangle$ and $|5\rangle$ are eigenstates with the eigenvalues $E_{1,2} = \lambda\hbar^2/2$. The states $|1\rangle$ and $|6\rangle$ form a two-dimensional subspace. The matrix representation is given by

$$\begin{pmatrix} 0 & \dfrac{\lambda\hbar^2}{\sqrt{2}} \\ \dfrac{\lambda\hbar^2}{\sqrt{2}} & -\dfrac{\lambda\hbar^2}{2} \end{pmatrix}.$$

The eigenvalues are

$$E_{3,4} = -\frac{\lambda\hbar^2}{2} \pm \frac{3\lambda\hbar^2}{4}.$$

Analogously, the states $|2\rangle$ and $|4\rangle$ form a two-dimensional subspace. The matrix representation is given by

$$\begin{pmatrix} -\dfrac{\lambda\hbar^2}{2} & \dfrac{\lambda\hbar^2}{\sqrt{2}} \\ \dfrac{\lambda\hbar^2}{\sqrt{2}} & 0 \end{pmatrix}.$$

The eigenvalues are

$$E_{5,6} = -\frac{\lambda\hbar^2}{2} \pm \frac{3\lambda\hbar^2}{4}.$$

The Hamilton operator (1) describes the *spin-orbit coupling*.

Chapter 3

Nambu Mechanics

In *Nambu mechanics* the equations of motion are given by

$$\frac{du_j}{dt} = \frac{\partial(u_j, I_1, \ldots, I_{n-1})}{\partial(u_1, u_2, \ldots, u_n)}, \qquad j = 1, 2, \ldots, n$$

where $\partial(u_j, I_1, \ldots, I_{n-1})/\partial(u_1, u_2, \ldots, u_n)$ denotes the *Jacobian determinant* and $I_k : \mathbb{R}^n \to \mathbb{R}$ $(k = 1, \ldots, n-1)$ are $n-1$ smooth functions. Equation (1) can also be written in summation convention as

$$\frac{du_j}{dt} = \varepsilon_{jl_1 \ldots l_{n-1}}(\partial_{l_1} I_1) \cdots (\partial_{l_{n-1}} I_{n-1}), \qquad j = 1, 2, \ldots, n$$

where $\varepsilon_{jl_1 \ldots l_{n-1}}$ is the generalized Levi-Civita symbol and $\partial_j \equiv \partial/\partial u_j$.

Problem 1. Show that I_k $(k = 1, \ldots, n-1)$ are first integrals of system

$$\frac{du_j}{dt} = \frac{\partial(u_j, I_1, \ldots, I_{n-1})}{\partial(u_1, u_2, \ldots, u_n)}, \qquad j = 1, 2, \ldots, n. \tag{1}$$

Solution 1. The proof that $I_1, \ldots, I_{n-1}$ are first integrals of system (1) is as follows. We have (summation convention is used)

$$\frac{dI_i}{dt} = \frac{\partial I_i}{\partial u_j}\frac{du_j}{dt} = \varepsilon_{jl_1 \ldots l_{n-1}}(\partial_j I_i)(\partial_{l_1} I_1) \cdots (\partial_{l_{n-1}} I_{n-1})$$

$$= \frac{\partial(I_i, I_1, \ldots, I_{n-1})}{\partial(u_1, \ldots, u_n)}.$$

Thus $dI_i/dt = 0$ as the Jacobian matrix has two equal rows and is therefore singular. If the first integrals are polynomials, then the dynamical system (1) is algebraically completely integrable.

Problem 2.   Let

$$\frac{du_j}{dt} = \frac{\partial(u_j, I_1, \ldots, I_{n-1})}{\partial(u_1, u_2, \ldots, u_n)}, \qquad j = 1, 2, \ldots, n. \tag{1}$$

From problem (1) we know that I_k $(k = 1, \ldots, n-1)$ are first integrals. Consider $n = 3$ and assume that

$$I_1(\mathbf{u}) = a_{11}u_1^2/2 + a_{12}u_1u_2 + \cdots + a_{33}u_3^2/2 + a_1u_1 + a_2u_2 + a_3u_3$$

$$I_2(\mathbf{u}) = b_{11}u_1^2/2 + b_{12}u_1u_2 + \cdots + b_{33}u_3^2/2 + b_1u_1 + b_2u_2 + b_3u_3.$$

Find the dynamical system.

Solution 2.   For $n = 3$ the equations of motion are given by

$$\frac{du_1}{dt} = \frac{\partial I_1}{\partial u_2}\frac{\partial I_2}{\partial u_3} - \frac{\partial I_1}{\partial u_3}\frac{\partial I_2}{\partial u_2}$$

$$\frac{du_2}{dt} = \frac{\partial I_1}{\partial u_3}\frac{\partial I_2}{\partial u_1} - \frac{\partial I_1}{\partial u_1}\frac{\partial I_2}{\partial u_3}$$

$$\frac{du_3}{dt} = \frac{\partial I_1}{\partial u_1}\frac{\partial I_2}{\partial u_2} - \frac{\partial I_1}{\partial u_2}\frac{\partial I_2}{\partial u_1}.$$

If I_1 and I_2 are quadratic polynomials given above then the equations of motion are given by

$$\frac{du_1}{dt} = (a_{12}b_{13} - a_{13}b_{12})u_1^2 + (a_{12}b_{23} + a_{22}b_{13} - a_{13}b_{22} - a_{23}b_{12})u_1u_2$$

$$+ (a_{12}b_{33} + a_{23}b_{13} - a_{13}b_{23} - a_{33}b_{12})u_1u_3 + (a_{22}b_{23} - a_{23}b_{22})u_2^2$$

$$+ (a_{22}b_{33} - a_{33}b_{22})u_2u_3 + (a_{23}b_{33} - a_{33}b_{23})u_3^2$$

$$+ (a_{12}b_3 + a_2b_{13} - a_{13}b_2 - a_3b_{12})u_1 + (a_{22}b_3 + a_2b_{23} - a_{23}b_2 - a_3b_{22})u_2$$

$$+ (a_{23}b_3 + a_2b_{33} - a_{33}b_2 - a_3b_{23})u_3 + a_2b_3 - a_3b_2$$

$$\frac{du_2}{dt} = (a_{13}b_{11} - a_{11}b_{13})u_1^2 + (a_{13}b_{12} + a_{23}b_{11} - a_{11}b_{23} - a_{12}b_{13})u_1u_2$$

$$+ (a_{33}b_{11} - a_{11}b_{33})u_1u_3 + (a_{23}b_{12} - a_{12}b_{23})u_2^2$$

$$+ (a_{23}b_{13} + a_{33}b_{12} - a_{12}b_{33} - a_{13}b_{23})u_2u_3 + (a_{33}b_{13} - a_{13}b_{33})u_3^2$$

$$+ (a_{13}b_1 + a_3b_{11} - a_{11}b_3 - a_1b_{13})u_1 + (a_{23}b_1 + a_3b_{12} - a_{12}b_3 - a_1b_{23})u_2$$

$$+ (a_{33}b_1 + a_3b_{13} - a_{13}b_3 - a_1b_{33})u_3 + a_3b_1 - a_1b_3$$

$$\frac{du_3}{dt} = (a_{11}b_{12} - a_{12}b_{11})u_1^2 + (a_{11}b_{22} - a_{22}b_{11})u_1u_2$$
$$+ (a_{11}b_{23} + a_{13}b_{12} - a_{12}b_{13} - a_{23}b_{11})u_1u_3$$
$$+ (a_{12}b_{22} - a_{22}b_{12})u_2^2 + (a_{12}b_{23} + a_{13}b_{22} - a_{22}b_{13} - a_{23}b_{12})u_2u_3$$
$$+ (a_{13}b_{23} - a_{23}b_{13})u_3^2 + (a_{11}b_2 + a_1b_{12} - a_{12}b_1 - a_2b_{11})u_1$$
$$+ (a_{12}b_2 + a_1b_{22} - a_{22}b_1 - a_2b_{12})u_2 + (a_{13}b_2 + a_1b_{23} - a_{23}b_1 - a_2b_{13})u_3$$
$$+ a_1b_2 - a_2b_1 \, .$$

Problem 3. Let

$$\Omega_1 \frac{du_1}{dt} = (\Omega_3 - \Omega_2)u_2u_3 \tag{1a}$$

$$\Omega_2 \frac{du_2}{dt} = (\Omega_1 - \Omega_3)u_3u_1 \tag{1b}$$

$$\Omega_3 \frac{du_3}{dt} = (\Omega_2 - \Omega_1)u_1u_2 \tag{1c}$$

where Ω_1, Ω_2 and Ω_3 are positive constants. This dynamical system describes *Euler's rigid body motion*. Show that this system can be derived within Nambu mechanics.

Solution 3. Using the result from problem 2 the equations of motion can be derived within Nambu mechanics from the first integrals

$$I_1(\mathbf{u}) = \frac{1}{2}(\Omega_1 u_1^2 + \Omega_2 u_2^2 + \Omega_3 u_3^2)$$

$$I_2(\mathbf{u}) = \frac{1}{2}\left(\frac{\Omega_1}{\Omega_2\Omega_3}u_1^2 + \frac{\Omega_2}{\Omega_3\Omega_1}u_2^2 + \frac{\Omega_3}{\Omega_1\Omega_2}u_3^2 \right).$$

The first integral I_1 represents the total kinetic energy. The general solution of system (1) can be expressed in terms of Jacobi elliptic functions.

Problem 4. The nonrelativistic motion of a charged particle with mass m and charge q in a constant electric field $\mathbf{E} = (E_1, E_2, E_3)$ and constant magnetic field $\mathbf{B} = (B_1, B_2, B_3)$ is given by

$$\frac{dv_1}{dt} = \frac{q}{m}E_1 + \frac{q}{m}(B_3v_2 - B_2v_3)$$

$$\frac{dv_2}{dt} = \frac{q}{m}E_2 + \frac{q}{m}(B_1v_3 - B_3v_1)$$

$$\frac{dv_3}{dt} = \frac{q}{m}E_3 + \frac{q}{m}(B_2v_1 - B_1v_2)$$

where $\mathbf{v} = (v_1, v_2, v_3)$ denotes the velocity of the particle. If $B_1 = B_2 = 0$, $B_3 \neq 0$ and $E_2 = E_3 = 0$, $E_1 \neq 0$, then we obtain

$$\frac{dv_1}{dt} = \frac{q}{m}E_1 + \frac{q}{m}B_3v_2$$

$$\frac{dv_2}{dt} = -\frac{q}{m}B_3v_1$$

$$\frac{dv_3}{dt} = 0.$$

Show that this dynamical system can be derived within Nambu mechanics.

Solution 4.   Using the result from problem (2) we find that the dynamical system can be derived from the first integrals

$$I_1(\mathbf{v}) = -v_3, \qquad I_2(\mathbf{v}) = \frac{1}{2}B_3^2v_1^2 + \frac{1}{2}B_3^2v_2^2 + E_1B_3v_2.$$

Problem 5.   Consider the hierarchy of autonomous system of first order ordinary differential equations

$$\frac{du_j}{dt} = cu_1\cdots u_{j-1}\hat{u}_ju_{j+1}\cdots u_n$$

where $n \geq 2$ and $j = 1, 2, \ldots, n$, $\hat{u}_j$ indicates omission and c is a nonzero constant. Find the first integrals. Can the dynamical system be derived from these first integrals using Nambu mechanics?

Solution 5.   We find $n - 1$ first integrals

$$\frac{1}{2}c(u_1^2 - u_2^2), \quad \frac{1}{2}c(u_2^2 - u_3^2), \quad \ldots, \quad \frac{1}{2}c(u_{n-1}^2 - u_n^2).$$

Using these first integrals and Nambu mechanics we obtain the dynamical system.

Chapter 4

Gateaux and Fréchet Derivatives

Let W_1 be some topological vector space and f a (nonlinear) mapping $f : W \to W_1$. We call f *Gateaux differentiable* in $u \in W$ if there exists a mapping $\theta \in L(W, W_1)$ such that for all $v \in W$

$$\lim_{\epsilon \to 0} \frac{1}{\epsilon}(f(u + \epsilon v) - f(u) - \epsilon \theta v) = 0$$

in the topology of W_1. The linear mapping $\theta \in L(W, W_1)$ is the called *Gateaux derivative* of f in u and is written as $\theta = f'(u)$. The Gateaux derivative is a linear operation.

Let V and W be two Banach spaces with the norms $\|\cdot\|_V$ and $\|\cdot\|_W$, respectively. Let $U \subset V$ be an open subset of V. A function $f : U \to W$ is called Fréchet differentiable at $p \in U$ if there exists a bounded operator $A_p : V \to W$ such that

$$\lim_{h \to \infty} \frac{\|f(p + h) - f(p) - A_p(h)\|_W}{\|h\|_V} = 0.$$

The Fréchet derivative is a linear operation. The chain rule also applies.

If the Fréchet derivative of a mapping exists, then the Gateaux derivative of this mapping also exists.

Problem 1. Let

$$\frac{du_1}{dt} - \sigma(u_2 - u_1) = 0$$

$$\frac{du_2}{dt} + u_1 u_3 - ru_1 + u_2 = 0$$

$$\frac{du_3}{dt} - u_1 u_2 + bu_3 = 0$$

where σ, b and r are positive constants. The left-hand side defines a map f. Calculate the Gateaux derivative. This system of differential equations is called the *Lorenz model*. The equation $\theta \mathbf{v} = \mathbf{0}$ is called the corresponding *variational equation* (or *linearized equation*). Calculate the variational equation of the Lorenz model.

Solution 1. Since

$$f(\mathbf{u} + \epsilon \mathbf{v}) - f(\mathbf{u}) = \begin{pmatrix} \epsilon dv_1/dt - \epsilon\sigma(v_2 - v_1) \\ \epsilon dv_2/dt + \epsilon u_1 v_3 + \epsilon u_3 v_1 + \epsilon^2 v_1 v_3 - \epsilon r v_1 + \epsilon v_2 \\ \epsilon dv_3/dt - \epsilon u_1 v_2 - \epsilon u_2 v_1 - \epsilon^2 v_1 v_2 + \epsilon b v_3 \end{pmatrix}$$

we obtain

$$\theta \mathbf{v} = \begin{pmatrix} dv_1/dt - \sigma(v_2 - v_1) \\ dv_2/dt + u_1 v_3 + u_3 v_1 - r v_1 + v_2 \\ dv_3/dt - u_1 v_2 - u_2 v_1 + b v_3 \end{pmatrix}.$$

Then the variational equation $\theta \mathbf{v} = \mathbf{0}$ takes the form

$$\frac{dv_1}{dt} = \sigma(v_2 - v_1)$$

$$\frac{dv_2}{dt} = -u_1 v_3 - u_3 v_1 + r v_1 - v_2$$

$$\frac{dv_3}{dt} = u_1 v_2 + u_2 v_1 - b v_3.$$

To solve this system we have find a solution of (1) to solve system (2).

Problem 2. The difference equation

$$u_{t+1} = 4u_t(1 - u_t), \qquad t = 0, 1, 2, \ldots \tag{1}$$

is called the *logistic equation*, where $u_0 \in [0, 1]$. It can be shown that $u_t \in [0, 1]$ for all $t \in \mathbb{N}$. Equation (1) can also be written as the map $f : [0, 1] \to [0, 1]$

$$f(u) = 4u(1 - u). \tag{2}$$

Then u_t is the t-th iterate of f. Calculate the Gateaux derivative of the map f. Find the variational equation of (1).

Solution 2. We obtain

$$f(u + \epsilon v) - f(u) = 4\epsilon v - 8\epsilon u v - 4\epsilon^2 v^2 .$$

Therefore

$$\frac{d}{d\epsilon}(f(u + \epsilon v) - f(u))\Big|_{\epsilon=0} = 4v - 8uv = 4(1 - 2u)v .$$

Thus we find that the Gateaux derivative of the the map is given by

$$\theta v = 4v - 8uv = 4(1 - 2u)v .$$

The variational equation of the nonlinear difference equation (1) is then

$$v_{t+1} = 4v_t - 8u_t v_t = 4(1 - 2u_t)v_t .$$

The stability of fixed points and periodic solutions is studied with the help of the variational equation.

Problem 3. Consider the map

$$f(u) := \frac{\partial u}{\partial t} - 6u\frac{\partial u}{\partial x} - \frac{\partial^3 u}{\partial x^3} .$$

Calculate the Gateaux derivative. Calculate the variational equation. The nonlinear partial differential equation $f(u) = 0$, i.e.

$$\frac{\partial u}{\partial t} = 6u\frac{\partial u}{\partial x} + \frac{\partial^3 u}{\partial x^3}$$

is called the *Korteweg-de Vries equation*.

Solution 3. We obtain

$$f(u + \epsilon v) - f(u) = \epsilon\frac{\partial v}{\partial t} - 6\epsilon v\frac{\partial u}{\partial x} - 6\epsilon u\frac{\partial v}{\partial x} - 6\epsilon^2 v\frac{\partial v}{\partial x} - \epsilon\frac{\partial^3 v}{\partial x^3} .$$

It follows that

$$\lim_{\epsilon \to 0} \frac{1}{\epsilon}(f(u + \epsilon v) - f(u)) = \frac{\partial v}{\partial t} - 6\frac{\partial u}{\partial x}v - 6u\frac{\partial v}{\partial x} - \frac{\partial^3 v}{\partial x^3} .$$

Therefore

$$\theta v \equiv f'(u)v = \frac{\partial v}{\partial t} - 6\frac{\partial u}{\partial x}v - 6u\frac{\partial v}{\partial x} - \frac{\partial^3 v}{\partial x^3} .$$

From this equation we find the variational equation $\theta v = 0$ as

$$\frac{\partial v}{\partial t} = 6\frac{\partial u}{\partial x}v + 6u\frac{\partial v}{\partial x} + \frac{\partial^3 v}{\partial x^3} .$$

This is a linear partial differential equation in v. A solution u from the Korteweg-de Vries equation has to be inserted. An example is the trivial solution $u(x, t) = 0$. Then we obtain the linear partial differential equation

$$\frac{\partial v}{\partial t} = \frac{\partial^3 v}{\partial x^3}.$$

Using this equation we can study the stability of the solution $u(x, t) = 0$.

Problem 4.   The Gateaux derivative of an operator-valued function $R(u)$ is defined as

$$R'(u)[v]w := \frac{\partial [R(u + \epsilon v)w]}{\partial \epsilon}\bigg|_{\epsilon=0} \qquad (1)$$

where ϵ is a real parameter. We say that $R'(u)[v]w$ is the derivative of $R(u)$ evaluated at v and then applied to w, where w, v and u are smooth functions of $x_1, \ldots, x_n$ and t. Let $n = 1$ and

$$R(u) := \frac{\partial}{\partial x} + u + \frac{\partial u}{\partial x} D_x^{-1}$$

where u is a smooth function of x and t and

$$D_x^{-1} f(x) := \int^x f(s) ds.$$

Calculate the Gateaux derivative of $R(u)$.

Solution 4.   From (1) we find

$$R(u + \epsilon v)w = \frac{\partial w}{\partial x} + (u + \epsilon v)w + \frac{\partial}{\partial x}(u + \epsilon v)D_x^{-1}w$$

or

$$R(u + \epsilon v)w = \frac{\partial w}{\partial x} + uw + \frac{\partial u}{\partial x}D_x^{-1}w + \epsilon vw + \epsilon \frac{\partial v}{\partial x}D_x^{-1}w.$$

Therefore

$$R'(u)[v]w = vw + \frac{\partial v}{\partial x}D_x^{-1}w \equiv \left(v + \frac{\partial v}{\partial x}D_x^{-1}\right)w.$$

Problem 5.   Let $f, g : W \longrightarrow W$ be two maps, where W is a topological vector space ($u \in W$). Assume that the Gateaux derivative of f and g exists, i.e.

$$f'(u)[v] := \frac{\partial f(u + \epsilon v)}{\partial \epsilon}\bigg|_{\epsilon=0}, \qquad g'(u)[v] := \frac{\partial g(u + \epsilon v)}{\partial \epsilon}\bigg|_{\epsilon=0}. \qquad (1)$$

The *Lie product* (or *commutator*) of f and g is defined by

$$[f, g] := f'(u)[g] - g'(u)[f].\qquad(2)$$

Let

$$f(u) := \frac{\partial u}{\partial t} - u\frac{\partial u}{\partial x} - \frac{\partial^3 u}{\partial x^3}, \qquad g(u) := \frac{\partial u}{\partial t} - \frac{\partial^2 u}{\partial x^2}.\qquad(3)$$

Calculate the commutator $[f, g]$.

Solution 5. Applying the definition (1) we find

$$f'(u)[g] \equiv \frac{\partial}{\partial \epsilon}\left[\frac{\partial}{\partial t}(u + \epsilon g(u))\right]_{\epsilon=0}$$

$$-\frac{\partial}{\partial \epsilon}\left[(u + \epsilon g(u))\frac{\partial}{\partial x}(u + \epsilon g(u))\right]_{\epsilon=0}$$

$$-\frac{\partial}{\partial \epsilon}\left[\frac{\partial^3}{\partial x^3}(u + \epsilon g(u))\right]_{\epsilon=0}.$$

Consequently,

$$f'(u)[g] = \frac{\partial g(u)}{\partial t} - g(u)\frac{\partial u}{\partial x} - u\frac{\partial g(u)}{\partial x} - \frac{\partial^3 g(u)}{\partial x^3}.\qquad(4)$$

For the second term on the right-hand side of (2) we find

$$g'(u)[f] = \frac{\partial}{\partial \epsilon}\left[\frac{\partial}{\partial t}(u + \epsilon f(u))\right]_{\epsilon=0} - \frac{\partial}{\partial \epsilon}\left[\frac{\partial^2}{\partial x^2}(u + \epsilon f(u))\right]_{\epsilon=0}.$$

Consequently

$$g'(u)[f] = \frac{\partial f(u)}{\partial t} - \frac{\partial^2 f(u)}{\partial x^2}.$$

Hence

$$[f, g] = \frac{\partial}{\partial t}(g(u) - f(u)) - g(u)\frac{\partial u}{\partial x} - u\frac{\partial g(u)}{\partial x} - \frac{\partial^3 g(u)}{\partial x^3} + \frac{\partial^2 f(u)}{\partial x^2}.$$

Inserting f and g we find

$$[f, g] = \frac{\partial}{\partial t}\left(-\frac{\partial^2 u}{\partial x^2} + u\frac{\partial u}{\partial x} + \frac{\partial^3 u}{\partial x^3}\right) - \left(\frac{\partial u}{\partial t} - \frac{\partial^2 u}{\partial x^2}\right)\frac{\partial u}{\partial x}$$

$$- u\left(\frac{\partial^2 u}{\partial t \partial x} - \frac{\partial^3 u}{\partial x^3}\right) - \frac{\partial^4 u}{\partial t \partial x^3} + \frac{\partial^5 u}{\partial x^5} + \frac{\partial^3 u}{\partial x^2 \partial t}$$

$$- 3\frac{\partial u}{\partial x}\frac{\partial^2 u}{\partial x^2} - u\frac{\partial^3 u}{\partial x^3} - \frac{\partial^5 u}{\partial x^5}.$$

It follows that
$$[f, g] = -2 \frac{\partial u}{\partial x} \frac{\partial^2 u}{\partial x^2}.$$

Problem 6.   Let
$$\frac{\partial u}{\partial t} = A(u) \tag{1}$$
be an evolution equation, where A is a differential expression, i.e. A depends on u and its partial derivatives with respect to $x_1, \ldots, x_n$ and on the variables $x_1, \ldots, x_n$ and t. A function
$$\sigma\left(x_1, \ldots, x_n, t, u(\mathbf{x}, t), \frac{\partial u}{\partial x_1}, \ldots\right) \tag{2}$$
is called a *symmetry generator* of (1) if
$$\frac{\partial \sigma}{\partial t} = A'(u)[\sigma] \tag{3}$$
where $A'(u)[\sigma]$ is the Gateaux derivative of A, i.e.
$$A'(u)[\sigma] = \frac{\partial}{\partial \epsilon} A(u + \epsilon \sigma)|_{\epsilon=0}. \tag{4}$$
Let $n = 1$ and consider the Korteweg-de Vries equation
$$\frac{\partial u}{\partial t} = 6u \frac{\partial u}{\partial x} + \frac{\partial^3 u}{\partial x^3}.$$
Show that
$$\sigma = 3t \frac{\partial u}{\partial x} + \frac{1}{2}$$
is a symmetry generator.

Solution 6.   First we calculate the Gateaux derivative of A, i.e.
$$A'(u)[\sigma] = \frac{\partial}{\partial \epsilon} A(u + \epsilon \sigma)|_{\epsilon=0}$$
$$= \frac{\partial}{\partial \epsilon} \left(6(u + \epsilon \sigma) \left(\frac{\partial(u + \epsilon \sigma)}{\partial x} \right) + \frac{\partial^3}{\partial x^3} (u + \epsilon \sigma) \right) \bigg|_{\epsilon=0}.$$
Therefore
$$A'(u)[\sigma] = 6 \frac{\partial u}{\partial x} \sigma + 6u \frac{\partial \sigma}{\partial x} + \frac{\partial^3 \sigma}{\partial x^3}.$$
From σ we find using differentiation with respect to t
$$\frac{\partial \sigma}{\partial t} = 3 \frac{\partial u}{\partial x} + 3t \frac{\partial^2 u}{\partial x \partial t}.$$

Inserting σ into $A'(u)[\sigma]$ gives

$$A'(u)[\sigma] = 6\frac{\partial u}{\partial x}\left(3t\frac{\partial u}{\partial x} + \frac{1}{2}\right) + 18ut\frac{\partial^2 u}{\partial x^2} + 3t\frac{\partial^4 u}{\partial x^4}$$

or

$$A'(u)[\sigma] = 18tu\frac{\partial^2 u}{\partial x^2} + 18t\left(\frac{\partial u}{\partial x}\right)^2 + 3\frac{\partial u}{\partial x} + 3t\frac{\partial^4 u}{\partial x^4}.$$

From the Korteweg-de Vries equation we obtain

$$\frac{\partial^2 u}{\partial x \partial t} = 6\left(\frac{\partial u}{\partial x}\right)^2 + 6u\frac{\partial^2 u}{\partial x^2} + \frac{\partial^4 u}{\partial x^4}.$$

Inserting the right-hand side of this equation into the equation for $\partial\sigma/\partial t$ leads to

$$\frac{\partial\sigma}{\partial t} = 3\frac{\partial u}{\partial x} + 18t\left(\frac{\partial u}{\partial x}\right)^2 + 18tu\frac{\partial^2 u}{\partial x^2} + 3t\frac{\partial^4 u}{\partial x^4}.$$

This proves (3).

Problem 7. Let

$$\frac{\partial u}{\partial t} = A(u)$$

be an evolution equation, where A is a nonlinear differential expression, i.e. A depends on u and its partial derivatives with respect to $x_1, \ldots, x_n$ and on the variables $x_1, \ldots, x_n$ and t. An operator $R(u)$ is called a *recursion operator* if

$$R'(u)[A(u)]v = [A'(u)v, R(u)v] \tag{1}$$

where

$$R'(u)[A(u)]v := \frac{\partial}{\partial\epsilon}(R(u + \epsilon A(u))v)|_{\epsilon=0} \tag{2}$$

and the commutator is defined as

$$[A'(u)v, R(u)v] := \frac{\partial}{\partial\epsilon}A'(u)(v+\epsilon R(u)v)|_{\epsilon=0} - \frac{\partial}{\partial\epsilon}R(u)(v+\epsilon A'(u)v)|_{\epsilon=0}. \tag{3}$$

Let $n = 1$. Consider the nonlinear diffusion equation

$$\frac{\partial u}{\partial t} = \frac{\partial^2 u}{\partial x^2} + 2u\frac{\partial u}{\partial x} \tag{4}$$

i.e.

$$A(u) \equiv \frac{\partial^2 u}{\partial x^2} + 2u\frac{\partial u}{\partial x}.$$

Show that

$$R(u) = \frac{\partial}{\partial x} + u + \frac{\partial u}{\partial x}D_x^{-1}$$

is a recursion operator of (4), where

$$D_x^{-1} f(x) := \int^x f(s) ds.$$

Solution 7.   First we calculate the left-hand side of (1). When we apply the operator $R(u)$ to v we obtain

$$R(u)v = \frac{\partial v}{\partial x} + uv + \frac{\partial u}{\partial x} D_x^{-1} v.$$

From definition (2) we find that the left-hand side is given by

$$R'(u)[A(u)]v = \frac{\partial}{\partial \epsilon} \left(\frac{\partial}{\partial x} + u + \epsilon A(u) + \frac{\partial(u + \epsilon A(u))}{\partial x} D_x^{-1} \right) v$$

$$= A(u)v + \frac{\partial A(u)}{\partial x} D_x^{-1} v$$

$$= \left(\frac{\partial^2 u}{\partial x^2} + 2u \frac{\partial u}{\partial x} \right) v + \left(\frac{\partial^3 u}{\partial x^3} + 2 \left(\frac{\partial u}{\partial x} \right)^2 + 2u \frac{\partial^2 u}{\partial x^2} \right) D_x^{-1} v.$$

To calculate the commutator we first have to find the Gateaux derivative of A. Since

$$A'(u)[v] := \left. \frac{\partial A(u + \epsilon v)}{\partial \epsilon} \right|_{\epsilon=0}$$

we find

$$A'(u)[v] = \left. \frac{\partial}{\partial \epsilon} \left(\frac{\partial^2}{\partial x^2} (u + \epsilon v) + 2(u + \epsilon v) \frac{\partial}{\partial x} (u + \epsilon v) \right) \right|_{\epsilon=0}$$

$$= \frac{\partial^2 v}{\partial x^2} + 2u \frac{\partial v}{\partial x} + 2 \frac{\partial u}{\partial x} v.$$

Therefore

$$A'(u)v = \frac{\partial^2 v}{\partial x^2} + 2u \frac{\partial v}{\partial x} + 2 \frac{\partial u}{\partial x} v.$$

For the first term of the commutator we find

$$\frac{\partial}{\partial \epsilon} A'(u)(v + \epsilon R(u)v)|_{\epsilon=0} = \frac{\partial}{\partial \epsilon} \left[\frac{\partial^2}{\partial x^2} (v + \epsilon R(u)v) + 2u \frac{\partial}{\partial x} (v + \epsilon R(u)v) \right.$$

$$\left. + 2 \frac{\partial u}{\partial x} (v + \epsilon R(u)v) \right]_{\epsilon=0}$$

$$= \frac{\partial^3 v}{\partial x^3} + 3 \frac{\partial^2 u}{\partial x^2} v + 5 \frac{\partial u}{\partial x} \frac{\partial v}{\partial x} + 3u \frac{\partial^2 v}{\partial x^2} + 6u \frac{\partial u}{\partial x} v$$

$$+ 2u^2 \frac{\partial v}{\partial x} + \left(\frac{\partial^3 u}{\partial x^3} + 2 \left(\frac{\partial u}{\partial x} \right)^2 + 2u \frac{\partial^2 u}{\partial x^2} \right) D_x^{-1} v.$$

For the second term of the commutator we obtain

$$\frac{\partial}{\partial \epsilon} R(u)(v + \epsilon A'(u)v)|_{\epsilon=0} = \frac{\partial}{\partial \epsilon} \left[\frac{\partial}{\partial x}(v + \epsilon A'(u)v) \right.$$

$$\left. + u(v + \epsilon A'(u)v) + \frac{\partial u}{\partial x} D_x^{-1}(v + \epsilon A'(u)v) \right]_{\epsilon=0}$$

$$= \frac{\partial A'(u)v}{\partial x} + uA'(u)v + \frac{\partial u}{\partial x} D_x^{-1} A'(u)v.$$

Using the identity $D_x^{-1}(v\partial u/\partial x + u\partial v/\partial x) = uv$, where D_x^{-1} is defined above we find

$$\frac{\partial}{\partial \epsilon} R(u)(v + \epsilon A'(u)v)|_{\epsilon=0} = \frac{\partial^3 v}{\partial x^3} + 2\frac{\partial^2 u}{\partial x^2}v + 5\frac{\partial u}{\partial x}\frac{\partial v}{\partial x} + 3u\frac{\partial^2 v}{\partial x^2} + 4u\frac{\partial u}{\partial x}v + 2u^2\frac{\partial v}{\partial x}.$$

Combining these equations we find that the commutator $[A'(u)v, R(u)v]$ is given by the right-hand side of $R'(u)[A(u)]v$ which proves (1).

Problem 8. Let f be a continuously differentiable real function on the interval $I = [a, b]$, i.e.

$$f : C^1[a, b] \to \mathbb{R}.$$

The norm is defined as

$$\|f\| := \max_{a \le t \le b}(|f(t)|, |f'(t)|)$$

where $f'(t) \equiv df/dt$. Let $\Phi(t, y, z)$ be a given real function for $t \in I$ and y, z arbitrary. Φ is assumed to possess partial derivatives up to second order with respect to all arguments. Find the *Fréchet derivative* of the functional

$$Tf(t) := \int_a^b \Phi(t, f, f')dt.$$

Solution 8. Taylor's theorem yields

$$T(f + g) - T(f) = \int_a^b [\Phi(t, f + g, f' + g') - \Phi(t, f, f')]dt$$

$$= \int_a^b [g\Phi_f + g'\Phi_{f'} + g^2\tilde{\Phi}_{ff} + gg'\tilde{\Phi}_{ff'} + g'^2\tilde{\Phi}_{f'f'}]dt$$

where g is continuously differentiable in I and

$$\frac{\partial \Phi}{\partial f} \equiv \Phi_f := \frac{\partial \Phi(t, y, z)}{\partial y}\bigg|_{y \to f(t), z \to f'(t)}$$

$$\frac{\partial \Phi}{\partial f'} \equiv \Phi_{f'} := \left. \frac{\partial \Phi(t, y, z)}{\partial z} \right|_{y \to f(t), z \to f'(t)} .$$

The tilde indicates that the function has to be taken at some intermediate argument. With the norm we have $|g|, |g'| \le \|g\|$. Therefore the contributions of the second derivatives under the integral sign may be estimated by $\|g\| \, \epsilon(\|g\|)$ where ϵ is a null function. Thus the Fréchet derivative of $T(f)$ becomes

$$T'_{(f)}g = \int_a^b \left[\frac{\partial \Phi}{\partial f} g(t) + \frac{\partial \Phi}{\partial f'} g'(t) \right] dt .$$

If $f(a) = f_a$ and $f(b) = f_b$, then $g(a) = g(b) = 0$. *Partial integration* in this case yields

$$\int_a^b \frac{\partial \Phi}{\partial f'} g'(t) dt = - \int_a^b \left(\frac{d}{dt} \frac{\partial \Phi}{\partial f'} \right) g(t) dt .$$

Consequently, the Fréchet derivative of T is given by

$$T'_{(f)}g = \int_a^b \left[\frac{\partial \Phi}{\partial f} - \frac{d}{dt} \frac{\partial \Phi}{\partial f'} \right] g(t) dt .$$

In *variational calculus* we find that an extremum of Tf with the function $f \in C^2[a, b]$ and $f(a) = f_a$, $f(b) = f_b$ is given by

$$\frac{\partial \Phi}{\partial f} - \frac{d}{dt} \frac{\partial \Phi}{\partial f'} = 0 .$$

This equation is called the *Euler-Lagrange equation*.

Using the Gateaux derivative we start from

$$I(t, f(t), f'(t)) = \int_a^b F(t, f(t), f'(t)) dt .$$

Then

$$I(t, f(t) + \epsilon g(t), f'(t) + \epsilon g'(t)) = \int_a^b F(t, f(t) + \epsilon g(t), f'(t) + \epsilon g'(t)) dt .$$

We assume that that F admits a Taylor expansion around f and f'. Then

$$\left. \frac{d}{d\epsilon} I(t, f(t) + \epsilon g(t), f'(t) + \epsilon g'(t)) \right|_{\epsilon=0} = \int_a^b \left(\frac{\partial F}{\partial f} g + \frac{\partial F}{\partial f'} g' \right) dt .$$

From the boundary conditions $g(a) = g(b) = 0$ and integration by parts we obtain

$$\int_a^b \left(\frac{\partial F}{\partial f} - \frac{d}{dt} \frac{\partial F}{\partial f'} \right) g(t) dt = 0 .$$

Problem 9. The *Euler-Lagrange equation* can be written as

$$\frac{d}{dt}\frac{\partial F(t, f(t), f'(t))}{\partial f'} - \frac{\partial F}{\partial f} = 0. \tag{1}$$

Show that if F does not depend explicitly on t the Euler-Lagrange equation can be written as

$$\frac{d}{dt}\left(f'\frac{\partial F}{\partial f'} - F\right) = 0. \tag{2}$$

Solution 9. From (1) we find

$$f''\frac{\partial^2 F}{\partial f'^2} + f'\frac{\partial^2 F}{\partial f \partial f'} + \frac{\partial^2 F}{\partial f' \partial t} - \frac{\partial F}{\partial f} = 0.$$

Since F does not depend explicitly on t, this differential equation simplifies to

$$f''\frac{\partial^2 F}{\partial f'^2} + f'\frac{\partial^2 F}{\partial f \partial f'} - \frac{\partial F}{\partial f} = 0.$$

On the other hand from (2) we obtain

$$f'\left(f''\frac{\partial^2 F}{\partial f'^2} + f'\frac{\partial^2 F}{\partial f \partial f'} - \frac{\partial F}{\partial f}\right) = 0$$

where we assume that $f'(t) \neq 0$. Thus this proves that the Euler-Lagrange equation can be written in the form (2). From (2) it follows that

$$f'\frac{\partial F}{\partial f'} - F = \text{const}.$$

Problem 10. Show that if $M \in C[a, b]$ and

$$\int_a^b M(x)h(x)dx = 0 \tag{1}$$

for all $h \in \mathcal{H}$, then it is necessary that $M(x) = 0$ for all $x \in [a, b]$, where

$$\mathcal{H} := \{h : h \in C^1[a, b], \ h(a) = h(b) = 0\}.$$

Solution 10. We prove it by contradiction. Suppose that there exists an $x_0 \in (a, b)$ such that $M(x_0) \neq 0$. Then we may assume without loss of generality that $M(x_0) > 0$. Since $M \in C[a, b]$ there exists a neighbourhood $N^\delta(x_0) \subset (a, b)$ such that $M(x) > 0$ for all $x \in N^\delta(x_0)$. Let

$$h(x) = \begin{cases} (x - x_0 - \delta)^2(x - x_0 + \delta)^2 & \text{for } x \in N^\delta(x_0) \\ 0 & \text{for } x \notin N^\delta(x_0) \end{cases}$$

Clearly, $h \in \mathcal{H}$. With this choice of h we find

$$\int_a^b M(x)h(x)dx = \int_{x_0-\delta}^{x_0+\delta} M(x)(x-x_0-\delta)^2(x-x_0+\delta)^2 dx > 0$$

with contradicts our hypothesis. Thus $M(x) = 0$ for all $x \in (a,b)$. That $M(x)$ also has to vanish at the endpoints of the interval follows from the continuity of M.

Problem 11.   Let $\rho \in C[a,b]$. Consider the function

$$J(u(x)) = \int_a^b \rho(x)\sqrt{1+u'(x)^2}dx \qquad (1)$$

which is defined for all $u \in C^1[a,b]$. Find the Gateaux derivative.

Solution 11.   From

$$J(u+\epsilon v) = \int_a^b \rho(x)\sqrt{1+(u+\epsilon v)'(x)^2}dx$$

it follows that

$$\frac{\partial}{\partial \epsilon}J(u+\epsilon v) = \int_a^b \frac{\rho(x)(u+\epsilon v)'(x)v'(x)}{\sqrt{1+(u+\epsilon v)'(x)^2}}dx.$$

This is justified by the continuity of this last integrand on $[a,b] \times \mathbb{R}$. Setting $\epsilon = 0$ we obtain

$$\left.\frac{\partial}{\partial \epsilon}J(u+\epsilon v)\right|_{\epsilon=0} = \int_a^b \frac{\rho(x)u'(x)v'(x)}{\sqrt{1+u'(x)^2}}dx.$$

Problem 12.   The *Fréchet derivative* of a matrix function $f : \mathbb{C}^{n \times n}$ at a point $X \in \mathbb{C}^{n \times n}$ is a linear mapping $L_X : \mathbb{C}^{n \times n} \to \mathbb{C}^{n \times n}$ such that for all $Y \in \mathbb{C}^{n \times n}$

$$f(X+Y) - f(X) - L_X(Y) = o(\|Y\|).$$

Calculate the Fréchet derivative of $f(X) = X^2$.

Solution 12.   We have

$$f(X+Y) - f(X) = XY + YX + Y^2.$$

Thus

$$L_X(Y) = XY + YX.$$

The right-hand side is the anti-commutator of X and Y.

Chapter 5

Stability and Bifurcations

Let S be an arbitrary set. Consider a map $f : S \to S$. Let $x^* \in S$. Then x^* is called a *fixed point* of f if $f(x^*) = x^*$. Consider the difference equation $u_{t+1} = f(u_t)$ with $t = 0, 1, \ldots$ and given u_0 (initial value). Then u^* is a fixed point if $f(u^*) = u^*$.

Let

$$\frac{du_j}{dt} = F_j(\mathbf{u}), \qquad j = 1, 2, \ldots, n$$

be an autonomous system of ordinary differential equations of first order, where F_j are smooth functions with $F_j : \mathbb{R}^n \to \mathbb{R}$. A vector $\mathbf{u}^*$ is called a *fixed point* or *stationary point* or *steady-state solution* of the autonomous system if

$$F_j(\mathbf{u}^*) = 0, \qquad j = 1, 2, \ldots, n.$$

The *variational equation* is defined by

$$\frac{dv_j}{dt} = \sum_{k=1}^{n} \frac{\partial F_j}{\partial u_k}(\mathbf{u}(t))v_k, \qquad j = 1, 2, \ldots, n.$$

Problem 1. The nonlinear difference equation

$$u_{t+1} = 4u_t(1 - u_t) \tag{1}$$

is called the *logistic equation*, where $t = 0, 1, 2, \ldots$ and $u_0 \in [0, 1]$.
(i) Find the fixed points.

(ii) Study the stability of the fixed points.

Solution 1.   (i) Since $u_0 \in [0,1]$ we find $u_t \in [0,1]$ for $t = 1, 2, \ldots$. The *fixed points* are given by the solutions of the equation

$$u^* = 4u^*(1 - u^*).$$

Solving this algebraic equation gives the two fixed points $u_1^* = 0$, $u_2^* = \frac{3}{4}$. The fixed points are time-independent solutions.
(ii) The variational equation of (1) is given by

$$v_{t+1} = (4 - 8u_t)v_t.$$

Inserting the fixed point $u_1^* = 0$ into the variational equation yields

$$v_{t+1} = 4v_t.$$

For $v_0 \neq 0$ the quantity $|v_t|$ is growing and therefore the fixed point $u_1^* = 0$ is unstable. Inserting the fixed point $u_2^* = 3/4$ into the variational equation yields

$$v_{t+1} = -2v_t.$$

Consequently, if $v_0 \neq 0$, then $|v_t|$ is growing and therefore the fixed point $u_2^* = 3/4$ is also unstable.

Problem 2.   Consider the logistic equation $(1 \leq r \leq 4)$

$$u_{t+1} = ru_t(1 - u_t), \qquad t = 0, 1, 2, \ldots \qquad u_0 \in [0,1] \qquad (1)$$

Show that $r = 3$ is a bifurcation point.

Solution 2.   The fixed points (time-independent solutions) are given by the solution of the quadratic equation

$$u^* = ru^*(1 - u^*).$$

We find

$$u_1^* = 0, \qquad u_2^* = \frac{r - 1}{r}.$$

From $f_r(u) = ru(1 - u)$ we find

$$\frac{df_r}{du} = r(1 - 2u).$$

Thus the variational equation is given by

$$v_{t+1} = r(1 - 2u_t)v_t$$

where u_t is a solution of (1). Inserting the fixed point u_2^* into the variational equation gives

$$v_{t+1} = (2 - r)v_t.$$

Thus, if $r > 3$, then $|2 - r| > 1$ and the fixed point u_2^* is unstable. If $r < 3$, then $|2 - r| < 1$ and the fixed point u_2^* is stable, i.e. v_t tends to zero as $t \to \infty$. Thus at $r = 3$ we have a qualitative change in the behaviour of the solution of (1).

Problem 3. The Henòn map $\mathbf{f}_{a,b} : \mathbb{R}^2 \to \mathbb{R}^2$ is defined by

$$f_1(x, y) = 1 + y - ax^2, \qquad f_2(x, y) = bx$$

where $a, b \geq 0$ and $b < 1$.
(i) Establish for which values of a and b this map has fixed points.
(ii) Find the values of the fixed points.
(iii) Find the values of a and b for which the fixed points are attractive, repellent or saddle points.

Solution 3. (i) Let (x^*, y^*) be a *fixed point* of $\mathbf{f}_{a,b}$. Then it must satisfy the simultaneous equations $x^* = f_1(x^*, y^*)$, $y^* = f_2(x^*, y^*)$ i.e.

$$x^* = 1 + y^* - a(x^*)^2, \qquad y^* = bx^*.$$

If y^* is eliminated, one finds a quadratic equation in x^*,

$$a(x^*)^2 + (1 - b)x^* - 1 = 0.$$

This equation has real roots if

$$a \geq -\frac{(1 - b)^2}{4} := a_0(b).$$

(ii) Solving the quadratic equation one finds the fixed points

$$(x_1, bx_1), \quad (x_2, bx_2)$$

where

$$x_{1,2} = \frac{(b - 1) \pm \sqrt{(b - 1)^2 + 4a}}{2a}.$$

It is obvious that $x_1 \geq x_2$, with the equality occurring for $a = a_0(b)$.
(iii) The stability of the fixed points are determined by the *Jacobian matrix*, defined as

$$D\mathbf{f}_{a,b}(x, y) := \begin{pmatrix} \dfrac{\partial f_1}{\partial x}(x, y) & \dfrac{\partial f_1}{\partial y}(x, y) \\ \dfrac{\partial f_2}{\partial x}(x, y) & \dfrac{\partial f_2}{\partial y}(x, y) \end{pmatrix} = \begin{pmatrix} -2ax & 1 \\ b & 0 \end{pmatrix}.$$

The eigenvalues λ of this matrix follow from the characteristic equation

$$(-2ax - \lambda)(-\lambda) - b = 0.$$

We obtain

$$\lambda_{+,-} = -ax \pm \sqrt{a^2 x^2 + b}.$$

From this equation follows that

$$|\lambda_+| < (>)1 \text{ if } 1 + 2ax > (<)b$$

and

$$|\lambda_-| < (>)1 \text{ if } 1 - 2ax > (<)b.$$

Now consider the eigenvalues for the two fixed points.
(a) $x = x_1$: We have $|\lambda_+| < 1$ if

$$1 + (b - 1) + \sqrt{(b - 1)^2 + 4a} > b,$$

i.e.

$$\sqrt{(b - 1)^2 + 4a} > 0,$$

which is true for all $a > a_0(b)$. We have $|\lambda_-| < 1$ if

$$1 - \left[(b - 1) + \sqrt{(b - 1)^2 + 4a} \right] < b,$$

i.e.

$$a < \frac{3}{4}(1 - b)^2 := a_1(b).$$

Therefore (x_1, bx_1) is an attractor for a and b such that $a_0(b) < a < a_1(b)$.
(b) $x = x_2$: We have $|\lambda_+| > 1$ if

$$1 + \left[(b - 1) - \sqrt{(b - 1)^2 + 4a} \right] < b,$$

i.e.

$$-\sqrt{(b - 1)^2 + 4a} < 0,$$

which is true for all $a > a_0(b)$. We have $|\lambda_-| < 1$ if

$$1 - \left[(b - 1) - \sqrt{(b - 1)^2 + 4a} \right] > b.$$

Since $|b| < 1$, this equality is satisfied if $a < a_1(b)$. Therefore, (x_2, bx_2) is a saddle point for a and b such that $a_0(b) < a < a_1(b)$. From (a) and (b) follows that a saddle node bifurcation occurs at $a = a_0(b)$.

Problem 4. Find the fixed points of the system of differential equations (*Lorenz model*)

$$\frac{du_1}{dt} = \sigma(u_2 - u_1)$$

$$\frac{du_2}{dt} = -u_1 u_3 + r u_1 - u_2$$

$$\frac{du_3}{dt} = u_1 u_2 - b u_3$$

where b, σ and r are positive constants. Study the stability of the fixed points.

Solution 4. The system of equations

$$\sigma(u_2^* - u_1^*) = 0, \qquad -u_1^* u_3^* + r u_1^* - u_2^* = 0, \qquad u_1^* u_2^* - b u_3^* = 0$$

determine the fixed points. For all $r > 0$ the this system of equations possesses the solution (fixed point) $u_1^* = u_2^* = u_3^* = 0$. The variational equation of Lorenz model can be written in matrix form as

$$\begin{pmatrix} dv_1/dt \\ dv_2/dt \\ dv_3/dt \end{pmatrix} = \begin{pmatrix} -\sigma & \sigma & 0 \\ r - u_3(t) & -1 & -u_1(t) \\ u_2(t) & u_1(t) & -b \end{pmatrix} \begin{pmatrix} v_1 \\ v_2 \\ v_3 \end{pmatrix}.$$

Inserting this solution into the variational equation leads to a linear system of differential equations with constant coefficients, i.e.

$$\begin{pmatrix} dv_1/dt \\ dv_2/dt \\ dv_3/dt \end{pmatrix} = \begin{pmatrix} -\sigma & \sigma & 0 \\ r & -1 & 0 \\ 0 & 0 & -b \end{pmatrix} \begin{pmatrix} v_1 \\ v_2 \\ v_3 \end{pmatrix}.$$

From the matrix of the right-hand side of this equation we obtain the characteristic equation

$$(\lambda + b)(\lambda^2 + (\sigma + 1)\lambda + \sigma(1 - r)) = 0.$$

This equation has three real roots when $r > 0$. All are negative when $r < 1$. One becomes positive when $r > 1$. Consequently, when $r > 1$ the system becomes unstable. For $r > 1$ the system of equations admits two additional fixed points, namely

$$u_1^* = u_2^* = \sqrt{b(r-1)}, \qquad u_3^* = r - 1$$

and

$$u_1^* = u_2^* = -\sqrt{b(r-1)}, \qquad u_3^* = r - 1.$$

For either of these solutions, the characteristic equation of the matrix in the variational equation is given by

$$\lambda^3 + (\sigma + b + 1)\lambda^2 + (r + \sigma)b\lambda + 2\sigma b(r - 1) = 0 .$$

This equation possesses one real negative root and two complex-conjugate roots when $r > 1$. The complex conjugate roots are purely imaginary if the product of the coefficients of λ^2 and λ equals the constant term, i.e.

$$r = \frac{\sigma(\sigma + b + 3)}{\sigma - b - 1} .$$

Thus for r larger then the right-hand side of this equation the two additional fixed points are unstable.

The Lorenz model shows so-called *chaotic behaviour* for certain parameter values, for example $b = 8/3$, $\sigma = 10$ and $r = 28$.

Problem 5.   Consider the autonomous system of ordinary differential equations

$$\frac{du_1}{dt} = \mu u_1 + u_2 - u_1(u_1^2 + u_2^2)$$

$$\frac{du_2}{dt} = -u_1 + \mu u_2 - u_2(u_1^2 + u_2^2)$$

$$\frac{du_3}{dt} = u_3(1 - u_1^2 - u_2^2 - u_3^2)$$

where $\mu \in \mathbb{R}$ is a bifurcation parameter.
(i) Find the fixed points.
(ii) Study the stability of the fixed points.
(iii) Does *Hopf bifurcation* occur for this system?

Solution 5.   (i) The fixed points are determined by the equations

$$0 = \mu u_1^* + u_2^* - u_1^*(u_1^{*2} + u_2^{*2})$$

$$0 = -u_1^* + \mu u_2^* - u_2^*(u_1^{*2} + u_2^{*2})$$

$$0 = u_3^*(1 - u_1^{*2} - u_2^{*2} - u_3^{*2}) .$$

We find the three fixed points

$$\mathbf{e}_0 = \begin{pmatrix} 0 & 0 & 0 \end{pmatrix}, \qquad \mathbf{e}_1 = \begin{pmatrix} 0 & 0 & 1 \end{pmatrix}, \qquad \mathbf{e}_2 = \begin{pmatrix} 0 & 0 & -1 \end{pmatrix}.$$

(ii) Let us now study the stability of the fixed points. The linearized equation (variational equation) of the autonomous system of differential equations is given by

$$\frac{dv_1}{dt} = (\mu - 3u_1^2 - u_2^2)v_1 + (1 - 2u_1 u_2)v_2$$

$$\frac{dv_2}{dt} = (-1 - 2u_1 u_2)v_1 + (\mu - u_1^2 - 3u_2^2)v_2$$

$$\frac{dv_3}{dt} = -2u_1 u_3 v_1 - 2u_2 u_3 v_2 + (1 - u_1^2 - u_2^2 - 3u_3^2)v_3 \,.$$

From the linearized equation we obtain for the fixed point e_0 the matrix

$$A_0 = \begin{pmatrix} \mu & 1 & 0 \\ -1 & \mu & 0 \\ 0 & 0 & 1 \end{pmatrix} \,.$$

The eigenvalues of A_0 are $\lambda_1^0 = 1$, $\lambda_{2,3}^0 = \mu \pm i$. From the linearized equation we obtain for the fixed point e_1 the matrix

$$A_1 = \begin{pmatrix} \mu & 1 & 0 \\ -1 & \mu & 0 \\ 0 & 0 & -2 \end{pmatrix} \,.$$

The eigenvalues of A_1 are $\lambda_1^1 = -2$, $\lambda_{2,3}^1 = \mu \pm i$. From the linearized equation we obtain for the fixed point e_2 the matrix

$$A_2 = \begin{pmatrix} \mu & 1 & 0 \\ -1 & \mu & 0 \\ 0 & 0 & -2 \end{pmatrix} \,.$$

The eigenvalues of A_2 are $\lambda_1^2 = -2$, $\lambda_{2,3}^2 = \mu \pm i$. Thus the fixed point e_0 is unstable for all $\mu \in \mathbb{R}$. The fixed points e_1 and e_2 are asymptotically stable for $\mu < 0$ and asymptotically unstable for $\mu > 0$.

(iii) The condition of the *Hopf bifurcation* is satisfied. This means

$$\Re(\lambda_{2,3}(\mu = 0)) = 0$$

$$\Re\left(\frac{d\lambda_{2,3}}{d\mu}\bigg|_{\mu=0}\right) = 1 \neq 0$$

for all three fixed points, where $\Re$ denotes the real part. Thus in the neighbourhood of the fixed points we find periodic solutions. For e_0 and $\mu > 0$ we obtain

$$v_1^0(\mu, t) = \sqrt{\mu}\sin t, \qquad v_2^0(\mu, t) = \sqrt{\mu}\cos t, \qquad v_3^0(\mu, t) = 0 \,.$$

For e_1 and $1 \geq \mu > 0$ we find

$$v_1^1(\mu, t) = \sqrt{\mu}\sin t, \qquad v_2^1(\mu, t) = \sqrt{\mu}\cos t, \qquad v_3^1(\mu, t) = \sqrt{1 - \mu} \,.$$

For e_2 and $1 \geq \mu > 0$ we obtain

$$v_1^2(\mu, t) = \sqrt{\mu}\sin t, \qquad v_2^2(\mu, t) = \sqrt{\mu}\cos t, \qquad v_3^2(\mu, t) = -\sqrt{1 - \mu} \,.$$

The matrix of the variational equation takes the form for $\mathbf{e}_0$

$$B_0 = \begin{pmatrix} -2v_1^2 & 1 - 2v_1v_2 & 0 \\ -1 - 2v_1v_2 & -2v_2^2 & 0 \\ 0 & 0 & 1 - \mu \end{pmatrix}$$

where $v_1(\mu, t) = \sqrt{\mu} \sin t$ and $v_2(\mu, t) = \sqrt{\mu} \cos t$. The eigenvalues are

$$1 - \mu, \qquad -\mu \pm \sqrt{\mu^2 - 1}.$$

For $\mathbf{e}_{1,2}$ and $0 < \mu \le 1$ we find

$$B_{1,2} = \begin{pmatrix} -2v_1^2 & 1 - 2v_1v_2 & 0 \\ -1 - 2v_1v_2 & -2v_2^2 & 0 \\ \mp 2v_1\sqrt{1-\mu} & \mp 2v_2\sqrt{1-\mu} & 2(\mu - 1) \end{pmatrix}.$$

The eigenvalues are $2(\mu - 1)$, $-\mu \pm \sqrt{\mu^2 - 1}$. The bifurcation phenomena can be understood as follows. Let

$$r^2 := u_1^2 + u_2^2, \qquad R^2 := u_1^2 + u_2^2 + u_3^2.$$

Then from (1) it follows that

$$\frac{d}{dt}(r^2) = 2r^2(\mu - r^2)$$

$$\frac{d}{dt}(R^2) = 2r^2(\mu - r^2) + 2(R^2 - r^2)(1 - R^2).$$

This system admits for $\mu > 0$ the time-independent solution

$$R^2(\mu, t) = r^2(\mu, t) = \mu.$$

Linearizing this system around this solution we obtain the matrix

$$A(\mu) = \begin{pmatrix} -2\mu & 0 \\ -2\mu & 2(1 - \mu) \end{pmatrix}.$$

The eigenvalues of $A(\mu)$ are given by $\lambda = -2\mu$, $\lambda = 2(1 - \mu)$. The matrix $A(\mu)$ is singular for $\mu = 0$ and $\mu = 1$. Consequently, the solution $R^2(\mu, t)$ is asymptotically stable with respect to $r^2(\mu, t)$ for $\mu > 1$ and unstable for $1 > \mu > 0$. The system of differential equation can be integrated The general solution is given by

$$r^2(\mu, t) = \frac{\mu r^2(0) \exp(2\mu t)}{r^2(0)(\exp(2\mu t) - 1) + \mu}$$

where $r^2(t = 0) = r^2(0)$. For $\mu < 0$ we find

$$\lim_{t \to \infty} r^2(\mu, t) = 0.$$

For $\mu > 0$ we find

$$\lim_{t \to \infty} r^2(\mu, t) = \mu.$$

Chapter 6

Nonlinear Ordinary Difference Equations

Problem 1. Consider the nonlinear difference equation

$$u_{t+1} = 2u_t(1 - u_t) \tag{1}$$

where $t = 0, 1, 2, \dots$ and $u_0 \in [0, 1]$.

(i) Show that the exact solution of the initial-value problem is given by

$$u_t = \frac{1}{2} - \frac{1}{2}(1 - 2u_0)^{2^t}. \tag{2}$$

(ii) Find the fixed points of the equation. Study the stability of the fixed points. Let

$$u_{t+1} = f(u_t), \qquad t = 0, 1, 2, \dots. \tag{3}$$

The fixed points are defined as the solutions of the equation $f(u^*) = u^*$.

Solution 1. (i) If $u_0 \in [0, 1]$, then $u_t \in [0, 1]$ for $t = 1, 2, \dots$. From (2) it follows that

$$u_{t+1} = \frac{1}{2} - \frac{1}{2}(1 - 2u_0)^{2^{t+1}} \equiv \frac{1}{2} - \frac{1}{2}(1 - 2u_0)^{2 \cdot 2^t}.$$

On the other hand we have

$$2u_t(1 - u_t) = 2 \left[\frac{1}{2} - \frac{1}{2}(1 - 2u_0)^{2^t} \right] \left[1 - \frac{1}{2} + \frac{1}{2}(1 - 2u_0)^{2^t} \right],$$

51

$$= \frac{1}{2} \left[1 - (1 - 2u_0)^{2^t} \right] \left[1 + (1 - 2u_0)^{2^t} \right]$$

$$= \frac{1}{2} \left[1 - (1 - 2u_0)^{2 \cdot 2^t} \right].$$

This proves that (2) is a solution of the initial-value problem of (1).
(ii) The fixed points are determined by the equation

$$u^* = 2u^*(1 - u^*).$$

It follows that $u_1^* = 0$ and $u_2^* = \frac{1}{2}$ are the fixed points. Let $0 < u_0 < 1$. Then from solution (2) we find that

$$u_t \to \frac{1}{2} \quad \text{as} \quad t \to \infty.$$

This means that the fixed point u_2^* is stable and the fixed point u_1^* is unstable.

Problem 2.   Let

$$x_{t+1} = \frac{ax_t + b}{cx_t + d} \tag{1}$$

where $c \neq 0$ and

$$D := \det \begin{pmatrix} a & b \\ c & d \end{pmatrix} = ad - bc \neq 0$$

with $t = 0, 1, 2, \ldots$.
(i) Let

$$x_t = y_t - \frac{d}{c}. \tag{2}$$

Find the difference equation for y_t.
(ii) Let

$$y_t = \frac{w_{t+1}}{w_t}.$$

Find the difference equation for w_t.

Solution 2.   (i) Inserting the transformation (2) into (1) yields

$$y_{t+1} = \frac{a + d}{c} - \frac{D}{c^2 y_t}.$$

This difference equation can be written in the form

$$y_{t+1}y_t - \frac{a + d}{c}y_t + \frac{D}{c^2} = 0.$$

(ii) The transformation (3) reduces this nonlinear difference equation to the linear second order difference equation

$$w_{t+2} - \frac{a+d}{c} w_{t+1} + \frac{D}{c^2} w_t = 0.$$

Problem 3. (i) Let r_t satisfy the nonlinear difference equation

$$r_{t+1}(1 + ar_t) = 1 \tag{1}$$

with $t = 0, 1, 2, \ldots$, $r_0 = 0$ and $a \neq 0$. Show that r_k can be expressed as the *continued fraction*

$$r_t = \cfrac{1}{1 + \cfrac{a}{1 + \cfrac{a}{1 + \cfrac{\cdot}{\cdot + \cfrac{\cdot}{\cdot + \cfrac{a}{1}}}}}} \tag{2}$$

which terminates at the k-th stage. Show that (1) can be linearized by

$$r_t = \frac{n_t}{n_{t+1}}. \tag{3}$$

(ii) Find the solution of the nonlinear difference equation

$$r_{t+1}(b + r_t) = 1 \tag{4}$$

with $r_0 = 0$ and $b \neq 0$.

Solution 3. (i) From (1) it follows that $r_1 = 1$ and

$$r_t = \frac{1}{1 + ar_{t-1}}$$

where $t \geq 1$. Since

$$r_{t-1} = \frac{1}{1 + ar_{t-2}}$$

etc., we obtain (2). By making the substitution (3) the nonlinear difference equation (1) reduces to the linear difference equation with constant coefficients

$$n_{t+1} - n_t - an_{t-1} = 0, \qquad t = 1, 2, \ldots$$

where $n_0 = 0$ and $n_1 = 1$.

(ii) From (4) we obtain

$$r_t = \frac{1}{b + r_{t-1}} \cdot$$

Since

$$r_{t-1} = \frac{1}{b + r_{t-2}}$$

etc., we find that the solution can be written as a continued fraction. The nonlinear difference equation (4) can be linearized using ansatz (3). We find

$$n_{t+2} - b n_{t+1} - n_t = 0$$

where $n_0 = 0$ and $n_1 = 1$.

The infinite continued fraction

$$\cfrac{1}{b + \cfrac{1}{b + \cfrac{1}{b + \cdots}}}$$

converges to

$$\frac{1}{2}(\sqrt{b^2 + 4} - b)$$

when $b > 0$, and to

$$-\frac{1}{2}(\sqrt{b^2 + 4} + b)$$

when $b < 0$. In particular, we have

$$\sqrt{2} = 1 + \cfrac{1}{2 + \cfrac{1}{2 + \cdots}}$$

Problem 4.   Find the solution of the nonlinear difference equation

$$y_{t+1} = 2y_t^2 - 1 \tag{1}$$

where $t = 0, 1, 2, \ldots$.

Solution 4.   By making the substitution $y_t = \cos u_t$ transforms (1) into

$$\cos u_{t+1} = \cos(2u_t)$$

where we have used the identity

$$\cos^2 \alpha \equiv \frac{1}{2}(1 + \cos(2\alpha)).$$

Hence the solution is either

$$u_{t+1} = 2u_t + 2m\pi$$

or

$$u_{t+1} = -2u_t + 2n\pi$$

where m and n are arbitrary integers. From the second alternative, we obtain the solution of (1)

$$y_t = \cos\left(2^t\theta + (-1)^t \frac{2n\pi}{3}\right)$$

where θ is an arbitrary constant and n an arbitrary integer. The solution corresponding to the first alternative is contained in this one.

Problem 5. Show that the solution of the initial-value problem of the nonlinear difference equation

$$x_{t+1} = \frac{4x_t(1-x_t)(1-k^2x_t)}{(1-k^2x_t^2)^2} \tag{1}$$

where $x_0 \in [0,1]$ and $0 \le k^2 \le 1$ is given by

$$x_t = \text{sn}^2(2^t \text{sn}^{-1}(\sqrt{x_0}, k), k) \tag{2}$$

where $t = 0, 1, 2, \ldots$ and sn denotes a *Jacobi elliptic function*. Consequently, we need the *addition theorems* for Jacobi elliptic functions. In the following the modulus k is omitted. The addition property of the Jacobi elliptic function sn is given by

$$\text{sn}(u \pm v) \equiv \frac{\text{sn}\,u\,\text{cn}\,v\,\text{dn}\,v \pm \text{cn}\,u\,\text{sn}\,v\,\text{dn}\,u}{1 - k^2\text{sn}^2 u\,\text{sn}^2 v}.$$

From this equation we obtain as special case

$$\text{sn}(2u) \equiv \frac{2\text{sn}\,u\,\text{cn}\,u\,\text{dn}\,u}{1 - k^2\text{sn}^4 u}.$$

The addition property of the Jacobi elliptic function cn is given by

$$\text{cn}(u \pm v) \equiv \frac{\text{cn}\,u\,\text{cn}\,v \mp \text{sn}\,u\,\text{sn}\,v\,\text{dn}\,u\,\text{dn}\,v}{1 - k^2\text{sn}^2 u\,\text{sn}^2 v}.$$

Thus we obtain as special case

$$\text{cn}(2u) \equiv \frac{\text{cn}^2 u - \text{sn}^2 u\,\text{dn}^2 u}{1 - k^2\text{sn}^4 u}.$$

The addition property of the Jacobi elliptic function dn is given by

$$\text{dn}(u \pm v) \equiv \frac{\text{dn}u\,\text{dn}v \mp k^2\text{sn}u\,\text{sn}v\,\text{cn}u\text{cn}v}{1 - k^2\text{sn}^2u\,\text{sn}^2v}.$$

From this identity we obtain as special case

$$\text{dn}(2u) \equiv \frac{\text{dn}^2u - k^2\text{sn}^2u\,\text{cn}^2u}{1 - k^2\text{sn}^4u}.$$

Solution 5.   Let

$$\alpha := 2^t\text{sn}^{-1}(\sqrt{x_0}, k).$$

Then

$$x_{t+1} = \text{sn}^2(2\alpha, k).$$

Applying the addition theorems given above we find that (2) is the solution to (1).

Problem 6.   Consider the *logistic equation*

$$u_{t+1} = 4u_t(1 - u_t) \tag{1}$$

which also can be written as map $f : [0, 1] \to [0, 1]$

$$f(u) = 4u(1 - u). \tag{2}$$

Show that the solution of the algebraic equation

$$f(f(u^*)) = u^* \tag{3}$$

gives the fixed points of f and a periodic orbit of f.

Solution 6.   The fixed points of f are given by

$$u_1^* = 0, \qquad u_2^* = \frac{3}{4}.$$

Since $f(f(u)) = 16u(1 - u)(1 - 4u(1 - u))$ we find that the solutions of (3) are

$$u_1^* = 0, \qquad u_2^* = \frac{3}{4}, \qquad u_3^* = \frac{5 - \sqrt{5}}{8}, \qquad u_4^* = \frac{5 + \sqrt{5}}{8}.$$

Thus u_1^* and u_2^* are the fixed points of f. Let

$$u_0 = \frac{5 - \sqrt{5}}{8}.$$

Then

$$u_1 = \frac{5 + \sqrt{5}}{8}, \qquad u_2 = \frac{5 - \sqrt{5}}{8} = u_0 \,.$$

Thus we have a periodic orbit with period 2. $f(f(u))$ is the second iterate of f. When we consider higher iterates we find other periodic orbits of f.

Problem 7. Consider a first-order difference equation

$$x_{t+1} = f(x_t), \qquad t = 0, 1, \ldots \tag{1}$$

and a second order difference equation

$$x_{t+2} = g(x_t, x_{t+1}), \qquad t = 0, 1, \ldots \tag{2}$$

Definition. If

$$g(x, f(x)) = f(f(x)) \tag{3}$$

then (1) is called an *invariant* of (2).
Show that

$$f(x) = 2x^2 - 1$$

is an invariant of

$$g(x, y) = y - 2x^2 + 2y^2 \,.$$

Solution 7. The left-hand side of (3) is given by

$$g(x, f(x)) = 2x^2 - 1 - 2x^2 + 2(2x^2 - 1)^2 = -1 + 2(2x^2 - 1)^2 \,.$$

The right-hand side of (3) is given by

$$f(f(x)) = 2(2x^2 - 1)^2 - 1 \,.$$

Thus we see that f is an invariant of g.

Problem 8. The *Fibonacci trace map* is given by

$$x_{t+3} = 2x_{t+2}x_{t+1} - x_t \,. \tag{1}$$

Show that the Fibonacci trace map admits the invariant

$$I(x_t, x_{t+1}, x_{t+2}) = x_t^2 + x_{t+1}^2 + x_{t+2}^2 - 2x_t x_{t+1} x_{t+2} - 1 \,. \tag{2}$$

Solution 8. From (1) we obtain by shifting $t \to t - 1$

$$x_{t+2} = 2x_{t+1}x_t - x_{t-1} \,.$$

Inserting this equation into (2) yields

$$x_t^2 + x_{t+1}^2 + (2x_{t+1}x_t - x_{t-1})^2 - 2x_t x_{t+1}(2x_{t+1}x_t - x_{t-1}) - 1 =$$

$$x_t^2 + x_{t+1}^2 - 4x_{t-1}x_t x_{t+1} + x_{t-1}^2 + 2x_{t-1}x_t x_{t+1} - 1.$$

Thus we find the expression

$$x_{t-1}^2 + x_t^2 + x_{t+1}^2 - 2x_{t-1}x_t x_{t+1} - 1.$$

Shifting $t \to t+1$ we obtain

$$x_t^2 + x_{t+1}^2 + x_{t+2}^2 - 2x_t x_{t+1}x_{t+2} - 1$$

which is the invariant (2).

Problem 9.   Find the exact solution of the initial value problem for the system of difference equations

$$x_{1t+1} = (2x_{1t} - 2x_{2t} - 1)(2x_{1t} + 2x_{2t} - 1), \qquad x_{2t+1} = 4x_{2t}(2x_{1t} - 1)$$

where $t = 0, 1, 2, \ldots$.

Solution 9.   Using the addition theorems for sine, cosine, cosh and sinh we obtain

$$x_{1t} = \cos^2(u_t) \cosh^2(v_t) - \sin^2(u_t) \sinh^2(v_t)$$
$$x_{2t} = -2 \sin(u_t) \cos(u_t) \sinh(v_t) \cosh(v_t)$$

where $u_t = 2^t u_0$ and $v_t = 2^t v_0$. The values u_0 and v_0 are determined by

$$x_{10} = \cos^2(u_0) \cosh^2(v_0) - \sin^2(u_0) \sinh^2(v_0)$$
$$x_{20} = -2 \sin(u_0) \cos(u_0) \sinh(v_0) \cosh(v_0).$$

Given x_{10} and x_{20} this is a system of transcendental equations.

Chapter 7

Nonlinear Ordinary Differential Equations

Problem 1. The nonlinear system of ordinary differential equations

$$\frac{du_1}{dt} = u_1 - u_1 u_2, \qquad \frac{du_2}{dt} = -u_2 + u_1 u_2$$

is a so-called *Lotka-Volterra model*, where $u_1 > 0$ and $u_2 > 0$.
(i) Give an interpretation of the system assuming that u_1 and u_2 are describing species.
(ii) Find the fixed points (time-independent solutions).
(iii) Find the variational equation. Study the stability of the fixed points.
(iv) Find the first integral of the system.
(v) Describe why the solution cannot be given explicitly.

Solution 1. (i) Since the quantities u_1 and u_2 are positive we find the following behaviour: owing to the first term u_1 on the right-hand side of the first equation u_1 is growing. The second term $-u_1 u_2$ describes the interaction of species 1 with species 2. Owing to the minus sign u_1 is decreasing. For the second equation we have the opposite behaviour. Owing to $-u_2$ on the right-hand side we find that u_2 is decreasing and the interacting part $u_1 u_2$ leads to a growing u_2. Consequently, one expects that the quantities u_1 and u_2 are oscillating.
(ii) The equations which determine the *fixed points* are given by

$$u_1^* - u_1^* u_2^* = 0, \qquad -u_2^* + u_1^* u_2^* = 0.$$

59

Since $u_1 > 0$ and $u_2 > 0$ we obtain only one fixed point, namely

$$u_1^* = u_2^* = 1.$$

The fixed points are also called *time-independent solutions* or *steady-state solutions* or *equilibrium solutions*.

(iii) The variational equation is given by

$$\frac{dv_1}{dt} = v_1 - u_1 v_2 - u_2 v_1, \quad \frac{dv_2}{dt} = -v_2 + u_1 v_2 + u_2 v_1.$$

Inserting the fixed point solution $(u_1^*, u_2^*) = (1, 1)$ into this system yields

$$\frac{dv_1}{dt} = -v_2, \quad \frac{dv_2}{dt} = v_1$$

or

$$\begin{pmatrix} dv_1/dt \\ dv_2/dt \end{pmatrix} = \begin{pmatrix} 0 & -1 \\ 1 & 0 \end{pmatrix} \begin{pmatrix} v_1 \\ v_2 \end{pmatrix}.$$

Thus in a sufficiently small neighbourhood of the fixed point $\mathbf{u}^* = (1, 1)$ we find a periodic solution, since the eigenvalues of the matrix on the right-hand side of this equation are given by i, $-i$.

(iv) From

$$\frac{du_1}{u_1(1 - u_2)} = \frac{du_2}{u_2(-1 + u_1)} = dt$$

we obtain

$$\frac{(-1 + u_1)du_1}{u_1} = \frac{(1 - u_2)du_2}{u_2}.$$

Integrating this equation leads to the first integral

$$I(\mathbf{u}) = u_1 u_2 e^{-u_1 - u_2}.$$

(v) To find the explicit solution we have to integrate

$$\frac{du_1}{u_1(1 - u_2)} = \frac{du_2}{u_2(-1 + u_1)} = \frac{dt}{1}.$$

Owing to the first integral the constant of motion is given by

$$u_1 u_2 e^{-u_1 - u_2} = c$$

where c is a positive constant. This equation cannot be solved with respect to u_1 or u_2. Since $c > 0$ we find that the Lotka-Volterra model oscillates around the fixed point $\mathbf{u}^* = (1, 1)$.

Problem 2.   Let

$$\frac{du_1}{dt} = 2u_1^3, \quad \frac{du_2}{dt} = -(1 + 6u_1^2)u_2. \tag{1}$$

Show that this autonomous system of ordinary differential equations has negative divergence and a particular solution with *exploding amplitude*.

Let $t_c < \infty$. If $u_j(t) \to \infty$ as $t \to t_c$, then $u_j(t)$ is said to have an exploding amplitude.

Solution 2. The divergence of system (1) is obviously

$$\frac{\partial}{\partial u_1}(2u_1^3) - \frac{\partial}{\partial u_2}(1 + 6u_1^2)u_2 = 6u_1^2 - 1 - 6u_1^2 = -1.$$

If $u_2(t) = 0$, then $du_2/dt = -(1 + 6u_1^2)u_2$ is satisfied and we obtain

$$\frac{du_1}{dt} = 2u_1^3.$$

This equation admits the particular solution

$$u_1(t) = \frac{1}{2}\frac{1}{\sqrt{1-t}}$$

with $u_1(0) = \frac{1}{2}$ and $0 \le t < 1$. If $t \to t_c = 1$, then $u_1(t) \to \infty$.

Problem 3. The linear diffusion equation in one-space dimension is given by

$$\frac{\partial u}{\partial t} = \frac{\partial^2 u}{\partial x^2}. \tag{1}$$

Insert the ansatz

$$u(x,t) = \prod_{j=1}^{n}(x - a_j(t)) \tag{2}$$

into (1) and show that the time dependent functions a_j satisfy the nonlinear autonomous system of ordinary differential equations

$$\frac{da_k}{dt} = -2\sum_{j}'\frac{1}{a_k - a_j}$$

where $\sum'$ means that $j \ne k$. Use the identity (see Volume I, chapter 2, problem 7)

$$\sum_{\substack{j,k=1 \\ j \ne k}}^{n}\frac{1}{x - a_j}\frac{1}{x - a_k} \equiv 2\sum_{\substack{j,k=1 \\ j \ne k}}^{n}\frac{1}{x - a_k}\frac{1}{a_k - a_j}. \tag{3}$$

Solution 3.   Using the result from problem (7) (Volume I, chapter 2) we obtain

$$\frac{\partial u}{\partial t} = -u \sum_{j=1}^{n} \frac{1}{x - a_j} \frac{da_j}{dt} \tag{4}$$

and

$$\frac{\partial u}{\partial x} = u \sum_{j=1}^{n} \frac{1}{x - a_j} \tag{5}$$

where u is given by ansatz (2). From (5) it follows that

$$\frac{\partial^2 u}{\partial x^2} = u \left(\sum_{k=1}^{n} \frac{1}{x - a_k} \right) \left(\sum_{j=1}^{n} \frac{1}{x - a_j} \right) - u \sum_{j=1}^{n} \frac{1}{(x - a_j)^2}$$

and therefore

$$\frac{\partial^2 u}{\partial x^2} = u \sum_{j \neq k}^{n} \frac{1}{x - a_j} \frac{1}{x - a_k} .$$

Using identity (3) we find

$$\frac{\partial^2 u}{\partial x^2} = 2u \sum_{j \neq k}^{n} \frac{1}{x - a_k} \frac{1}{a_k - a_j} . \tag{6}$$

Inserting (4) and (6) into the linear diffusion equation (1) gives

$$\sum_{k=1}^{n} \frac{1}{x - a_k} \frac{da_k}{dt} = -2 \sum_{j \neq k}^{n} \frac{1}{x - a_k} \frac{1}{a_k - a_j} .$$

Consequently,

$$\frac{da_k}{dt} = -2 \sum_{j}^{n} {}' \frac{1}{a_k - a_j} .$$

Problem 4.   Let

$$\frac{d\mathbf{u}}{dt} = L\mathbf{u} + (\mathbf{a} \cdot \mathbf{u})\mathbf{u} \tag{1}$$

be a nonlinear autonomous system of n first-order ordinary differential equations, where L is an $n \times n$ matrix with constant coefficients and

$$\mathbf{a} \cdot \mathbf{u} := a_1 u_1 + a_2 u_2 + \cdots + a_n u_n .$$

Find the solution of the initial-value problem, where $\mathbf{u}(0) = \mathbf{u}_0$.

Solution 4. Method 1. We start from the ansatz

$$\mathbf{u}(\mathbf{u}_0, t) = f(t)\mathbf{v}(\mathbf{u}_0, t). \tag{2}$$

Differentiating (2) with respect to t gives

$$\frac{d\mathbf{u}}{dt} = \frac{df}{dt}\mathbf{v} + f\frac{d\mathbf{v}}{dt}.$$

Inserting this expression into (1) yields

$$\frac{df}{dt}\mathbf{v} + f\frac{d\mathbf{v}}{dt} = Lf\mathbf{v} + f(\mathbf{a}\cdot\mathbf{v})f\mathbf{v} = f(L\mathbf{v}) + f^2(\mathbf{a}\cdot\mathbf{v})\mathbf{v}.$$

Therefore

$$\frac{d\mathbf{v}}{dt} = L\mathbf{v}, \qquad \mathbf{v}(0) = \mathbf{u}_0 \tag{3}$$

and

$$\frac{df}{dt} = f^2\mathbf{a}\cdot\mathbf{v}, \qquad f(0) = 1. \tag{4}$$

The method of the solution of system (1) is equivalent to the well known method of *variation of constants*. Equation (4) is a *Riccati equation*. Solving first the system of linear differential equations (3) and then the nonlinear differential equation (4), we obtain

$$\mathbf{u}(\mathbf{u}_0, t) = \frac{\mathbf{v}(\mathbf{u}_0, t)}{1 - \int_0^t \mathbf{a}\cdot\mathbf{v}(\mathbf{u}_0, \tau)d\tau}.$$

Method 2. The transformation

$$\mathbf{u}(\mathbf{u}_0, t) = \frac{1}{1 - \mathbf{a}\cdot\mathbf{B}}\frac{d\mathbf{B}}{dt}$$

where $\mathbf{a}\cdot\mathbf{B} = a_1 B_1 + a_2 B_2 + \cdots + a_n B_n$ with

$$\frac{d\mathbf{B}(0)}{dt} = \mathbf{u}_0, \qquad \mathbf{B}(0) = 0$$

is the linearization of system (1) to the system of linear differential equations with constant coefficients

$$\frac{d^2\mathbf{B}}{dt^2} = L\frac{d\mathbf{B}}{dt}.$$

Problem 5. The mathematical *pendulum* is described by the nonlinear second order ordinary differential equation

$$\frac{d^2\theta}{dt^2} + \omega^2 \sin\theta = 0 \tag{1}$$

where $\omega^2 = g/L$ (L length of the pendulum, g acceleration due to gravity) and θ is the angular displacement of the pendulum from its position of equilibrium. Find the solution of the initial-value problem $\theta(t = 0) = \alpha$ and $d\theta(t = 0)/dt = 0$.

Solution 5.   The constant of motion of equation (1) is given by

$$\frac{1}{2}\left(\frac{d\theta}{dt}\right)^2 - \omega^2 \cos\theta = C \tag{2}$$

where C is a constant of integration. The constant of integration is given by the initial condition. We find

$$C = -\left(\frac{g}{L}\cos\alpha\right).$$

We solve (2) for $d\theta/dt$, and thus obtain

$$\frac{d\theta}{dt} = \sqrt{\frac{2g}{L}}\sqrt{\cos\theta - \cos\alpha}. \tag{3}$$

It follows that

$$dt = \sqrt{\frac{L}{2g}}\frac{d\theta}{\sqrt{\cos\theta - \cos\alpha}}. \tag{4}$$

In order to reduce the right-hand member of this equation to standard form, we introduce

$$\cos\theta =: 1 - 2k^2 \sin^2\phi, \qquad k := \sin(\alpha/2)$$

and use the following equations

$$\cos\theta - \cos\alpha = 2k^2 \cos^2\phi$$

$$\sin\theta = 2k\sin\phi\sqrt{1 - k^2 \sin^2\phi}$$

$$\sin\theta d\theta = 4k^2 \sin\phi\cos\phi d\phi.$$

When these equations are substituted into (4), we obtain

$$dt = \sqrt{\frac{L}{g}}\frac{d\phi}{\sqrt{1 - k^2 \sin^2\phi}}. \tag{5}$$

Therefore the time T required for the pendulum to swing from its position of equilibrium at $\theta = 0$ to a displacement of $\theta = \theta_0$ is given by

$$T = \sqrt{\frac{L}{g}}\int_0^{\phi_0}\frac{d\phi}{\sqrt{1 - k^2 \sin^2\phi}}$$

where ϕ_0 is obtained from

$$\sin^2 \phi_0 = \frac{1 - \cos \theta_0}{2k^2} = \frac{\sin^2(\theta_0/2)}{k^2}.$$

Thus

$$\phi_0 = \arcsin\left(\frac{\sin(\theta_0/2)}{k}\right).$$

In terms of elliptic integrals, we can write

$$T = \sqrt{\frac{L}{g}} F(\phi_0, k).$$

The period of the simple pendulum is defined to be the time required to make a complete oscillation between positions of maximum displacement. To determine this position of maximum displacement, we use (3). Consequently

$$\frac{d\theta}{dt} = 2k\sqrt{\frac{g}{L}} \cos \phi.$$

Since the desired value of θ is that for which $d\theta/dt = 0$, we see that this corresponds to $\phi = \frac{1}{2}\pi$. If we let $P(k)$ be the period of the pendulum, we obtain

$$P(k) = 4\sqrt{\frac{L}{g}} \int_0^{\pi/2} \frac{d\phi}{\sqrt{1 - k^2 \sin^2 \phi}} = 4\sqrt{\frac{L}{g}} K(k)$$

where $K(k)$ is the *complete elliptic integral* of the first kind. When $k = 0$, this reduces to

$$P = 2\pi\sqrt{\frac{L}{g}}.$$

To find the displacement θ as a function of t, we integrate (5). We find

$$t = \sqrt{\frac{L}{g}} \int_0^{\phi} \frac{d\phi}{\sqrt{1 - k^2 \sin^2 \phi}}.$$

This equation can be written as

$$\text{sn}\left(t\sqrt{\frac{g}{L}}, k\right) = \sin \phi = \frac{1}{k} \sin \frac{1}{2}\theta$$

from which we obtain

$$\theta(t) = 2\arcsin\left(k\,\text{sn}\left(t\sqrt{\frac{g}{L}}, k\right)\right)$$

with $k = \sin(\alpha/2)$.

Problem 6.   Consider the nonlinear autonomous system of ordinary differential equations

$$\frac{du_1}{dt} = u_2 + u_1(1 - u_1^2 - u_2^2), \qquad \frac{du_2}{dt} = -u_1 + u_2(1 - u_1^2 - u_2^2).$$

(i) Find the fixed points.
(ii) Show that system (1) admits a periodic solution of the form

$$u_1(t) = \sin(\omega t), \qquad u_2(t) = \cos(\omega t).$$

Determine ω.
(iii) Find the solution for $t \to \infty$.

Solution 6.   (i) The fixed points are determined by

$$u_2^* + u_1^*(1 - u_1^{*2} - u_2^{*2}) = 0, \qquad -u_1^* + u_2^*(1 - u_1^{*2} - u_2^{*2}) = 0.$$

We find one fixed point, namely $u_1^* = u_2^* = 0$.
(ii) Inserting the periodic solution (2) into system (1) and applying the identity $\sin^2 \omega t + \cos^2 \omega t \equiv 1$ yields $\omega = 1$.
(iii) Introducing the quantity

$$r^2 := u_1^2 + u_2^2$$

and differentiating we obtain the equation of motion

$$\frac{d}{dt} r^2 = 2r^2(1 - r^2).$$

Obviously, we find the time-independent solution $r^2 = 0$ (i.e. $u_1^* = u_2^* = 0$) and the *limit cycle* $r^2 = 1$ as solutions of the original differential equations (1). Let $r^2 \neq 1$ and $r^2 \neq 0$. For $0 < r^2 < 1$, the solution of the above equation is given by

$$r^2(t, r_0^2) = \frac{1}{1 + \left(\frac{1}{r_0^2} - 1\right) e^{-2t}}$$

with $0 < r_0^2 < 1$ and $-\infty < t < \infty$. For $1 < r^2 < \infty$, the solution of the initial value problem is given by

$$r^2(t, r_0^2) = \frac{1}{1 + \left(\frac{1}{r_0^2} - 1\right) e^{-2t}}$$

with $1 < r_0^2 < \infty$ and

$$\frac{1}{2} \ln\left(1 - \frac{1}{r_0^2}\right) < t < \infty.$$

For both cases as $t \to \infty$, we have $r^2(t \to \infty) = 1$. Consequently, the limit cycle is stable. This can also be seen from applying the *Ljapunov theory*. Let $V : \mathbb{R}^2 \to \mathbb{R}$ with

$$V(\mathbf{u}) = \frac{r^2}{2}.$$

Then

$$\frac{dV}{dt} = r\frac{dr}{dt} = r^2(1 - r^2).$$

Thus $dV/dt > 0$ for $0 < r^2 < 1$ and $dV/dt < 0$ for $r^2 > 1$.

Problem 7.   Let

$$\{w, z\} := \frac{w'''}{w'} - \frac{3}{2}\left(\frac{w''}{w'}\right)^2$$

be the *Schwarzian derivative* of w, where $w' \equiv dw/dz$. Let y_1 and y_2 be two linearly independent solutions of the equation

$$y'' + Q(z)y = 0 \tag{1}$$

which are defined and holomorphic in some simply connected domain D in the complex plane.
(i) Show that

$$w(z) = \frac{y_1(z)}{y_2(z)}$$

satisfies the third order differential equation

$$\{w, z\} = 2Q(z) \tag{2}$$

at all points of D where $y_2(z) \neq 0$.
(ii) Show that if $w(z)$ is a solution of (2) and is holomorphic in some neighbourhood of a point $z_0 \in D$, then one can find two linearly independent solutions, $u(z)$ and $v(z)$ of (2) defined in D so that

$$w(z) = \frac{u(z)}{v(z)}.$$

If $v(z_0) = 1$ the solutions u and v are uniquely defined.

Solution 7.   (i) The expression

$$y_1 y_2' - y_2 y_1'$$

is called the *Wronskian*. We may assume that the Wronskian of y_1 and y_2 is identically one. Then

$$w'(z) = (y_2(z))^{-2}$$

so that

$$\frac{w''(z)}{w'(z)} = -2\frac{y_2'(z)}{y_2(z)}$$

and

$$\left(\frac{w''(z)}{w'(z)}\right)' = -2\frac{y_2''(z)}{y_2(z)} + 2\left(\frac{y_2'(z)}{y_2(z)}\right)^2 = 2Q(z) + \frac{1}{2}\left(\frac{w''(z)}{w'(z)}\right)^2$$

from which the first assertion follows.

(ii) Suppose that a solution of equation (2) is given by its initial values

$$w(z_0), \qquad w'(z_0), \qquad w''(z_0)$$

at a point z_0 in D. We may assume that $w'(z_0) \neq 0$ since otherwise $Q(z)$ could not be holomorphic at $z = z_0$. We can now choose two linearly independent solutions, $u(z)$ and $v(z)$, of (2) so that at $z = z_0$ the quotient

$$w_1(z) = \frac{u(z)}{v(z)}$$

has the initial values $w(z_0), w'(z_0), w''(z_0)$ the same as $w(z)$. If $v(z_0) = 1$, then the solutions $u(z)$ and $v(z)$ are uniquely determined, and we must have

$$w(z) \equiv w_1(z).$$

Problem 8. The nonlinear system of ordinary differential equations

$$A\frac{dp}{dt} + (C - B)qr = Mg(y_0\gamma'' - z_0\gamma')$$

$$B\frac{dq}{dt} + (A - C)rp = Mg(z_0\gamma - x_0\gamma'')$$

$$C\frac{dr}{dt} + (B - A)pq = Mg(x_0\gamma' - y_0\gamma)$$

$$\frac{d\gamma}{dt} = r\gamma' - q\gamma''$$

$$\frac{d\gamma'}{dt} = p\gamma'' - r\gamma$$

$$\frac{d\gamma''}{dt} = q\gamma - p\gamma'$$

describes the motion of a *heavy rigid body* about a fixed point, where M denotes the mass and A, B and C are the principle moments of inertia.

(i) Show that

$$I_1 = Ap^2 + Bq^2 + Cr^2 - 2Mg(x_0\gamma + y_0\gamma' + z_0\gamma'')$$
$$I_2 = Ap\gamma + Bq\gamma' + Cr\gamma''$$
$$I_3 = \gamma^2 + \gamma'^2 + \gamma''^2$$

are first integrals of the nonlinear dynamical system.

(ii) Find the conditions on the constants A, B, C, x_0, y_0 and z_0 under which

$$I_4 = A^2 p^2 + B^2 q^2 + C^2 r^2$$
$$I_5 = r$$
$$I_6 = x_0 p + y_0 q + z_0 r$$
$$I_7 = (p^2 - q^2 + c\gamma)^2 + (2pq + c\gamma')^2$$

are first integrals of the nonlinear dynamical system, where $c = Mgx_0/C$.

Solution 8.   (i) We recall that I is a *first integral* if $dI/dt = 0$. Straightforward calculation yields

$$\frac{dI_1}{dt} = 2Ap\frac{dp}{dt} + 2Bq\frac{dq}{dt} + 2Cr\frac{dr}{dt} - 2Mg\left(x_0\frac{d\gamma}{dt} + y_0\frac{d\gamma'}{dt} + z_0\frac{d\gamma''}{dt}\right).$$

Inserting the dynamical system yields

$$\begin{aligned}
\frac{dI_1}{dt} = {} & 2p[(B-C)qr + Mg(y_0\gamma'' - z_0\gamma')] \\
& + 2q[(C-A)rp + Mg(z_0\gamma - x_0\gamma'')] \\
& + 2r[(A-B)pq + Mg(x_0\gamma' - y_0\gamma)] \\
& - 2Mg[x_0(r\gamma' - q\gamma'') + y_0(p\gamma'' - r\gamma) + z_0(q\gamma - p\gamma')].
\end{aligned}$$

Hence we arrive at

$$\frac{dI_1}{dt} = 0.$$

Analogously, for I_2 and I_3 we find

$$\frac{dI_2}{dt} = 0, \qquad \frac{dI_3}{dt} = 0.$$

(ii) The condition that I_4 is a first integral of the nonlinear dynamical system, i.e., $dI_4/dt = 0$ leads to

$$\frac{dI_4}{dt} = 2Mg[\gamma(z_0 Bq - y_0 Cr) + \gamma'(x_0 Cr - z_0 Ap) + \gamma''(y_0 Ap - x_0 Bq)] = 0.$$

Therefore we obtain the condition $x_0 = y_0 = z_0 = 0$. We find that the condition that $I_5 = r$ is a first integral is given by

$$A = B, \qquad x_0 = y_0 = 0.$$

The condition that I_6 is a first integral leads to

$$\frac{dI_6}{dt} = x_0\frac{dp}{dt} + y_0\frac{dq}{dt} + z_0\frac{dr}{dt} = 0.$$

Thus

$$\frac{dI_6}{dt} = \frac{x_0 Mg}{A}(y_0\gamma'' - z_0\gamma') + \frac{y_0 Mg}{B}(z_0\gamma - x_0\gamma'') + \frac{z_0 Mg}{C}(x_0\gamma' - y_0\gamma) = 0.$$

Thus we find $A = B = C$. Taking the time derivative of I_7 we find terms such as

$$4p^3 qr\left(\frac{B-C}{A} - \frac{C-A}{B} + 2\frac{C-A}{B}\right) + 4pq^3 r\left(\frac{C-B}{A} + \frac{C-A}{B} + 2\frac{B-C}{A}\right).$$

From this we find the condition $A = B = 2C$. Furthermore we find the term

$$4\frac{Mg}{A}p^3(y_0\gamma'' - z_0\gamma').$$

From this we conclude that $z_0 = y_0 = 0$. All other terms vanish if the these conditions are satisfied.

Problem 9.   Consider the eigenvalue equation

$$\frac{d^2\psi}{dx^2} = (V(x) - E)\psi(x) \tag{1}$$

where $V(x) = V(-x)$. Let

$$\Phi(x) := x^{-s}\psi(x) \tag{2}$$

where $s = 0$ or $s = 1$ for even or odd states, respectively. The logarithmic derivative

$$f(x) := -\frac{1}{\Phi(x)}\frac{d\Phi(x)}{dx} \tag{3}$$

is regular at the origin for all eigenstates of (1). Thus we have

$$\Phi(x) = \exp\left(-\int^x f(s)ds\right).$$

(i) Find the differential equation that f satisfies.
(ii) Assume that

$$V(x) = \sum_{j=1}^{K} v_j x^{2j}, \qquad v_K > 0. \tag{4}$$

We expand f in a Taylor series around the origin

$$f(x) = \sum_{j=0}^{\infty} f_j x^{2j+1}. \tag{5}$$

Find the condition on the coefficients f_j.

Solution 9. (i) Inserting (3) into (1) yields

$$\frac{df}{dx} - f^2(x) + 2s\frac{f(x)}{x} = E - V(x) \tag{6}$$

This is a *Riccati differential equation*.

(ii) Inserting the Taylor expansion (5) and the expansion (4) into (6) we find that the coefficients f_j satisfy the condition

$$f_j = \frac{1}{2j + 2s + 1}\left(\sum_{i=0}^{j-1} f_i f_{j-i-1} + E\delta_{j0} - \sum_{i=1}^{K} v_i \delta_{ij}\right)$$

where δ_{ij} denotes the Kronecker delta. The function f can be approximated by a sequence of rational functions

$$g(x) = \frac{A(x)}{B(x)}$$

where

$$A(x) = \sum_{j=0}^{M} a_j x^{2j+1}, \qquad B(x) = \sum_{j=0}^{N} b_j x^{2j}, \qquad b_0 = 1.$$

If g is exactly a *Padé approximant*, then

$$f(x) - g(x) = O(x^{2(M+N)+3}).$$

Problem 10. Consider the second-order ordinary differential equation

$$\frac{d^2x}{dt^2} + f(x)\frac{dx}{dt} + g(x) = 0. \tag{1}$$

This equation will have a unique periodic solution provided that

$$\oint F(x)dy = 0 \tag{2}$$

where

1) f is even, g is odd, $xg(x) > 0$ for all $x \neq 0$, and $f(0) < 0$

2) f and g are continuous and g is Lipschitzian

3) $F(x) \to \pm\infty$ as $x \to \pm\infty$, where

$$F(x) := \int_0^x f(s)ds$$

4) F has one single positive zero $x = a$, for $x \geq a$ the function F increases monotonically with x.

5) The variable y is defined by $y := dx/dt + F(x)$.

Give a physical interpretation of these conditions.

Solution 10. Equation (1) can be written as

$$\frac{d^2 x}{dt^2} = -f(x)\frac{dx}{dt} - g(x)$$

and may be thought of as representing a unit mass acted on by a restoring force $-g(x)$ and a damping force $-f(x)dx/dt$. As the unit mass moves a distance dx, the damper does an amount of work dW given by

$$dW = -f(x)\frac{dx}{dt}dx \, .$$

Since $f(0) < 0$, we find that $dW/dx > 0$ when the unit mass passes through the origin with a positive velocity. Thus at this point the damper is putting energy into the system. As x increases, $F(x)$ reaches a minimum, at which point $f(x) = 0$. After this point has been reached f becomes positive and the damper removes energy from the system. Thus, from physical considerations, we realize the possibility of a limit cycle. When the limit cycle is reached, the sum of all of the elements of work done by the damper in one cycle must be zero. Thus, if the system is in a steady-state oscillation

$$W_{cycle} = \int_{cycle} dW = \int_{cycle} \left(-f(x)\frac{dx}{dt}\right) dx = 0 \, . \tag{3}$$

If we let $dx/dt = v$, then (3) can be integrated by parts

$$(vF(x))_{cycle} - \int_{cycle} F(x)dv = 0 \, .$$

The first term of this equation is zero because the integration is over a cycle. Thus, the condition for a steady-state oscillation becomes

$$\int_{cycle} F(x)dv = 0 \, .$$

This equation can be converted into (2) by introducing the variable y. Thus we can write

$$\int F(x)(dy - f(x)dx) = \int_{cycle} F(x)dy - \int_{cycle} F(x)f(x)dx \, .$$

The second integral is zero because the integration is over a cycle. Hence we obtain

$$\oint F(x)dy = 0$$

which is the usual curvilinear integral taken along a trajectory.

Problem 11. Consider the differential equation

$$\frac{du}{dt} = f(t, u(t)).$$

Let (t_0, u_0) be a particular pair of values assigned to the real variable (t, u) such that within a rectangular domain D surrounding the point (t_0, u_0) and defined by the inequalities

$$|t - t_0| \leq a, \qquad |u - u_0| \leq b.$$

If (t, u) and (t, U) are two points within D, then the *Lipschitz condition* is

$$|f(t, U) - f(t, u)| < K|(U - u)|$$

where K is a constant. Discuss the solution of the one-dimensional nonlinear differential equation

$$\frac{du}{dt} = -u^{1/3} \tag{1}$$

where $u(t) \geq 0$. Show that the Lipschitz condition is violated.

Solution 11. The differential equation (1) has a fixed point at $u^* = 0$. Let $f(u) = -u^{1/3}$. Then

$$\frac{df}{du} = -\frac{1}{3}u^{-2/3}.$$

Thus

$$\frac{df}{du} \to -\infty \qquad \text{at} \quad u \to 0.$$

Thus the Lipschitz condition is violated. The fixed point $u^* = 0$ is an attractor with "infinite" local stability. Let $u(t = 0) = u_0 > 0$. Then the time t_f to reach the fixed point (attractor) is finite, i.e.

$$t_f = -\int_{u_0}^0 \frac{du}{u^{1/3}} = \frac{3}{2}u_0^{3/2} < \infty.$$

Problem 12. Consider the hypothetical chemical reaction mechanism (*Lotka-Volterra model*)

$$A + X \xrightarrow{k_1} 2X$$
$$X + Y \xrightarrow{k_2} 2Y$$
$$Y \xrightarrow{k_3} B$$

where X and Y are intermediaries, k_1, k_2, and k_3 are the reaction rate constants, and the concentrations of the reactants A and B are kept constant. Find the kinetic equations.

Solution 12.   Denoting the concentrations of A, B, X, and Y by the same letters for convenience, the law of mass action then gives

$$\frac{dA}{dt} = k_1 AX$$

$$\frac{dX}{dt} = -k_1 AX + k_1 X^2 - k_2 XY$$

$$\frac{dY}{dt} = -k_2 XY + k_2 Y^2 - k_3 Y$$

$$\frac{dB}{dt} = k_3 Y \, .$$

Since A and B are kept constant the conditions on A and B mean that the system is open and so there must be an exchange of matter with the surroundings. Under this condition the equations reduce to two equations

$$\frac{dX}{dt} = -k_1 AX + k_1 X^2 - k_2 XY$$

$$\frac{dY}{dt} = -k_2 XY + k_2 Y^2 - k_3 Y$$

where A, k_1, k_2 and k_3 are constants.

Problem 13.   Consider the Hamilton function

$$H(p_x, p_y, x, y) = \frac{1}{2} \left(p_x^2 + p_y^2 + (x^2 y^2)^{1/\alpha} \right) \tag{1}$$

where $\alpha \in [0, 1]$. Discuss the equation of motion for $x \gg y$.

Solution 13.   In the limit $\alpha \to 0$ we obtain the hyperbola billard. Increasing the parameter α means gradual softening of the billard walls and when $\alpha = 1$ we recover the $x^2 y^2$ potential. The symmetry group of this family is C_{4v}. For $x \gg y$ the motion in the x-direction will be much slower than the motion in the y-direction. Thus x may be regarded as a slowly varying parameter. In the *adiabatic approximation*, the motion in the y-direction is described by the Hamilton function

$$H_y = \frac{1}{2} \left(p_y^2 + (x^2 y^2)^{1/\alpha} \right) . \tag{2}$$

By the *adiabatic theorem* the action integral in the y-direction is given by

$$J_y = \oint p_y dy = \frac{(2H_y)^{(1+\alpha)/2} f(2/\alpha)}{x} \tag{3}$$

where

$$f(s) := \frac{1}{2\pi} \oint \sqrt{1 - |z|^s} dz. \tag{4}$$

The quantity J_y is approximately a constant of motion. The full Hamilton function may now be written as

$$H \approx \frac{1}{2} p_x^2 + H_y(x) \tag{5}$$

giving (approximately) the x motion for any J_y. If we transform the (x, p_x) pair to action-angle variables we obtain the following expression for H

$$H \approx \frac{1}{2} \left[\frac{J_x J_y}{f(2/\alpha) f(2/(\alpha + 1))} \right]^{2/(\alpha+2)}.$$

Although this expression was derived under the asymmetric assumption $x \gg y$, the Hamilton function (1) is symmetric in x and y. These adiabatic expressions are not valid in the central region.

Problem 14. (i) Let $f : \mathbb{R} \to \mathbb{R}$ be an analytic function. Solve the functional equation

$$f(x + y) = f(x) + f(y) \tag{1}$$

using differentiation.

(ii) Let $f : \mathbb{R} \to \mathbb{R}$ be an analytic function. Solve the functional equation

$$f(x + y) = f(x)f(y) \tag{2}$$

using differentiation.

(iii) Let $f : \mathbb{R} \to \mathbb{R}$ be an analytic function. Solve the functional equation

$$f(x + y) + f(x - y) = 2f(x)f(y) \tag{3}$$

using differentiation.

Solution 14. (i) If we set $y = 0$ in (1) we obtain $f(0) = 0$. If we differentiate (1) with respect to x we obtain

$$f'(x + y) = f'(x).$$

Thus

$$f'(x) = c, \qquad f(x) = cx + b$$

where c and b are constants. Owing to $f(0) = 0$ we obtain $b = 0$ and thus

$$f(x) = cx.$$

(ii) Inserting $y = 0$ into (2) yields $f(x) = 0$ or $f(0) = 1$. We find the f with the condition $f(0) = 1$. If we differentiate (2) with respect to x we obtain

$$f'(x + y) = f'(x)f(y).$$

Setting $x = 0$ yields

$$\frac{f'(y)}{f(y)} = f'(0) = c.$$

Thus

$$f(x) = \exp(cx).$$

(iii) Setting $y = 0$ we find

$$f(x) = f(x)f(0)$$

we see that (3) admits the trivial solution $f(x) = 0$. For a non-trival solution of (3) we find that $f(0) = 1$. Setting $x = 0$ we obtain $f(-x) = f(x)$. Taking the second derivative of (3) with respect to x we obtain

$$f''(x + y) + f''(x - y) = 2f''(x)f(y).$$

Taking the second derivative of (3) with respect to y we obtain

$$f''(x + y) + f''(x - y) = 2f(x)f''(y).$$

Thus

$$f''(x)f(y) = f(x)f''(y).$$

It follows that

$$f''(x) = kf(x).$$

For $k = 0$ we obtain $f(x) = 1$, for $k > 0$ we obtain

$$f(x) = \cosh(\sqrt{k}x).$$

For $k < 0$ we obtain

$$f(x) = \cos(\sqrt{-k}x).$$

Problem 15.   Study the singularity structure of the nonlinear ordinary differential equation

$$\frac{d^2u}{dt^2} + \frac{1}{2}u^3 = 0$$

where $u(t)$ is a real-valued function.

Solution 15.   We consider the differential equations in the complex domain

$$\frac{d^2w}{dz^2} + \frac{1}{2}w^3 = 0.$$

Inserting the ansatz (locally represented as a *Laurent expansion*)

$$w(z) = \sum_{j=0}^{\infty} a_j (z - z_0)^{j-1}$$

we find the recursion relation for the expansion coefficients a_j

$$a_j(j+1)(j-4) = -\frac{1}{2} \sum_k \sum_l a_{j-k-l} a_k a_l, \qquad 0 < k+l \le j, \quad 0 \le k, l < j$$

where

$$a_0 = 2i, \quad a_1 = a_2 = a_3 = 0, \quad a_4 = \text{arbitrary}.$$

Owing to the arbitrary pole position z_0 and coefficient a_4 we have a local representation of the general solution to this second-order differential equation. The nonlinear differential equation can be solved exactly in terms of the Jacobi elliptic functions and the movable singularities in the complex z-plane form a regular lattice of first-order poles.

Problem 16. Study the singularity structure of the nonlinear ordinary differential equation

$$\frac{d^2 x}{dt^2} + \lambda \frac{dx}{dt} + \frac{1}{2} x^3 = \epsilon f(t)$$

where $u(t)$ is a real-valued function and $f(t)$ is an analytic function of t.

Solution 16. We consider the differential equation in the complex plane, i.e.,

$$\frac{d^2 w}{dz^2} + \lambda \frac{dw}{dz} + \frac{1}{2} w^3 = \epsilon f(z).$$

The introduction of either the damping term or driving term leads to a breakdown of the Laurent series

$$w(z) = \sum_{j=0}^{\infty} a_j (z - z_0)^{j-1}$$

discussed in the previous problem since it is not possible to introduce an arbitrary coefficient at $j = 4$. We have to add logarithmic terms

$$w(z) = \sum_{j=0}^{\infty} \sum_{k=0}^{\infty} a_{jk} (z - z_0)^{j-1} ((z - z_0)^4 \ln(z - z_0))^k.$$

Computation of the recursion relation for the a_{jk} yields

$$a_{jk}((j-1)(j-2) + 4k(2j+4k-3)) + a_{j-4,k+1}(k+1)(2j+8k-3)$$
$$+ a_{j-8,k+2}(k+1)(k+2) + \lambda a_{j-1,k}(j+4k-2) + \lambda a_{j-5,k+1}$$
$$= -\frac{1}{2} \sum_{p,q,r,s} a_{j-r,k-s} a_{r-p,s-q} a_{pq} + \epsilon f_{j-3} \delta_{k0}$$

where the summation is for $0 \le p \le r \le j$ and $0 \le q \le s \le k$ and

$$f_j = \frac{1}{j!} \frac{\partial^j f}{\partial z^j}.$$

The values of the first few coefficients are

$$a_{00} = 2i, \quad a_{10} = -\frac{\lambda}{3}, \quad a_{20} = -\frac{i\lambda^2}{18}, \quad a_{30} = -\frac{i\lambda^3}{27} - \frac{\epsilon f_0}{4}.$$

The coefficient a_{40} is arbitrary and therefore a_{01} is given by

$$a_{01} = \frac{4}{135} i\lambda^4 + \frac{1}{5}\epsilon(\lambda f_0 + f_1).$$

The sets of coefficients a_{0k}, $k = 0, 1, 2, \ldots$, satisfy

$$4k(k-1)a_{0k} + ka_{0k} + \frac{1}{2}a_{0k} = -\frac{1}{8} \sum_s \sum_q a_{0,k-s} a_{0,s-q} a_{0q}$$

where the summation is for $0 \le q \le s \le k$. Introducing the generating function

$$\Theta(s) = \sum_{k=0}^{\infty} a_{0k} s^k$$

where s is some independent variable, the nonlinear differential equation for Θ is obtained

$$16s^2 \frac{d^2\Theta}{ds^2} + 4s\frac{d\Theta}{ds} + 2\Theta + \frac{1}{2}\Theta^3 = 0.$$

The differential equation can also be obtained by substituting

$$w(z) = \frac{1}{z - z_0}\Theta_0(s)$$

where $s = (z - z_0)^4 \ln(z - z_0)$ into this second order differential equation. In the limit $z \to z_0$ we find that Θ_0 satisfies this differential equation provided that there is an ordering in which $|z - z_0| \ll |s|$. This second order differential equation can be solved in terms of elliptic functions by making the substitution

$$\Theta_0(s) = s^{1/4} g(s^{1/4})$$

which leads to the equation

$$\frac{d^2 g(y)}{dy^2} + \frac{1}{2} g^3(y) = 0$$

where $y = s^{1/4} = z(\ln z)^{1/4}$.

Problem 17. Suppose that an elementary magnet is situated at the origin and that its axis corresponds to the z-axis. The trajectory

$$(x(t), y(t), z(t))$$

of an electrical particle in this magnetic field is then given as the autonomous system of second order ordinary differential equations

$$\frac{d^2 x}{dt^2} = \frac{1}{r^5} \left(3yz \frac{dz}{dt} - (3z^2 - r^2) \frac{dy}{dt} \right)$$

$$\frac{d^2 y}{dt^2} = \frac{1}{r^5} \left((3z^2 - r^2) \frac{dx}{dt} - 3xz \frac{dz}{dt} \right)$$

$$\frac{d^2 z}{dt^2} = \frac{1}{r^5} \left(3xz \frac{dy}{dt} - 3yz \frac{dx}{dt} \right)$$

where $r^2 := x^2 + y^2 + z^2$. Show that the system can be simplified by introducing *polar coordinates*

$$x(t) = R(t) \cos(\phi(t)), \qquad y(t) = R(t) \sin(\phi(t)).$$

Solution 17. Since

$$\frac{dx}{dt} = \frac{dR}{dt} \cos \phi - R \frac{d\phi}{dt} \sin \phi$$

$$\frac{dy}{dt} = \frac{dR}{dt} \sin \phi + R \frac{d\phi}{dt} \cos \phi$$

and

$$\frac{d^2 x}{dt^2} = \frac{d^2 R}{dt^2} \cos \phi - 2 \frac{dR}{dt} \frac{d\phi}{dt} \sin \phi - R \frac{d^2 \phi}{dt^2} \sin \phi - R \left(\frac{d\phi}{dt} \right)^2 \cos \phi$$

$$\frac{d^2 y}{dt^2} = \frac{d^2 R}{dt^2} \sin \phi + 2 \frac{dR}{dt} \frac{d\phi}{dt} \cos \phi + R \frac{d^2 \phi}{dt^2} \cos \phi - R \left(\frac{d\phi}{dt} \right)^2 \cos \phi$$

we obtain

$$\frac{d^2 R}{dt^2} = \left(\frac{2C}{R} + \frac{R}{r^3} \right) \left(\frac{2C}{R^2} + \frac{3R^2}{r^5} - \frac{1}{r^3} \right)$$

$$\frac{d^2 z}{dt^2} = \left(\frac{2C}{R} + \frac{R}{r^3} \right) \frac{3Rz}{r^5}$$

$$\frac{d\phi}{dt} = \left(\frac{2C}{R} + \frac{R}{r^3} \right) \frac{1}{R}$$

where $r^2 = R^2 + z^2$ and C is some constant originating from the integration of $d^2\phi/dt^2$.

Problem 18.   Consider the pair of coupled nonlinear differential equations relevant to the quantum field theory of charged solitons

$$\frac{d^2\sigma}{dx^2} = -\sigma + \sigma^3 + d\rho^2\sigma, \qquad \frac{d^2\rho}{dx^2} = f\rho + \lambda\rho^3 + d\rho(\sigma^2 - 1) \qquad (1)$$

where σ and ρ are real scalar fields and d, f, λ are constants. Try to find an exact solution of (1) with the ansatz

$$\rho(x) = b_1 \tanh(\lambda_0(x + c_0)), \qquad \sigma(x) = \sum_{n=1} a_n \tanh^n(\lambda_0(x + c_0)). \qquad (2)$$

Solution 18.   Inserting (2) into the first equation of (1) and equating with the terms of the same order in $\tanh(\lambda_0(x+c_0))$ we obtain a recurrence relation for the coefficients a_n, namely

$$(n+1)(n+2)\lambda_0^2 - (2n^2\lambda_0^2 - 1)a_n + ((n-1)(n-2)\lambda_0^2 - db_1^2)a_{n-2} - c_n = 0 \quad (3)$$

where

$$c_n = \sum_{i+j+k=n} a_i a_j a_k. \qquad (4)$$

Inserting (2) into the second equation of (1), we obtain

$$2\lambda_0^2 + f - d = 0, \qquad 2\lambda_0^2 - \lambda b_1^2 - dc_2' = 0, \qquad (5a)$$

$$c_n' = 0, \qquad n \neq 2, \qquad (5b)$$

where

$$c_n' = \sum_{i+j=n} a_i a_j. \qquad (6)$$

Using $c_m' = 0$ $(n \neq 2)$, we have $a_n = 0$ $(n \neq 1)$. Let $n = 0, 1, 2, 3 \ldots$. Then from (3), (4) and (6) we find $\lambda_0 = \sqrt{1/2}$ and

$$a_1 = \pm\sqrt{(d-\lambda)/(d^2-\lambda)}, \qquad b_1 = \pm\sqrt{(d-1)/(d^2-\lambda)}.$$

Thus we find an exact static soliton solutions for (1)

$$\rho(x) = \pm\sqrt{(d-1)/(d^2-\lambda)} \tanh\left(\sqrt{\frac{1}{2}}(x+c_0)\right)$$

$$\sigma(x) = \pm\sqrt{(d-\lambda)/(d^2-\lambda)} \tanh\left(\sqrt{\frac{1}{2}}(x+c_0)\right).$$

Chapter 8

Lax Representations in Classical Mechanics

Let

$$\frac{d\mathbf{u}}{dt} = \mathbf{V}(\mathbf{u}), \qquad \mathbf{u} \equiv (u_1, u_2, \ldots, u_m)^T$$

be an autonomous system of first-order ordinary differential equations. Assume that the functions $V_k : \mathbb{R}^m \to \mathbb{R}$ are smooth. Assume that this system can be written in the form (the so-called *Lax representation*)

$$\frac{dL}{dt} = [A, L](t) \equiv [A(t), L(t)]$$

where A and L are $n \times n$ matrices and $[A, L] \equiv AL - LA$. Then the $n \times n$ matrices L and A are called a *Lax pair*.

Problem 1. (i) Show that

$$\frac{dL^k}{dt} = [A, L^k](t). \tag{1}$$

(ii) Show that $\operatorname{tr}(L^k)$ $(k = 1, 2, \ldots)$ are first integrals, where $\operatorname{tr}(.)$ denotes the trace.
(iii) Assume that L^{-1} exists. Show that $\operatorname{tr}(L^{-1})$ is a first integral.

Solution 1. (i) The formula (1) is true by assumption for $k = 1$. If it is

81

true for k, then

$$\frac{dL^{k+1}}{dt} = \frac{d}{dt}(L^k L) = \frac{dL^k}{dt}L + L^k\frac{dL}{dt}$$

$$= ([A, L^k]L + L^k[A, L])(t) \qquad \text{by the induction hypothesis}$$

$$= (AL^k L - L^k AL + L^k AL - L^k LA)(t) = (AL^{k+1} - L^{k+1}A)(t)$$

$$= [A, L^{k+1}](t).$$

(ii) We have

$$\frac{d}{dt}\mathrm{tr}L^k = \mathrm{tr}\left(\frac{dL^k}{dt}\right) = \mathrm{tr}([A, L^k]) = \mathrm{tr}(AL^k - L^k A) = \mathrm{tr}(AL^k) - \mathrm{tr}(L^k A) = 0$$

since $\mathrm{tr}(XY) = \mathrm{tr}(YX)$ for arbitrary $n \times n$ matrices X and Y. Consequently, $\mathrm{tr}L^k$ ($k = 1, 2, \ldots$) are first integrals of system (1).

(iii) From $L^{-1}L = I$ where I is the unit matrix it follows that

$$\frac{dL^{-1}}{dt} = -L^{-1}\frac{dL}{dt}L^{-1}.$$

Then

$$\frac{d}{dt}\mathrm{tr}L^{-1} = \mathrm{tr}\left(\frac{d}{dt}L^{-1}\right) = -\mathrm{tr}\left(L^{-1}\frac{dL}{dt}L^{-1}\right)$$

$$= -\mathrm{tr}(L^{-1}[A, L]L^{-1}) = -\mathrm{tr}(L^{-1}ALL^{-1}) + \mathrm{tr}(L^{-1}LAL^{-1})$$

$$= -\mathrm{tr}(L^{-1}A) + \mathrm{tr}(AL^{-1}) = 0.$$

Problem 2.   Let $H : \mathbb{R}^4 \to \mathbb{R}$ be the Hamilton function

$$H(\mathbf{p}, \mathbf{q}) = \frac{1}{2}(p_1^2 + p_2^2) + e^{q_2 - q_1}. \tag{1}$$

(i) Find the equations of motion. The Hamilton function (1) is a first integral, i.e. $dH/dt = 0$. Find the second first integral on inspection of the equations of motion.

(ii) Define

$$a := \frac{1}{2}e^{(q_2 - q_1)/2}, \quad b_1 := \frac{1}{2}p_1, \quad b_2 := \frac{1}{2}p_2.$$

Give the equations of motion for a, b_1 and b_2.

(iii) Show that the equations of motion for a, b_1 and b_2 can be written in Lax form, i.e.,

$$\frac{dL}{dt} = [A, L](t)$$

where

$$L := \begin{pmatrix} b_1 & a \\ a & b_2 \end{pmatrix}, \qquad A := \begin{pmatrix} 0 & a \\ -a & 0 \end{pmatrix}.$$

Solution 2. (i) The *Hamilton equations of motion* are given by

$$\frac{dq_1}{dt} = \frac{\partial H}{\partial p_1} = p_1, \qquad \frac{dp_1}{dt} = -\frac{\partial H}{\partial q_1} = e^{q_2 - q_1}$$

$$\frac{dq_2}{dt} = \frac{\partial H}{\partial p_2} = p_2, \qquad \frac{dp_2}{dt} = -\frac{\partial H}{\partial q_2} = -e^{q_2 - q_1}.$$

Adding the last two equations provides the first integral $I(\mathbf{p}, \mathbf{q}) = p_1 + p_2$ since

$$\frac{dI}{dt} = \frac{dp_1}{dt} + \frac{dp_2}{dt} = e^{q_2 - q_1} - e^{q_2 - q_1} = 0.$$

Obviously, the Hamilton function H is the other first integral.

(ii) The time derivative of $a(t)$ yields

$$\frac{da}{dt} = \frac{1}{4} e^{(q_2 - q_1)/2} \left(\frac{dq_2}{dt} - \frac{dq_1}{dt} \right) = \frac{1}{4} e^{(q_2 - q_1)/2} (p_2 - p_1) = \frac{1}{2} e^{(q_2 - q_1)/2} (b_2 - b_1).$$

Thus

$$\frac{da}{dt} = a(b_2 - b_1).$$

The time derivative of $b_1(t)$ yields

$$\frac{db_1}{dt} = \frac{1}{2} \frac{dp_1}{dt} = \frac{1}{2} e^{q_2 - q_1} = 2a^2.$$

Analogously,

$$\frac{db_2}{dt} = \frac{1}{2} \frac{dp_2}{dt} = -\frac{1}{2} e^{q_2 - q_1} = -2a^2.$$

To summarize: the equations of motion for a, b_1 and b_2 are given by

$$\frac{da}{dt} = a(b_2 - b_1), \qquad \frac{db_1}{dt} = 2a^2, \qquad \frac{db_2}{dt} = -2a^2.$$

(iii) The Lax representation is given by L and A since

$$[A, L] \equiv AL - LA$$

$$= \begin{pmatrix} 0 & a \\ -a & 0 \end{pmatrix} \begin{pmatrix} b_1 & a \\ a & b_2 \end{pmatrix} - \begin{pmatrix} b_1 & a \\ a & b_2 \end{pmatrix} \begin{pmatrix} 0 & a \\ -a & 0 \end{pmatrix}$$

$$= \begin{pmatrix} 2a^2 & a(b_2 - b_1) \\ a(b_2 - b_1) & -2a^2 \end{pmatrix}$$

and

$$\frac{dL}{dt} = \begin{pmatrix} db_1/dt & da/dt \\ da/dt & db_2/dt \end{pmatrix}.$$

The first integrals follow as

$$I_1(a, b_1, b_2) = \text{tr} L = b_1 + b_2, \qquad I_2(a, b_1, b_2) = \text{tr} L^2 = 2a^2 + b_1^2 + b_2^2.$$

Problem 3.   Consider the nonlinear system of ordinary differential equations

$$\frac{du_1}{dt} = (\lambda_3 - \lambda_2)u_2u_3 \tag{1a}$$

$$\frac{du_2}{dt} = (\lambda_1 - \lambda_3)u_3u_1 \tag{1b}$$

$$\frac{du_3}{dt} = (\lambda_2 - \lambda_1)u_1u_2 \tag{1c}$$

where $\lambda_j \in \mathbb{R}$. This dynamical system describes *Euler's rigid body motion*.

(i) Show that the first integrals are given by

$$I_1(\mathbf{u}) = u_1^2 + u_2^2 + u_3^2, \qquad I_2(\mathbf{u}) = \lambda_1 u_1^2 + \lambda_2 u_2^2 + \lambda_3 u_3^2. \tag{2}$$

(ii) A Lax representation is given by

$$\frac{dL}{dt} = [L, \lambda L](t) \tag{3}$$

where

$$L := \begin{pmatrix} 0 & -u_3 & u_2 \\ u_3 & 0 & -u_1 \\ -u_2 & u_1 & 0 \end{pmatrix}, \qquad \lambda L := \begin{pmatrix} 0 & -\lambda_3 u_3 & \lambda_2 u_2 \\ \lambda_3 u_3 & 0 & -\lambda_1 u_1 \\ -\lambda_2 u_2 & \lambda_1 u_1 & 0 \end{pmatrix}. \tag{4}$$

Show that $\text{tr}(L^k)$ ($k = 1, 2, \ldots$) gives only one first integral of system (1).

(iii) Instead of (3) we consider now

$$\frac{d(L + Ay)}{dt} = [L + Ay, \lambda L + By](t) \tag{5}$$

where y is a dummy variable and A and B are time-independent diagonal matrices, i.e., $A = \text{diag}(A_1, A_2, A_3)$ and $B = \text{diag}(B_1, B_2, B_3)$ with $A_j, B_j \in \mathbb{R}$. Equation (5) decomposes into various powers of y, namely

$$y^0: \quad \frac{dL}{dt} = [L, \lambda L](t), \qquad y^1: \quad 0 = [L, B] + [A, \lambda L], \qquad y^2: \quad [A, B] = 0. \tag{6}$$

Equation (6c) is satisfied identically since A and B are diagonal matrices. Equation (6b) leads to

$$\lambda_i = \frac{B_j - B_k}{A_j - A_k}$$

where (i, j, k) are permutations of $(1, 2, 3)$. This equation can be satisfied by setting

$$B_j = A_j^2, \qquad \lambda_i = A_j + A_k.$$

Consequently the original Lax pair L, λL satisfies the *extended Lax pair*

$$L + Ay, \qquad \lambda L + By.$$

Show that $\mathrm{tr}((L + Ay)^2)$ and $\mathrm{tr}((L + Ay)^3)$ provide the first integrals (2).

Solution 3. (i) Straightforward calculation yields

$$\frac{d}{dt} I_1(\mathbf{u}) = 2u_1 \frac{du_1}{dt} + 2u_2 \frac{du_2}{dt} + 2u_3 \frac{du_3}{dt}.$$

Thus

$$\frac{d}{dt} I_1(\mathbf{u}) = 2u_1(\lambda_3 - \lambda_2)u_2 u_3 + 2u_2(\lambda_1 - \lambda_3)u_3 u_1 + 2u_3(\lambda_2 - \lambda_1)u_1 u_2 = 0.$$

Analogously, $dI_2(\mathbf{u})/dt = 0$.
(ii) We obtain $\mathrm{tr} L = 0$ and

$$\mathrm{tr} L^2 = -2(u_1^2 + u_2^2 + u_3^2) = -2I_1(\mathbf{u}).$$

Since L does not depend on λ we cannot find I_2.
(iii) Straightforward calculation yields

$$\mathrm{tr}((L + Ay)^2) = -2I_1(\mathbf{u}) + C_1, \qquad \mathrm{tr}((L + Ay)^3) = -3y I_2(\mathbf{u}) + C_2$$

where C_1 and C_2 are constants. Thus the extended Lax pair provides both first integrals.

Problem 4. The semisimple Lie algebra $sl(2, \mathbb{R})$ is defined by the commutator rules

$$[X_0, X_+] = 2X_+ \qquad [X_0, X_-] = -2X_- \qquad [X_+, X_-] = X_0 \qquad (1)$$

where X_0, X_+, X_- denote the generators. Let

$$\{A(\mathbf{p}, \mathbf{q}), B(\mathbf{p}, \mathbf{q})\} := \sum_{j=1}^{N} \left(\frac{\partial A}{\partial q_j} \frac{\partial B}{\partial p_j} - \frac{\partial B}{\partial q_j} \frac{\partial A}{\partial p_j} \right) \qquad (2)$$

be the *Poisson bracket*. Let

$$t_+ := \frac{1}{2} \sum_{k=1}^{N} p_k^2, \qquad t_- := -\frac{1}{2} \sum_{k=1}^{N} q_k^2, \qquad t_0 := -\sum_{k=1}^{N} p_k q_k.$$

(i) Show that the functions $-t_+$, $-t_-$, $-t_0$ form a basis of a Lie algebra under the Poisson bracket and that the Lie algebra is isomorphic to $sl(2,\mathbb{R})$.

(ii) Let

$$H(\mathbf{p},\mathbf{q}) = \frac{1}{2}\sum_{k=1}^{N} p_k^2 + U(\mathbf{q}) \equiv t_+ + U(\mathbf{q}).$$

Find the condition on U such that $\{-H, -t_-, -t_0\}$ forms a basis of a Lie algebra which is isomorphic to $sl(2,\mathbb{R})$.

(iii) The Hamilton system

$$H(\mathbf{p},\mathbf{q}) = \frac{1}{2}\sum_{k=1}^{N} p_k^2 + \sum_{j=2}^{N}\sum_{k=1}^{j-1} \frac{a^2}{(q_j - q_k)^2}$$

admits a Lax representation with

$$L := \begin{pmatrix} p_1 & \dfrac{ia}{(q_1 - q_2)} & \cdots & \dfrac{ia}{(q_1 - q_N)} \\ \dfrac{ia}{(q_2 - q_1)} & p_2 & \cdots & \dfrac{ia}{(q_2 - q_N)} \\ \vdots & & & \vdots \\ \dfrac{ia}{(q_N - q_1)} & & \cdots & p_N \end{pmatrix}$$

where a is a nonzero real constant. Let $N = 3$. Find the first integrals.

Solution 4.   (i) Straightforward calculation yields

$$\{-t_0, -t_+\} = \{t_0, t_+\} = -2t_+$$
$$\{-t_0, -t_-\} = \{t_0, t_-\} = 2t_-$$
$$\{-t_+, -t_-\} = \{t_+, t_-\} = -t_0.$$

Consider the map $X_0 \to -t_0$, $X_- \to -t_-$, $X_+ \to -t_+$. Then the two Lie algebras are isomorphic.

(ii) Since

$$\{-t_0, -H\} = -\sum_{j=1}^{N}\left(p_j^2 - \frac{\partial U}{\partial q_j} q_j\right)$$

$$\{-H, -t_-\} = \sum_{j=1}^{N} p_j q_j = -t_0$$

we find from the condition $\{-t_0, -H\} = -2H$ that the potential U satisfies the linear partial differential equation

$$\sum_{j=1}^{N} q_j \frac{\partial U}{\partial q_j} = -2U.$$

This is a homogeneous equation (of rank -2) for the potential U. This equation admits the two solutions

$$U(\mathbf{q}) = \sum_{j=2}^{N} \sum_{k=1}^{j-1} \frac{a^2}{(q_j - q_k)^2}$$

and

$$U(\mathbf{q}) = \frac{a^2}{q_1^2 + q_2^2 + \cdots + q_N^2}.$$

(iii) The first integrals are the invariants of L. From L we find (with $N = 3$) that $\mathrm{tr}(L) = p_1 + p_2 + p_3$ and

$$\mathrm{tr}(L^2) = p_1^2 + p_2^2 + p_3^2 + 2a^2 \left(\frac{1}{(q_1 - q_2)^2} + \frac{1}{(q_3 - q_1)^2} + \frac{1}{(q_2 - q_3)^2} \right).$$

Obviously, $2H = \mathrm{tr}L^2$. The calculation of $\mathrm{tr}(L^3)$ is very lengthy. To find the third first integral we use

$$\det(L) = \lambda_1 \lambda_2 \lambda_3 = p_1 p_2 p_3 - a^2 \left(\frac{p_1}{(q_2 - q_3)^2} + \frac{p_2}{(q_3 - q_1)^2} + \frac{p_3}{(q_1 - q_2)^2} \right)$$

where λ_1, λ_2 and λ_3 denote the eigenvalues of L.

Problem 5. Assume that

$$L := \begin{pmatrix} p_1 & if_{12} & 0 & \cdots & 0 & 0 & -if_{N1} \\ -if_{12} & p_2 & if_{23} & \cdots & 0 & 0 & 0 \\ 0 & -if_{23} & p_3 & \cdots & & & \\ \vdots & & & & & & if_{N-1,N} \\ if_{N1} & 0 & 0 & \cdots & 0 & -if_{N-1,N} & p_N \end{pmatrix} \tag{1}$$

where $f_{n,n+1} := f(q_n - q_{n+1})$ with $f(\mathbf{q})$ a certain given smooth real function. Assume that the Hamilton function H is given by

$$H(\mathbf{p}, \mathbf{q}) = \frac{1}{2} \mathrm{tr} L^2. \tag{2}$$

Let

$$A := i \begin{pmatrix} h_1 & g_{12} & 0 & \cdots & 0 & g_{N1} \\ g_{12} & h_2 & g_{23} & \cdots & 0 & 0 \\ 0 & g_{23} & h_3 & & & \\ 0 & \cdots & & & & g_{N-1,N} \\ g_{N1} & 0 & & \cdots & g_{N-1,N} & h_N \end{pmatrix} \tag{3}$$

where $g(\mathbf{q})$ and $h_k(\mathbf{p}, \mathbf{q})$ are real-valued smooth functions. Find the condition on f, g and h such that L, A are a Lax pair of the Hamilton function H.

Solution 5. From the Hamilton function we obtain

$$H(\mathbf{p}, \mathbf{q}) = \frac{1}{2} \mathrm{tr} L^2 = \frac{1}{2} \sum_{k=1}^{N} p_k^2 + \sum_{k=1}^{N} f_{k,k+1}^2 \tag{4}$$

with $N + 1 \equiv 1$, i.e. modulo N. Then the Hamilton equations of motion are given by

$$\frac{dq_k}{dt} = \frac{\partial H}{\partial p_k} = p_k, \quad \frac{dp_k}{dt} = -\frac{\partial H}{\partial q_k} = 2\left(-f_{k,k+1}' f_{k,k+1} + f_{k-1,k}' f_{k-1,k}\right)$$

where $k = 1, \ldots, N$ and f' means differentiation with respect to the argument. For f we obtain

$$\frac{df_{k,k+1}}{dt} = (p_k - p_{k+1}) f_{k,k+1}' \tag{6}$$

where we used $f_{k,k+1} = f(q_k - q_{k+1})$. From the requirement that L, A are a Lax pair, i.e. L, A satisfies $dL/dt = [L, A](t)$ we obtain

$$\frac{dp_k}{dt} = 2(f_{k-1,k} g_{k-1,k} - f_{k,k+1} g_{k,k+1})$$

$$\frac{df_{k,k+1}}{dt} = g_{k,k+1}(p_k - p_{k+1}) + i f_{k,k+1}(h_{k+1} - h_k) \quad \text{modulo} \quad N$$

$$0 = f_{k,k+1} g_{k,k+2} - g_{k,k+1} f_{k,k+2} \cdot$$

The choice $h_k = 0$ and

$$g_{k,k+1} = f_{k,k+1}', \quad 0 = f_{k,k+1} g_{k,k+2} - g_{k,k+1} f_{k,k+2}$$

provides the consistency of this system and the Hamilton equations of motion. From the last two equations we can eliminate $g_{k,k+1}$ and obtain

$$\frac{f_{k,k+1}'}{f_{k,k+1}} = \frac{f_{k,k+2}'}{f_{k,k+2}} \cdot$$

Consequently,

$$\frac{f_{k,k+1}'}{f_{k,k+1}} = c \equiv \text{const.}$$

Therefore, the solution of this equation is given by

$$f_{k,k+1} \equiv f(q_k - q_{k+1}) = A \exp(c(q_k - q_{k+1}))$$

where A is a constant.

Problem 6.   We consider systems of $(2n + 1)$ ordinary nonlinear differential equations. These are the multiple three-wave interaction system describing triads (a_j, b_j, u), $j = 1, \ldots, n$, evolving in time alone and interacting with each other through the single common member u. These systems can be derived from a Hamilton function

$$H(\mathbf{b}, \mathbf{c}, u) = \frac{1}{2} i \sum_{j=1}^{n} \epsilon_j (c_j c_j^* - b_j b_j^*) + i \sum_{j=1}^{n} \alpha_j (u b_j^* c_j + u^* b_j c_j^*) \quad (1)$$

and with *Poisson bracket* defined as

$$\{f, g\} := \frac{\partial f}{\partial u} \frac{\partial g}{\partial u^*} - \frac{\partial f}{\partial u^*} \frac{\partial g}{\partial u} + \sum_{j=1}^{n} \left(\frac{\partial f}{\partial b_j} \frac{\partial g}{\partial b_j^*} - \frac{\partial f}{\partial b_j^*} \frac{\partial g}{\partial b_j} + \frac{\partial f}{\partial c_j} \frac{\partial g}{\partial c_j^*} - \frac{\partial f}{\partial c_j^*} \frac{\partial g}{\partial c_j} \right).$$
$$(2)$$

Thus (u, u^*), (b_j, b_j^*), (c_j, c_j^*) are pairs of canonical variables (where * means complex conjugate). The arbitrary real parameters α_j, ϵ_j play the role of frequencies.

(i) Let $\alpha_j = 1$ for $j = 1, 2, \ldots, n$. Find the Hamilton's equation of motion from (1) and (2).

(ii) Find the Lax representation.

Solution 6. From the Poisson bracket we find

$$\frac{du}{dt} = \{u, H\} = \frac{\partial H}{\partial u^*}$$

$$\frac{db_j}{dt} = \{b_j, H\} = \frac{\partial H}{\partial b_j^*}, \qquad \frac{dc_j}{dt} = \{c_j, H\} = \frac{\partial H}{\partial c_j^*}$$

and *c.c.*, where *c.c.* stands for complex conjugate. Thus the Hamilton's equations of motion are

$$\frac{du}{dt} = i \sum_{j=1}^{n} b_j c_j^*, \quad \frac{db_j}{dt} = -\frac{1}{2} i \epsilon_j b_j + i u c_j, \quad \frac{dc_j}{dt} = \frac{1}{2} i \epsilon_j c_j + i u^* b_j, \text{ and c.c.}$$

(ii) We have

$$\frac{dL}{dt} = \{L, H\}(t) \equiv [A, L](t)$$

in which L and A are the $(2n + 2) \times (2n + 2)$ matrices

$$L := \frac{1}{2} \begin{pmatrix} \pi & \sigma_1 & \cdots & \sigma_n \\ \tau_1 & \epsilon_1 I_2 & & 0 \\ \vdots & & \ddots & \\ \tau_n & 0 & & \epsilon_n I_2 \end{pmatrix}, \qquad A := \frac{1}{2} i \begin{pmatrix} 0 & \omega \sigma_1 & \cdots & \omega \sigma_n \\ \tau_1 \omega & & & \\ \vdots & & 0 & \\ \tau_n \omega & & & \end{pmatrix}.$$

The π, σ_j and τ_j are the 2×2 matrices

$$\pi := \begin{pmatrix} 0 & -2u \\ 2u^* & 0 \end{pmatrix}, \qquad \tau_j := \begin{pmatrix} b_j^* & -c_j^* \\ -c_j & -b_j \end{pmatrix}, \qquad \sigma_j := \begin{pmatrix} -b_j & -c_j^* \\ -c_j & b_j^* \end{pmatrix}$$

and $\omega = \text{diag}(-1, 1)$. I_2 is the 2×2 identity matrix. The Hamilton function (1) is given by

$$H = \frac{2}{3}i\text{tr}(L^3) + \text{ const.}$$

where tr denotes the trace.

Problem 7.   Consider the autonomous first order system

$$\frac{du_1}{dt} = u_1(u_2 - u_n), \quad \frac{du_2}{dt} = u_2(u_3 - u_1), \ldots, \quad \frac{du_n}{dt} = u_n(u_1 - u_{n-1}).$$

(i) Find a Lax representation for the case $n = 3$ and the first integrals using $\text{tr}L^n$.
(ii) Find the Lax representation for the case $n = 4$ and the first integrals.
(iii) Find the Lax representation for arbitrary n.

Solution 7.   (i) For $n = 3$ we find the Lax pair

$$L = \begin{pmatrix} 0 & 1 & u_1 \\ u_2 & 0 & 1 \\ 1 & u_3 & 0 \end{pmatrix}, \quad A = \begin{pmatrix} u_1 + u_2 & 0 & 1 \\ 1 & u_2 + u_3 & 0 \\ 0 & 1 & u_1 + u_2 \end{pmatrix}.$$

The first integrals are given by $\text{tr}L^2 = u_1 + u_2 + u_3$, $\text{tr}L^3 = u_1 u_2 u_3$.
(ii) For $n = 4$ we have the Lax pair

$$L = \begin{pmatrix} 0 & 1 & 0 & u_1 \\ u_2 & 0 & 1 & 0 \\ 0 & u_3 & 0 & 1 \\ 1 & 0 & u_4 & 0 \end{pmatrix}, \quad A = \begin{pmatrix} u_1 + u_2 & 0 & 1 & 0 \\ 0 & u_2 + u_3 & 0 & 1 \\ 1 & 0 & u_3 + u_4 & 0 \\ 0 & 1 & 0 & u_4 + u_1 \end{pmatrix}.$$

The first integrals are given by

$$I_1(\mathbf{u}) = \sum_{j=1}^{4} u_j, \quad I_2(\mathbf{u}) = \prod_{j=1}^{4} u_j, \quad I_3(\mathbf{u}) = u_2 u_4 + u_1 u_3$$

where

$$\text{tr}L = 0, \quad \text{tr}L^2 = 2I_1, \quad \text{tr}L^3 = 0, \quad \text{tr}L^4 = 4 + 4I_3 + 2I_1^2 - 4I_2.$$

(iii) For arbitrary n the Lax representation is given by

$$\phi_{j+1} + u_j \phi_{j-1} = \lambda \phi_j$$

$$\frac{d\phi_j}{dt} = \phi_{j+2} + (u_{j+1} + u_j)\phi_j$$

where $j = 1, \ldots, n$ and $0 \equiv n$, $1 \equiv n + 1$ etc. The parameter λ is time-independent.

Chapter 9

Hilbert Spaces

A *Hilbert space* is a set, $\mathcal{H}$ of elements, or vectors, $(f, g, h, \ldots)$ which satisfies the following conditions (1) to (5).

(1) If f and g belong to $\mathcal{H}$, then there is a unique element of $\mathcal{H}$, denoted by $f + g$, the operation of addition (+) being invertible, commutative and associative.

(2) If c is a complex number, then for any f in $\mathcal{H}$, there is an element cf of $\mathcal{H}$; and the multiplication of vectors by complex numbers thereby defined satisfies the distributive conditions

$$c(f + g) = cf + cg, \qquad (c_1 + c_2)f = c_1 f + c_2 f.$$

(3) Hilbert spaces $\mathcal{H}$ possess a zero element, 0, characterized by the property that $0 + f = f$ for all vectors f in $\mathcal{H}$.

(4) For each pair of vectors f, g in $\mathcal{H}$, there is a complex number $\langle f|g \rangle$, termed the inner product or scalar product of f with g, such that

$$\langle f|g \rangle = \overline{\langle g|f \rangle}$$

$$\langle f|g + h \rangle = \langle f|g \rangle + \langle f|h \rangle$$

$$\langle f|cg \rangle = c \langle f|g \rangle$$

and

$$\langle f|f \rangle \geq 0.$$

91

Equality in the last formula occurs only if $f = 0$. The scalar product defines the norm $\|f\| = \langle f|f \rangle^{1/2}$.

(5) If $\{\, f_n \,\}$ is a sequence in $\mathcal{H}$ satisfying the Cauchy condition that

$$\| f_m - f_n \| \to 0$$

as m and n tend independently to infinity, then there is a unique element f of $\mathcal{H}$ such that $\| f_n - f \| \to 0$ as $n \to \infty$.

Let $B = \{\, \phi_n \,:\, n \in I \,\}$ be an orthonormal basis in the Hilbert space $\mathcal{H}$. I is the countable index set. Then

$$(1) \qquad \bigwedge_{f \in \mathcal{H}} \qquad f = \sum_{n \in I} \langle f|\phi_n \rangle \phi_n$$

$$(2) \qquad \bigwedge_{f,g \in \mathcal{H}} \qquad \langle f|g \rangle = \sum_{n \in I} \overline{\langle f|\phi_n \rangle} \langle g|\phi_n \rangle$$

Problem 1.   Consider the Hilbert space $L_2[0, \pi]$ and the function $f(x) = \sin(x)$. Find a and b $(a, b \in \mathbb{R})$ such that

$$\| f(x) - (ax^2 + bx) \|$$

is a minimum. The norm in the Hilbert space $L_2[0, \pi]$ is induced by the scalar product. Hence

$$\| f(x) - (ax^2 + bx) \|^2 = \int_0^\pi (f(x) - (ax^2 + bx))^2 dx \,.$$

Solution 1.   We have

$$h(a,b) := \int_0^\pi (\sin(x) - (ax^2 + bx))^2 dx \,.$$

From the conditions

$$\frac{\partial h}{\partial a} = 0, \qquad \frac{\partial h}{\partial b} = 0$$

we obtain

$$\int_0^\pi x^2 (\sin(x) - (ax^2 + bx)) dx = 0$$

$$\int_0^\pi x(\sin(x) - (ax^2 + bx)) dx = 0 \,.$$

From these two equations we obtain

$$a \int_0^\pi x^4 dx + b \int_0^\pi x^3 dx = \int_0^\pi x^2 \sin(x) dx$$

$$a \int_0^\pi x^3 dx + b \int_0^\pi x^2 dx = \int_0^\pi x \sin(x) dx .$$

Integrating yields

$$\frac{\pi^5 a}{5} + \frac{\pi^4 b}{4} = \pi^2 - 4$$

$$\frac{\pi^4 a}{4} + \frac{\pi^3 b}{3} = \pi .$$

This is a system of linear equations for a and b. Solving for a and b we find

$$a = \frac{20}{\pi^3} - \frac{320}{\pi^5}, \qquad b = \frac{240}{\pi^4} - \frac{12}{\pi^2} .$$

Problem 2. Consider the Hilbert space $L_2(\mathbb{R})$. Let $\phi \in L_2(\mathbb{R})$ be a real-valued function with

$$\int_\mathbb{R} \phi(x) dx = 1$$

and

$$\int_\mathbb{R} \phi(x) \phi(x - n) dx = \delta_{n0}, \quad n \in \mathbb{Z} .$$

Assume that

$$\phi(x) = \sum_{k=0}^{M-1} c_k \phi(2x - k)$$

where $c_k \in \mathbb{R}$.
(i) Show that

$$\sum_{k=0}^{M-1} c_k = 2 .$$

(ii) Show that

$$\sum_{k=0}^{M-1} c_k^2 = 2 .$$

(iii) Give a function that satisfies these conditions.

Solution 2. (i) We have

$$1 = \int_\mathbb{R} \phi(x) dx = \int_\mathbb{R} \sum_{k=0}^{M-1} c_k \phi(2x - k) dx = \sum_{k=0}^{M-1} c_k \int_\mathbb{R} \phi(2x - k) dx .$$

Using the transformation $y = 2x - k$ and $dy = 2dx$ we find

$$1 = \frac{1}{2} \sum_{k=0}^{M-1} c_k \int_{\mathbb{R}} \phi(y) dy = \frac{1}{2} \sum_{k=0}^{M-1} c_k \,.$$

Thus equation (1) follows.

(ii) To prove equation (2) we start from

$$1 = \int_{\mathbb{R}} \phi(x)\phi(x) dx = \int_{\mathbb{R}} \left(\sum_{k=0}^{M-1} c_k \phi(2x - k) \right) \left(\sum_{j=0}^{M-1} c_j \phi(2x - j) \right) dx \,.$$

Thus

$$1 = \sum_{k=0}^{M-1} \sum_{j=0}^{M-1} c_k c_j \int_{\mathbb{R}} \phi(2x - k)\phi(2x - j) dx \,.$$

Using $y = 2x - k$ and $dy = 2dx$ we obtain

$$1 = \frac{1}{2} \sum_{k=0}^{M-1} \sum_{j=0}^{M-1} c_k c_j \int_{\mathbb{R}} \phi(y)\phi(y + k - j) dy = \frac{1}{2} \sum_{k=0}^{M-1} c_k^2 \,.$$

(iii) The function

$$\phi(x) = \begin{cases} 1 \text{ for } & 0 \le x \le 1 \\ 0 & \text{otherwise} \end{cases}$$

satisfies the conditions given above.

The function ϕ plays an important role in wavelet theory. The function ϕ is called the *scaling function*. What is desired is a function ψ which is also orthogonal to its dilations, or scales, i.e.,

$$\int_{\mathbb{R}} \psi(x)\psi(2x - k) dx = 0 \,.$$

Such a function ψ (the so-called associated *wavelet function*) does exist and is given by

$$\psi(x) = \sum_{k=1}^{M} (-1)^k c_{1-k} \phi(2x - k)$$

which is dependent on the solution of ϕ. Periodic boundary conditions are used

$$c_k \equiv c_{k+nM} \,.$$

Problem 3. Consider the function $H \in L_2(\mathbb{R})$

$$H(x) := \begin{cases} 1 & 0 \le x < 1/2 \\ -1 & 1/2 \le x \le 1 \\ 0 & \text{otherwise} \end{cases} \tag{1}$$

Let

$$H_{mn}(x) := 2^{-m/2} H(2^{-m}x - n)$$

where $m, n \in \mathbb{Z}$ and n is the translation parameter and m the dilation parameter.
(i) Find H_{11}, H_{12}.
(ii) Show that

$$\langle H_{mn}(x), H_{kl}(x) \rangle = \delta_{mk}\delta_{nl}, \qquad k, l \in \mathbb{Z} \tag{2}$$

where $\langle . \rangle$ denotes the scalar product in $L_2(\mathbb{R})$.
(iii) Expand the function ($f \in L_2(\mathbb{R})$)

$$f(x) = \exp(-|x|)$$

with respect to H_{mn}. The functions H_{mn} form an orthonormal basis in $L_2(\mathbb{R})$.

Solution 3. (i) We have

$$H_{mn}(x) = 2^{-\frac{m}{2}} H(2^{-m}x - n) = \begin{cases} 2^{-\frac{m}{2}} & 0 \le 2^{-m}x - n < \frac{1}{2} \\ -2^{-\frac{m}{2}} & \frac{1}{2} \le 2^{-m}x - n \le 1 \\ 0 & \text{otherwise} \end{cases}$$

Thus

$$H_{mn}(x) = \begin{cases} 2^{-\frac{m}{2}} & 2^m n \le x < 2^m(n + \frac{1}{2}) \\ -2^{-\frac{m}{2}} & 2^m(n + \frac{1}{2}) \le x \le 2^m(n + 1) \\ 0 & \text{otherwise} \end{cases}.$$

It follows that

$$H_{11}(x) = \begin{cases} \frac{1}{\sqrt{2}} & 2 \le x < 3 \\ -\frac{1}{\sqrt{2}} & 3 \le x \le 4 \\ 0 & \text{otherwise} \end{cases} \qquad H_{12}(x) = \begin{cases} \frac{1}{\sqrt{2}} & 4 \le x < 5 \\ -\frac{1}{\sqrt{2}} & 5 \le x \le 6 \\ 0 & \text{otherwise} \end{cases}$$

(ii) We have

$$I_{mnkl} := \langle H_{mn}(x), H_{kl}(x) \rangle = \int_{-\infty}^{\infty} H_{mn}(x) H_{kl}(x) dx.$$

The intervals on which H_{mn} and H_{kl} are non-zero are

$$I_{mn} := (2^m n, 2^m(n + 1)), \qquad I_{kl} := (2^k l, 2^k(l + 1)).$$

We consider the different cases:

Case 1. $m = k, n = l$ with

$$I_{mnkl} = \int_{2^m n}^{2^m (n+1)} 2^{-m} dx = 1.$$

Case 2. $m = k, n \neq l$ with

$$I_{mnkl} = 0$$

since $I_{mn} \cap I_{kl} = \emptyset$.

Case 3. $m \neq k$. Suppose without loss of generality that $m < k$. Either $I_{mn} \cap I_{kl} = \emptyset$ ($I_{mnkl} = 0$), or $I_{mn} \subset I_{kl}$ (as shown below). We have

$$2^k l \leq 2^m n < 2^k (l + \frac{1}{2}) \Rightarrow 2^{k-m} l \leq n < 2^{k-m} (l + \frac{1}{2})$$

$$\Rightarrow 2^{k-m} l \leq n + 1 \leq 2^{k-m} (l + \frac{1}{2}) \Rightarrow 2^k l \leq 2^m (n+1) \leq 2^k (l + \frac{1}{2})$$

Also

$$2^k (l + \frac{1}{2}) \leq 2^m n < 2^k (l+1) \Rightarrow 2^k (l + \frac{1}{2}) \leq 2^m (n+1) \leq 2^k (l+1)$$

$$2^k l < 2^m (n+1) \leq 2^k (l + \frac{1}{2}) \Rightarrow 2^k l \leq 2^m n \leq 2^k (l + \frac{1}{2})$$

$$2^k (l + \frac{1}{2}) < 2^m (n+1) \leq 2^k (l+1) \Rightarrow 2^k (l + \frac{1}{2}) \leq 2^m n \leq 2^k (l+1)$$

which gives

$$\langle H_{mn}(x), H_{kl}(x) \rangle = \pm \int_{2^m n}^{2^m (n+\frac{1}{2})} 2^{-\frac{1}{2}(m+k)} dx \mp \int_{2^m (n+\frac{1}{2})}^{2^m (n+1)} 2^{-\frac{1}{2}(m+k)} dx = 0.$$

Thus $I_{mnkl} = \delta_{mk} \delta_{nl}$.

(iii) The scalar product $\langle f(x), H_{mn}(x) \rangle$ is given by

$$\langle f(x), H_{mn}(x) \rangle = \int_{-\infty}^{\infty} f(x) H_{mn}(x) dx$$

$$= \int_{2^m n}^{2^m (n+\frac{1}{2})} 2^{-\frac{m}{2}} e^{-|x|} dx - \int_{2^m (n+\frac{1}{2})}^{2^m (n+1)} 2^{-\frac{m}{2}} e^{-|x|} dx$$

$$= 2^{-\frac{m}{2}} \begin{cases} -e^{-x} \Big|_{2^m n}^{2^m (n+\frac{1}{2})} + e^{-x} \Big|_{2^m (n+\frac{1}{2})}^{2^m (n+1)} & n \geq 0 \\[2mm] e^x \Big|_{2^m n}^{2^m (n+\frac{1}{2})} - e^x \Big|_{2^m (n+\frac{1}{2})}^{2^m (n+1)} & n < 0 \end{cases}$$

$$= -\frac{n + \delta_{n,0}}{|n| + \delta_{n,0}} 2^{-\frac{m}{2}} (2e^{-2^m |n+\frac{1}{2}|} - e^{-2^m |n|} - e^{-2^m |n+1|}).$$

The expansion is given by

$$f(x) = \sum_{m,n\in\mathbf{Z}} \langle f(x), H_{mn}(x)\rangle H_{mn}(x).$$

Problem 4. Consider the Hilbert space $\mathcal{H}$ of the 2×2 matrices over the complex numbers with the scalar product

$$\langle A, B\rangle := \operatorname{tr}(AB^*), \qquad A, B \in \mathcal{H}.$$

Show that the rescaled Pauli matrices $\mu_j := \frac{1}{\sqrt{2}}\sigma_j$, $j = 1, 2, 3$

$$\mu_1 = \frac{1}{\sqrt{2}}\begin{pmatrix} 0 & 1 \\ 1 & 0 \end{pmatrix}, \quad \mu_2 = \frac{1}{\sqrt{2}}\begin{pmatrix} 0 & -i \\ i & 0 \end{pmatrix}, \quad \mu_3 = \frac{1}{\sqrt{2}}\begin{pmatrix} 1 & 0 \\ 0 & -1 \end{pmatrix}$$

plus the rescaled 2×2 identity matrix

$$\mu_0 = \frac{1}{\sqrt{2}}\begin{pmatrix} 1 & 0 \\ 0 & 1 \end{pmatrix}$$

form an orthonormal basis in the Hilbert space $\mathcal{H}$.

Solution 4. Let $c_0, c_1, c_2, c_3 \in \mathbb{C}$. Then the equation

$$c_0\mu_0 + c_1\mu_1 + c_2\mu_2 + c_3\mu_3 = \begin{pmatrix} 0 & 0 \\ 0 & 0 \end{pmatrix}$$

only admits the zero solution, i.e. $c_0 = c_1 = c_2 = c_3 = 0$. Thus the four matrices are linearly independent. Since

$$\operatorname{tr}(\mu_j\mu_k) = 0, \qquad j \neq k$$

and

$$\operatorname{tr}(\mu_j\mu_j) = 1$$

we have an orthonormal basis in this Hilbert space.

Problem 5. Let $\mathbb{C}^n$ denote the complex Euclidean space. Let $\mathbf{z} = (z_1, \ldots, z_n) \in \mathbb{C}^n$ and $\mathbf{w} = (w_1, \ldots, w_n) \in \mathbb{C}^n$ then the scalar product (inner product) is given by

$$\mathbf{z} \cdot \mathbf{w} := \mathbf{z}\mathbf{w}^* = \mathbf{z}\overline{\mathbf{w}}^T$$

where $\overline{\mathbf{z}} = (\overline{z}_1, \ldots, \overline{z}_n)$. Let E_n denote the set of entire functions in $\mathbb{C}^n$. Let F_n denote the set of $f \in E_n$ such that

$$\|f\|^2 := \frac{1}{\pi^n}\int_{\mathbb{C}^n} |f(\mathbf{z})|^2 \exp(-|\mathbf{z}|^2)dV$$

is finite. Here dV is the volume element (Lebesgue measure)

$$dV = \prod_{j=1}^{n} dx_j dy_j = \prod_{j=1}^{n} r_j dr_j d\theta_j$$

with $z_j = r_j e^{i\theta_j}$. The norm follows from the scalar product of two functions $f, g \in F_n$

$$\langle f, g \rangle := \frac{1}{\pi^n} \int_{\mathbb{C}^n} f(\mathbf{z})\overline{g(\mathbf{z})} \exp(-|\mathbf{z}|^2) dV .$$

Let

$$\mathbf{z}^m := z_1^{m_1} \cdots z_n^{m_n}$$

where the multiindex m is defined by $m! = m_1! \cdots m_n!$ and $|m| = \sum_{j=1}^{n} m_j$. Find the scalar product

$$\langle \mathbf{z}^m, \mathbf{z}^p \rangle .$$

Solution 5.   We obtain

$$\langle \mathbf{z}^m, \mathbf{z}^p \rangle = \langle z_1^{m_1} \cdots z_n^{m_n}, z_1^{p_1} \cdots z_n^{p_n} \rangle = 0 \quad \text{for } m \neq p$$

and

$$\langle \mathbf{z}^m, \mathbf{z}^m \rangle = \langle z_1^{m_1} \cdots z_n^{m_n}, z_1^{m_1} \cdots z_n^{m_n} \rangle m! = m_1! \cdots m_n! .$$

Thus the monomials

$$\frac{1}{\sqrt{m!}} z^m$$

are orthonormal in this Hilbert space.

Problem 6.   Let Ψ be a complex-valued differentiable function of ϕ in the interval $[0, 2\pi]$ and $\Psi(0) = \Psi(2\pi)$, i.e. Ψ is an element of the Hilbert space $L_2([0, 2\pi])$. Assume that (normalization condition)

$$\int_0^{2\pi} \Psi^*(\phi)\Psi(\phi)d\phi = 1 .$$

Calculate

$$\Im \frac{\hbar}{i} \int_0^{2\pi} \Psi^*(\phi)\phi \frac{d}{d\phi} \Psi(\phi)d\phi$$

where $\Im$ denotes the imaginary part.

Solution 6.   Using integration by parts, $\Psi(2\pi) = \Psi(0)$ and the normalization condition we find

$$\int_0^{2\pi} \Psi^*(\phi)\phi \frac{d}{d\phi} \Psi(\phi)d\phi = \Psi^*(\phi)\phi\Psi(\phi)\big|_0^{2\pi} - \int_0^{2\pi} \Psi(\phi) \left(\frac{d}{d\phi}(\Psi^*(\phi)\phi) \right) d\phi$$

$$= 2\pi\Psi^*(2\pi)\Psi(2\pi) - \int_0^{2\pi} \Psi(\phi)\phi\frac{d}{d\phi}\Psi^*(\phi)d\phi$$

$$- \int_0^{2\pi} \Psi(\phi)\Psi^*(\phi)d\phi$$

$$= 2\pi\Psi^*(0)\Psi(0) - 1 - \int_0^{2\pi} \Psi(\phi)\phi\frac{d}{d\phi}\Psi^*(\phi)d\phi.$$

Thus

$$\int_0^{2\pi} \Psi^*(\phi)\phi\frac{d}{d\phi}\Psi(\phi)d\phi + \int_0^{2\pi} \Psi(\phi)\phi\frac{d}{d\phi}\Psi^*(\phi)d\phi = 2\pi\Psi^*(0)\Psi(0) - 1$$

or

$$\int_0^{2\pi} \left(\Psi^*(\phi)\phi\frac{d}{d\phi}\Psi(\phi) + \Psi(\phi)\phi\frac{d}{d\phi}\Psi^*(\phi) \right) d\phi = 2\pi|\Psi(0)|^2 - 1.$$

It follows that

$$\Im\frac{\hbar}{i} \int_0^{2\pi} \Psi^*(\phi)\phi\frac{d}{d\phi}\Psi(\phi)d\phi = \frac{\hbar}{2}(1 - 2\pi|\Psi(0)|^2).$$

Problem 7. Consider a complex Hilbert space $\mathcal{H}$ and $|\phi_1\rangle, |\phi_2\rangle \in \mathcal{H}$. Let $c_1, c_2 \in \mathbb{C}$. An *antilinear operator* K in this Hilbert space $\mathcal{H}$ is characterized by

$$K(c_1|\phi_1\rangle + c_2|\phi_2\rangle) = c_1^*K|\phi_1\rangle + c_2^*K|\phi_2\rangle.$$

A *comb* is an antilinear operator K with zero expectation value for all states $|\psi\rangle$ of a certain complex Hilbert space $\mathcal{H}$. This means

$$\langle\psi|K|\psi\rangle = \langle\psi|LC|\psi\rangle = \langle\psi|L|\psi^*\rangle = 0$$

for all states $|\psi\rangle \in \mathcal{H}$, where L is a linear operator and C is the complex conjugation.

(i) Consider the two-dimensional Hilbert space $\mathcal{H} = \mathbb{C}^2$. Find a unitary 2×2 matrix such that

$$\langle\psi|UC|\psi\rangle = 0.$$

(ii) Consider the Pauli spin matrices with $\sigma_0 = I_2$, $\sigma_1 = \sigma_x$, $\sigma_2 = \sigma_y$, $\sigma_3 = \sigma_z$. Find

$$\sum_{\mu=0}^{3}\sum_{\nu=0}^{3}\langle\psi|\sigma_\mu C|\psi\rangle g^{\mu,\nu}\langle\psi|\sigma_\nu C|\psi\rangle$$

where $g^{\mu,\nu} = \text{diag}(-1,1,0,1)$.

Solution 7. (i) We find $U = \sigma_y$ since

$$\langle \psi | \sigma_y C | \psi \rangle = \langle \psi | \sigma_y | \psi^* \rangle$$

$$= (\psi_1^* \quad \psi_2^*) \begin{pmatrix} 0 & -i \\ i & 0 \end{pmatrix} \begin{pmatrix} \psi_1^* \\ \psi_2^* \end{pmatrix}$$

$$= (\psi_1^* \quad \psi_2^*) \begin{pmatrix} -i\psi_2^* \\ i\psi_1^* \end{pmatrix}$$

$$= 0.$$

(ii) We have

$$\sum_{\mu=0}^{3} \sum_{\nu=0}^{3} \langle \psi | \sigma_\mu C | \psi \rangle g^{\mu,\nu} \langle \psi | \sigma_\nu C | \psi \rangle = -\langle \psi | \sigma_0 | \psi^* \rangle^2 + \langle \psi | \sigma_1 | \psi^* \rangle + \langle \psi | \sigma_3 | \psi^* \rangle$$

$$= -(\psi_1^* \psi_1^* + \psi_2^* \psi_2^*)^2 + (\psi_1^* \psi_2^* + \psi_2^* \psi_1^*)^2$$

$$+ (\psi_1^* \psi_2^* - \psi_2^* \psi_2^*)^2$$

$$= 0.$$

Problem 8. Consider the Hilbert space $L_2([0,1])$. The *shifted Legrendre polynomials*, defined on the interval $[0, 1]$, are obtained from the Legrendre polynomial by the transformation $y = 2x - 1$. The shifted Legrendre polynomials are given by the recurrence formula

$$P_j(x) = \frac{(2j+1)(2x-1)}{j+1} P_j(x) - \frac{j}{j+1} P_{j-1}(x) \qquad j = 1, 2, \ldots$$

and $P_0(x) = 1$, $P_1(x) = 2x - 1$. They are elements of the Hilbert space $L_2([0,1])$. A function u in the Hilbert space $L_2([0,1])$ can be approximated in the form of a series with $n+1$ terms

$$u(x) = \sum_{j=0}^{n} c_j P_j(x)$$

where the coeffients $c_j \in \mathbb{R}$, $j = 0, 1, \ldots, n$. Consider the Volterra integral equation of first kind

$$\lambda \int_0^x \frac{y(t)}{(x-t)^\alpha} dt = f(x), \qquad 0 \le t \le x \le 1$$

with $0 < \alpha < 1$ and $f \in L_2([0,1])$. Consider the ansatz

$$y_n(x) = a_0 x^\alpha + \sum_{j=0}^{n} c_j P_j(x)$$

to find an approximate solution to the *Volterra integral equation* of first kind ($\alpha = 1/2$)

$$\lambda \int_0^x \frac{y(t)}{\sqrt{x-t}} dt = f(x)$$

where

$$f(x) = \frac{2}{105} \sqrt{x}(105 - 56x^2 + 48x^3).$$

Solution 8. Inserting the ansatz in the Volterra integral equation yields

$$\lambda a_0 \int_0^x \frac{t^\alpha}{(x-t)^\alpha} dt + \lambda \sum_{j=0}^n c_j \int_0^x \frac{P_j(t)}{(x-t)^\alpha} dt = f(x)$$

where $\alpha = 1/2$. Since

$$\int_0^x \frac{t^n}{(x-t)^\alpha} dt = \frac{\Gamma(n+1)\Gamma(1-\alpha)}{\Gamma(n+2-\alpha)} x^{n+1-\alpha}$$

and

$$\int_0^x \frac{t^\alpha}{(x-t)^\alpha} = \frac{\pi\alpha}{\sin(\pi\alpha)} x$$

we find

$$\int_0^x \frac{P_j(t)}{(x-t)^{(\alpha)}} dt = \sum_{k=0}^j a_{jk}^\alpha x^{k+1-\alpha}$$

where

$$a_{jk}^{(\alpha)} = (-1)^{j+k} \frac{(j+k)!\Gamma(1-\alpha)}{k!(j-k)!\Gamma(k+2-\alpha)}.$$

Thus we find

$$\lambda a_0 \frac{\pi\alpha}{\sin(\pi\alpha)} x + \lambda \sum_{j=0}^n \sum_{k=0}^j c_j a_{jk}^{(\alpha)} x^{k+1-\alpha} = f(x).$$

Since $\alpha = 1/2$ we have

$$\sin(\pi/2) = 1, \qquad \Gamma(1/2) = \sqrt{\pi}$$

and the equation simplifies to

$$\frac{1}{2} \lambda a_0 \pi x + \lambda \sum_{j=0}^n \sum_{k=0}^j c_j a_{jk}^{(\alpha)} x^{k+1-\alpha} = f(x).$$

We consider this equation at the roots y_j : $(j = 1, 2, \ldots, n + 1)$ of the shifted Legrende polynmials and $y_0 = 0$. Thus we find $n + 2$ equations for the coeffients c_j $(j = 0, 1, \ldots, n)$ and a_0. We have

$$P_2(x) = 6x^2 - 6x + 1, \qquad P_3(x) = 20x^3 - 30x^2 + 12x - 1.$$

We find $a_0 = 0$ and

$$c_0 = \frac{11}{12}, \quad c_1 = -\frac{1}{20}, \quad c_2 = \frac{1}{12}, \quad c_3 = \frac{1}{20}.$$

Problem 9. (i) Consider the Hilbert space $L_2[0, 1]$ with the scalar product $\langle \cdot, \cdot \rangle$. Let $f : [0, 1] \to [0, 1]$

$$f(x) := \begin{cases} 2x & \text{if } x \in [0, 1/2) \\ 2(1 - x) & \text{if } x \in [1/2, 1] \end{cases}$$

Thus $f \in L_2[0, 1]$. Calculate the moments μ_k, $k = 0, 1, 2, \ldots$ defined by

$$\mu_k := \langle f(x), x^k \rangle \equiv \int_0^1 f(x) x^k \, dx.$$

(ii) Show that

$$\sum_{k=0}^{\infty} |\mu_k|^2 < \pi \int_0^1 |f(x)|^2 \, dx.$$

Solution 9. (i) For $k = 0$ we have

$$\mu_0 = \int_0^1 f(x) dx = \frac{1}{2}.$$

For $k \geq 1$ we obtain

$$\mu_k = \int_0^{1/2} 2xx^k \, dx + \int_{1/2}^1 (2 - 2x) x^k \, dx$$

$$= 2 \int_0^{1/2} x^{k+1} \, dx + 2 \int_{1/2}^1 x^k \, dx - 2 \int_{1/2}^1 x^{k+1} \, dx$$

$$= 2 \frac{(1/2)^{k+2}}{k+2} + 2 \frac{1}{k+1} - 2 \frac{(1/2)^{k+1}}{k+1} - 2 \frac{1}{k+2} + 2 \frac{(1/2)^{k+2}}{k+2}$$

$$= -\left(\frac{1}{2}\right)^{k+1} \frac{1}{(k+2)(k+1)} + \frac{2}{(k+1)(k+2)}$$

(ii) For the right-hand side we find

$$\int_0^1 |f(x)|^2 dx = \int_0^{1/2} 4x^2 dx + \int_0^1 2(1-x)2(1-x)dx$$

$$= \frac{1}{3}.$$

Problem 10. Let $L_2(\mathbb{R}^2)$ be the Hilbert space of the square-integrable functions over $\mathbb{R}^2$. Let $C^2(\mathbb{R}^2, \mathbb{R})$ be the set of $\mathbb{R}$ valued functions having two continuous derivatives on $\mathbb{R}^2$. Assume that

$$f, g \in C^2(\mathbb{R}^2, \mathbb{R}) \cap L_2(\mathbb{R}^2) \qquad \Delta f, \Delta g \in L_2(\mathbb{R}^2)$$

and that

$$\nabla f, \ \nabla g \in L_2(\mathbb{R}^2).$$

This means each component of each vector is in $L_2(\mathbb{R}^2)$. Show that

$$\int_{\mathbb{R}^2} f(x,y)\Delta g(x,y)dxdy = \int_{\mathbb{R}^2} g(x,y)\Delta f(x,y)dxdy. \qquad (1)$$

Solution 10. Owing to the assumptions we know that all functions are in the Hilbert space $L_2(\mathbb{R}^2)$. It follows that all integrals exist. We apply *Green's theorem* to solve the problem. Let P and Q be two real valued functions that are continuous on a region $E \subset \mathbb{R}^2$ bounded by a rectifiable Jordan curve Γ and such that $\partial P/\partial y$ and $\partial Q/\partial x$ exist and are bounded in their interior of E and that

$$\int\int_E \frac{\partial Q}{\partial x} dxdy \qquad \text{and} \qquad \int\int_E \frac{\partial P}{\partial y} dxdy$$

exist. Then, if Γ is positively orientated, the line integral exists and we have

$$\int_\Gamma (Pdx + Qdy) = \int\int_E \left(\frac{\partial Q}{\partial x} - \frac{\partial P}{\partial y} \right) dxdy.$$

Green's theorem implies that for all positive R,

$$\int_{B(0,R)} (f\Delta g - g\Delta f)dxdy = \int_{\partial B(0,R)} \left(f\frac{\partial g}{\partial x} - g\frac{\partial f}{\partial x} \right) dy$$

$$+ \int_{\partial B(0,R)} \left(g\frac{\partial f}{\partial y} - f\frac{\partial g}{\partial y} \right) dx.$$

An estimate for the integrals in these equations is

$$\left| \int_{\partial B(0,R)} f\frac{\partial g}{\partial x} dy \right| \leq \int_0^{2\pi} |f(R\cos\phi, R\sin\phi)| \left| \frac{\partial g(R\cos\phi, R\sin\phi)}{\partial x} \right| |R\sin\phi|d\phi$$

$$= A(R)$$

where we have introduced polar coordinates. Let $\|.\|_2$ be the norm of the Hilbert space $L_2(\mathbb{R}^2)$. Since

$$\int_{B(0,R)} \left| f(r\cos\phi, r\sin\phi) \frac{\partial g(r\cos\phi, r\sin\phi)}{\partial x} \right| r\,dr\,d\phi = \int_0^R A(r)r\,dr$$

$$\leq \|f\|_2 \|\partial g/\partial x\|_2$$

it follows that there is a sequence $\{R_n\}_{n=1}^\infty$ such that $R_n \to \infty$ and

$$A(R_n) \to 0$$

as $n \to \infty$. In a similar manner the other three integrals may be estimated. Thus

$$\left| \int_{B(0,R_n)} (f\Delta g - g\Delta f)\,dx\,dy \right|$$

is dominated by the sum of four quantities, each of which tends to zero as $n \to \infty$. Consequently, (1) follows.

Problem 11.   Consider the Hilbert space $L_2(\mathbb{R})$. Let

$$\hat{\psi}(\omega) = \begin{cases} 1 & \text{if } 1/2 \leq |\omega| \leq 1 \\ 0 & \text{otherwise} \end{cases}$$

and

$$\hat{\phi}(\omega) = e^{-\alpha|\omega|}, \qquad \alpha > 0.$$

(i) Calculate the inverse Fourier transforms of $\hat{\psi}(\omega)$ and $\hat{\phi}(\omega)$, i.e.

$$\psi(t) = \frac{1}{2\pi} \int_\mathbb{R} e^{-i\omega t} \hat{\psi}(\omega)\,d\omega, \quad \phi(t) = \frac{1}{2\pi} \int_\mathbb{R} e^{-i\omega t} \hat{\phi}(\omega)\,d\omega.$$

(ii) Calculate the scalar product $\langle \psi(t)|\phi(t) \rangle$ by utilizing the identity

$$2\pi \langle \psi(t)|\phi(t) \rangle = \langle \hat{\psi}(\omega)|\hat{\phi}(\omega) \rangle.$$

Solution 11.   (i) For the inverse Fourier transforms we find

$$\psi(t) = \frac{1}{2\pi} \left(\int_{-1}^{-1/2} e^{-i\omega t}\,d\omega + \int_{1/2}^1 e^{-i\omega t}\,d\omega \right) = \frac{1}{\pi t}(\sin(t) - \sin(t/2)).$$

$$\phi(t) = \frac{1}{2\pi} \left(\int_{-\infty}^0 e^{-i\omega t} e^{\alpha\omega}\,d\omega + \int_0^\infty e^{-i\omega t} e^{-\alpha\omega}\,d\omega \right) = \frac{1}{\pi} \frac{\alpha}{t^2 + \alpha^2}.$$

(ii) We find

$$\langle \psi(t)|\phi(t)\rangle = \frac{1}{2\pi}\langle \hat{\psi}(\omega)|\hat{\phi}(\omega)\rangle = \frac{1}{2\pi}\left(\int_{-1}^{-1/2} e^{\alpha\omega}d\omega + \int_{1/2}^{1} e^{-\alpha\omega}d\omega\right) = 0.$$

Problem 12. Consider the two-body problem in quantum mechanics with the coordinates

$$\mathbf{r}_1 = (x_1, y_1, z_1)^T, \qquad \mathbf{r}_2 = (x_2, y_2, z_2)^T$$

and the operators

$$\Delta_1 = \frac{\partial^2}{\partial x_1^2} + \frac{\partial^2}{\partial y_1^2} + \frac{\partial^2}{\partial z_1^2}, \qquad \Delta_2 = \frac{\partial^2}{\partial x_2^2} + \frac{\partial^2}{\partial y_2^2} + \frac{\partial^2}{\partial z_2^2}$$

where $\mathbf{r}_1$ is the position vector of particle 1 and $\mathbf{r}_2$ is the position vector of particle 2. The eigenvalue equation takes the form

$$\hat{H}u(\mathbf{r}_1,\mathbf{r}_2) \equiv \left(-\frac{\hbar^2}{2m_1}\Delta_1 - \frac{\hbar^2}{2m_2}\Delta_2 + V(\mathbf{r}_1 - \mathbf{r}_2)\right)u(\mathbf{r}_1,\mathbf{r}_2) = Eu(\mathbf{r}_1,\mathbf{r}_2)$$

where, on physical grounds, only potentials depending on $\mathbf{r}_1 - \mathbf{r}_2$ are considered. The function u is an element of the Hilbert space $L_2(\mathbb{R}^6)$. Consider the linear invertible transformation $(\mathbf{r}_1, \mathbf{r}_2) \to (\mathbf{r}, \mathbf{R})$

$$\mathbf{r}(\mathbf{r}_1,\mathbf{r}_2) = \mathbf{r}_2 - \mathbf{r}_1$$
$$\mathbf{R}(\mathbf{r}_1,\mathbf{r}_2) = \frac{m_1\mathbf{r}_1 + m_2\mathbf{r}_2}{m_1 + m_2}$$
$$U(\mathbf{r}(\mathbf{r}_1,\mathbf{r}_2), \mathbf{R}(\mathbf{r}_1,\mathbf{r}_2)) = u(\mathbf{r}_1,\mathbf{r}_2)$$

with

$$\mathbf{r} = (x, y, z)^T, \qquad \mathbf{R} = (X, Y, Z)^T$$

where $\mathbf{R}$ is the centre of mass position vector and $\mathbf{r}$ is the relative position vector.

(i) Find the inverse transformation.

(ii) Express the Hamilton operator in terms these new coordinates.

Solution 12. (i) Straightforward calculation yields

$$\mathbf{r}_1 = \frac{(m_1 + m_2)\mathbf{R} - m_2\mathbf{r}}{m_1 + m_2}, \qquad \mathbf{r}_2 = \frac{(m_1 + m_2)\mathbf{R} + m_1\mathbf{r}}{m_1 + m_2}$$

(ii) We set $M := m_1 + m_2$. Using the chain rule we find

$$\frac{\partial U}{\partial x_1} = -\frac{\partial U}{\partial x} + \frac{m_1}{M}\frac{\partial U}{\partial X} = \frac{\partial u}{\partial x_1}$$

$$\frac{\partial U}{\partial y_1} = -\frac{\partial U}{\partial y} + \frac{m_1}{M}\frac{\partial U}{\partial Y} = \frac{\partial u}{\partial y_1}$$

$$\frac{\partial U}{\partial z_1} = -\frac{\partial U}{\partial z} + \frac{m_1}{M}\frac{\partial U}{\partial Z} = \frac{\partial u}{\partial z_1}$$

$$\frac{\partial U}{\partial x_2} = \frac{\partial U}{\partial x} + \frac{m_2}{M}\frac{\partial U}{\partial X} = \frac{\partial u}{\partial x_2}$$

$$\frac{\partial U}{\partial y_2} = \frac{\partial U}{\partial y} + \frac{m_2}{M}\frac{\partial U}{\partial Y} = \frac{\partial u}{\partial y_2}$$

$$\frac{\partial U}{\partial z_2} = \frac{\partial U}{\partial z} + \frac{m_2}{M}\frac{\partial U}{\partial Z} = \frac{\partial u}{\partial z_2}.$$

For the second order derivatives we have

$$\frac{\partial^2 U}{\partial x_1^2} = \frac{\partial^2 U}{\partial x^2} - 2\frac{m_1}{M}\frac{\partial^2 U}{\partial x\partial X} + \frac{m_1^2}{M^2}\frac{\partial^2 U}{\partial X^2} = \frac{\partial^2 u}{\partial x_1^2}$$

$$\frac{\partial^2 U}{\partial x_2^2} = \frac{\partial^2 U}{\partial x^2} + 2\frac{m_2}{M}\frac{\partial^2 U}{\partial x\partial X} + \frac{m_2^2}{M^2}\frac{\partial^2 U}{\partial X^2} = \frac{\partial^2 u}{\partial x_2^2}$$

and analogously for y_1, z_1, y_2 and z_2. Thus we obtain the new Hamilton operator

$$\hat{H}_S = -\frac{\hbar^2}{2M}\Delta_R - \frac{\hbar^2}{2m}\Delta + V(\mathbf{r})$$

where

$$\Delta_R := \frac{\partial^2}{\partial X^2} + \frac{\partial^2}{\partial Y^2} + \frac{\partial^2}{\partial Z^2}, \quad \Delta := \frac{\partial^2}{\partial x^2} + \frac{\partial^2}{\partial y^2} + \frac{\partial^2}{\partial z^2}$$

with

$$m = \frac{m_1 m_2}{m_1 + m_2}$$

the reduced mass.

Chapter 10

Generalized Functions

The vector space $D(\mathbb{R}^n)$ denotes all infinitely-differentiable functions in $\mathbb{R}$ with compact support. The functions $\phi \in D(\mathbb{R}^n)$ are called test functions. Each linear continuous functional over the vector space of test functions $D(\mathbb{R}^n)$ is called a generalized function.

Instead of considering the space $D(\mathbb{R})$ for the test functions we can also consider the vector space $S(\mathbb{R})$. The vector space $S(\mathbb{R}^n)$ is the set of all infinitely differentiable functions which decrease as $|\mathbf{x}| \to \infty$, together with all their derivatives, faster than any power of $|\mathbf{x}|^{-1}$. These functions are called test functions. Each linear continuous functional over the vector space $S(\mathbb{R}^n$ is called a generalized function (of slow growth). Note that $S(\mathbb{R}^n) \subset L_2(\mathbb{R}^n)$ and the vector space $S(\mathbb{R}^n)$ is dense in the Hilbert space $L_2(\mathbb{R}^n)$.

The *delta function* is defined by

$$(\delta(\mathbf{x}), \phi(\mathbf{x})) := \phi(\mathbf{0})$$
$$(\delta(\mathbf{x} - \mathbf{x}_0), \phi(\mathbf{x})) := \phi(\mathbf{x}_0) .$$

The delta function δ is a singular generalized function.

Let ϕ be a test function. Then the derivative of a generalized function T is defined as

$$\left(\frac{\partial T}{\partial x_j}, \phi \right) := - \left(T, \frac{\partial \phi}{\partial x_j} \right) .$$

Problem 1.   Let $a > 0$. We define

$$f_a(x) := \begin{cases} \dfrac{1}{2a} & \text{for} \quad |x| \le a \\ 0 & \text{for} \quad |x| > a \,. \end{cases}$$

(i) Calculate the derivative of f_a in the sense of generalized functions.
(ii) What happens if $a \to +0$?

Solution 1.   (i) Using the definition given in the introduction we find

$$\left(\frac{df_a(x)}{dx}, \phi(x) \right) = - \left(f_a(x), \frac{d\phi(x)}{dx} \right) \,.$$

Thus

$$\left(\frac{df_a(x)}{dx}, \phi(x) \right) = - \int_{\mathbb{R}} f_a(x) \frac{d\phi}{dx} dx = - \frac{1}{2a} \int_{-a}^{a} \frac{d\phi}{dx} dx \,.$$

Finally

$$\left(\frac{df_a(x)}{dx}, \phi(x) \right) = \frac{1}{2a} (-\phi(a) + \phi(-a)) = \frac{1}{2a} ((\delta(x+a), \phi(x)) - (\delta(x-a), \phi(x))) \,.$$

We can write

$$\frac{df_a(x)}{dx} = \frac{1}{2a} (\delta(x + a) - \delta(x - a))$$

where δ denotes the *delta function*.
(ii) In the sense of generalized functions the function f_a tends to the delta function for $a \to +0$.

Problem 2.   Show that

$$\frac{1}{2\pi} \sum_{k=-\infty}^{\infty} e^{ikx} \equiv \sum_{k=-\infty}^{\infty} \delta(x - 2k\pi) \tag{1}$$

in the sense of generalized functions.

Hint. Expand the 2π-periodic function

$$f(x) = \frac{1}{2} - \frac{x}{2\pi} \tag{2}$$

into a Fourier series. An orthonormal basis in the Hilbert space $L_2(0, 2\pi)$ is given by

$$\left\{ \phi_k(x) := \frac{1}{\sqrt{2\pi}} \exp(ikx) \; : \; k \in \mathbb{Z} \right\} \,.$$

Solution 2. Since we assume that the function f is periodic with respect to 2π we can expand it into a *Fourier expansion*

$$f(x) = \sum_{k \in \mathbb{Z}} \langle f, \phi_k \rangle \phi_k \,.$$

Calculating the scalar product $\langle f, \phi_k \rangle$ in the Hilbert space $L_2(0, 2\pi)$ we obtain

$$f(x) = -\frac{i}{2\pi} \sum_{\substack{k=-\infty \\ k \neq 0}}^{\infty} \frac{1}{k} e^{ikx}$$

where

$$\langle f, \phi_0 \rangle = \int_0^{2\pi} \left(\frac{1}{2} - \frac{x}{2\pi} \right) \frac{1}{\sqrt{2\pi}} dx = 0 \,.$$

In the sense of generalized functions we can differentiate the right-hand side of (2) term by term. We obtain

$$f' = -\frac{1}{2\pi} + \sum_{k=-\infty}^{\infty} \delta(x - 2k\pi) = \frac{1}{2\pi} \sum_{\substack{k=-\infty \\ k \neq 0}}^{\infty} e^{ikx} \,.$$

Thus we obtain identity (1).

The identity

$$\delta(f(x)) \equiv \sum_n \frac{1}{|f'(x_n)|} \delta(x - x_n)$$

can also be used to find identity (1). Here the sum runs over all the zeros of f and $f'(x_n) \equiv df(x = x_n)/dx$.

Problem 3. Calculate the Fourier transform of

$$f(t) = \sin(\Omega t), \qquad g(t) = \cos(\Omega t)$$

in the sense of generalized functions. The Fourier transform for generalized functions is given by

$$(F[f](\omega), \psi(\omega)) := 2\pi (f(t), \phi(t))$$

where

$$\psi(\omega) := \int_{-\infty}^{+\infty} e^{i\omega t} \phi(t) \, dt.$$

Solution 3. Inserting $f(t) = \sin(\Omega t)$ into the equation for the Fourier transform for generalized function we obtain

$$(F[f](\omega), \psi(\omega)) = 2\pi \int_{-\infty}^{+\infty} \sin(\Omega t) \phi(t) \, dt.$$

Using the identity

$$\sin(\Omega t) \equiv \frac{e^{i\Omega t} - e^{-i\Omega t}}{2i}$$

we find

$$(F[f](\omega), \psi(\omega)) = \frac{\pi}{i} \int_{-\infty}^{+\infty} e^{i\Omega t} \phi(t) \, dt - \frac{\pi}{i} \int_{-\infty}^{+\infty} e^{-i\Omega t} \phi(t) \, dt.$$

From $\psi(\omega)$ we find

$$\psi(-\omega) = \int_{-\infty}^{+\infty} e^{-i\omega t} \phi(t) \, dt.$$

Therefore

$$(F[f](\omega), \psi(\omega)) = \frac{\pi}{i} [\psi(\Omega) - \psi(-\Omega)].$$

Consequently,

$$(F[f](\omega), \psi(\omega)) = \frac{\pi}{i} ((\delta(\omega - \Omega), \psi(\omega)) - (\delta(\omega + \Omega), \psi(\omega))).$$

Thus

$$F[f](\omega) = \frac{\pi}{i} (\delta(\omega - \Omega) - \delta(\omega + \Omega)).$$

Analogously we find the Fourier transform of g

$$F[g](\omega) = \pi(\delta(\omega + \Omega) + \delta(\omega - \Omega)).$$

The Fourier transform of $\exp(bx)$ in the sense of generalized functions is

$$F[\exp(bx)](\omega) = 2\pi\delta(\omega - ib).$$

Using the identities

$$\exp(i\Omega t) \equiv \cos(\Omega t) + i\sin(\Omega t), \quad \exp(-i\Omega t) \equiv \cos(\Omega t) - i\sin(\Omega t)$$

and $b = i\Omega$ and $b = -i\Omega$, we can also find $F[f](\omega)$ and $F[g](\omega)$ from $F[\exp(bx)](\omega)$.

Problem 4.   Let $\mathbf{x} = (x_1, x_2, \ldots, x_n)$. Show that

$$E(\mathbf{x}, t) := \frac{1}{(4\pi a^2 t)^{n/2}} \exp\left(-\frac{|\mathbf{x}|^2}{4a^2 t}\right) \to \delta(\mathbf{x}), \qquad t \to +0 \qquad (1)$$

in the sense of generalized functions, where $n = 1, 2, \ldots$.

Solution 4.   Let $\phi \in S(\mathbb{R}^n)$. Then we have the estimates

$$\left| \int_{\mathbb{R}^n} E(\mathbf{x}, t)[\phi(\mathbf{x}) - \phi(0)]d\mathbf{x} \right| \leq \frac{K}{(4\pi a^2 t)^{n/2}} \int_{\mathbb{R}^n} \exp\left(-\frac{|\mathbf{x}|^2}{4a^2 t}\right)|\mathbf{x}|d\mathbf{x}$$

$$= \frac{K\sigma_n}{(4\pi a^2 t)^{n/2}} \int_0^\infty \exp\left(-\frac{r^2}{4a^2 t}\right)r^n dr$$

$$= K'\sqrt{t} \int_0^\infty \exp(-u^2)u^n du$$

$$= C\sqrt{t}$$

where

$$\sigma_n := \frac{2\pi^{n/2}}{\Gamma(n/2)} \tag{2}$$

is the area of the surface of a unit sphere in $\mathbb{R}^n$ and

$$\Gamma(z) := \int_0^\infty t^{z-1}e^{-t}dt, \qquad \Re z > 0 \tag{3}$$

where Γ denotes the *Gamma function*. For $t > 0$ we have

$$\int_{\mathbb{R}^n} E(\mathbf{x}, t)d\mathbf{x} = \frac{1}{(2a\sqrt{\pi t})^n} \int_{\mathbb{R}^n} \exp\left(-\frac{|\mathbf{x}|^2}{4a^2 t}\right)d\mathbf{x}$$

$$= \prod_{j=1}^n \frac{1}{\sqrt{\pi}} \int_{-\infty}^\infty \exp(-\xi_j^2)d\xi_j^2$$

$$= 1.$$

Thus we obtain

$$(E(\mathbf{x}, t), \phi) = \int_{\mathbb{R}^n} E(\mathbf{x}, t)\phi(\mathbf{x})d\mathbf{x}$$

$$= \phi(0) \int_{\mathbb{R}^n} E(\mathbf{x}, t)d\mathbf{x} + \int_{\mathbb{R}^n} E(\mathbf{x}, t)[\phi(\mathbf{x}) - \phi(0)]d\mathbf{x}$$

$$\rightarrow \phi(0)$$

$$= (\delta, \phi).$$

Problem 5. Consider the function $H : \mathbb{R} \rightarrow \mathbb{R}$

$$H(x) := \begin{cases} 1 & 0 \leq x \leq 1/2 \\ -1 & 1/2 \leq x \leq 1 \\ 0 & \text{otherwise} \end{cases}$$

(i) Find the derivative of H in the sense of generalized functions. Obviously H can be considered as a regular functional

$$\int_{\mathbb{R}} H(x)\phi(x)dx .$$

(ii) Find the Fourier transform of H. Draw a picture of the Fourier transform.

Solution 5.   (i) We have

$$\left(\frac{dH}{dx}, \phi\right) := -\left(H, \frac{d\phi}{dx}\right)$$

$$= -\int_0^{\frac{1}{2}} \frac{d\phi}{dx}dx + \int_{\frac{1}{2}}^1 \frac{d\phi}{dx}dx$$

$$= -\phi(1/2) + \phi(0) + \phi(1) - \phi(1/2)$$

$$= (\delta(x) + \delta(x-1) - 2\delta(x-1/2), \phi) .$$

(ii) For the Fourier transform we find

$$\int_{-\infty}^{\infty} e^{ikx} H dx = \int_0^{\frac{1}{2}} e^{ikx} dx - \int_{\frac{1}{2}}^1 e^{ikx} dx$$

$$= \frac{e^{ikx}}{ik}\Big|_0^{\frac{1}{2}} - \frac{e^{ikx}}{ik}\Big|_{\frac{1}{2}}^1$$

$$= \frac{1}{ik}(2e^{\frac{ik}{2}} - 1 - e^{ik})$$

$$= \frac{e^{\frac{ik}{2}}}{ik}(2 - (e^{\frac{ik}{2}} + e^{-\frac{ik}{2}}))$$

$$= \frac{2ie^{\frac{k}{2}}}{k}(\cos(k/2) - 1).$$

Problem 6.   Find a particular solution to the differential equation

$$\frac{d^2u}{dt^2} + a\frac{du}{dt} + bu = \delta(t) \qquad a, b \in \mathbb{R} \tag{1}$$

in the sense of generalized functions.

Solution 6.   Consider the ansatz

$$u(t) = \theta(t)f(t) \tag{2}$$

where

$$\theta(t) := \begin{cases} 1 & t \geq 0 \\ 0 & \text{otherwise} \end{cases} \tag{3}$$

and f is a differentiable function. Let ϕ be a test function $\phi \in S(\mathbb{R})$. Then from (1) we obtain

$$\left(\frac{d^2u}{dt^2}, \phi\right) + a\left(\frac{du}{dt}, \phi\right) + b\left(u, \phi\right) = (\delta(t), \phi) = \phi(0). \tag{4}$$

Applying differentiation in the sense of generalized functions we obtain from (4)

$$\left(u, \frac{d^2\phi}{dt^2}\right) - a\left(u, \frac{d\phi}{dt}\right) + b(u, \phi) = \phi(0). \tag{5}$$

Inserting the ansatz (2) into (5) it follows that

$$\int_{-\infty}^{\infty} \theta(t)f(t)\frac{d^2\phi}{dt^2}dt - a\int_{-\infty}^{+\infty} \theta(t)f(t)\frac{d\phi}{dt}dt + b\int_{-\infty}^{\infty} \theta(t)f(t)\phi(t)dt = \phi(0).$$

Applying the property of the step-function (3) leads to

$$\int_0^{\infty} f(t)\frac{d^2\phi}{dt^2}dt - a\int_0^{\infty} f(t)\frac{d\phi}{dt}dt + b\int_0^{\infty} f(t)\phi(t)dt = \phi(0) \tag{6}$$

Applying integration by parts and taking into account the boundary conditions for ϕ

$$\phi(|\infty|) = 0, \qquad \phi'(|\infty|) = 0$$

we find

$$\int_0^{\infty} f(t)\frac{d\phi}{dt}dt = -f(0)\phi(0) - \int_0^{\infty} \frac{df}{dt}\phi dt$$

and

$$\int_0^{\infty} f(t)\frac{d^2\phi}{dt^2}dt = -f(0)\frac{d\phi(0)}{dt} - \int_0^{\infty} \frac{df}{dt}\frac{d\phi}{dt}dt$$

$$= -f(0)\frac{d\phi(0)}{dt} + \frac{df(0)}{dt}\phi(0) + \int_0^{\infty} \frac{d^2f}{dt^2}\phi dt.$$

Inserting these two equations into (6) we find

$$\int_0^{\infty} \left(\frac{d^2f}{dt^2} + a\frac{df}{dt} + bf\right)\phi(t)dt - f(0)\frac{d\phi(0)}{dt} + \frac{df(0)}{dt}\phi(0) + af(0)\phi(0) = \phi(0).$$

To satisfy this equation we have to fulfill the following three conditions

$$\frac{d^2f}{dt^2} + a\frac{df}{dt} + bf = 0$$

$$f(0) = 0$$

$$\frac{df(0)}{dt} + af(0) = 1.$$

Problem 7.   Use the identity

$$\sum_{k=-\infty}^{\infty} \delta(x - 2\pi k) \equiv \frac{1}{2\pi} \sum_{k=-\infty}^{\infty} e^{ikx} \tag{1}$$

to show that

$$2\pi \sum_{k=-\infty}^{\infty} \phi(2\pi k) = \sum_{k=-\infty}^{\infty} F[\phi](k) \tag{2}$$

where F denotes the Fourier transform and ϕ is a test function. Equation (2) is called *Poisson's summation formula*.

Solution 7.   We rewrite (1) in the form

$$2\pi \sum_{k=-\infty}^{\infty} \delta(x - 2\pi k) = \sum_{k=-\infty}^{\infty} F[\delta(x - k)].$$

Applying this equation to $\phi \in S(\mathbb{R})$, we obtain

$$2\pi \left(\sum_{k=-\infty}^{\infty} \delta(x - 2\pi k), \phi \right) = 2\pi \sum_{k=-\infty}^{\infty} \phi(2\pi k) = \left(\sum_{k=-\infty}^{\infty} F[\delta(x - k)], \phi \right).$$

Finally we arrive at

$$2\pi \left(\sum_{k=-\infty}^{\infty} \delta(x - 2\pi k), \phi \right) = \sum_{k=-\infty}^{\infty} (\delta(x - k), F[\phi]) = \sum_{k=-\infty}^{\infty} F[\phi](k).$$

This proves Poisson's summation formula.

Problem 8.   The *Hermite polynomials* are defined by

$$H_n(x) := (-1)^n e^{x^2} \frac{d^n}{dx^n} e^{-x^2}, \qquad n = 0, 1, 2, \ldots \tag{1}$$

where $H_0(x) = 1$, $H_1(x) = 2x$, $H_2(x) = 4x^2 - 2$ etc. An orthonormal basis in the Hilbert space $L_2(\mathbb{R})$ is given by

$$B := \left\{ \frac{(-1)^n}{2^{n/2}\sqrt{n!}\,\sqrt[4]{\pi}} e^{x^2/2} \frac{d^n}{dx^n} e^{-x^2} \; : \; n = 0, 1, 2, \ldots \right\}. \tag{2}$$

Use the completeness relation to find an expansion of a function $f \in L_2(\mathbb{R})$.

Solution 8.   Using the basis (2) and the Hermite polynomials (1) the completeness relation is given by

$$\sum_{n=0}^{\infty} \frac{1}{2^n n! \sqrt{\pi}} e^{-(x^2 + x'^2)/2} H_n(x) H_n(x') = \delta(x - x'), \qquad x, x' \in \mathbb{R}.$$

Using the completeness relation and the properties of the delta function we obtain

$$f(x) = \sum_{n=0}^{\infty} \frac{1}{2^n n! \sqrt{\pi}} H_n(x) e^{-x^2/2} \int_{-\infty}^{\infty} f(x') H_n(x') e^{-x'^2/2} dx'. \quad (4)$$

Problem 9. The probability amplitude for the three-dimensional transition of a particle with mass m from point $\mathbf{r}_0$ at the time t_0 to the point $\mathbf{r}_N$ at time t_N is given by the kernel

$$K(\mathbf{r}_N, t_N; \mathbf{r}_0, t_0) =$$

$$\lim_{N \to \infty} \prod_{n=1}^{N} \left(\frac{m}{2\pi i \hbar (t_n - t_{n-1})} \right)^{3/2} \int \cdots \int d^3 \mathbf{r}_1 \ldots d^3 \mathbf{r}_{N-1} \exp((i/\hbar) S(\mathbf{r}_N, \mathbf{r}_0)).$$

$$(1)$$

In a compact form one writes

$$K(\mathbf{r}_N, t_N; \mathbf{r}_0, t_0) = \int_0^N \exp((i/\hbar) S(\mathbf{r}_N, \mathbf{r}_0) D\mathbf{r}(t). \quad (2)$$

The integrals extend over all paths in three-dimensional coordinate space and S is the *action* of the corresponding classical system. Thus

$$S(\mathbf{r}_N, \mathbf{r}_0) = \int_{t_0}^{t_N} L(\dot{\mathbf{r}}, \mathbf{r}, t) dt \quad (3)$$

where $L(\dot{\mathbf{r}}, \mathbf{r}, t)$ represents the *Lagrange function*. We can also write (3) as

$$S(\mathbf{r}_N, \mathbf{r}_0) = \sum_{n=1}^{N} (t_n - t_{n-1}) L\left(\frac{\mathbf{r}_n - \mathbf{r}_{n-1}}{t_n - t_{n-1}}, \mathbf{r}_n, t_n \right). \quad (4)$$

The wave function ψ satisfies the integral equation

$$\psi(\mathbf{r}, t) = \int K(\mathbf{r}, t; \mathbf{r}_0, t_0) \psi(\mathbf{r}_0, t_0) d^3 \mathbf{r}_0. \quad (5)$$

The momentum amplitude is given by the Fourier transform of the wave function

$$\phi(\mathbf{p}, t) = \int \exp(-(i/\hbar) \mathbf{p} \cdot \mathbf{r}) \psi(\mathbf{r}, t) d^3 \mathbf{r}. \quad (6)$$

With $\mathbf{p} = \hbar \mathbf{k}$ we obtain

$$\phi(\mathbf{k}, t) = \int \exp(-2\pi i \mathbf{k} \cdot \mathbf{r}) \psi(\mathbf{r}, t) d^3 \mathbf{r}. \quad (7)$$

The momentum amplitude satisfies a similar integral equation as the wave function

$$\phi(\mathbf{k}, t) = \int \kappa(\mathbf{k}, t; \mathbf{k}_0, t_0)\phi(\mathbf{k}_0, t_0)d^3k_0 \,.$$

The momentum kernel κ represents the probability amplitude of the transition from the wave vector $\mathbf{k}_0$ at the time t_0 to the wave vector $\mathbf{k}$ at the time t. We have

$$\kappa(\mathbf{k}, t; \mathbf{k}_0, t_0) = \int \int \exp(-2\pi i \mathbf{k} \cdot \mathbf{r})K(\mathbf{r}, t; \mathbf{r}_0, t_0)\exp(2\pi \mathbf{k}_0 \cdot \mathbf{r}_0)d^3r_0 d^3r \,.$$

Describe the motion of a particle in a local potential $V(\mathbf{r}, t)$. The Lagrangian function is given by

$$L(\dot{\mathbf{r}}, \mathbf{r}, t) = \frac{1}{2}m\dot{\mathbf{r}}^2 - V(\mathbf{r}, t)$$

where $V(\mathbf{r}, t)$ represents the potential energy.

Solution 9.   The action S becomes

$$S(\mathbf{r}_N, \mathbf{r}_0) = \sum_{n=1}^{N}(t_n - t_{n-1})\left(\frac{1}{2}m\left(\frac{\mathbf{r}_n - \mathbf{r}_{n-1}}{t_n - t_{n-1}}\right)^2 - V(\mathbf{r}_n, t_n)\right) \,.$$

Inserting this expression into (1) yields

$$K(\mathbf{r}_N, t_N; \mathbf{r}_0, t_0) = \lim_{N \to \infty} \prod_{n=1}^{N}\left(\frac{m}{ih(t_n - t_{n-1})}\right)^{3/2} \int \cdots \int d^3r_1 \ldots d^3r_{N-1}$$

$$\times \exp\left((i\pi m/h)\sum_{n=1}^{N}\frac{(\mathbf{r}_n - \mathbf{r}_{n-1})^2}{(t_n - t_{n-1})}\right)$$

$$\times \exp\left(-(2\pi i/h)\sum_{n=1}^{N}(t_n - t_{n-1})V(\mathbf{r}_n, t_n)\right) \,.$$

Using the identity

$$\left(\frac{m}{iht}\right)^{3/2}\exp(i\pi m\mathbf{r}^2/ht) \equiv \int d^3k \exp(2\pi i(\mathbf{k} \cdot \mathbf{r} - ht\mathbf{k}^2/2m))$$

where the integration extends over the three-dimensional momentum space, $K(\mathbf{r}_N, t_N; \mathbf{r}_0, t_0)$ can be written as

$$K(\mathbf{r}_N, t_N; \mathbf{r}_0, t_0) = \lim_{N \to \infty}\int \cdots \int d^3r_1 \cdots d^3r_{N-1}d^3k_0 \cdots d^3k_{N-1}$$

$$\times \exp\left(2\pi i \sum_{n=1}^{N}\mathbf{k}_{n-1} \cdot (\mathbf{r}_n - \mathbf{r}_{n-1}) - \left(\frac{h}{2m}\right)(t_n - t_{n-1})\mathbf{k}_{n-1}^2\right)$$

$$\times \exp\left(-2\pi i/h\sum_{n=1}^{N}(t_n - t_{n-1})V(\mathbf{r}_n, t_n)\right) \,.$$

From (3) we obtain

$$\kappa(\mathbf{k}_N, t_N; \mathbf{k}'_0, t_0) = \lim_{N \to \infty} \int \cdots \int d^3\mathbf{r}_0 \cdots d^3\mathbf{r}_N d^3\mathbf{k}_0 \cdots d^3\mathbf{k}_{N-1}$$

$$\times \exp(-2\pi i(\mathbf{k}_0 - \mathbf{k}'_0) \cdot \mathbf{r}_0) \exp(-2\pi i \sum_{n=1}^{N} (\mathbf{k}_n - \mathbf{k}_{n-1}) \cdot \mathbf{r}_n)$$

$$\times \exp(-(2\pi i/h) \sum_{n=1}^{N} (t_n - t_{n-1})V(\mathbf{r}_n, t_n))$$

$$\times \exp(-(\pi i h/m) \sum_{n=1}^{N} (t_n - t_{n-1})\mathbf{k}_{n-1}^2).$$

The integration over $\mathbf{r}_0$ can be carried out giving the delta function

$$\int d^3\mathbf{r}_0 \exp(-2\pi i(\mathbf{k}_0 - \mathbf{k}'_0) \cdot \mathbf{r}_0) = \delta(\mathbf{k}_0 - \mathbf{k}'_0).$$

Using the property of the delta function we can also calculate the integral over $\mathbf{k}_0$. We obtain

$$\kappa(\mathbf{k}_N, t_N; \mathbf{k}_0, t_0) = \lim_{N \to \infty} \int \cdots \int d^3\mathbf{r}_1 \cdots d^3\mathbf{r}_N d^3\mathbf{k}_1 \cdots \mathbf{k}_{N-1}$$

$$\times \exp(-2\pi i \sum_{n=1}^{N} (\mathbf{k}_n - \mathbf{k}_{n-1}) \cdot \mathbf{r}_n)$$

$$\times \exp(-(2\pi i/h) \sum_{n=1}^{N} (t_n - t_{n-1})V(\mathbf{r}_n, t_n))$$

$$\times \exp(-(i\pi h/m) \sum_{n=1}^{N} (t_n - t_{n-1})\mathbf{k}_{n-1}^2).$$

Defining

$$F(\mathbf{k}_n - \mathbf{k}_{n-1}) :=$$

$$\int d^3\mathbf{r}_n \exp(-2\pi i(\mathbf{k}_n - \mathbf{k}_{n-1}) \cdot \mathbf{r}_n) \exp(-(2\pi i/h)(t_n - t_{n-1})V(\mathbf{r}_n, t_n))$$

we can write

$$\kappa(\mathbf{k}_N, t_N; \mathbf{k}_0, t_0) =$$

$$\lim_{N \to \infty} \int \cdots \int d^3\mathbf{k}_1 \cdots d^3\mathbf{k}_{N-1} \prod_{n=1}^{N} F(\mathbf{k}_n - \mathbf{k}_{n-1}) e^{-i\pi h/m}(t_n - t_{n-1})\mathbf{k}_{n-1}^2).$$

The physical interpretation of this formula is that the transmission of the wave, with wave vector $\mathbf{k}_{n-1}$ through the n-th slice can be described by the propagation factor exp multiplied by the probability amplitude

$$F(\mathbf{k}_n - \mathbf{k}_{n-1})$$

for the scattering into the new wave vector $\mathbf{k}_n$. Integration over all possible wave vectors and time slices gives the expression for the momentum kernel.

Problem 10.   Given a function (signal) $f(\mathbf{t}) = f(t_1, t_2, \ldots, t_n) \in L_2(\mathbb{R}^n)$ of n real variables $\mathbf{t} = (t_1, t_2, \ldots, t_n)$. We define the *symplectic tomogram* associated with the square integrable function f

$$
w(\mathbf{X}, \boldsymbol{\mu}, \boldsymbol{\nu}) = \prod_{k=1}^{n} \frac{1}{2\pi|\nu_k|} \left| \int_{\mathbb{R}^n} dt_1 \cdots dt_n f(\mathbf{t}) \exp\left(\sum_{j=1}^{n} \left(\frac{i\mu_j}{2\nu_j} t_j^2 - \frac{iX_j}{\nu_j} t_j \right) \right) \right|^2
$$

where $(\nu_j \neq 0$ for $j = 1, 2, \ldots, n)$

$$
\mathbf{X} = (X_1, X_2, \ldots, X_n), \quad \boldsymbol{\mu} = (\mu_1, \mu_2, \ldots, \mu_n), \quad \boldsymbol{\nu} = (\nu_1, \nu_2, \ldots, \nu_n).
$$

(i) Prove the equality

$$
\int_{\mathbb{R}^n} w(\mathbf{X}, \boldsymbol{\mu}, \boldsymbol{\nu}) d\mathbf{X} = \int_{\mathbb{R}^n} |f(\mathbf{t})|^2 d\mathbf{t} \tag{1}
$$

for the special case $n = 1$. The tomogram is the probabilty distribution function of the random variable $\mathbf{X}$. This probability distribution function depends on $2n$ extra real parameters $\boldsymbol{\mu}$ and $\boldsymbol{\nu}$.
(ii) The map of the function $f(\mathbf{t})$ onto the tomogram $w(\mathbf{X}, \boldsymbol{\mu}, \boldsymbol{\nu})$ is invertible. The square integrable function $f(\mathbf{t})$ can be associated with the density matrix

$$
\rho_f(\mathbf{t}, \mathbf{t}') = f(\mathbf{t}) f^*(\mathbf{t}').
$$

This density matrix can be mapped onto the *Ville-Wigner function*

$$
W(\mathbf{q}, \mathbf{p}) = \int_{\mathbb{R}^n} \rho_f \left(\mathbf{q} + \frac{\mathbf{u}}{2}, \mathbf{q} - \frac{\mathbf{u}}{2} \right) e^{-i\mathbf{p}\cdot\mathbf{u}} d\mathbf{u}.
$$

Show that this map is invertible.
(iii) How is the tomogram $w(\mathbf{X}, \boldsymbol{\mu}, \boldsymbol{\nu})$ related to the Ville-Wigner function?
(iv) Show that the Ville-Wigner function can be reconstructed from the function $w(\mathbf{X}, \boldsymbol{\mu}, \boldsymbol{\nu})$.
(v) Show that the density matrix $f(\mathbf{t}) f^*(\mathbf{t}')$ can be found from $w(\mathbf{X}, \boldsymbol{\mu}, \boldsymbol{\nu})$.

Solution 10.   (i) Since

$$
\frac{1}{2\pi\nu} \left| \int_{\mathbb{R}} dt f(t) e^{i\mu t^2/(2\nu) - iXt/\nu} \right|^2
$$

$$
= \frac{1}{2\pi\nu} \int_{\mathbb{R}} \int_{\mathbb{R}} dt dt' f(t) f^*(t') e^{i\mu(t^2 - (t')^2)/(2\nu)} e^{iX(t'-t)/\nu}
$$

and in the sense of generalized functions ($\nu \neq 0$)

$$\int_{\mathbb{R}} e^{iX(t'-t)/\nu} dX = 2\pi\delta((t'-t)/\nu) = 2\pi\nu\delta(t'-t)$$

we find, applying the property of the delta function, that equality (1) holds.
(ii) Since we have a Fourier transform the inverse transform is given by

$$\rho(\mathbf{t}, \mathbf{t}') = \frac{1}{(2\pi)^n} \int_{\mathbb{R}^n} W\left(\frac{\mathbf{t}+\mathbf{t}'}{2}, \mathbf{p}\right) e^{i\mathbf{p}\cdot(\mathbf{t}-\mathbf{t}')} d\mathbf{p}.$$

(iii) Using a delta function we can write

$$w(\mathbf{X}, \boldsymbol{\mu}, \boldsymbol{\nu}) = \int_{\mathbb{R}^n} W(\mathbf{q}, \mathbf{q}) \prod_{k=1}^{n} \delta(X_k - \mu_k q_k - \nu_k p_k) \frac{dp_k dq_k}{2\pi}.$$

(iv) The Ville-Wigner function can be reconstructed from the function $w(\mathbf{X}, \boldsymbol{\mu}, \boldsymbol{\nu})$ via

$$W(\mathbf{p}, \mathbf{q}) = \frac{1}{(2\pi)^n} \int_{\mathbb{R}^n} w(\mathbf{X}, \boldsymbol{\mu}, \boldsymbol{\nu}) \prod_{k=1}^{n} e^{i(X_k - \mu_k q_k - \nu_k p_k)} dX_k d\mu_k d\nu_k.$$

(v) The density matrix $f(\mathbf{t})f^*(\mathbf{t}')$ can be found from $w(\mathbf{X}, \boldsymbol{\mu}, \boldsymbol{\nu})$ as

$$f(\mathbf{t})f^*(\mathbf{t}') = \frac{1}{(2\pi)^n} \int_{\mathbb{R}^n} w(\mathbf{X}, \boldsymbol{\mu}, \mathbf{t}-\mathbf{t}') \prod_{k=1}^{n} \exp\left(i\left(X_k - \mu_k \frac{t_k + t_k'}{2}\right)\right) dX_k d\mu_k.$$

Problem 11. Let $c > 0$. Consider the Schrödinger equation

$$-\frac{\hbar^2}{2m} \frac{d^2\psi}{dx^2} + c\delta^{(n)}(x)\psi = E\psi$$

where $\delta^{(n)}$ ($n = 0, 1, 2, \ldots$) denotes the n-th derivative of the delta function. Derive the joining conditions on the wave function ψ.

Solution 11. Let $\epsilon > 0$. We define

$$\overline{f}(0) := \frac{1}{2}(f(0^+) + f(0^-)).$$

For functions f that are discontinuous at the origin the expression

$$\int_{-\epsilon}^{\epsilon} \delta(x)f(x)dx$$

is in general not well-defined. If we assume that the delta function is the limit of a sequence of even functions, then

$$\int_{-\epsilon}^{\epsilon} \delta(x)f(x)dx = \overline{f}(0).$$

This we assume in the following. Integrating from $-\epsilon$ to ϵ the eigenvalue equation yields

$$-\frac{\hbar^2}{2m}[\psi'(\epsilon) - \psi'(-\epsilon)) + c\int_{-\epsilon}^{\epsilon}\left(\left(\frac{d}{dx}\right)^n \delta(x)\right)\psi(x)dx = E\int_{-\epsilon}^{\epsilon}\psi(x)dx.$$

Using integration by parts n times and using the fact that all boundary terms vanish yields

$$\int_{-\epsilon}^{\epsilon}\left(\left(\frac{d}{dx}\right)^n \delta(x)\right)\psi(x)dx = (-1)^n\int_{-\epsilon}^{\epsilon}\delta(x)\left(\left(\frac{d}{dx}\right)^n\psi(x)\right)dx$$

$$= (-1)^n\overline{\psi}^{(n)}(0).$$

In the limit $\epsilon \to 0$ we have

$$\lim_{\epsilon\to 0}\int_{-\epsilon}^{\epsilon}\psi(\epsilon)dx = 0.$$

Thus we have the first boundary condition

$$\Delta\psi' = (-1)^n\frac{2mc}{\hbar^2}\overline{\psi}^{(n)}(0).$$

Furthermore integrating the eigenvalue equation from $-L$ (L positive) to x yields

$$-\frac{\hbar^2}{2m}(\psi'(x)-\psi'(-L))+c\int_{-L}^{x}\left(\left(\frac{d}{ds}\right)^n \delta(s)\right)\psi(s)ds = E\int_{-L}^{x}\psi(s)ds. \quad (1)$$

We integrate by parts and find

$$\int_{-L}^{x}\left(\left(\frac{d}{ds}\right)^n \delta(s)\right)\psi(s)ds = \delta^{(n-1)}(x)\psi(x) - \delta^{(n-2)}(x)\psi'(x)$$

$$+ \delta^{(n-3)}(x)\psi^{(2)}(x) - \cdots + (-1)^{n-1}\delta(x)\psi^{(n-1)}(x) + (-1)^n\theta(x)\overline{\psi}^{(n)}(0)$$

where now the upper boundary terms are nonzero. Integrating (1) from $-\epsilon$ to ϵ and taking the limit $\epsilon \to 0$ we obtain

$$-\frac{\hbar^2}{2m}\Delta\psi + c\left(\int_{-\epsilon}^{\epsilon}\delta^{(n-1)}(x)\psi(x)dx - \int_{-\epsilon}^{\epsilon}\delta^{(n-2)}(x)\psi'(x)dx\right.$$

$$\left. + \int_{-\epsilon}^{\epsilon}\delta^{(n-3)}(x)\psi^{(2)}(x)dx - \cdots + (-1)^{n-1}\int_{-\epsilon}^{\epsilon}\delta(x)\psi^{(n-1)}(x)dx\right) = 0.$$

Using once more integration by parts we obtain for the expression in the parenthesis

$$(-1)^{n-1}n\overline{\psi}^{(n-1)}(0)$$

Thus we have the second boundary condition

$$\Delta\psi = (-1)^{n-1}\frac{2mc}{\hbar^2}n\overline{\psi}^{(n-1)}(0).$$

For $n = 0$ we have

$$\Delta\psi = 0, \qquad \Delta\psi' = \frac{2mc}{\hbar^2}\psi(0)$$

and for $n = 1$ we have

$$\Delta\psi = \frac{2mc}{\hbar^2}\overline{\psi}(0), \qquad \Delta\psi' = -\frac{2mc}{\hbar^2}\overline{\psi}'(0).$$

Problem 12. Any $SU(2)$ matrix A can be written as $(x_0, x_1, x_2, x_3 \in \mathbb{R})$

$$A = \begin{pmatrix} x_0 - ix_3 & -ix_1 - x_2 \\ x_2 - ix_1 & x_0 + ix_3 \end{pmatrix}, \qquad x_0^2 + x_1^2 + x_2^2 + x_3^2 = 1 \qquad (1)$$

i.e., $\det A = 1$. Using *Euler angles* α, β, γ the matrix can also be written as

$$A = \begin{pmatrix} \cos(\beta/2)e^{i(\alpha+\gamma)/2} & -\sin(\beta/2)e^{i(\alpha-\gamma)/2} \\ \sin(\beta/2)e^{-i(\alpha-\gamma)/2} & \cos(\beta/2)e^{-i(\alpha+\gamma)/2} \end{pmatrix}. \qquad (2)$$

(i) Show that the invariant measure dg of $SU(2)$ can be written as

$$dg = \frac{1}{\pi^2}\delta(x_0^2 + x_1^2 + x_2^2 + x_3^2 - 1)dx_0 dx_1 dx_2 dx_3$$

where δ is the Dirac delta function.
(ii) Show that dg is normalized, i.e.

$$\int dg = 1.$$

(iii) Using (1) and (2) find $x_1(\alpha, \beta, \gamma)$, $x_2(\alpha, \beta, \gamma)$, $x_3(\alpha, \beta, \gamma)$. Find the Jacobian determinant.
(iv) Using the results from (iii) show that the invariant measure can be written as

$$\frac{1}{16\pi^2}\sin\beta d\alpha d\beta d\gamma.$$

Solution 12. (i) Let $B \in SU(2)$. Under left multiplication $A' = BA$ is expressed in terms of the new variables x_0', x_1', x_2' and x_3'. By regarding $x_0 + ix_3$ and $x_2 - ix_1$ as two independent complex variables, we find that the Jacobian determinant of changing variables from (x_0, x_1, x_2, x_3) to (x_0', x_1', x_2', x_3') is simply $\det(B) = 1$. It follows that dg is invariant.

(ii) Let $r^2 := x_1^2 + x_2^2 + x_3^2$. We have

$$\int dg = \frac{1}{2\pi^2} \int_{|r| \leq 1} \frac{dx_1 dx_2 dx_3}{\sqrt{1 - r^2}}$$

$$= \frac{2}{\pi} \int_0^1 \frac{r^2 dr}{\sqrt{1 - r^2}}$$

$$= 1$$

where we used spherical coordinates for the angle integration which provides 4π and

$$\int \frac{x^2 dx}{\sqrt{a^2 - x^2}} = -\frac{x\sqrt{a^2 - x^2}}{2} + \frac{a^2}{2} \arcsin\left(\frac{x}{a}\right).$$

(iii) We find

$$x_1(\alpha, \beta, \gamma) = \sin\frac{\beta}{2} \sin\frac{\alpha - \gamma}{2}$$

$$x_2(\alpha, \beta, \gamma) = \sin\frac{\beta}{2} \cos\frac{\alpha - \gamma}{2}$$

$$x_3(\alpha, \beta, \gamma) = -\cos\frac{\beta}{2} \sin\frac{\alpha + \gamma}{2}.$$

For the Jacobian determinant we obtain

$$\left| \frac{\partial(x_1, x_2, x_3)}{\partial(\alpha, \beta, \gamma)} \right| = -\frac{1}{4} \sin\frac{\beta}{2} \cos^2\frac{\beta}{2} \cos\frac{\alpha + \gamma}{2}.$$

(iv) Since

$$r^2 = x_1^2 + x_2^2 + x_3^2 = 1 - \cos^2\left(\frac{\alpha + \gamma}{2}\right) \cos^2\frac{\beta}{2}$$

we have

$$\sqrt{1 - r^2} = \sqrt{1 - (x_1^2 + x_2^2 + x_3^2)} = \cos\frac{\alpha + \gamma}{2} \cos\frac{\beta}{2}.$$

Furthermore $\sin(\beta/2)\cos(\beta/2) = \frac{1}{2}\sin\beta$. Using these results and the results from (ii) and (iii) we obtain

$$dg = \frac{1}{2\pi^2} \frac{dx_1 dx_2 dx_3}{\sqrt{1 - r^2}} = \frac{1}{16\pi^2} \sin\beta d\alpha d\beta d\gamma.$$

Chapter 11

Linear Partial Differential Equations

Problem 1. Show that

$$u(x,t) = \sum_{n=1}^{\infty} c_n e^{-n^2 t} \sin(nx)$$

satisfies the one-dimensional linear heat equation

$$\frac{\partial u}{\partial t} = \frac{\partial^2 u}{\partial x^2} \tag{1}$$

together with the initial and boundary conditions

$$u(x,0) = f(x), \quad 0 < x < \pi$$
$$u(0,t) = 0, \quad t > 0$$
$$u(\pi,t) = 0, \quad t > 0.$$

Start with the *separation ansatz* $u(x,t) = X(x)T(t)$.

Solution 1. Substituting the separation ansatz into (1) yields

$$\frac{1}{X(x)} \frac{d^2 X}{dx^2} = \frac{1}{T(t)} \frac{dT}{dt} = -\lambda$$

where λ is an arbitrary real constant. The eigenvalue problem

$$-\frac{d^2 X}{dx^2} = \lambda X(x), \qquad X(0) = X(\pi) = 0$$

123

gives an infinite countable number of solutions

$$X_n(x) = C \sin(nx)$$

with the eigenvalues $\lambda_n = n^2$ where $n = 1, 2, \ldots, \infty$. Therefore, for the function T it follows the linear differential equation

$$\frac{dT}{dt} + n^2 T = 0, \qquad t > 0$$

with the solution $T(t) = Ce^{-n^2 t}$. Since $(2/\pi)^{1/2} \sin(nx)$ with $n = 1, 2, \ldots, \infty$ is a basis in the Hilbert space $L_2(0, \pi)$ and $f \in L_2(0, \pi)$ we find

$$f(x) = \sqrt{\frac{2}{\pi}} \sum_{n=1}^{\infty} d_n \sin(nx)$$

with the expansion coefficients given by

$$d_n := \left(f(x), \sqrt{\frac{2}{\pi}} \sin(nx) \right) = \sqrt{\frac{2}{\pi}} \int_0^\pi f(x) \sin(nx) dx$$

where $(\,,\,)$ denotes the scalar product in the Hilbert space $L_2(0, \pi)$. Since

$$u(x, 0) = \sum_{n=1}^{\infty} c_n \sin(nx) = f(x)$$

we obtain

$$c_n = \frac{2}{\pi} \int_0^\pi f(x) \sin(nx) dx.$$

Problem 2. Show that

$$E(\mathbf{x}) = -\frac{e^{ik|\mathbf{x}|}}{4\pi |\mathbf{x}|}$$

satisfies the partial differential equation

$$\Delta E + k^2 E = \delta$$

in the sense of generalized functions, where $\mathbf{x} = (x_1, x_2, x_3)$ and

$$|\mathbf{x}| \equiv r = \sqrt{x_1^2 + x_2^2 + x_3^2}$$

and δ is the delta function.

Solution 2. First we have to show that

$$\Delta \frac{1}{|\mathbf{x}|} = -4\pi\delta(\mathbf{x})$$

in the sense of generalized functions. Let ϕ be a test function. Without loss of generality we can assume that ϕ is spherically symmetric. From the definition of the derivative of a generalized function and using spherical coordinates we obtain

$$\left(\Delta \frac{1}{r}, \phi(r)\right) := \left(\frac{1}{r}, \Delta\phi(r)\right) = \int_{\mathbb{R}^3} \frac{\Delta\phi}{r} r^2 \sin\theta \, dr d\theta d\phi. \tag{1}$$

Since ϕ depends only on r, the angle part of Δ can be omitted. We find that the Laplace operator takes the form

$$\Delta := \frac{1}{r^2} \frac{\partial}{\partial r} \left(r^2 \frac{\partial}{\partial r}\right).$$

Inserting this equation into (1) gives

$$\left(\frac{1}{r}, \Delta\phi\right) = 4\pi \int_{r=0}^{\infty} \frac{1}{r} \left(2r \frac{d}{dr}\phi + r^2 \frac{d^2}{dr^2}\phi\right) dr.$$

Thus

$$\left(\frac{1}{r}, \Delta\phi\right) = -4\pi\phi(0) = -4\pi(\delta, \phi)$$

where we have used integration by parts and that $\phi(r) \to 0$ as $r \to \infty$. We can write

$$\Delta \frac{1}{r} = -4\pi\delta.$$

For $r > 0$ we have

$$\frac{\partial}{\partial x_j} \frac{1}{|\mathbf{x}|} = -\frac{x_j}{|\mathbf{x}|^3}, \qquad \frac{\partial}{\partial x_j} e^{ik|\mathbf{x}|} = ik \frac{x_j}{|\mathbf{x}|} e^{ik|\mathbf{x}|}$$

and

$$\Delta e^{ik|\mathbf{x}|} = \left(\frac{2ik}{|\mathbf{x}|} - k^2\right) e^{ik|\mathbf{x}|}.$$

It follows that

$$(\Delta + k^2) \frac{1}{|\mathbf{x}|} e^{ik|\mathbf{x}|} = -4\pi\delta(r).$$

Problem 3. Consider the linear partial differential equation

$$\frac{\partial u}{\partial t} + c_0 \frac{\partial u}{\partial x} + \beta \frac{\partial^3 u}{\partial x^3} = 0 \tag{1}$$

where β and c_0 are positive constants.

(i) Find the *dispersion relation* (i.e. the relation between the frequency ω and the wave vector k) using the ansatz

$$u(x,t) = u_0 \exp(i(kx - \omega(k)t))$$

(ii) Let

$$x'(x,t) = x - c_0 t, \qquad t'(x,t) = t$$
$$u'(x'(x,t), t'(x,t)) = u(x,t).$$

Find the partial differential equation for $u'(x',t')$.

(iii) Show that the solution to the initial-value problem of the partial differential equation

$$\frac{\partial u}{\partial t} + \beta \frac{\partial^3 u}{\partial x^3} = 0 \tag{2}$$

is given by

$$u(x,t) = (\pi)^{-1/2}(3\beta t)^{-1/3} \int_{-\infty}^{\infty} \text{Ai}\left(\frac{x-s}{(3\beta t)^{1/3}}\right) u(s, t=0)ds \tag{3}$$

where

$$\text{Ai}(s) := \frac{1}{\sqrt{\pi}} \int_0^{\infty} \cos\left(\frac{v^3}{3} + vs\right) dv$$

is the *Airy function*.

Solution 3.   (i) Inserting the ansatz into (1) leads to the dispersion relation

$$\omega(k) = c_0 k - \beta k^3.$$

(ii) Straightforward application of the chain rule yields

$$\frac{\partial u'}{\partial t'} + \beta \frac{\partial^3 u'}{\partial x'^3} = 0.$$

(iii) The Airy function has the asymptotic representation

$$\text{Ai}(s) = \begin{cases} \frac{1}{2}s^{-1/4} \exp\left(-\frac{2}{3}s^{3/2}\right) & \text{for } s \to \infty \\ |s|^{-1/4} \cos\left(\frac{2}{3}|s|^{3/2} - \frac{\pi}{4}\right) & \text{for } s \to -\infty. \end{cases}$$

If we assume that the initial distribution $u(x, t = 0)$ tends to zero sufficiently fast as $x \to \pm\infty$ we find by differentiating that (3) is the solution of the initial-value problem of the partial differential equation (2).

The Airy functions $\text{Ai}(z)$ and $\text{Bi}(z)$ can be expressed in terms of Bessel functions. Vice versa the Bessel functions can be expressed in terms of the Airy functions $\text{Ai}(z)$ and $\text{Bi}(z)$.

Problem 4. Consider a wave function $u(x, y)$ which is confined within a regular triangle. The regular triangle is bound by the three lines

$$x = -A, \qquad x = \sqrt{3}y + 2A, \qquad x = -\sqrt{3}y + 2A$$

where $A > 0$. The eigenvalue problem is given by

$$\Delta u(x, y) \equiv \frac{\partial^2 u}{\partial x^2} + \frac{\partial^2 u}{\partial y^2} = -\frac{mE}{\hbar^2} u(x, y) \tag{1}$$

where at the boundary

$$u|_B = 0. \tag{2}$$

(i) Find the symmetry group of the system.
(ii) Find the eigenvalues and eigenfunctions for an invariant subspace.

Solution 4. (i) Diffraction in a corner of a polygon does not occur whenever the angle of the corner is π divided by an integer. In our case we have $\pi/3$. Therefore it is possible to express the wave function in terms of elementary functions, i.e. sin and cos. The triangle is invariant under the following six symmetry operations:

identity, labelled I

rotation by $\frac{2}{3}\pi$ around the origin, labelled C_3

rotation by $\frac{4}{3}\pi$ around the origin, labelled C_3^2

reflection in the line $x = (1/\sqrt{3})y$, labelled σ

reflection in the x axis, labelled σ'

reflection in the line $x = -(1/\sqrt{3})y$, labelled σ''.

These operations form the group C_{3v}. The *character table* is given by

	$\{I\}$	$\{2C_3\}$	$\{3\sigma\}$
A1	1	1	1
A2	1	1	-1
E	2	-1	0

(ii) To apply group theory we first have to find the projection operators. From the character table we find that the projection operators are given by

$$\Pi_{A1} = I + C_3 + C_3^2 + \sigma + \sigma' + \sigma''$$
$$\Pi_{A2} = I + C_3 + C_3^2 - \sigma - \sigma' - \sigma''$$
$$\Pi_E = 2I - C_3 - C_3^2.$$

We construct the eigenfunctions as linear combinations of

$$f(x, y) := \sin(P(x + x_0)) \sin(Qy)$$
$$g(x, y) := \sin(P(x + x_0)) \cos(Qy).$$

Inserting these functions into (1) yields

$$E = \frac{\hbar^2}{m}(P^2 + Q^2).$$

Now we have to apply the projection operators to f and g and then impose the boundary condition. We find $\Pi_{A1} f = 0$ and the eigenfunctions of the symmetry type $A1$ are given by

$$\phi(x, y) := \frac{1}{2}\Pi_{A1} g(x, y).$$

Thus

$$\phi(x, y) = \sin(P(x + x_0)) \cos(Qy)$$
$$+ \sin(P(-\frac{1}{2}x - \frac{\sqrt{3}}{2}y + x_0)) \cos(Q(-\frac{1}{2}y + \frac{\sqrt{3}}{2}x))$$
$$+ \sin(P(-\frac{1}{2}x + \frac{\sqrt{3}}{2}y + x_0)) \cos(Q(-\frac{1}{2}y - \frac{\sqrt{3}}{2}x)).$$

Analogously, we find $\Pi_{A2} g = 0$ and the eigenfunctions of the symmetry type $A2$ are given by

$$\zeta(x, y) := \frac{1}{2}\Pi_{A2} f(x, y)$$
$$= \sin(P(x + x_0)) \sin(Qy)$$
$$+ \sin(P(-\frac{1}{2}x - \frac{\sqrt{3}}{2}y + x_0)) \sin(Q(-\frac{1}{2}y + \frac{\sqrt{3}}{2}x))$$
$$+ \sin(P(-\frac{1}{2}x + \frac{\sqrt{3}}{2}y + x_0)) \sin(Q(-\frac{1}{2}y - \frac{\sqrt{3}}{2}x)).$$

Since the functions of the symmetry class $A1$ are symmetric under C_3 it is sufficient to require that $\phi(x, y) = 0$ for $x = -A$ and any y. Then the functions automatically satisfy the boundary conditions along the two other sides of the triangle. Thus we obtain

$$P = \frac{k\pi}{3A}, \qquad Q = \frac{n\pi}{\sqrt{3}A}$$

where k and n are integers which are either both even or both odd. We can restrict the values of k and n as follows: k and n are positive integers with

$k \leq n$ and are either both even or both odd. For the symmetry class $A2$ it is sufficient to require that

$$\zeta(x, y) = 0 \quad \text{for} \quad x = -A$$

and any y. The function ζ is identically zero for $k = n$. Therefore we can restrict the values of k as follows: k and n are positive integers with $k < n$ and are either both even or both odd. Consequently, we have an infinite countable number of eigenfunctions of the symmetry class $A1$ and $A2$. Inserting these eigenfunctions into the eigenvalue equation we find the eigenvalues

$$E_{n,k} = \frac{\pi^2 \hbar^2}{mA^2} \left(\frac{k^2}{9} + \frac{n^2}{3} \right).$$

Problem 5. The *telegraph equation* is given by

$$\frac{\partial^2 w}{\partial t^2} + a \frac{\partial w}{\partial t} + bw = c^2 \frac{\partial^2 w}{\partial x^2}. \tag{1}$$

Show that this equation can be transformed into the canonical form

$$\frac{\partial^2 u}{\partial \eta \partial \xi} + ku = 0 \tag{2}$$

where $k = (a^2 - 4b^2)/(16c^2)$ by applying the transformation

$$\eta(x, t) = x - ct, \qquad \xi(x, t) = x + ct \tag{3a}$$

$$w(x(\eta, \xi), t(\eta, \xi)) = u(\eta, \xi) \exp(-at(\eta, \xi)/2). \tag{3b}$$

Solution 5. From (3a) and (3b) we find by applying the chain rule

$$\frac{\partial w}{\partial \eta} = \frac{\partial w}{\partial x} \frac{\partial x}{\partial \eta} + \frac{\partial w}{\partial t} \frac{\partial t}{\partial \eta}$$

$$= \frac{\partial u}{\partial \eta} \exp(-at(\eta, \xi)/2) - \frac{a}{2} u(\eta, \xi) \exp(-at(\eta, \xi)/2) \frac{\partial t}{\partial \eta}.$$

Since $\partial x/\partial \eta = 1/2$, $\partial t/\partial \eta = -1/(2c)$, $\partial x/\partial \xi = 1/2$, $\partial t/\partial \xi = 1/(2c)$ and setting $s := -at(\beta, \xi)/2$ we obtain

$$\frac{1}{2} \frac{\partial w}{\partial x} - \frac{1}{2c} \frac{\partial w}{\partial t} = \frac{\partial u}{\partial \eta} \exp(s) + \frac{a}{4c} u(\eta, \xi) \exp(s). \tag{4}$$

Analogously when we take the derivative of w with respect to ξ we find

$$\frac{1}{2} \frac{\partial w}{\partial x} + \frac{1}{2c} \frac{\partial w}{\partial t} = \frac{\partial u}{\partial \xi} \exp(s) - \frac{a}{4c} u(\eta, \xi) \exp(s). \tag{5}$$

Subtracting (4) and (5) yields

$$\frac{\partial w}{\partial t} = -c\frac{\partial u}{\partial \eta}\exp(s) + c\frac{\partial u}{\partial \xi}\exp(s) - \frac{a}{2}u\exp(s).$$

Taking the derivative of (4) with respect to ξ yields

$$\frac{\partial^2 w}{\partial x^2} - \frac{1}{c^2}\frac{\partial^2 w}{\partial t^2} = 4\frac{\partial^2 u}{\partial \eta \partial \xi}\exp(s) - \frac{a}{c}\frac{\partial u}{\partial \eta}\exp(s) + \frac{a}{c}\frac{\partial u}{\partial \xi}\exp(s) - \frac{a^2}{4c^2}u\exp(s).$$

Taking the derivative of (5) with respect to η yields

$$\frac{\partial^2 w}{\partial x^2} - \frac{1}{c^2}\frac{\partial^2 w}{\partial t^2} = 4\frac{\partial^2 u}{\partial \eta \partial \xi}\exp(s) + \frac{a}{c}\frac{\partial u}{\partial \xi}\exp(s) - \frac{a}{c}\frac{\partial u}{\partial \eta}\exp(s) - \frac{a^2}{4c^2}u\exp(s).$$

Inserting these equations into (1) yields (2).

Problem 6.   Consider the *linear diffusion equation*

$$\frac{\partial u}{\partial t} = \frac{\partial^2 u}{\partial x^2}$$

with

$$u(x,0) = f(x), \qquad x \in [0,\infty)$$
$$u(0,t) = U, \qquad t \geq 0$$

The exact solution is given by

$$u(x,t) = U\,\mathrm{erfc}(\frac{1}{2}x/t^{1/2})$$
$$+ \frac{1}{2\sqrt{\pi t}}\int_0^\infty f(x_1)(\exp(-(x-x_1)^2/4t) - \exp(-(x+x_1)^2/4t))dx_1\,.$$

Find the asymptotic expansion of $u(x,t)$ as $t \to \infty$, $\eta = O(1)$ with

$$\eta = \frac{1}{2}x/t^{1/2}\,.$$

Solution 6.   The solution $u(x,t)$ can be written as

$$u(x,t) \sim U\,\mathrm{erfc}(\frac{1}{2}\frac{x}{\sqrt{t}}) + \frac{1}{\sqrt{\pi}}\left(\frac{I_2(0)\eta e^{-\eta^2}}{t} + \frac{I_4(0)(\eta^3 - \frac{3}{2}\eta)e^{-\eta^2}}{t^2} + \cdots\right).$$

The coefficients being proportional to *Hermite polynomials*. If we assume that the integrals converge, then the numbers $I_2(0)$, $I_4(0)$, $\ldots$, are given by

$$I_2(x) = \int_x^\infty dx_1 \int_{x_1}^\infty f(x_2)dx_2, \qquad I_4(x) = \int_x^\infty dx_1 \int_{x_1}^\infty I_2(x_2)dx_2\,, \qquad \ldots$$

at $x = 0$. The expansion depends on the initial condition only via these coefficients.

Problem 7. Let u be a differentiable function of x, y, and z which satisfies the partial differential equation

$$(y - z)\frac{\partial u}{\partial x} + (z - x)\frac{\partial u}{\partial y} + (x - y)\frac{\partial u}{\partial z} = 0.$$

Show that u contains x, y and z only in combinations $x + y + z$ and $x^2 + y^2 + z^2$.

Solution 7. The auxiliary equations of the partial differential equation are

$$\frac{dx}{y - z} = \frac{dy}{z - x} = \frac{dz}{x - y} = \frac{du}{0}.$$

They are equivalent to the three differential relations

$$du = 0, \qquad dx + dy + dz = 0, \qquad x\,dx + y\,dy + z\,dz = 0.$$

Thus the integrals are

$$u = c_1, \qquad x + y + z = c_2, \qquad x^2 + y^2 + z^2 = c_3$$

where c_1, c_2, c_3 are constants. Therefore the general solution is given by

$$u(x, y, z) = f(x + y + z, x^2 + y^2 + z^2)$$

where f is an arbitrary differentiable function.

Problem 8. Consider the initial value problem

$$\frac{\partial^2 u}{\partial x^2} + \frac{\partial^2 u}{\partial t^2} = 0, \qquad -\infty < x < +\infty, \quad t \geq 0$$
$$u(x, 0) = f(x), \qquad -\infty < x < +\infty$$
$$\frac{\partial u(x, 0)}{\partial x} = g(x), \qquad -\infty < x < +\infty.$$

(i) Find the solution of the initial problem for the case $f(x) = f_1(x) = 0$ and $g(x) = g_1(x) = 0$.
(ii) Find the solution of the initial value problem for the case $f(x) = f_2(x) = 0$ and $g(x) = g_2(x) = \frac{1}{n}\sin(nx)$, where $n = 1, 3, \ldots$.
(iii) Compare the two solutions from (i) and (ii) for $x = \pi/2$ and $n \to \infty$.

Solution 8. (i) Obviously we have $u(x, t) = u_1(x, t) = 0$.

(ii) Since

$$\frac{\partial^2}{\partial x^2} \sin(nx) = -n^2 \sin(nx), \qquad \frac{\partial^2}{\partial t^2} \sinh(nt) = n^2 \sinh(nt)$$

we find

$$u(x,t) = u_2(x,t) = \frac{1}{n^2} \sinh(nt) \sin(nx)$$

with $u_2(x,0) = 0$ and $\partial u_2(x,0)/\partial t = \frac{1}{n} \sin(nx)$.

(iii) We have $f_1(x) = f_2(x) = 0$ and

$$\lim_{n \to \infty} |g_1(x) - g_2(x)| = 0$$

uniformly in x since $|(1/n)\sin(nx)| \leq 1/n$. Thus the data of the two problems, f_1, g_1 and f_2, g_2 can be made arbitrarily close for $n \to \infty$. However, if we compare the two solutions at $x = \pi/2$, we obtain

$$|u_1(\pi/2, t) - u_2(\pi/2, t)| = |u_2(\pi/2, t)| = \frac{\sinh(nt)}{n^2} \equiv \frac{e^{nt} - e^{-nt}}{2n^2}$$

since $\sin(n\pi/2) = -(-1)^n$. For t positive, e^{nt} approaches infinity faster than n^2, as n tends to infinity. Therefore

$$\lim_{n \to \infty} |u_1(\pi/2, t) - u_2(\pi/2, t)| = \infty.$$

Therefore as the data for the two problems become more alike, the solution become inreasingly different. This is called the failure to depend continuously on the problem data.

Problem 9.   Consider an electron of mass m confined to the $x - y$ plane and a constant magnetic field **B** parallel to the z-axis, i.e.

$$\mathbf{B} = \begin{pmatrix} 0 \\ 0 \\ B \end{pmatrix}.$$

The Hamilton operator for this two-dimensional electron is given by

$$\hat{H} = \frac{(\hat{\mathbf{p}} + e\mathbf{A})^2}{2m} = \frac{1}{2m}((\hat{p}_x + eA_x)^2 + (\hat{p}_y + eA_y)^2)$$

where **A** is the vector potential with

$$\mathbf{B} = \nabla \times \mathbf{A}$$

and

$$\hat{p}_x = -i\hbar \frac{\partial}{\partial x}, \qquad \hat{p}_y = -i\hbar \frac{\partial}{\partial y}.$$

(i) Show that **B** can be obtained from the vector potential

$$\mathbf{A} = \begin{pmatrix} 0 \\ xB \\ 0 \end{pmatrix} \quad \text{or} \quad \mathbf{A} = \begin{pmatrix} -yB \\ 0 \\ 0 \end{pmatrix}.$$

(ii) Use the second choice for **A** to find the Hamilton operator $\hat{H}$.

(iii) Show that

$$[\hat{H}, \hat{p}_x] = 0.$$

(iv) Let $k = p_x/\hbar$. Make the ansatz for the wave function

$$\psi(x, y) = e^{ikx} \phi(y)$$

and show that the eigenvalue equation $\hat{H}\psi = E\psi$ reduces to

$$\left(-\frac{\hbar^2}{2m}\frac{d^2}{dy^2} + \frac{m\omega_c^2}{2}(y - y_0)^2 \right) \phi(y) = E\phi(y)$$

where

$$\omega_c := \frac{eB}{m}, \qquad y_0 := \frac{\hbar k}{eB}.$$

(v) Show that the eigenvalues are given by

$$E_n = \left(n + \frac{1}{2} \right)\hbar\omega_c, \qquad n = 0, 1, 2, \ldots.$$

Solution 9.   (i) Since

$$\nabla \times \mathbf{A} := \begin{pmatrix} \partial A_z/\partial y - \partial A_y/\partial z \\ \partial A_x/\partial z - \partial A_z/\partial x \\ \partial A_y/\partial x - \partial A_x/\partial y \end{pmatrix}$$

we obtain the desired result.

(ii) Inserting $A_x = -yB$, $A_y = 0$, $A_z = 0$ into the Hamilton operator provides

$$\hat{H} = \frac{1}{2m}(\hat{p}_x - eyB)^2 + \frac{1}{2m}\hat{p}_y^2.$$

(iii) Since coordinate x does not appear in the Hamilton operator $\hat{H}$ and

$$[\hat{y}, \hat{p}_x] = [\hat{p}_y, \hat{p}_x] = 0$$

it follows that $[\hat{H}, \hat{p}_x] = 0$.

(iv) From (iii) we have

$$\hat{p}_x\psi(x, y) \equiv -i\hbar\frac{\partial}{\partial x}\psi(x, y) = \hbar k\psi(x, y), \qquad \hat{H}\psi(x, y) = E\psi(x, y).$$

Inserting the ansatz $\psi(x, y) = e^{ikx}\psi(y)$ into the first equation we find

$$\psi(x, y) = e^{ikx}\phi(y).$$

For $\hat{H}\psi$ we obtain

$$\hat{H}\psi = \frac{1}{2m}((\hat{p}_x - eyB)^2 + \hat{p}_y)e^{ikx}\phi(y)$$

$$= \frac{1}{2m}e^{ikx}((\hbar k - eyB)^2 + \hat{p}_y^2)\phi(y)$$

$$= \frac{1}{2m}e^{ikx}(e^2B^2(y - \hbar k/eB)^2 + \hat{p}_y^2)\phi(y)$$

$$= e^{ikx}\left(\frac{m\omega_c^2}{2}(y - y_0)^2 + \frac{\hat{p}_y^2}{2m}\right)\phi(y).$$

Now the right-hand side must be equal to $E\psi(x, y) = Ee^{ikx}\phi(y)$. Thus since

$$\hat{p}_y^2 = -\hbar^2 \frac{\partial^2}{\partial y^2}$$

the second order ordinary differential equation follows

$$\left(-\frac{\hbar^2}{2m}\frac{d^2}{dy^2} + \frac{m\omega_c^2}{2}(y - y_0)^2\right)\phi(y) = E\phi(y).$$

(v) The eigenvalue problem of (iv) is essentially the one-dimensional harmonic oscillator except the term is $(y - y_0)^2$ instead of y^2. This means that the centre of oscillation is at $y = y_0$ instead of 0. This has no influence on the eigenvalues which are the same as for the harmonic oscillator, namely

$$E_n = \left(n + \frac{1}{2}\right)\hbar\omega_c, \quad n = 0, 1, 2, \ldots.$$

Problem 10. Consider the eigenvalue equation

$$-\frac{d^2\psi}{dx^2} + V(x)\psi(x) = E\psi(x). \tag{1}$$

(i) Let the potential V be given by

$$V(x) = a^2x^6 - 3ax^2 \tag{2}$$

where $a > 0$. Show that for this potential the ground-state wave function ψ_0 is

$$\psi_0(x) = \exp(-ax^4/4). \tag{3}$$

(ii) What happens when we analytically continue the eigenvalue $E_0(a)$ from positive to negative values of the parameter a?

Solution 10.   (i) Inserting (3) into (1) yields the energy eigenvalue

$$E(a) = 0.$$

The eigenfunction corresponds to the lowest-lying energy eigenvalue because it has no nodes.

(ii) When we analytically continue the function $E_0(a)$ from positive to negative values, the result is still $E_0(a) = 0$. However, this conclusion is wrong. The spectrum of the potential (2) for $a < 0$ is strictly positive. When we continue the wave function ψ_0 to a negative value of a we find a nonnormalizable solution of the eigenvalue equation (1), which indicates that a simple replacement of a by $-a$ is not allowed.

Problem 11.   Consider the *Dirac-Hamilton operator*

$$\hat{H} = mc^2\beta + c\boldsymbol{\alpha}\cdot\mathbf{p} \equiv mc^2\beta + c(\alpha_1 p_1 + \alpha_2 p_2 + \alpha_3 p_3) \tag{1}$$

where m is the rest mass, c is the speed of light and

$$\beta := \begin{pmatrix} 1 & 0 & 0 & 0 \\ 0 & 1 & 0 & 0 \\ 0 & 0 & -1 & 0 \\ 0 & 0 & 0 & -1 \end{pmatrix}, \quad \alpha_1 := \begin{pmatrix} 0 & 0 & 0 & 1 \\ 0 & 0 & 1 & 0 \\ 0 & 1 & 0 & 0 \\ 1 & 0 & 0 & 0 \end{pmatrix},$$

$$\alpha_2 := \begin{pmatrix} 0 & 0 & 0 & -i \\ 0 & 0 & i & 0 \\ 0 & -i & 0 & 0 \\ i & 0 & 0 & 0 \end{pmatrix}, \quad \alpha_3 := \begin{pmatrix} 0 & 0 & 1 & 0 \\ 0 & 0 & 0 & -1 \\ 1 & 0 & 0 & 0 \\ 0 & -1 & 0 & 0 \end{pmatrix},$$

$$\hat{p}_1 := -i\hbar\frac{\partial}{\partial x_1}, \quad \hat{p}_2 := -i\hbar\frac{\partial}{\partial x_2}, \quad \hat{p}_3 := -i\hbar\frac{\partial}{\partial x_3}.$$

Thus the Dirac-Hamilton operator takes the form

$$\hat{H} = c\hbar \begin{pmatrix} mc/\hbar & 0 & -i\frac{\partial}{\partial x_3} & -i\frac{\partial}{\partial x_1} - \frac{\partial}{\partial x_2} \\ 0 & mc/\hbar & -i\frac{\partial}{\partial x_1} + \frac{\partial}{\partial x_2} & i\frac{\partial}{\partial x_3} \\ -i\frac{\partial}{\partial x_3} & -i\frac{\partial}{\partial x_1} - \frac{\partial}{\partial x_2} & -mc/\hbar & 0 \\ -i\frac{\partial}{\partial x_1} + \frac{\partial}{\partial x_2} & i\frac{\partial}{\partial x_3} & 0 & -mc/\hbar \end{pmatrix}.$$

Let I_4 be the 4×4 unit matrix. Use the *Heisenberg equation of motion* to find the time evolution of β, α_j and $I_4 p_j$, where $j = 1, 2, 3$. The *Heisenberg equation of motion* is given by

$$i\hbar\frac{d\hat{A}}{dt} = [\hat{A}, \hat{H}](t).$$

Solution 11. We find for the commutator

$$[\beta, \hat{H}] = 2c\hbar \begin{pmatrix} 0 & 0 & -i\frac{\partial}{\partial x_3} & -i\frac{\partial}{\partial x_1} - \frac{\partial}{\partial x_2} \\ 0 & 0 & -i\frac{\partial}{\partial x_1} + \frac{\partial}{\partial x_2} & i\frac{\partial}{\partial x_3} \\ i\frac{\partial}{\partial x_3} & i\frac{\partial}{\partial x_1} + \frac{\partial}{\partial x_2} & 0 & 0 \\ i\frac{\partial}{\partial x_1} - \frac{\partial}{\partial x_2} & -i\frac{\partial}{\partial x_3} & 0 & 0 \end{pmatrix}.$$

Thus the Heisenberg equation of motion takes the form

$$i\hbar\frac{d\beta}{dt} = 2c\beta\alpha_1(p_1 - ip_2)(t) + 2c\alpha_3 p_3(t).$$

Since

$$[\alpha_1, \hat{H}] = -2\hat{H}\alpha_1 + 2cI_4 p_1$$

we obtain

$$i\hbar\frac{d\alpha_1}{dt} = -(2\hat{H}\alpha_1 + 2cI_4 p_1)(t).$$

Analogously,

$$i\hbar\frac{d\alpha_2}{dt} = -(2\hat{H}\alpha_2 + 2cI_4 p_2)(t)$$

$$i\hbar\frac{d\alpha_3}{dt} = -(2\hat{H}\alpha_3 + 2cI_4 p_3)(t).$$

Furthermore we obviously have

$$[I_4 p_j, \hat{H}] = 0, \qquad j = 1, 2, 3.$$

Thus

$$i\hbar\frac{d}{dt}I_4 p_j = 0, \qquad j = 1, 2, 3.$$

From these equations we obtain

$$i\hbar\frac{d^2\alpha}{dt^2} = 2\frac{d\alpha}{dt}\hat{H} = -2\hat{H}\frac{d\alpha}{dt}.$$

This equation can be integrated once and we find

$$\frac{d\alpha}{dt} = \frac{d\alpha(0)}{dt}\exp(-2i\hat{H}t/\hbar).$$

We also have the identities

$$[\hat{H}, \dot{\alpha}]_+ = 0, \qquad [\hat{H}, \dot{\alpha}] = 2\hat{H}\alpha.$$

Chapter 12

Nonlinear Partial Differential Equations

Problem 1. The nonlinear partial differential equation

$$\frac{\partial \mathbf{S}}{\partial t} = \mathbf{S} \times \frac{\partial^2 \mathbf{S}}{\partial x^2} \tag{1}$$

where (constraint) $S_1^2 + S_2^2 + S_3^2 = 1$ is called the *Heisenberg ferromagnet equation* in one-space dimension. Here $\times$ denotes the vector product and

$$\mathbf{S} = \begin{pmatrix} S_1 \\ S_2 \\ S_3 \end{pmatrix}.$$

(i) Write equation (1) in components.
(ii) Let

$$Q := 1 + u^2 + v^2$$

and set

$$S_1 := \frac{2u}{Q}, \qquad S_2 := \frac{2v}{Q}, \qquad S_3 := \frac{-1 + u^2 + v^2}{Q}. \tag{2}$$

Show that constraint (2) is satisfied identically. Find the time evolution of u and v. This transformation is called *stereographic projection*.

Solution 1. (i) Since the *vector product* is given by

$$\mathbf{a} \times \mathbf{b} := \begin{pmatrix} a_2 b_3 - a_3 b_2 \\ a_3 b_1 - a_1 b_3 \\ a_1 b_2 - a_2 b_1 \end{pmatrix}$$

137

we have

$$\frac{\partial S_1}{\partial t} = S_2 \frac{\partial^2 S_3}{\partial x^2} - S_3 \frac{\partial^2 S_2}{\partial x^2} \tag{3a}$$

$$\frac{\partial S_2}{\partial t} = S_3 \frac{\partial^2 S_1}{\partial x^2} - S_1 \frac{\partial^2 S_3}{\partial x^2} \tag{3b}$$

$$\frac{\partial S_3}{\partial t} = S_1 \frac{\partial^2 S_2}{\partial x^2} - S_2 \frac{\partial^2 S_1}{\partial x^2}. \tag{3c}$$

(ii) From (2) we find

$$\frac{\partial S_1}{\partial t} = \frac{2}{Q} \frac{\partial u}{\partial t} - \frac{2u}{Q^2} \frac{\partial Q}{\partial t} = \frac{1}{Q^2} \left(2Q \frac{\partial u}{\partial t} - 2u \frac{\partial Q}{\partial t} \right) \tag{4a}$$

$$\frac{\partial S_2}{\partial t} = \frac{2}{Q} \frac{\partial v}{\partial t} - \frac{2v}{Q^2} \frac{\partial Q}{\partial t} = \frac{1}{Q^2} \left(2Q \frac{\partial v}{\partial t} - 2v \frac{\partial Q}{\partial t} \right). \tag{4b}$$

Since

$$S_3 = \frac{-1 + u^2 + v^2}{Q} \equiv \frac{Q - 2}{Q} = 1 - \frac{2}{Q}$$

we have

$$\frac{\partial S_3}{\partial t} = \frac{2}{Q^2} \frac{\partial Q}{\partial t}. \tag{4c}$$

Now

$$\frac{\partial^2 S_1}{\partial x^2} = \frac{2}{Q} \frac{\partial^2 u}{\partial x^2} - \frac{4}{Q^2} \frac{\partial u}{\partial x} \frac{\partial Q}{\partial x} - \frac{2u}{Q^2} \frac{\partial^2 Q}{\partial x^2} + \frac{4u}{Q^3} \left(\frac{\partial Q}{\partial x} \right)^2 \tag{5a}$$

$$\frac{\partial^2 S_2}{\partial x^2} = \frac{2}{Q} \frac{\partial^2 v}{\partial x^2} - \frac{4}{Q^2} \frac{\partial v}{\partial x} \frac{\partial Q}{\partial x} - \frac{2v}{Q^2} \frac{\partial^2 Q}{\partial x^2} + \frac{4v}{Q^3} \left(\frac{\partial Q}{\partial x} \right)^2 \tag{5b}$$

and

$$\frac{\partial^2 S_3}{\partial x^2} = \frac{2}{Q^2} \frac{\partial^2 Q}{\partial x^2} - \frac{4}{Q^3} \left(\frac{\partial Q}{\partial x} \right)^2. \tag{5c}$$

Inserting (4) through (5) into (3c) yields

$$\frac{2}{Q^2} \frac{\partial Q}{\partial t} = \frac{2u}{Q} \left(\frac{2}{Q} \frac{\partial^2 v}{\partial x^2} - \frac{4}{Q^2} \frac{\partial v}{\partial x} \frac{\partial Q}{\partial x} - \frac{2v}{Q^2} \frac{\partial^2 Q}{\partial x^2} + \frac{4v}{Q^3} \left(\frac{\partial Q}{\partial x} \right)^2 \right)$$

$$- \frac{2v}{Q} \left(\frac{2}{Q} \frac{\partial^2 u}{\partial x^2} - \frac{4}{Q^2} \frac{\partial u}{\partial x} \frac{\partial Q}{\partial x} - \frac{2u}{Q^2} \frac{\partial^2 Q}{\partial x^2} + \frac{4u}{Q^3} \left(\frac{\partial Q}{\partial x} \right)^2 \right)$$

or

$$\frac{\partial Q}{\partial t} = u \left(2 \frac{\partial^2 v}{\partial x^2} - \frac{4}{Q} \frac{\partial v}{\partial x} \frac{\partial Q}{\partial x} \right) - v \left(2 \frac{\partial^2 u}{\partial x^2} - \frac{4}{Q} \frac{\partial u}{\partial x} \frac{\partial Q}{\partial x} \right).$$

Inserting (4) through (5) and this equation into (3a) and (3b) yields

$$Q \frac{\partial v}{\partial t} + Q \frac{\partial^2 u}{\partial x^2} - 2 \left(\left(\frac{\partial u}{\partial x} \right)^2 - \left(\frac{\partial v}{\partial x} \right)^2 \right) u - 4v \frac{\partial u}{\partial x} \frac{\partial v}{\partial x} = 0$$

$$-Q\frac{\partial u}{\partial t} + Q\frac{\partial^2 v}{\partial x^2} + 2\left(\left(\frac{\partial u}{\partial x}\right)^2 - \left(\frac{\partial v}{\partial x}\right)^2\right)v - 4u\frac{\partial u}{\partial x}\frac{\partial v}{\partial x} = 0.$$

Problem 2. (i) Solve the partial differential equation

$$F\left(u, \frac{\partial u}{\partial t}, \frac{\partial u}{\partial x}\right) \equiv \frac{\partial u}{\partial t} + u\frac{\partial u}{\partial x} = 0 \tag{1}$$

with the initial condition

$$u(x, t = 0) = x. \tag{2}$$

(ii) Show that the nonlinear partial differential equation

$$\frac{\partial u}{\partial t} + (\alpha + \beta u)\frac{\partial u}{\partial x} = 0 \tag{3}$$

admits the general solution

$$u(x, t) = f(x - (\alpha + \beta u(x, t))t) \tag{4}$$

to the initial-value problem, where $u(x, t = 0) = f(x)$ and $\alpha, \beta \in \mathbb{R}$.
(iii) Simplify case (ii) to case (i).

Solution 2. (i) From (1) we obtain the surface

$$F(u, p, q) \equiv p + uq = 0.$$

Then we obtain the autonomous system of first-order ordinary differential equations

$$\frac{dt}{ds} = \frac{\partial F}{\partial p} = 1$$

$$\frac{dx}{ds} = \frac{\partial F}{\partial q} = u$$

$$\frac{du}{ds} = p\frac{\partial F}{\partial p} + q\frac{\partial F}{\partial q} = p + uq = 0$$

where we have used $F(u, p, q) \equiv p + uq$. These equations determine the *characteristic strip*. The solution of the initial-value problem of this autonomous system of differential equations is given by

$$x(s) = u_0 s + x_0, \quad t(s) = s + t_0, \quad u(s) = u_0.$$

To impose the initial condition (2) we have to set

$$x_0(\alpha) = \alpha, \quad t_0(\alpha) = 0, \quad u_0(\alpha) = \alpha.$$

Thus we obtain as solution of the autonomous system

$$x(s, \alpha) = \alpha s + \alpha, \quad t(s, \alpha) = s, \quad u(s, \alpha) = \alpha.$$

Since

$$D := \frac{\partial(t, x)}{\partial(s, \alpha)} = 1 + s$$

we can solve these equations with respect to s and α if $D \neq 0$. We find

$$s(x, t) = t, \quad \alpha(x, t) = \frac{x}{1 + t}.$$

Inserting s and α into $u(s, \alpha) = \alpha$ gives the solution $u(x, t) = x/(1 + t)$ of the initial-value problem.

(ii) Since

$$\frac{\partial u}{\partial t} = \left(-\alpha - \beta u - \beta t \frac{\partial u}{\partial t} \right) f', \qquad \frac{\partial u}{\partial x} = \left(1 - \beta t \frac{\partial u}{\partial x} \right) f'$$

where f' is the derivative of f with respect to the argument we obtain

$$\left(1 - \beta t \frac{\partial u}{\partial x} \right) \frac{\partial u}{\partial t} = \left(-\alpha - \beta u - \beta t \frac{\partial u}{\partial t} \right) \frac{\partial u}{\partial x}$$

or

$$\frac{\partial u}{\partial t} = (-\alpha - \beta u) \frac{\partial u}{\partial x}$$

which is (3).

(iii) Let us now simplify (3) to case (i). Let $\alpha = 0$ and $\beta = 1$. Then we obtain from (4) that

$$u(x, t) = f(x - u(x, t)t).$$

Since $u(x, t = 0) = f(x) = x$ it follows that

$$u(x, t) = f(x - u(x, t)t) = x - u(x, t)t.$$

From this solution we see that the solution $u(x, t) = x/(1 + t)$ follows.

Problem 3.   (i) Show that the nonlinear partial differential equation

$$\frac{\partial^2 u}{\partial t^2} - \frac{\partial^2 u}{\partial x^2} = \mu^2 u - \lambda u^3 \tag{1}$$

can be derived from the *Lagrange density*

$$\mathcal{L} = \frac{1}{2} \left(\left(\frac{\partial u}{\partial t} \right)^2 - \left(\frac{\partial u}{\partial x} \right)^2 \right) + \frac{1}{2} \mu^2 u^2 - \frac{1}{4} \lambda u^4 \tag{2}$$

where $\mu^2 > 0$ and $\lambda > 0$. From the Lagrange density $\mathcal{L}$ it follows that we can introduce the Hamilton density and therefore the *energy functional*

$$E(u) = \int_{-\infty}^{+\infty} dx \left(\frac{1}{2}\left(\frac{\partial u}{\partial t}\right)^2 + \frac{1}{2}\left(\frac{\partial u}{\partial x}\right)^2 - \frac{1}{2}\mu^2 u^2 + \frac{1}{4}\lambda u^4 \right).$$

(ii) Find the space-time independent solutions of (1).
(iii) Calculate the energy difference between the different solutions of (i).
(iv) Show that (1) admits the time-independent solution

$$u_{\text{kink}}(x) = \pm\frac{\mu}{\sqrt{\lambda}} \tanh\left(\frac{\mu}{\sqrt{2}}(x - x_0)\right) \tag{3}$$

and calculate the energy difference. Owing to its shape, the solution (4) is called the *kink solution*.

Solution 3. (i) The *Euler-Lagrange equation* is given by

$$\frac{\partial \mathcal{L}}{\partial u} - \frac{\partial}{\partial x}\left(\frac{\partial \mathcal{L}}{\partial \left(\frac{\partial u}{\partial x}\right)}\right) - \frac{\partial}{\partial t}\left(\frac{\partial \mathcal{L}}{\partial \left(\frac{\partial u}{\partial t}\right)}\right) = 0.$$

Inserting the Lagrange density (2) into (4) gives (1).
(ii) The space-time independent solutions are determined by

$$\mu^2 u - \lambda u^3 = 0.$$

We obtain as solution $u = 0$, $u = \pm\mu/\sqrt{\lambda}$.
(iii) Straightforward calculation shows that

$$E(u = 0) - E\left(u = \pm\frac{\mu}{\sqrt{\lambda}}\right) = \lim_{L\to\infty} \int_{-L}^{+L} dx \frac{1}{4}\frac{\mu^4}{\lambda} = +\infty$$

which implies that the difference between the energy densities is positive and moreover that $u = \pm\mu/\sqrt{\lambda}$ are the minima ($u = 0$ is a maximum). The set of vacua is degenerate.
(iv) If u is independent of t we obtain from (1)

$$\frac{d^2 u}{dx^2} + \mu^2 u - \lambda u^3 = 0.$$

Consequently integration yields

$$\frac{1}{2}\left(\frac{du}{dx}\right)^2 + \mu^2\frac{u^2}{2} - \frac{\lambda u^4}{4} = \text{const.}$$

It follows that

$$x - x_0 = \int_0^u \frac{du'}{\sqrt{\lambda u'^4/2 - \mu^2 u'^2 + k}}, \qquad x_0, k = \text{const}$$

which represents an *elliptic integral*. Therefore $u(x)$ is periodic on the complex x-plane. Hence $E[u]$ is of infinite energy, unless the two zeroes of the square root coalesce. This happens for $k = \mu^4/2\lambda$. Then we obtain the solution (3). Straightforward calculation gives

$$E(u_{\text{kink}}) - E(u_{\text{vac}}) = \frac{2\sqrt{2}}{3}\frac{\mu^3}{\lambda} < +\infty$$

where $u_{\text{vac}} = \pm\mu/\sqrt{\lambda}$. A space-time dependent solution can be found from (3) by applying the Lorentz transformation.

Problem 4.   The *Navier-Stokes equation* of a viscous incompressible fluid is given by

$$\left(\frac{\partial}{\partial t} + \mathbf{u} \cdot \nabla\right)\mathbf{u} = -\frac{1}{\rho}\nabla p + \nu\Delta\mathbf{u}. \tag{1}$$

We assume that the fluid is incompressible, i.e.

$$\nabla \cdot \mathbf{u} = \mathbf{0}. \tag{2}$$

(i) Express the Navier-Stokes equation in cylindrical coordinates r, ϕ, z.
(ii) Consider a flow between concentric rotating cylinders. The cylinders are infinitely long. Let r_1, r_2 and Ω_1, Ω_2 denote the radii and angular velocities of the inner and outer cylinders, respectively. We denote the velocity components in the increasing r, ϕ and z directions by u, v and w. The boundary conditions are

$$u = w = 0 \quad \text{at} \quad r = r_1 \quad \text{and} \quad r = r_2$$
$$v(r_1, \phi, z, t) = r_1\Omega_1$$
$$v(r_2, \phi, z, t) = r_2\Omega_2.$$

These conditions are called *no-slip conditions*. Find an exact solution of the form
$$u = w = 0, \qquad v = V(r).$$

Solution 4.   Writing the Navier-Stokes equation in cylindrical coordinates

$$x(\phi, r) = r\cos\phi, \quad y(\phi, r) = r\sin\phi, \quad z = z$$

yields

$$\left(\frac{\partial}{\partial t} + \mathbf{u} \cdot \nabla\right)u - \frac{v^2}{r} = -\frac{\partial}{\partial r}\frac{p}{\rho} + \nu\left(\Delta - \frac{1}{r^2}\right)u - \frac{2\nu}{r^2}\frac{\partial}{\partial\phi}v$$

$$\left(\frac{\partial}{\partial t} + \mathbf{u} \cdot \nabla\right)v + \frac{uv}{r} = -\frac{1}{r}\frac{\partial}{\partial\phi}\frac{p}{\rho} + \nu\left(\Delta - \frac{1}{r^2}\right)v + \frac{2\nu}{r^2}\frac{\partial}{\partial\phi}u$$

$$\left(\frac{\partial}{\partial t} + \mathbf{u} \cdot \nabla\right) w = -\frac{\partial}{\partial z}\frac{p}{\rho} + \nu \Delta w$$

$$\left(\frac{\partial}{\partial r} + \frac{1}{r}\right) u = -\frac{1}{r}\frac{\partial}{\partial \phi}v - \frac{\partial}{\partial z}w$$

where

$$\mathbf{u} \cdot \nabla := u\frac{\partial}{\partial r} + \frac{1}{r}v\frac{\partial}{\partial \phi} + w\frac{\partial}{\partial z}, \quad \Delta := \frac{\partial^2}{\partial r^2} + \frac{1}{r}\frac{\partial}{\partial r} + \frac{1}{r^2}\frac{\partial^2}{\partial \phi^2} + \frac{\partial^2}{\partial z^2}.$$

Inserting $u = w = 0$ and $v = V(r)$ into the first equation yields

$$-\frac{V^2(r)}{r} = -\frac{1}{\rho}\frac{\partial p}{\partial r}.$$

Thus we can assume that the pressure p depends only on r. The third equation is satisfied identically. For the second equation we find

$$\nu\left(\frac{\partial^2}{\partial r^2} + \frac{1}{r}\frac{\partial}{\partial r} - \frac{1}{r^2}\right)V(r) = 0.$$

Therefore the solution is given by

$$V(r) = Ar + \frac{B}{r}$$

where A and B are the constants of integration. Imposing the boundary conditions gives

$$A = \frac{\Omega_2 r_2^2 - \Omega_1 r_1^2}{r_2^2 - r_1^2}, \qquad B = \frac{r_1^2 r_2^2 (\Omega_1 - \Omega_2)}{r_2^2 - r_1^2}.$$

Problem 5. Consider the nonlinear diffusion equation

$$\frac{\partial u}{\partial t} = D\frac{\partial}{\partial x}\left(\frac{1}{u}\frac{\partial u}{\partial x}\right)$$

where $u(x,t)$ is the density and x, t are the space, time coordinates and $x \in [0,1]$. D is a positive constant. We set at the boundaries

$$u(0,t) = u(1,t) = u_0 \geq 0.$$

We assume that $u(x,0) \geq u_0$.
(i) Show that the problem is well-posed.
(ii) Introduce the new independent and dependent variables

$$\tau(t,x) = \frac{D}{u_0}t, \qquad y(t,x) = x \tag{1a}$$

$$v(\tau(t,x), y(t,x)) = \ln(u(x,t)/u_0).$$
(1b)

Find the partial differential equation for these variables.
(iii) How do the boundary conditions change?

Solution 5.   (i) Since

$$\frac{d}{dt} \int_0^1 u(x,t)dx = D \left(\frac{1}{u} \frac{\partial u}{\partial x} \right) \Big|_0^1$$
(2)

the flux (i.e. the right-hand side of (2)) will be finite and the problem is well posed.
(ii) From (1b) we find

$$u(x,t) = u_0 \exp(v(\tau(t,x), y(t,x))).$$
(3)

Using (3) and (1b) we find, by applying the chain rule,

$$\frac{\partial^2 v}{\partial y^2} = \exp(v) \frac{\partial v}{\partial \tau} = \frac{\partial}{\partial \tau} \exp(v).$$

(iii) The boundary conditions change to

$$v(0,t) = v(1,t) = 0.$$

The quantity v is nonnegative and the new time scale differs from the old by a factor of u_0/D.

Problem 6.   Consider the complex *Ginzburg-Landau equation*

$$\frac{\partial A}{\partial t} = (1 + ic_1) \frac{\partial^2 A}{\partial x^2} + A - (-1 + ic_2)|A|^2 A.$$
(1)

Find a solution of the form

$$A(x,t) = R(x,t) \exp(i\Theta(x,t))$$
(2)

where R and Θ are real-valued functions.

Solution 6.   Inserting (2) into (1) we obtain two real equations

$$\frac{1}{c_1^2} \frac{\partial R}{\partial t} = \frac{1}{c_1^2} \frac{\partial^2 R}{\partial x^2} - \frac{1}{c_1^2} R \left(\frac{\partial \Theta}{\partial x} \right)^2 - \frac{2}{c_1} \frac{\partial R}{\partial x} \frac{\partial \Theta}{\partial x} - \frac{1}{c_1} R \frac{\partial^2 \Theta}{\partial x^2} + \frac{1}{c_1^2} R + \frac{1}{c_1^2} R^3$$

$$\frac{1}{c_1} R \frac{\partial \Theta}{\partial t} = \frac{2}{c_1} \frac{\partial R}{\partial x} \frac{\partial \Theta}{\partial x} + \frac{1}{c_1} R \frac{\partial^2 \Theta}{\partial x^2} + \frac{\partial^2 R}{\partial x^2} - R \left(\frac{\partial \Theta}{\partial x} \right)^2 - \frac{c_2}{c_1} R^3.$$

Adding the two equation provides

$$\epsilon^2 \frac{\partial R}{\partial t} + \epsilon R \frac{\partial \Theta}{\partial t} = (1 + \epsilon^2) \left(\frac{\partial^2 R}{\partial x^2} - \left(\frac{\partial \Theta}{\partial x} \right)^2 R \right) + \epsilon^2 R + (\beta + \epsilon^2) R^3$$

where $c_2 = -\beta c_1$ and $\epsilon = 1/c_1$. We also obtain

$$-\frac{1}{2} \epsilon \frac{\partial R^2}{\partial t} + \epsilon^2 R^2 \frac{\partial \Theta}{\partial t} = (1 + \epsilon^2) \frac{\partial}{\partial x} \left(R^2 \frac{\partial \Theta}{\partial x} \right) - \epsilon (1 + (1 - \beta) R^2) R^2 .$$

For sufficiently large values of c_1 an expansion in ϵ becomes meaningful

$$R = R_0 + \epsilon^2 R_2 + \cdots, \qquad \Theta = \frac{1}{\epsilon} (\Theta_{-1} + \epsilon^2 \Theta_1 + \cdots) .$$

For the order ϵ^{-2} we find

$$R_0 \left(\frac{\partial \Theta_{-1}}{\partial x} \right)^2 = 0 .$$

For the order ϵ^{-1} we find

$$\frac{\partial}{\partial x} \left(R_0^2 \frac{\partial \Theta_{-1}}{\partial x} \right) = 0 .$$

Excluding $R_0 = 0$ we obtain $\partial \Theta_{-1}/\partial x = 0$, so that Θ_{-1} only depends on t. Setting

$$\gamma(t) := \frac{\partial \Theta_{-1}}{\partial t}$$

we obtain for the ϵ^0

$$0 = \frac{\partial^2 R_0}{\partial x^2} - \gamma(t) R_0 + \beta R_0^3 . \tag{3a}$$

For ϵ^1 we find

$$-\frac{\partial}{\partial x} \left(R_0^2 \frac{\partial \Theta_1}{\partial x} \right) = \frac{1}{2} \frac{\partial R_0^2}{\partial t} - (1 + \gamma(t)) R_0^2 - (1 - \beta) R_0^4 . \tag{3b}$$

For $\beta, \gamma > 0$ we obtain from (3a) the spatially periodic solution

$$R_0(x, t) = \left(\frac{2\gamma(t)}{(2 - m(t))\beta} \right)^{1/2} \mathrm{dn} \left(\left(\frac{\gamma(t)}{2 - m(t)} \right)^{1/2} x, m(t) \right) . \tag{4}$$

Here $\mathrm{dn}(u, m)$ is a Jacobian elliptic function that varies between $(1 - m)$ and 1 with period $2K(m)$. The parameter m is between 0 and 1, and $K(m)$

is the complete elliptic integral of the first kind. For $m \to 1$ the period of the function dn goes to infinity and (3) degenerates into the pulse

$$(2\gamma/\beta)^{1/2}\text{sech}(\gamma^{1/2}x)$$

while for $m \to 0$ one has small, harmonic oscillations.

Problem 7. In *plasma physics* the differential equations for the density of the ion-fluid n and its velocity u, in dimensionless form, is given by

$$\frac{\partial n}{\partial t} + \frac{\partial(nu)}{\partial x} = 0 \qquad\qquad \text{equation of continuity}$$

$$\frac{\partial u}{\partial t} + u\frac{\partial u}{\partial x} = E \qquad\qquad \text{equation of motion}$$

$$\frac{\partial n_e}{\partial x} = -n_e E \qquad\qquad \text{balance of pressure and electric force}$$

$$\frac{\partial E}{\partial x} = n - n_e \qquad\qquad \text{Poisson equation}$$

where n_e is the electron density, E the electric field, and the inertia term is neglected because of the small mass of electrons.
(i) Eliminate n and E to find the differential equations for u and n_e.
(ii) Introduce

$$\xi := \epsilon^{1/2}(x - t), \qquad \eta := \epsilon^{3/2}x$$

and find a solution of the form

$$u = \epsilon u^{(1)} + \epsilon^2 u^{(2)} + \cdots, \qquad n_e = 1 + \epsilon n_e^{(1)} + \epsilon^2 n_e^{(2)} + \cdots . \qquad (1)$$

Solution 7. (i) Eliminating n and E yields

$$\frac{\partial u}{\partial t} + u\frac{\partial u}{\partial x} + \frac{1}{n_e}\frac{\partial n_e}{\partial x} = 0 \qquad\qquad (2a)$$

$$\frac{\partial n_e}{\partial t} + \frac{\partial(n_e u)}{\partial x} + \frac{\partial P}{\partial x} = 0 \qquad\qquad (2b)$$

where

$$P := -\left(\frac{\partial}{\partial t} + u\frac{\partial}{\partial x}\right)\left(\frac{1}{n_e}\frac{\partial n_e}{\partial x}\right) .$$

(ii) Inserting expansion (1) into (2) we obtain $u^{(1)} = n_e^{(1)}$ and

$$\frac{\partial u^{(1)}}{\partial \eta} + u^{(1)}\frac{\partial u^{(1)}}{\partial \xi} + \frac{\partial^3 u^{(1)}}{\partial \xi^3} = 0, \qquad \frac{\partial n_e^{(1)}}{\partial \eta} + n_e^{(1)}\frac{\partial n_e^{(1)}}{\partial \xi} + \frac{\partial^3 n_e^{(1)}}{\partial \xi^3} = 0.$$

We see that u, the ion-fluid velocity, and n_e, the electron density, obey the same *Korteweg-de Vries equation* and move with the same phase $u^{(1)} = n_e^{(1)}$. The nonlinear term

$$u^{(1)} \frac{\partial u^{(1)}}{\partial \xi}$$

comes from the interaction of the ions with the electrons which affects the ions themselves, and

$$n_e^{(1)} \frac{\partial n_e^{(1)}}{\partial \xi}$$

expresses similar effects on the electrons through the interaction with the ions. We have

$$n^{(1)}(\eta, \xi) = -\int^{\xi} E^{(1)}(\eta, s)ds = u^{(1)}(\eta, \xi) = n_e^{(1)}(\eta, \xi).$$

Problem 8. Consider the *Korteweg-de Vries-Burgers equation*

$$\frac{\partial u}{\partial t} + a_1 u \frac{\partial u}{\partial x} + a_2 \frac{\partial^2 u}{\partial x^2} + a_3 \frac{\partial^3 u}{\partial x^3} = 0 \tag{1}$$

where a_1, a_2 and a_3 are non-zero constants. It contains dispersive, dissipative and nonlinear terms.
(i) Find a solution of the form

$$u(x,t) = \frac{b_1}{(1 + \exp(b_2(x + b_3 t + b_4)))^2} \tag{2}$$

where b_1, b_2, b_3 and b_4 are constants determined by a_1, a_2, a_3 and a_4.
(ii) Study the case $t \to \infty$ and $t \to -\infty$.

Solution 8. (i) This is a typical problem for computer algebra. Inserting the ansatz (2) into (1) yields the conditions

$$b_3 = -(a_1 b_1 + a_2 b_2 + a_3 b_2^2) \tag{3}$$

and

$$(2a_1 b_1 + 3a_2 b_2 + 9a_3 b_2^2) \exp((a_1 b_1 b_2 + a_2 b_2^2 + a_3 b_2^3)t) + (a_1 b_1 + 3a_2 b_2 - 3a_3 b_2^2) = 0.$$

The quantity b_4 is arbitrary. Thus

$$2a_1 b_1 = -3a_2 b_2 - 9a_3 b_2^2, \qquad a_1 b_1 = -3a_2 b_2 + 3a_3 b_2^2. \tag{4}$$

Solving these system of equations with respect to b_1 and b_2 yields

$$b_1 = -\frac{12a_2^2}{25a_1 a_3}, \qquad b_2 = \frac{a_2}{5a_3}.$$

Inserting b_1 and b_2 into (3) gives

$$b_3 = \frac{6}{25} \frac{a_2^2}{a_3}.$$

Thus the solution is

$$u(x,t) = -\frac{12a_2^2}{25a_1a_3} \left(1 + \exp\left(\frac{a_2}{5a_3} \left(x + \frac{6a_2^2}{25a_3}t + b_4 \right) \right) \right)^{-2}.$$

(ii) Two asymptotic values exist which are for $t \to -\infty$, $u_I = 2v_c/a$ and for $t \to \infty$, $u_{II} = 0$.

Problem 9. For a barotropic fluid of index γ the *Navier-Stokes equation* in one space dimension is given by

$$\frac{\partial c}{\partial t} + v\frac{\partial c}{\partial x} + \frac{(\gamma-1)}{2}c\frac{\partial v}{\partial x} = 0 \qquad \text{continuity equation} \qquad (1a)$$

$$\frac{\partial v}{\partial t} + v\frac{\partial v}{\partial x} + \frac{2}{(\gamma-1)}c\frac{\partial c}{\partial x} - \frac{\partial^2 v}{\partial x^2} = 0 \qquad \text{Euler's equation} \qquad (1b)$$

where v represents the fluid's velocity and c the speed of sound. For the class of solutions characterized by a vanishing presure (i.e., $c = 0$), the above system reduces to the Burgers equation. We assume for simplicity the value $\gamma = 3$ in what follows.

(i) Show that the *velocity potential* Φ exists, and it is given by the following pair of equations

$$\frac{\partial \Phi}{\partial x} = v, \qquad \frac{\partial \Phi}{\partial t} = \frac{\partial v}{\partial x} - \frac{1}{2}(v^2 + c^2). \qquad (2)$$

(ii) Consider the *similarity ansatz*

$$v(x,t) = f(s)\frac{x}{t}, \qquad c(x,t) = g(s)\frac{x}{t} \qquad (3)$$

where the *similarity variable* s is given by

$$s(x,t) := \frac{x}{\sqrt{t}}. \qquad (4)$$

Show that the Navier-Stokes equation yields the following system of ordinary differential equations

$$s\frac{df}{ds}g + s\left(f - \frac{1}{2}\right)\frac{dg}{ds} + g(2f-1) = 0$$

$$2\frac{d^2f}{ds^2} + \frac{df}{ds}\left(\frac{4}{s} + s(1-2f)\right) + 2f(1-f) = 2g\left(s\frac{dg}{ds} + g\right).$$

(iii) Show that the continuity equation (1a) admits a first integral, expressing the law of mass conservation, namely

$$s^2 g(2f - 1) = C,$$

where C is a constant. (iv) Show that g can be eliminated, and we obtain a second-order ordinary differential equation for the function f.

Solution 9. (i) From (2) with

$$\frac{\partial^2 \Phi}{\partial t \partial x} = \frac{\partial^2 \Phi}{\partial x \partial t}$$

the Euler equation (1b) follows. Thus the condition of integrability of Φ precisely coincides with the Euler equation (1b). Conversely inserting $\partial \Phi / \partial x = v$ into (1b) yields the expression for $\partial \Phi / \partial t$.
(ii) From (3) we obtain

$$\frac{\partial v}{\partial t} = \frac{df}{ds}\frac{ds}{dt}\frac{x}{t} - f\frac{x}{t^2}, \qquad \frac{\partial v}{\partial x} = \frac{df}{ds}\frac{ds}{dx}\frac{x}{t} + f\frac{1}{t}$$

$$\frac{\partial c}{\partial t} = \frac{dg}{ds}\frac{ds}{dt}\frac{x}{t} - g\frac{x}{t^2}, \qquad \frac{\partial c}{\partial x} = \frac{dg}{ds}\frac{ds}{dx}\frac{x}{t} + g\frac{1}{t}$$

and

$$\frac{\partial^2 v}{\partial x^2} = \frac{d^2 f}{ds^2}\left(\frac{ds}{dx}\right)^2\frac{x}{t} + \frac{df}{ds}\frac{d^2 s}{dx^2}\frac{x}{t} + \frac{df}{ds}\frac{ds}{dx}\frac{1}{t} + \frac{df}{ds}\frac{ds}{dx}\frac{1}{t}.$$

Sinçe

$$\frac{ds}{dx} = \frac{1}{\sqrt{t}}, \qquad \frac{ds}{dt} = -\frac{1}{2}xt^{-3/2}$$

the system of ordinary differential equation follows.

Problem 10. For the functions $u, v : [0, \infty) \times \mathbb{R} \to \mathbb{R}$ we consider the Cauchy problem

$$\frac{\partial u}{\partial t} + \frac{\partial u}{\partial x} = v^2 - u^2, \qquad \frac{\partial v}{\partial t} - \frac{\partial v}{\partial x} = u^2 - v^2 \tag{1a}$$

$$u(0, x) = u_0(x), \qquad v(0, x) = v_0(x). \tag{1b}$$

This is the *Carleman model* introduced above.
(i) Define

$$S := u + v, \qquad D := u - v \tag{2}$$

Find the partial differential equations for S and D.
(ii) Find explicit solutions assuming that S and D are conjugate harmonic functions.

Solution 10.   (i) We find that S and D satisfy the system of partial differential equations

$$\frac{\partial S}{\partial t} + \frac{\partial D}{\partial x} = 0, \qquad \frac{\partial D}{\partial t} + \frac{\partial S}{\partial x} = -2DS. \tag{3}$$

and the conditions $u \geq 0$, $v \geq 0$ take the form $S \geq 0$ and $S^2 - D^2 \geq 0$.
(ii) Assume that S und D conjugate harmonic functions, i.e.,

$$\frac{\partial S}{\partial t} + \frac{\partial D}{\partial x} = 0, \qquad \frac{\partial S}{\partial x} - \frac{\partial D}{\partial t} = 0. \tag{4}$$

Let $z := x + it$ and $f(z) := S + iD$. Then

$$\frac{df}{dz} = -\frac{i}{2}(f^2(z) + c), \qquad c \in \mathbb{R} \tag{5}$$

since

$$\frac{df}{dz} = \frac{\partial S}{\partial x} + i\frac{\partial D}{\partial x} = -\frac{i}{2}(S^2 - D^2 + 2iDS + c), \qquad \frac{\partial S}{\partial x} = DS.$$

Owing to $\partial S/\partial x = \partial D/\partial t$ the second equation is also satsified. Let

$$g(z) := \alpha f(\alpha z), \qquad \alpha \in \mathbb{R}$$

then

$$\frac{dg(z)}{dz} = \alpha^2 f'(\alpha z) = -\frac{i\alpha^2}{2}(f^2(\alpha z) + c) = -\frac{i}{2}(g^2(z) + \alpha^2 c).$$

For the solution

$$f(z) = \begin{cases} \dfrac{2f(z_0)}{2 + if(z_0)z_0 - if(z_0)z} & c = 0 \\[4mm] id\dfrac{(f(z_0) + id)e^{d(z-z_0)} + f(z_0) - id}{(f(z_0) + id)e^{d(z-z_0)} - f(z_0) + id} & d = \sqrt{c} \neq 0 \end{cases}$$

of (5) we only have to study the three cases $c = -1, 0, 1$. Since the problem is analytic we can consider

$$c = -1, \quad z_0 = 0, \quad f(z_0) = \frac{1 + \theta}{1 - \theta}, \qquad \theta \in \mathbb{R}, \quad \theta \neq 1$$

$$f(z) = -\frac{\theta e^{iz} + 1}{\theta e^{iz} - 1} = \frac{1 - \theta^2 e^{-2t} + i2\theta e^{-t}\sin x}{1 - 2\theta e^{-t}\cos x + \theta^2 e^{-2t}}.$$

We can easily check that for $|\theta| \leq \sqrt{2} - 1$ the conditions $S \geq 0$ and $S^2 - D^2 \geq 0$ are satisfied. Thus a particular solution of (1) is given by

$$u(t, x) = \frac{1}{2}\frac{\operatorname{sgn}\theta \sinh(t - \log|\theta|) + \sin x}{\operatorname{sgn}\theta \cosh(t - \log|\theta|) - \cos x}$$

$$v(t, x) = \frac{1}{2} \frac{\text{sgn}\theta \sinh(t - \log|\theta|) - \sin x}{\text{sgn}\theta \cosh(t - \log|\theta|) - \cos x}.$$

Problem 11. Consider the partial differential equation

$$\frac{\partial u}{\partial t} + u\frac{\partial u}{\partial x} + \delta\frac{\partial^3 u}{\partial x^3} + \frac{\partial^2 u}{\partial x^2} + \frac{\partial^4 u}{\partial x^4} = 0 \tag{1}$$

subject to periodic boundary conditions in the interval $[0, L]$, with initial conditions $u(x, 0) = u_0(x)$. We only consider solutions with zero spatial average. We recall that for $L \leq 2\pi$ all initial conditions evolve into $u(x, t) = 0$. We expand the solution for u in the Fourier series

$$u(x, t) = \sum_{n=-\infty}^{\infty} a_n(t) \exp(ik_n x) \tag{2}$$

where $k_n := 2n\pi/L$ and the complex expansion coefficients satisfy

$$a_{-n}(t) = \bar{a}_n(t). \tag{3}$$

Here $\bar{a}$ denotes the complex conjugate of a. Since we choose solutions with zero average we have $a_0 = 0$.
(i) Show that inserting the series expansion (2) into (1) we obtain the following system for the time evolution of the Fourier amplitudes $a_n(t)$

$$\frac{da_n}{dt} + (k_n^4 - k_n^2 - i\delta k_n^3)a_n + \frac{1}{2}ik_n \sum_{m=0}^{\infty}(a_m a_{n-m} + \bar{a}_m a_{n+m}) = 0. \tag{4}$$

(ii) Find the autonomous system when we only keeping the first five modes. where

$$k := 2\frac{\pi}{L}, \qquad \mu_n := k_n^4 - k_n^2.$$

Solution 11. (i) Since

$$\frac{\partial u}{\partial x} = \sum_{m=-\infty}^{\infty} a_m(t)ik_m e^{ik_m x}$$

we have

$$u\frac{\partial u}{\partial x} = \sum_{q=-\infty}^{\infty}\sum_{p=-\infty}^{\infty} a_p(t)a_q(t)ik_q e^{i(k_q+k_p)x}.$$

Furthermore

$$\frac{\partial u}{\partial t} = \sum_{n=-\infty}^{\infty} \frac{da_n}{dt} e^{ik_n x}$$

and the higher order derivatives with respect to x are given by

$$\frac{\partial^2 u}{\partial x^2} = -\sum_{m=-\infty}^{\infty} a_m(t) k_m^2 e^{ik_m x}, \quad \frac{\partial^3 u}{\partial x^3} = -i \sum_{m=-\infty}^{\infty} a_m(t) k_m^3 e^{ik_m x}$$

and

$$\frac{\partial^4 u}{\partial x^4} = \sum_{m=-\infty}^{\infty} a_m(t) k_m^4 e^{ik_m x}.$$

Thus

$$\sum_{m=-\infty}^{\infty} \left(\frac{da_m}{dt} + (k_m^4 - k_m^2 - i\delta k_m^3) a_m \right) e^{ik_m t} + \sum_{\substack{q=-\infty \\ p=-\infty}}^{\infty} a_p(t) a_q(t) ik_q e^{i(k_q + k_p)x} = 0.$$

Applying the Kronecker delta $\delta_{m,n}$ and $\delta_{p+q,n}$ to this equation we obtain (4) where we used (3).

(ii) We obtain the autonomous system of first order

$$\frac{da_1}{dt} + (\mu_1 - i\delta k^3) a_1 + ik(\bar{a}_1 a_2 + \bar{a}_2 a_3 + \bar{a}_3 a_4 + \bar{a}_4 a_5) = 0$$

$$\frac{da_2}{dt} + (\mu_2 - 8i\delta k^3) a_2 + ik(a_1^2 + 2\bar{a}_1 a_3 + 2\bar{a}_2 a_4 + 2\bar{a}_3 a_5) = 0$$

$$\frac{da_3}{dt} + (\mu_3 - 27i\delta k^3) a_3 + 3ik(a_1 a_2 + \bar{a}_1 a_4 + \bar{a}_2 a_5) = 0$$

$$\frac{da_4}{dt} + (\mu_4 - 64i\delta k^3) a_4 + 2ik(a_2^2 + 2a_1 a_3 + 2\bar{a}_1 a_5) = 0$$

$$\frac{da_5}{dt} + (\mu_5 - 125i\delta k^3) a_5 + 5ik(a_1 a_4 + a_2 a_3) = 0.$$

Problem 12. The spherically symmetric $SU(2)$ *Yang-Mills equations* can be written in the form

$$\frac{\partial \phi_1}{\partial t} - \frac{\partial \phi_2}{\partial r} = -A_0 \phi_2 - A_1 \phi_1 \tag{1a}$$

$$\frac{\partial \phi_2}{\partial t} + \frac{\partial \phi_1}{\partial r} = -A_1 \phi_2 + A_0 \phi_1 \tag{1b}$$

$$r^2 \left(\frac{\partial A_1}{\partial t} - \frac{\partial A_0}{\partial r} \right) = 1 - (\phi_1^2 + \phi_2^2) \tag{1c}$$

where r is the spatial radius vector and t the time. Introduce the new variables

$$z(r,t) = \phi_1(r,t) + i\phi_2(r,t) \tag{2a}$$

$$z(r,t) = R(r,t) \exp(i\theta(r,t)) \tag{2b}$$

to find particular solutions of system (1). Find the differential equation for R.

Solution 12.   Using (2a) we obtain from (1a) and (1b) that

$$\frac{\partial z}{\partial t} + i\frac{\partial z}{\partial r} = (iA_0 - A_1)z. \tag{3}$$

Using (2b) and (1a), (1b) we obtain after separating the imaginary and real parts

$$A_0 = \frac{\partial \theta}{\partial t} + \frac{\partial \ln R}{\partial r}, \qquad A_1 = \frac{\partial \theta}{\partial r} - \frac{\partial \ln R}{\partial t} \tag{4}$$

Substitution of this equation into (1c) yields

$$r^2 \left(\frac{\partial^2}{\partial r^2} + \frac{\partial^2}{\partial t^2} \right) \ln R = R^2 - 1.$$

Thus θ does not arise in this differential equation. Changing the variables

$$R(r,t) = r\exp(g(r,t))$$

we find the *Liouville equation*

$$\left(\frac{\partial^2}{\partial r^2} + \frac{\partial^2}{\partial t^2} \right) g = \exp(2g)$$

which has the general solution in terms of two harmonic functions $a(r,t)$ and $b(r,t)$ related by the *Cauchy-Riemann condition*

$$\exp(2g) = 4 \left(\left(\frac{\partial a}{\partial r} \right)^2 + \left(\frac{\partial a}{\partial t} \right)^2 \right) \frac{1}{(1 - a^2 - b^2)^2}$$

$$\left(\frac{\partial^2}{\partial r^2} + \frac{\partial^2}{\partial t^2} \right) a = 0, \qquad \left(\frac{\partial^2}{\partial r^2} + \frac{\partial^2}{\partial t^2} \right) b = 0$$

$$\frac{\partial a}{\partial r} = \frac{\partial b}{\partial t}, \qquad \frac{\partial a}{\partial t} = -\frac{\partial b}{\partial r}.$$

Equation (3) was obtained with an arbitrary smooth function $\theta(r,t)$.

Problem 13.   Let $-\infty < t, x, y, z < +\infty$. Show that the gravitational field (*metric tensor field*)

$$g = a^2 (dx \otimes dx - \frac{1}{2}e^{2x}\,dy \otimes dy + dz \otimes dz - ce^x\,dy \otimes dt - ce^x\,dt \otimes dy - c^2\,dt \otimes dt) \tag{1}$$

is produced by the *energy-momentum tensor* $m, n = 1, 2, 3, 4)$

$$T^{mn} := \frac{1}{2ka^2}g^{mn} + \frac{u^m u^n}{kc^2 a^2} \tag{2}$$

where

$$u^m := \left(0, 0, 0, \frac{c}{a}\right) \tag{3}$$

and a, k, c are constants. This means show that the Einstein field equation is satisfied. The *Einstein field equations* are given by

$$R_{mn} - \frac{1}{2} R g_{mn} = k T_{mn} \tag{4}$$

or

$$R_m^n - \frac{1}{2} R g_m^n = k T_m^n. \tag{5}$$

We have

$$T_{ik} = g_{il} g_{km} T^{lm}, \qquad T_k^i = g_{km} T^{im}, \qquad g_{nm} g^{mr} = \delta_n^r \tag{6}$$

where δ_n^r denotes the Kronecker delta. We use sum convention and $m, n = 1, 2, 3, 4$. Therefore from (4) it follows that

$$R^{mn} = \frac{1}{2} R g^{mn} - k T^{mn}. \tag{7}$$

We set $1 = x$, $2 = y$, $3 = z$, and $4 = ct$.

Solution 13. The metric tensor field (1) can be written as a 4×4 matrix

$$g_{nm} = a^2 \begin{pmatrix} 1 & 0 & 0 & 0 \\ 0 & -\frac{1}{2} e^{2x} & 0 & -e^x \\ 0 & 0 & 1 & 0 \\ 0 & -e^x & 0 & -1 \end{pmatrix}. \tag{8}$$

We consider ct as the fourth coordinate. Therefore the inverse of (g_{nm}) is given by

$$g^{nm} = \frac{1}{a^2} \begin{pmatrix} 1 & 0 & 0 & 0 \\ 0 & 2e^{-2x} & 0 & -2e^{-x} \\ 0 & 0 & 1 & 0 \\ 0 & -2e^{-x} & 0 & 1 \end{pmatrix}.$$

From (2) and

$$u^m u^n = u^m \otimes u^n = \frac{1}{a^2} \begin{pmatrix} 0 & 0 & 0 & 0 \\ 0 & 0 & 0 & 0 \\ 0 & 0 & 0 & 0 \\ 0 & 0 & 0 & c^2 \end{pmatrix}$$

we find

$$T^{mn} = \frac{1}{ka^4} \begin{pmatrix} \frac{1}{2} & 0 & 0 & 0 \\ 0 & e^{-2x} & 0 & -e^{-x} \\ 0 & 0 & \frac{1}{2} & 0 \\ 0 & -e^{-x} & 0 & \frac{3}{2} \end{pmatrix}.$$

The *Christoffel symbols* Γ^a_{mn} are defined as

$$\Gamma^a_{mn} := \frac{1}{2} g^{ab}(g_{bm,n} + g_{bn,m} - g_{mn,b}), \qquad \Gamma^a_{mn} = \Gamma^a_{nm}.$$

We have $g_{11,b} = 0$, for $b = 1, 2, 3, 4$. Therefore

$$\Gamma^1_{11} = \frac{1}{2} g^{1b}(g_{b1,1} + g_{b1,1} - g_{11,b}) = 0.$$

Furthermore we find

$$\Gamma^1_{12} = \Gamma^1_{13} = \Gamma^1_{14} = \Gamma^1_{23} = \Gamma^1_{34} = \Gamma^1_{44} = \Gamma^4_{11} = \Gamma^4_{44} = 0.$$

The nonzero Christoffel symbols are

$$\Gamma^1_{22} = \frac{1}{2} g^{11}(g_{12,2} + g_{12,2} - g_{22,1}) = \frac{1}{2} e^{2x}$$
$$\Gamma^1_{24} = \Gamma^1_{42} = \frac{1}{2} e^x$$
$$\Gamma^2_{14} = \Gamma^2_{41} = e^{-x}$$
$$\Gamma^4_{12} = \Gamma^4_{21} = \frac{1}{2} e^x$$
$$\Gamma^4_{14} = \Gamma^4_{41} = 1$$

The *curvature* is defined as

$$R^b_{msq} := \Gamma^b_{mq,s} - \Gamma^b_{ms,q} + \Gamma^b_{ns}\Gamma^n_{mq} - \Gamma^b_{nq}\Gamma^n_{ms}$$

and $R_{mq} := R^s_{msq}$. Thus

$$R_{mq} = R^s_{msq} = \Gamma^s_{mq,s} - \Gamma^s_{ms,q} + \Gamma^s_{ns}\Gamma^n_{mq} - \Gamma^s_{nq}\Gamma^n_{ms}.$$

From the definition we find

$$R_{11} = 0, \quad R_{22} = e^{2x}, \quad R_{33} = 0, \quad R_{44} = 1, \quad R_{24} = R_{42} = e^x$$

All the other R_{nm} are equal to zero. Thus we find

$$R_{nm} = \begin{pmatrix} 0 & 0 & 0 & 0 \\ 0 & e^{2x} & 0 & e^x \\ 0 & 0 & 0 & 0 \\ 0 & e^x & 0 & 1 \end{pmatrix}.$$

Now we consider the right-hand side of the Einstein field equations (4). From

$$T_{ik} = g_{il}g_{km}T^{lm}$$

we obtain

$$T_{11} = \frac{1}{2k}, \quad T_{22} = \frac{3}{4}\frac{e^{2x}}{k}, \quad T_{33} = \frac{1}{2k}, \quad T_{44} = \frac{1}{2k}, \quad T_{24} = T_{42} = \frac{e^x}{2k}$$

All the other T_{jk} are equal to zero. To find R we have to find R_k^i. We have

$$R_k^i = g^{im} R_{km}, \qquad R = R_i^i.$$

Consequently

$$R_1^1 = 0, \qquad R_2^2 = 0, \qquad R_3^3 = 0, \qquad R_4^4 = -\frac{1}{a^2}.$$

Thus

$$R = -\frac{1}{a^2}.$$

The right-hand side of the Einstein field equations (4) is given by

$$kT_{mn} = \begin{pmatrix} \frac{1}{2} & 0 & 0 & 0 \\ 0 & \frac{3}{4}e^{2x} & 0 & \frac{e^x}{2} \\ 0 & 0 & \frac{1}{2} & 0 \\ 0 & \frac{e^x}{2} & 0 & \frac{1}{2} \end{pmatrix}.$$

The left-hand side follows from R_{nm} and R

$$-\frac{1}{2}Rg_{mn} = \begin{pmatrix} \frac{1}{2} & 0 & 0 & 0 \\ 0 & -\frac{e^{2x}}{4} & 0 & -\frac{e^x}{2} \\ 0 & 0 & \frac{1}{2} & 0 \\ 0 & -\frac{e^x}{2} & 0 & -\frac{1}{2} \end{pmatrix}.$$

This proves that the Einstein field equations are satisfied by the metric tensor field given by (1) and the energy momentum tensor field given by (2).

Chapter 13

Symmetries and Group Theoretical Reductions

If a partial differential equation admits continuous symmetries (Lie symmetries) we can reduce it to an ordinary differential equation using a similarity variable which is constructed from the Lie symmetries.

Problem 1. The motion of an incompressible constant-property fluid in two-space dimension is described by the *stream function equation*

$$\nabla^2 \frac{\partial \psi}{\partial t} + \frac{\partial \psi}{\partial y} \nabla^2 \frac{\partial \psi}{\partial x} - \frac{\partial \psi}{\partial x} \nabla^2 \frac{\partial \psi}{\partial y} = \nu \nabla^4 \psi \tag{1}$$

where $\nabla^2 := \partial^2/\partial x^2 + \partial^2/\partial y^2$ and ν is a positive constant. The *stream function* ψ is defined by $u = \partial \psi/\partial y$, $v = -\partial \psi/\partial x$ in order to satisfy the *continuity equation* identically, i.e. $\partial u/\partial x + \partial v/\partial y = 0$.
(i) Show that (1) is invariant under the transformation

$$x'(x, y, t, \epsilon) = x \cos(\epsilon t) + y \sin(\epsilon t)$$
$$y'(x, y, t, \epsilon) = -x \sin(\epsilon t) + y \cos(\epsilon t)$$
$$t'(x, y, t, \epsilon) = t$$
$$\psi'(x'(x, y, t), y'(x, y, t), t'(x, y, t), \epsilon) = \psi(x, y, t) + \frac{1}{2}\epsilon(x^2 + y^2)$$

where ϵ is a real parameter.
(ii) Find the infinitesimal generator of this transformation.

Solution 1.   (i) Applying the chain rule we find

$$\frac{\partial \psi'}{\partial x'} \cos(\epsilon t) - \frac{\partial \psi'}{\partial y'} \sin(\epsilon t) = \frac{\partial \psi}{\partial x} + \epsilon x \qquad (2a)$$

$$\frac{\partial \psi'}{\partial x'} \sin(\epsilon t) + \frac{\partial \psi'}{\partial y'} \cos(\epsilon t) = \frac{\partial \psi}{\partial y} + \epsilon y \qquad (2b)$$

and

$$\frac{\partial \psi'}{\partial x'}(-\epsilon x \sin(\epsilon t) + \epsilon y \cos(\epsilon t)) + \frac{\partial \psi'}{\partial y'}(-\epsilon x \cos(\epsilon t) - \epsilon y \sin(\epsilon t)) + \frac{\partial \psi'}{\partial t'} = \frac{\partial \psi}{\partial t}. \qquad (2c)$$

From (2a) and (2b) it follows that

$$\frac{\partial^2 \psi'}{\partial x'^2} \cos^2(\epsilon t) - 2\frac{\partial^2 \psi'}{\partial x' \partial y'} \cos(\epsilon t) \sin(\epsilon t) + \frac{\partial^2 \psi'}{\partial y'^2} \sin^2(\epsilon t) = \frac{\partial^2 \psi}{\partial x^2} + \epsilon \qquad (3a)$$

$$\frac{\partial^2 \psi'}{\partial x'^2} \sin^2(\epsilon t) + 2\frac{\partial^2 \psi'}{\partial x' \partial y'} \cos(\epsilon t) \sin(\epsilon t) + \frac{\partial^2 \psi'}{\partial y'^2} \cos^2(\epsilon t) = \frac{\partial^2 \psi}{\partial y^2} + \epsilon. \qquad (3b)$$

Therefore

$$\frac{\partial^2 \psi'}{\partial x'^2} + \frac{\partial^2 \psi'}{\partial y'^2} = \frac{\partial^2 \psi}{\partial x^2} + \frac{\partial^2 \psi}{\partial y^2} + 2\epsilon.$$

Thus it is obvious that $\nu \nabla'^4 \psi' = \nu \nabla^4 \psi$. From (2c) we find that

$$\nabla'^2 \frac{\partial \psi'}{\partial t'} = \nabla^2 \frac{\partial \psi}{\partial t} + \epsilon x \nabla^2 \frac{\partial \psi}{\partial y} - \epsilon y \nabla^2 \frac{\partial \psi}{\partial x}.$$

From (2a) and (2b) we also find that

$$\frac{\partial \psi'}{\partial y'} \nabla'^2 \frac{\partial \psi'}{\partial x'} - \frac{\partial \psi'}{\partial x'} \nabla'^2 \frac{\partial \psi'}{\partial y'} = \frac{\partial \psi}{\partial y} \nabla^2 \frac{\partial \psi}{\partial x} - \frac{\partial \psi}{\partial x} \nabla^2 \frac{\partial \psi}{\partial y} - \epsilon x \nabla^2 \frac{\partial \psi}{\partial y} + \epsilon y \nabla^2 \frac{\partial \psi}{\partial x}.$$

Thus we arrive at

$$\nabla'^2 \frac{\partial \psi'}{\partial t'} + \frac{\partial \psi'}{\partial y'} \nabla'^2 \frac{\partial \psi'}{\partial x'} - \frac{\partial \psi'}{\partial x'} \nabla'^2 \frac{\partial \psi'}{\partial y'} - \nu \nabla'^4 \psi' = 0.$$

(ii) From the transformation we obtain the mapping

$$x'(x, y, t, \epsilon) = x \cos(\epsilon t) + y \sin(\epsilon t)$$
$$y'(x, y, t, \epsilon) = -x \sin(\epsilon t) + y \cos(\epsilon t)$$
$$t'(x, y, t, \epsilon) = t$$
$$\psi'(x, y, t, \psi, \epsilon) = \psi + \frac{1}{2}\epsilon(x^2 + y^2).$$

It follows that

$$\frac{dx'}{d\epsilon}\bigg|_{\epsilon=0} = yt, \quad \frac{dy'}{d\epsilon}\bigg|_{\epsilon=0} = -xt, \quad \frac{dt'}{d\epsilon}\bigg|_{\epsilon=0} = 0, \quad \frac{d\psi'}{d\epsilon}\bigg|_{\epsilon=0} = \frac{1}{2}(x^2 + y^2).$$

Thus the infinitesimal symmetry generator is given by

$$S = yt\frac{\partial}{\partial x} - xt\frac{\partial}{\partial y} + \frac{1}{2}(x^2 + y^2)\frac{\partial}{\partial \psi}.$$

We find the transformation from the symmetry generator S when we apply the exponential map, i.e.,

$$\begin{pmatrix} x'(x, y, t, \epsilon) \\ y'(x, y, t, \epsilon) \\ t'(x, y, t, \epsilon) \\ \psi'(x'(x, y, t), y'(x, y, t), t'(x, y, t), \epsilon) \end{pmatrix} = e^{\epsilon S} \begin{pmatrix} x \\ y \\ t \\ \psi \end{pmatrix}\bigg|_{\psi \rightarrow \psi(x, y, t)}.$$

Problem 2. (i) Show that the linear one-dimensional diffusion equation

$$\frac{\partial u}{\partial t} = \frac{\partial^2 u}{\partial x^2} \tag{1}$$

is invariant under the transformation group

$$t'(x, t, \epsilon) = t, \quad x'(x, t, \epsilon) = t\epsilon + x, \quad u'(x'(x, t), t'(x, t), \epsilon) = u(x, t)e^{-\frac{1}{2}(\frac{1}{2}t\epsilon^2 + x\epsilon)}$$

where ϵ is a real parameter.

(ii) Find the infinitesimal generator of this transformation.

(iii) Show that the transformation given by (2) can be derived from the infinitesimal generator.

(iv) Find a *similarity ansatz* and *similarity solution* from the transformation.

Solution 2. (i) Applying the chain rule and inserting the transformation yields

$$\frac{\partial u'}{\partial t} = \frac{\partial u'}{\partial x'}\frac{\partial x'}{\partial t} + \frac{\partial u'}{\partial t'}\frac{\partial t'}{\partial t} = \frac{\partial u}{\partial t}e^{-\frac{1}{2}(\frac{1}{2}t\epsilon^2 + x\epsilon)} - \frac{1}{4}\epsilon^2 u e^{-\frac{1}{2}(\frac{1}{2}t\epsilon^2 + x\epsilon)}$$

$$\frac{\partial u'}{\partial x} = \frac{\partial u'}{\partial x'}\frac{\partial x'}{\partial x} + \frac{\partial u'}{\partial t'}\frac{\partial t'}{\partial x} = \frac{\partial u}{\partial x}e^{-\frac{1}{2}(\frac{1}{2}t\epsilon^2 + x\epsilon)} - \frac{1}{2}\epsilon u e^{-\frac{1}{2}(\frac{1}{2}t\epsilon^2 + x\epsilon)}.$$

Since $\partial x'/\partial t = \epsilon$, $\partial x'/\partial x = 1$, $\partial t'/\partial t = 1$ and $\partial t'/\partial x = 0$ we find

$$\frac{\partial u'}{\partial x'}\epsilon + \frac{\partial u'}{\partial t'} = \frac{\partial u}{\partial t}e^{-\frac{1}{2}(\frac{1}{2}t\epsilon^2 + x\epsilon)} - \frac{1}{4}\epsilon^2 u e^{-\frac{1}{2}(\frac{1}{2}t\epsilon^2 + x\epsilon)}$$

$$\frac{\partial u'}{\partial x'} = \frac{\partial u}{\partial x} e^{-\frac{1}{2}(\frac{1}{2}t\epsilon^2 + x\epsilon)} - \frac{1}{2}\epsilon u e^{-\frac{1}{2}(\frac{1}{2}t\epsilon^2 + x\epsilon)}.$$

From this equation we obtain

$$\frac{\partial^2 u'}{\partial x'^2} = \frac{\partial^2 u}{\partial x^2} e^{-\frac{1}{2}(\frac{1}{2}t\epsilon^2 + x\epsilon)} - \epsilon\frac{\partial u}{\partial x} e^{-\frac{1}{2}(\frac{1}{2}t\epsilon^2 + x\epsilon)} + \frac{1}{4}\epsilon^2 u e^{-\frac{1}{2}(\frac{1}{2}t\epsilon^2 + x\epsilon)}.$$

Therefore we find from (1) that

$$\frac{\partial u'}{\partial t'} = \frac{\partial^2 u'}{\partial x'^2}.$$

(ii) From the transformation we obtain the mapping

$$t'(x, t, \epsilon) = t, \quad x'(x, t, \epsilon) = t\epsilon + x, \quad u'(x, t, u, \epsilon) = ue^{-\frac{1}{2}(\frac{1}{2}t\epsilon^2 + x\epsilon)}.$$

The transformation is called the *Galilean transformation*. From this equation we find

$$\frac{dt'}{d\epsilon}\Big|_{\epsilon=0} = 0, \qquad \frac{dx'}{d\epsilon}\Big|_{\epsilon=0} = t, \qquad \frac{du'}{d\epsilon}\Big|_{\epsilon=0} = -\frac{1}{2}xu.$$

Therefore the infinitesimal generator is given by

$$G = t\frac{\partial}{\partial x} - \frac{1}{2}xu\frac{\partial}{\partial u}.$$

(iii) The autonomous system associated with the symmetry generator G is

$$\frac{dt}{d\epsilon} = 0, \qquad \frac{dx}{d\epsilon} = t, \qquad \frac{du}{d\epsilon} = -\frac{1}{2}xu.$$

Solving the first equation yields $t(\epsilon) = t_0$. Inserting this solution into the second equation and integrating, we find $x(\epsilon) = t_0\epsilon + x_0$. Inserting this solution into the third equation and integrating, we arrive at

$$u(\epsilon) = u_0 e^{-\frac{1}{2}(\frac{1}{2}t_0\epsilon^2 + x_0\epsilon)}$$

which is the general solution to this autonomous system. When we set

$$t \to t,' \quad t_0 \to t, \qquad x \to x', \qquad x_0 \to x, \qquad u \to u', \quad u_0 \to u$$

we obtain the mapping. To find the transformation we have to calculate

$$\begin{pmatrix} x'(x, t, \epsilon) \\ t'(x, t, \epsilon) \\ u'(x'(x, t), t'(x, t), \epsilon) \end{pmatrix} = e^{\epsilon G}\begin{pmatrix} x \\ t \\ u \end{pmatrix}\Bigg|_{u \to u(x,t)}$$

where

$$e^{\epsilon G} x = x + \epsilon \left(t \frac{\partial}{\partial x} - \frac{1}{2} x u \frac{\partial}{\partial u} \right) x + \frac{\epsilon^2}{2!} \left(t \frac{\partial}{\partial x} - \frac{1}{2} x u \frac{\partial}{\partial u} \right)^2 x + \cdots = x + \epsilon t$$

$$e^{\epsilon G} t = t$$

$$e^{\epsilon G} u = u - \frac{\epsilon}{2} x u + \frac{\epsilon^2}{2!} \left(-\frac{1}{2} t u + \frac{1}{4} x^2 u \right) + \cdots = u e^{-\frac{1}{2}(\frac{1}{2} t \epsilon^2 + x \epsilon)}.$$

Thus we find the transformation group.

(iv) To find a similarity ansatz we write the mapping as

$$t(x_0, t_0, \epsilon) = t_0, \quad x(x_0, t_0, \epsilon) = t_0 \epsilon + x_0, \quad u(x_0, t_0, u_0, \epsilon) = u_0 e^{-\frac{1}{2}(\frac{1}{2} t_0 \epsilon^2 + x_0 \epsilon)}.$$

We set $t_0 = s$ and $x_0 = 0$, where s is the *similarity variable*. Then we find $s = t$, $\epsilon = x/s$. Therefore the similarity ansatz is given by

$$u(x, t) = f(s) e^{-\frac{1}{4} \frac{x^2}{s}}.$$

Inserting the similarity ansatz into (1) leads to the linear ordinary differential equation

$$\frac{df}{ds} + \frac{1}{2s} f = 0.$$

The general solution of this differential equation is given by $f(s) = C/\sqrt{s}$, where C is the constant of integration. Since $s = t$ is the similarity variable, we find an exact solution of the partial differential equation (1)

$$u(x, t) = \frac{C}{\sqrt{t}} e^{-\frac{1}{4} \frac{x^2}{t}}.$$

Problem 3. Consider the *Harry-Dym equation*

$$\frac{\partial u}{\partial t} - u^3 \frac{\partial^3 u}{\partial x^3} = 0. \tag{1}$$

(i) Find the Lie symmetry vector fields.

(ii) Compute the flow for one of the Lie symmetry vector fields.

Solution 3. (i) The Lie symmetry vector field is given by

$$V = \eta_x(x, t, u) \frac{\partial}{\partial x} + \eta_t(x, t, u) \frac{\partial}{\partial t} + \phi(x, t, u) \frac{\partial}{\partial u}. \tag{2}$$

There are eight determining equations for η_x, η_t, ϕ

$$\frac{\partial \eta_t}{\partial u} = 0, \quad \frac{\partial \eta_t}{\partial x} = 0, \quad \frac{\partial \eta_x}{\partial u} = 0, \quad \frac{\partial^2 \phi}{\partial u^2} = 0,$$

$$\frac{\partial^2 \phi}{\partial u \partial x} - \frac{\partial^2 \eta_x}{\partial x^2} = 0, \qquad \frac{\partial \phi}{\partial t} - u^3 \frac{\partial^3 \phi}{\partial x^3} = 0$$

$$3u^3 \frac{\partial^3 \phi}{\partial u \partial x^2} + \frac{\partial \eta_x}{\partial t} - u^3 \frac{\partial^3 \eta_x}{\partial x^3} = 0, \qquad u \frac{\partial \eta_t}{\partial t} - 3u \frac{\partial \eta_x}{\partial x} + 3\phi = 0.$$

These determining equations can easily be solved explicitly. The general solution is

$$\eta_x = k_1 + k_3 x + k_5 x^2, \quad \eta_t = k_2 - 3k_4 t, \quad \phi = (k_3 + k_4 + 2k_5 x)u$$

where $k_1, \ldots, k_5$ are arbitrary constants. The five infinitesimal generators then are

$$G_1 = \frac{\partial}{\partial x}, \qquad G_2 = \frac{\partial}{\partial t}$$

$$G_3 = x\frac{\partial}{\partial x} + u\frac{\partial}{\partial u}, \qquad G_4 = -3t\frac{\partial}{\partial t} + u\frac{\partial}{\partial u}, \qquad G_5 = x^2\frac{\partial}{\partial x} + 2xu\frac{\partial}{\partial u}.$$

Thus (1) is invariant under translations (G_1 and G_2) and scaling (G_3 and G_4). The flow corresponding to each of the infinitesimal generators can be obtained via simple integration.

(ii) Let us compute the flow corresponding to G_5. This requires integration of the first order system

$$\frac{d\bar{x}}{d\epsilon} = \bar{x}^2, \qquad \frac{d\bar{t}}{d\epsilon} = 0, \qquad \frac{d\bar{u}}{d\epsilon} = 2\bar{x}\bar{u}$$

together with the initial conditions $x(0) = x$, $\bar{t}(0) = t$, $\bar{u}(0) = u$, where ϵ is the parameter of the transformation group. One obtains

$$\bar{x}(\epsilon) = \frac{x}{(1 - \epsilon x)}, \qquad \bar{t}(\epsilon) = t, \qquad \bar{u}(\epsilon) = \frac{u}{(1 - \epsilon x)^2}.$$

We therefore conclude that for any solution $u = f(x, t)$ of (1), the transformed solution

$$\bar{u}(\bar{x}, \bar{t}) = (1 + \epsilon\bar{x})^2 f\left(\frac{\bar{x}}{1 + \epsilon\bar{x}}, \bar{t}\right)$$

will solve

$$\frac{\partial \bar{u}}{\partial \bar{t}} - \bar{u}^3 \frac{\partial^3 \bar{u}}{\partial \bar{x}^3} = 0.$$

Problem 4.   The nonlinear partial differential equation

$$\frac{\partial^2 u}{\partial x_1 \partial x_2} + \frac{\partial u}{\partial x_1} + u^2 = 0 \tag{1}$$

describes the relaxation to a *Maxwell distribution*. The symmetry vector fields are given by

$$Z_1 = \frac{\partial}{\partial x_1}, \quad Z_2 = \frac{\partial}{\partial x_2}, \quad Z_3 = -x_1\frac{\partial}{\partial x_1} + u\frac{\partial}{\partial u}, \quad Z_4 = e^{x_2}\frac{\partial}{\partial x_2} - e^{x_2}u\frac{\partial}{\partial u}.$$

Construct a similarity ansatz from the symmetry vector field

$$Z = c_1 \frac{\partial}{\partial x_1} + c_2 \frac{\partial}{\partial x_2} + c_3 \left(-x_1 \frac{\partial}{\partial x_1} + u \frac{\partial}{\partial u} \right)$$

and find the corresponding ordinary differential equation. Here $c_1, c_2, c_3 \in \mathbb{R}$.

Solution 4. The corresponding initial value problem of the symmetry vector field Z is given by

$$\frac{dx_1'}{d\epsilon} = c_1 - c_3 x_1', \quad \frac{dx_2'}{d\epsilon} = c_2, \quad \frac{du'}{d\epsilon} = c_3 u' .$$

The solution to this system provides the transformation group

$$x_1'(\mathbf{x}, u, \epsilon) = \frac{c_1}{c_3} - \frac{c_1 - c_3 x_1}{c_3} e^{-c_3 \epsilon}, \quad x_2'(\mathbf{x}, u, \epsilon) = c_2 \epsilon + x_2, \quad x_3'(\mathbf{x}, u, \epsilon) = u e^{c_3 \epsilon}$$

where $c_3 \neq 0$. Now let $x_2 = s/c$ and $x_1 = 1$ with the constant $c \neq 0$. The *similarity variable* s follows as

$$s = cx_2' + \frac{c_2 c}{c_3} \ln \frac{c_3 x_1' - c_1}{c_3 - c_1}$$

and the *similarity ansatz* is

$$u'(x_1', x_2') = v(s) \frac{c_1 - c_3}{c_1 - c_3 x_1'} .$$

Inserting this equation into

$$\frac{\partial^2 u'}{\partial x_1' \partial x_2'} + \frac{\partial u'}{\partial x_1'} + u'^2 = 0$$

leads to the ordinary differential equation

$$c \frac{d^2 v}{ds^2} + (1 - c) \frac{dv}{ds} - (1 - v)v = 0.$$

Problem 5. Consider the partial differential equation

$$\frac{\partial^2 u}{\partial x \partial t} = f(u)$$

where f is an analytic function of u. Find the Lie symmetry vector field

$$V = a(x, t, u) \frac{\partial}{\partial x} + b(x, t, u) \frac{\partial}{\partial t} + c(x, t, u) \frac{\partial}{\partial u}$$

with the corresponding vertical vector field

$$V_v = (-au_x - bu_t + c)\frac{\partial}{\partial u}.$$

Solution 5. The prolongation of the vertical vector field V_v is given by

$$\tilde{V}_v = (-au_x - bu_t + c)\frac{\partial}{\partial u}$$

$$\left(-\frac{\partial^2 a}{\partial x \partial t}u_x - \frac{\partial^2 a}{\partial x \partial u}u_t u_x - \frac{\partial a}{\partial x}u_{xt} - \frac{\partial^2 a}{\partial u \partial t}u_x^2 - \frac{\partial^2 a}{\partial u^2}u_t u_x^2 - 2\frac{\partial a}{\partial u}u_x u_{xt} \right.$$

$$-\frac{\partial a}{\partial t}u_{xx} - \frac{\partial a}{\partial u}u_t u_{xx} - au_{xxt} - \frac{\partial^2 b}{\partial x \partial t}u_t - \frac{\partial^2 b}{\partial x \partial u}u_t^2 - \frac{\partial b}{\partial x}u_{tt}$$

$$-\frac{\partial^2 b}{\partial u \partial t}u_x u_t - \frac{\partial^2 b}{\partial u^2}u_x u_t^2 - \frac{\partial b}{\partial u}u_{xt}u_t - \frac{\partial b}{\partial u}u_x u_{tt}$$

$$\left. +\frac{\partial^2 c}{\partial x \partial t} + \frac{\partial^2 c}{\partial x \partial u}u_t + \frac{\partial^2 c}{\partial u \partial t}u_x + \frac{\partial^2 c}{\partial u^2}u_t u_x + \frac{\partial c}{\partial u}u_{xt} \right)\frac{\partial}{\partial u_{xt}}$$

Using that

$$u_{xt} = f(u), \quad u_{xxt} = \frac{\partial f}{\partial u}u_x, \quad u_{xtt} = \frac{\partial f}{\partial u}u_t$$

we find form the condition that

$$L_{\tilde{V}_v}(u_{xt} - f(u)) = 0$$

where $L_{\tilde{V}_v}$ twelve conditions for the coefficients $1, u_x, u_t, u_x u_t, u_x^2, u_t^2, u_{xx},$ $u_{tt}, u_x^2 u_t, u_x u_t^2, u_x x u_t, u_x u_{tt}$. For the coefficient 1 we have

$$-c\frac{\partial f}{\partial u} - \frac{\partial a}{\partial x}f - \frac{\partial b}{\partial t} + \frac{\partial^2 c}{\partial x \partial t} + \frac{\partial c}{\partial u}f = 0.$$

From the other conditions we can conclude that the function a depends only on x and the function b depends only on t. Then we find that the function c is of the form

$$c(x, t, u) = Ku + h(x, t)$$

for some smooth function h and K is a constant. The equation for the coefficient 1 reduces to

$$-Ku\frac{\partial f}{\partial u} + h\frac{\partial f}{\partial u} - \left(\frac{da}{dx} + \frac{db}{dt} - K\right)f + \frac{\partial^2 h}{\partial x \partial t} = 0.$$

In general the functions udf/du, df/du, f, 1 will be linearly independent. Then it follows that $K = 0$, $h = 0$ and $da/dx + db/dt = 0$. Thus $c = 0$ and $a(x) = Ax + B$, $b(t) = -At + B$. We find the four symmetry vector fields

$$V_1 = x\frac{\partial}{\partial x} - t\frac{\partial}{\partial t}, \quad V_2 = \frac{\partial}{\partial x}, \quad V_3 = \frac{\partial}{\partial t}.$$

Chapter 14

Soliton Equations

Soliton equations are special partial differential equations. They have a number of interesting properties such as a Lax representation, solvable by the inverse scattering method, infinite many conservation laws. They also admit Bäcklund transformations and Lie-Bäcklund symmetries. Most of them also pass the Painlevé test.

Problem 1. The *Korteweg-de Vries equation* is given by

$$\frac{\partial u}{\partial t} - 6u\frac{\partial u}{\partial x} + \frac{\partial^3 u}{\partial x^3} = 0. \tag{1}$$

Find a solution of the form (*traveling wave solution*)

$$u(x,t) = -2f(x - ct). \tag{2}$$

Solution 1. From ansatz (2) we find

$$\frac{\partial u}{\partial t} = 2cf', \quad \frac{\partial u}{\partial x} = -2f', \quad \frac{\partial^3 u}{\partial x^3} = -2f''' \tag{3}$$

where f' denotes differentiation with respect to the argument $s := x - ct$. Inserting (3) into (1) yields

$$2cf' - 24ff' - 2f''' = 0. \tag{4}$$

Integrating (4) once gives

$$cf - 6f^2 - f'' + \frac{a}{2} = 0.$$

This second order differential equation can be integrated again and we find

$$(f')^2 = -4f^3 + cf^2 + af + b \tag{5}$$

where a and b are the constants of integration. This differential equation can written as

$$\left(\frac{df}{ds}\right)^2 = -4(f - \alpha_1)(f - \alpha_2)(f - \alpha_3)$$

with real roots α_1, α_2 and α_3 and $\alpha_1 < \alpha_2 < \alpha_3$. Let $f - \alpha_3 =: -g$ then we obtain

$$\left(\frac{dg}{dx}\right)^2 = 4g(\alpha_3 - \alpha_2 - g)(\alpha_3 - \alpha_1 - g).$$

Setting

$$g = (\alpha_3 - \alpha_2)v^2$$

we arrive at

$$\left(\frac{dv}{dx}\right)^2 = (\alpha_3 - \alpha_1)(1 - v^2)(1 - k^2v^2)$$

where $k^2 = (\alpha_3 - \alpha_2)/(\alpha_3 - \alpha_1)$. The first order differential equation

$$(dw/dz)^2 = (1 - w^2)(1 - k^2w^2)$$

is the equation that defines the *Jacobi elliptic function* sn(z,k). The constant k is the modulus of the elliptic function. Therefore we obtain for v the solution

$$v(x) = \mathrm{sn}(\sqrt{\alpha_3 - \alpha_1}\, x, k).$$

Consequently

$$f(x - ct) = \alpha_3 - (\alpha_3 - \alpha_2)\mathrm{sn}^2(\sqrt{\alpha_3 - \alpha_1}(x - ct), k).$$

Problem 2. The *Korteweg-de Vries equation* is given by

$$\frac{\partial u}{\partial t} + u\frac{\partial u}{\partial x} + \beta\frac{\partial^3 u}{\partial x^3} = 0. \tag{1}$$

A *conservation law* is given by

$$\frac{\partial Q}{\partial t} + \frac{\partial P}{\partial x} = 0.$$

(i) Show that (1) can be written as a conservation law.
(ii) Show that

$$\frac{\partial}{\partial t}\left(\frac{u^2}{2}\right) + \frac{\partial}{\partial x}\left(\frac{u^3}{3} + \beta\left(u\frac{\partial^2 u}{\partial x^2} - \frac{1}{2}\left(\frac{\partial u}{\partial x}\right)^2\right)\right) = 0 \tag{2}$$

and

$$\frac{\partial}{\partial t}\left(\frac{u^3}{3} - \beta\frac{\partial^2 u}{\partial x^2}\right) + \frac{\partial}{\partial x}\left(\frac{u^4}{4} + \beta\left(u^2\frac{\partial^2 u}{\partial x^2} + 2\frac{\partial u}{\partial t}\frac{\partial u}{\partial x}\right) + \beta^2\left(\frac{\partial^2 u}{\partial x^2}\right)^2\right) = 0$$
$$(3)$$

are conservation laws of the Korteweg-de Vries equation (1).

Solution 2. (i) Since $\partial u^2/\partial x \equiv 2u\partial u/\partial x$ we find that the Korteweg-de Vries equation (1) can be written as

$$\frac{\partial u}{\partial t} + \frac{\partial}{\partial x}\left(\frac{u^2}{2} + \beta\frac{\partial^2 u}{\partial x^2}\right) = 0.$$

(ii) Multiplying (1) by u and using the identity

$$\frac{\partial}{\partial x}\left(u\frac{\partial^2 u}{\partial x^2} - \frac{1}{2}\left(\frac{\partial u}{\partial x}\right)^2\right) \equiv u\frac{\partial^3 u}{\partial x^3}$$

we obtain (2). Multiplying (1) by u^2 we obtain (3).

Problem 3. Consider the *Schrödinger eigenvalue equation*

$$\frac{d^2 y}{dx^2} + (k^2 - u(x))y = 0 \tag{1}$$

where k is an arbitrary real parameter.
(i) Find a solution of the form

$$y(x) = e^{ikx}(2k + if(x)) \tag{2}$$

i.e. determine f and u.
(ii) Find the condition that $u(ax + bt)$ satisfies the Korteweg-de Vries equation

$$\frac{\partial u}{\partial t} - 6u\frac{\partial u}{\partial x} + \frac{\partial^3 u}{\partial x^3} = 0. \tag{3}$$

Solution 3. (i) Since

$$\frac{d^2 y}{dx^2} = -k^2 e^{ikx}(2k + if(x)) - 2ke^{ikx}\frac{df}{dx} + ie^{ikx}\frac{d^2 f}{dx^2} \tag{4}$$

we obtain

$$i\frac{d^2 f}{dx^2} - iu(x)f(x) + k\left(-2\frac{df}{dx} - 2u(x)\right) = 0. \tag{5}$$

Separating terms according to powers of k, we arrive at

$$\frac{d^2 f}{dx^2} = fu$$

$$\frac{df}{dx} = -u.$$

Elimination of u and integration yields

$$\frac{df}{dx} + \frac{1}{2}f^2 = 2c^2$$

where $2c^2$ is the constant of integration. The substitution

$$f = \frac{2}{w}\frac{dw}{dx}$$

leads to the linear second order differential equation

$$\frac{d^2 w}{dx^2} = c^2 w$$

with the general solution

$$w(x) = Ae^{cx} + Be^{-cx}$$

where A, B are the constants of integration. Using $df/du = -u$ and $dw/dx = fw$ we obtain

$$u(x) = -2\frac{d^2}{dx^2}\ln w(x).$$

Inserting the $w(x)$ into this expression leads to

$$u(x) = -2c^2\text{sech}^2(cx - \phi)$$

where $\phi = \frac{1}{2}\ln(B/A)$ and $\text{sech}\,\alpha := 1/\cosh\alpha$.
(ii) Inserting the ansatz

$$u(x,t) = -2c^2\text{sech}^2(cx - \phi(t))$$

into the Korteweg-de Vries equation provides

$$\frac{d\phi}{dt} = 4c^3.$$

Therefore

$$u(x,t) = -2c^2\text{sech}^2(cx - 4c^3 t).$$

Problem 4. The function

$$u(x,t) = -12\frac{4\cosh(2x - 8t) + \cosh(4x - 64t) + 3}{(3\cosh(x - 28t) + \cosh(3x - 36t))^2} \tag{1}$$

is a so-called *two-soliton* solution of the Korteweg-de Vries equation

$$\frac{\partial u}{\partial t} - 6u\frac{\partial u}{\partial x} + \frac{\partial^3 u}{\partial x^3} = 0. \tag{2}$$

(i) Find the solution for $t = 0$.
(ii) Find the solution for $|2x - 32t| \gg 0$ and $|x - 4t| < 1$.
(iii) Find the solution for $|x - 4t| \gg 0$ and $|2x - 32t| < 1$.

Solution 4. For $t = 0$ the function (1) simplifies to

$$u(x, t = 0) = -12\frac{4\cosh(2x) + \cosh(4x) + 3}{(3\cosh(x) + \cosh(3x))^2} = -6\mathrm{sech}^2 x \tag{3}$$

where we have used the identities

$$\cosh(2x) \equiv \sinh^2 x + \cosh^2 x$$
$$\cosh(4x) \equiv 8\cosh^4(x) - 8\cosh^2(x) + 1$$
$$\sinh(3x) \equiv 3\sinh(x) + 4\sinh^3(x).$$

Thus the two solitons completely overlap.
(ii) For $|2x - 32t| \gg 1$ and $|x - 4t| < 1$ we find

$$u(x,t) = -2\mathrm{sech}^2\left(x - 4t \mp \tanh^{-1}\frac{1}{2}\right).$$

(iii) For $|x - 4t| \gg 1$ and $|2x - 32t| < 1$ we find

$$u(x,t) = -8\mathrm{sech}^2\left(2x - 32t \mp \tanh^{-1}\frac{1}{2}\right).$$

Problem 5. Consider the *Korteweg-de Vries-Burgers equation*

$$\frac{\partial u}{\partial t} + u\frac{\partial u}{\partial x} + b\frac{\partial^3 u}{\partial x^3} - a\frac{\partial^2 u}{\partial x^2} = 0 \tag{1}$$

where u is a *conserved quantity*

$$\frac{d}{dt}\int_{-\infty}^{\infty} u(x,t)dx = 0$$

i.e. the area under u is conserved for all t. Find solutions of the form

$$u(x,t) = U(\xi(x,t)), \qquad \xi(x,t) := c(x-vt)$$

where

$$U(\xi) = S(Y(\xi)) = \sum_{n=0}^{N} a_n Y(\xi)^n, \qquad Y(\xi) = \tanh(\xi(x,t)) = \tanh(c(x-vt)).$$
(2)

Solution 5. Inserting (2) into (1) yields the ordinary differential equation

$$-cvU(\xi) + \frac{1}{2}cU(\xi)^2 + bc^3\frac{d^2U}{d\xi^2} - ac^2\frac{dU(\xi)}{d\xi} = C.$$

Requiring that

$$U(\xi) \to 0, \qquad \frac{dU(\xi)}{d\xi} \to 0, \qquad \frac{d^2U(\xi)}{d\xi^2} \to 0 \quad \text{as} \quad \xi \to \infty$$
(3)

we find that the integration constant C must be set to zero. Then the ordinary differential equation can be expressed in the variable Y as

$$-vS + \frac{1}{2}S^2 + bc^2(1-Y^2)\left(-2Y\frac{dS}{dY} + (1-Y^2)\frac{d^2S}{dY^2}\right) - ac(1-Y^2)\frac{dS}{dY} = 0.$$
(4)

Substitution of the expansion (2) into (4) and balancing the highest degree in Y, yields $N = 2$. The boundary condition $U(\xi) \to 0$ for $\xi \to +\infty$ or $\xi \to -\infty$ implies that $S(Y) \to 0$ for $Y \to +1$ or $Y \to -1$. Without loss of generality we only consider the limit $Y \to 1$. Two possible solutions arise

$$S(Y) = F(Y) = b_0(1-Y)(1+b_1Y)$$

or

$$S(Y) = G(Y) = d_0(1-Y)^2.$$

In the first case $S(Y)$ decays as $\exp(2\xi)$, whereas in the second case, $S(Y)$ decays as $\exp(-4\xi)$ as $\xi \to +\infty$. We first take

$$F(Y) = (1-Y)b_0(1+b_1Y).$$

Upon substituting this ansatz into (4) and subsequent cancellation of a common factor $(1-Y)$, we take the limit $Y \to 1$, which gives

$$v = 4bc^2 + 2ac.$$

The remaining constants b_0 and b_1 can be found through simple algebra. We obtain

$$c = \frac{a}{10b}, \qquad v = 24bc^2.$$

Thus

$$F(Y) = 36bc^2(1 - Y)(1 + \frac{1}{3}Y).$$

The second possible solution $G(Y) = d_0(1-Y)^2$ also leads to a real solution.

Problem 6.   The equation which describes small amplitude waves in a dispersive medium with a slight deviation from one-dimensionality is

$$\frac{\partial}{\partial x}\left(4\frac{\partial u}{\partial t} + 6u\frac{\partial u}{\partial x} + \frac{\partial^2 u}{\partial x^2}\right) \pm 3\frac{\partial^2 u}{\partial y^2} = 0. \tag{1}$$

The $+$ refers to the two-dimensional Korteweg-de Vries equation. The $-$ refers to the two-dimensional Kadomtsev-Petviashvili equation. The formulation of the *inverse scattering transform* is as follows. Let

$$u(x, y, t) := 2\frac{\partial}{\partial x}K(x, x; y, t) \tag{2}$$

where

$$K(x, z; y, t) + F(x, z; y, t) + \int_x^\infty K(x, s; y, t)F(s, z; y, t)ds = 0 \tag{3}$$

and F satsifies the system of linear partial differential equations

$$\frac{\partial^3 F}{\partial x^3} + \frac{\partial^3 F}{\partial z^3} + \frac{\partial F}{\partial t} = 0, \qquad \frac{\partial^2 F}{\partial x^2} - \frac{\partial^2 F}{\partial z^2} + \sigma\frac{\partial F}{\partial y} = 0 \tag{4}$$

with $\sigma = 1$ for the two-dimensional Korteweg-de Vries equation and $\sigma = i$ for the Kadomtsev-Petviashvili equation. Find solutions of the product form

$$F(x, z; y, t) = \alpha(x; y, t)\beta(z; y, t) \tag{5}$$

and

$$K(x, z; y, t) = L(x; y, t)\beta(z; y, t). \tag{6}$$

Solution 6.   Inserting (5) and (6) into (3) yields

$$L(x; y, t) = -\frac{\alpha(x; y, t)}{\left(1 + \int_x^\infty \alpha(s; y, t)\beta(s; y, t)ds\right)}. \tag{7}$$

From (2) we then have the solution of (1)

$$u(x, y, t) = 2\frac{\partial^2}{\partial x^2}\ln\left(1 + \int_x^\infty \alpha(s; y, t)\beta(s; y, t)ds\right) \tag{8}$$

provided functions α and β can be found. From (4) we obtain

$$\frac{\partial \alpha}{\partial t} + \frac{\partial^3 \alpha}{\partial x^3} = 0, \qquad \frac{\partial \beta}{\partial t} + \frac{\partial^3 \beta}{\partial z^3} = 0$$

$$\sigma \frac{\partial \alpha}{\partial y} + \frac{\partial^2 \alpha}{\partial x^2} = 0, \qquad \sigma \frac{\partial \beta}{\partial y} - \frac{\partial^2 \beta}{\partial z^2} = 0.$$

Then α and β admit the solution

$$\alpha(x, y, t) = \exp(-lx - (l^2/\sigma)y + l^3 t + \delta)$$

and similarly

$$\beta(x, y, t) = \exp(-Lz + (L^2/\sigma)y + L^3 t + \Delta)$$

where δ, Δ are arbitrary shifts and l, L are constants. This form gives the oblique solitary wave solution of the two-dimensional Korteweg-de Vries equation,

$$u(x, y, t) = 2a^2 \text{sech}^2(a(x + 2my - (a^2 + 3m^2)t)) \tag{9}$$

where

$$a = \frac{1}{2}(l + L), \qquad m = \frac{1}{2}(l - L).$$

For the Kadomtsev-Petviashvili equation we have $\sigma = i$, and so if we regard l, L as complex constants a real solution is just (9) with $m \to -im$ and $l = \bar{L}$.

Chapter 15

Lax Pairs for Partial Differential Equations

Let L and M be two linear differential operators. Assume that

$$Lv = \lambda v$$

$$\frac{\partial v}{\partial t} = Mv$$

where L is the linear differential operator of the *spectral problem*, M is the linear differential operator of an associated *time-evolution equation* and λ is a parameter. The differential operators L and M are called a *Lax pair*.

Problem 1. Show that

$$L_t = [M, L](t) \tag{1}$$

where $[L, M] := LM - ML$ is the commutator. Equation (1) is called the *Lax representation*. It contains a nonlinear evolution equation if L and M are correctly chosen.

Solution 1. From $Lv = \lambda v$ we obtain

$$\frac{\partial}{\partial t}(Lv) = \frac{\partial}{\partial t}(\lambda v). \tag{2}$$

Since λ does not depend on t it follows that

$$L_t v + L\frac{\partial v}{\partial t} = \lambda \frac{\partial v}{\partial t} \tag{3}$$

173

where we assume that λ does not depend on t. Inserting $Lv = \lambda v$ and $\partial b/\partial t = Mv$ into (3) gives

$$L_t v + LMv = \lambda Mv.$$

From $Lv = \lambda v$ we obtain $MLv = \lambda Mv$. Inserting this expression into this equation yields

$$L_t v = -LMv + MLv.$$

Consequently, $L_t v = [M, L]v$. Since the smooth function v is arbitrary we obtain $L_t = [M, L](t)$.

Problem 2.   Let

$$L := \frac{\partial^2}{\partial x^2} + u(x,t), \qquad M := 4\frac{\partial^3}{\partial x^3} + 6u(x,t)\frac{\partial}{\partial x} + 3\frac{\partial u}{\partial x}.$$

Find the evolution equation for u.

Solution 2.   Since

$$L_t v = \frac{\partial}{\partial t}(Lv) - L\left(\frac{\partial v}{\partial t}\right)$$

we obtain

$$L_t v = \frac{\partial}{\partial t}\left(\left(\frac{\partial^2}{\partial x^2} + u\right)v\right) - \left(\frac{\partial^2}{\partial x^2} + u\right)\frac{\partial v}{\partial t} = \frac{\partial u}{\partial t}v.$$

Now $[M, L]v = (ML)v - (LM)v = M(Lv) - L(Mv)$. Thus

$$[M, L]v = \left(4\frac{\partial^3}{\partial x^3} + 6u\frac{\partial}{\partial x} + 3\frac{\partial u}{\partial x}\right)\left(\frac{\partial^2 v}{\partial x^2} + uv\right)$$
$$- \left(\frac{\partial^2}{\partial x^2} + u\right)\left(4\frac{\partial^3 v}{\partial x^3} + 6u\frac{\partial v}{\partial x} + 3\frac{\partial u}{\partial x}v\right).$$

Finally

$$[M, L]v = 6u\frac{\partial u}{\partial x}v + \frac{\partial^3 u}{\partial x^3}v = \left(6u\frac{\partial u}{\partial x} + \frac{\partial^3 u}{\partial x^3}\right)v.$$

Since v is arbitrary we obtain the *Korteweg-de Vries equation*

$$\frac{\partial u}{\partial t} = 6u\frac{\partial u}{\partial x} + \frac{\partial^3 u}{\partial x^3}.$$

Problem 3.   Let

$$L := \frac{\partial^2}{\partial x^2} + bu(x,y,t) + \frac{\partial}{\partial y} \tag{1}$$

and

$$T := -4\frac{\partial^3}{\partial x^3} - 6bu(x,y,t)\frac{\partial}{\partial x} - 3b\frac{\partial u}{\partial x} - 3b\partial_x^{-1}\frac{\partial u}{\partial y} + \frac{\partial}{\partial t} \tag{2}$$

where $b \in \mathbb{R}$ and

$$\partial_x^{-1} f := \int_{-\infty}^{x} f(s)ds. \tag{3}$$

Calculate the nonlinear evolution equation which follows from the condition $[T, L]v = 0$, where v is a smooth function of x, y, t and $[,]$ denotes the commutator. Use the property that

$$\frac{\partial}{\partial x}\left(\partial_x^{-1} f\right) = f(x).$$

Solution 3.   Since $[T, L]v = (TL)v - (LT)v = T(Lv) - L(Tv)$ we obtain after a lengthly calculation

$$[T, L]v = b\frac{\partial u}{\partial t}v - b\frac{\partial^3 u}{\partial x^3}v - 6b^2u\frac{\partial u}{\partial x}v - 3b\left(\partial_x^{-1}\frac{\partial^2 u}{\partial y^2}\right)v.$$

Since v is arbitrary, the condition $[T, L]v = 0$ yields

$$\frac{\partial u}{\partial t} = \frac{\partial^3 u}{\partial x^3} + 6bu\frac{\partial u}{\partial x} + 3\partial_x^{-1}\frac{\partial^2 u}{\partial y^2}.$$

To find the local form we differentiate this equation with respect to x and find

$$\frac{\partial^2 u}{\partial t \partial x} = \frac{\partial^4 u}{\partial x^4} + 6b\left(\frac{\partial u}{\partial x}\right)^2 + 6bu\frac{\partial^2 u}{\partial x^2} + 3\frac{\partial^2 u}{\partial y^2}.$$

This equation is called the *Kadomtsev-Petviashvili equation*.

Problem 4.   Consider the system of partial differential equations

$$\frac{\partial \psi_1}{\partial x} = \lambda \psi_1 + q\psi_2 \tag{1a}$$

$$\frac{\partial \psi_2}{\partial x} = r\psi_1 - \lambda \psi_2 \tag{1b}$$

$$\frac{\partial \psi_1}{\partial t} = A\psi_1 + B\psi_2 \tag{1c}$$

$$\frac{\partial \psi_2}{\partial t} = C\psi_1 - A\psi_2 \tag{1d}$$

where λ is a real parameter and q and r are smooth functions of the independent variables x and t. The coefficients A, B and C are one-parameter

(λ) families of x, t and q and r with their derivatives. This system is called the *AKNS system*.

(i) Find the equation which follows from the *compatibility condition* (also called *integrability condition*)

$$\frac{\partial^2 \psi_1}{\partial x \partial t} = \frac{\partial^2 \psi_1}{\partial t \partial x}, \qquad \frac{\partial^2 \psi_2}{\partial x \partial t} = \frac{\partial^2 \psi_2}{\partial t \partial x}. \tag{2}$$

(ii) Let w be a complex-valued function of x and t. Assume that

$$A := 2i\lambda^2 + i|w|^2, \qquad B := i\frac{\partial w}{\partial x} + 2i\lambda w, \qquad C := i\frac{\partial \bar{w}}{\partial x} - 2i\lambda \bar{w}$$

$$q := w, \qquad r := -\bar{q} \equiv -\bar{w}. \tag{3}$$

Find the partial differential equation of w.

Solution 4.   Taking the derivative of (1a) and (1b) with respect to t yields

$$\frac{\partial^2 \psi_1}{\partial t \partial x} = \lambda \frac{\partial \psi_1}{\partial t} + \frac{\partial q}{\partial t} \psi_2 + q \frac{\partial \psi_2}{\partial t} = (\lambda A + qC)\psi_1 + \left(\lambda B - qA + \frac{\partial q}{\partial t}\right)\psi_2$$

and

$$\frac{\partial^2 \psi_2}{\partial t \partial x} = -\lambda \frac{\partial \psi_2}{\partial t} + \frac{\partial r}{\partial t} \psi_1 + r \frac{\partial \psi_1}{\partial t} = \left(-\lambda C + \frac{\partial r}{\partial t} + rA\right)\psi_1 + (\lambda A + rB)\psi_2$$

where we have used (1a) through (1d). Taking the derivative of (1c) and (1d) with respect to x yields

$$\frac{\partial^2 \psi_1}{\partial x \partial t} = \left(\frac{\partial A}{\partial x} + \lambda A + rB\right)\psi_1 + \left(Aq + \frac{\partial B}{\partial x} - \lambda B\right)\psi_2$$

and

$$\frac{\partial^2 \psi_2}{\partial x \partial t} == \left(\frac{\partial C}{\partial x} + \lambda C - rA\right)\psi_1 + \left(qC - \frac{\partial A}{\partial x} + \lambda A\right)\psi_2$$

where we have used (1a) through (1d). From these equations we obtain

$$\frac{\partial A}{\partial x} = qC - rB, \qquad \frac{\partial B}{\partial x} = 2\lambda B + \frac{\partial q}{\partial t} - 2Aq$$

and

$$\frac{\partial C}{\partial x} = -2\lambda C + \frac{\partial r}{\partial t} + 2rA, \qquad \frac{\partial A}{\partial x} = qC - rB.$$

(ii) Using (3) and these equations we obtain the nonlinear partial differential equation

$$i\frac{\partial w}{\partial t} + \frac{\partial^2 w}{\partial x^2} + 2|w|^2 w = 0.$$

The differential equation $\partial A / \partial x = qC - rB$ is satisfied identically. We used that $w = w_1 + iw_2$, where w_1 and w_2 are real-valued functions and $|w|^2 = w_1^2 + w_2^2$. This nonlinear partial differential equation is the so-called one-dimensional *cubic Schrödinger equation*.

Problem 5.   Consider a system consisting of a linear equation for an eigenfunction and an evolution equation

$$L(x, \partial)\psi(x, \lambda) = \lambda\psi(x, \lambda) \tag{1}$$

$$\frac{\partial\psi(x, \lambda)}{\partial x_n} = B_n(x, \partial)\psi(x, \lambda). \tag{2}$$

(i) Show that

$$\frac{\partial L}{\partial x_n} = [B_n, L] \equiv B_n L - LB_n \tag{3}$$

and

$$\frac{\partial B_m}{\partial x_n} - \frac{\partial B_n}{\partial x_m} = [B_n, B_m]. \tag{4}$$

(ii) Let

$$L := \partial + u_2(x)\partial^{-1} + u_3(x)\partial^{-2} + u_4(x)\partial^{-3} + \cdots$$

where $x = (x_1, x_2, x_3, \ldots)$, $\partial \equiv \partial/\partial x_1$ and

$$\partial^{-1}f(x) := \int^{x_1} f(s, x_2, x_3, \ldots)ds.$$

Define $B_n(x, \partial)$ as the differential part of $(L(x, \partial))^n$. Show that

$$B_1 = \partial$$
$$B_2 = \partial^2 + 2u_2$$
$$B_3 = \partial^3 + 3u_2\partial + 3u_3 + 3\frac{\partial u_2}{\partial x_1}$$
$$B_4 = \partial^4 + 4u_2\partial^2 + \left(4u_3 + 6\frac{\partial u_2}{\partial x_1}\right)\partial + 4u_4 + 6\frac{\partial u_3}{\partial x_1} + 4\frac{\partial^2 u_2}{\partial x_1^2} + 6u_2^2.$$

(iii) Find the equations of motion from (4) for $n = 2$ and $m = 3$.

Solution 5.   (i) From (1) we obtain

$$\frac{\partial}{\partial x_n}(L\psi) = \frac{\partial L}{\partial x_n}\psi + L\frac{\partial\psi}{\partial x_n} = \frac{\partial}{\partial x_n}(\lambda\psi) = \lambda\frac{\partial\psi}{\partial x_n}. \tag{5}$$

Inserting (2) into (5) yields

$$\frac{\partial L}{\partial x_n}\psi + LB_n\psi = \lambda B_n\psi = B_n\lambda\psi$$

or

$$\frac{\partial L}{\partial x_n}\psi = -LB_n\psi + B_nL\psi = [B_n, L]\psi. \tag{6}$$

(ii) From this equation we obtain

$$\frac{\partial}{\partial x_n}(L\psi) - L\frac{\partial\psi}{\partial x_n} = B_nL\psi - LB_n\psi \tag{7}$$

and

$$\frac{\partial}{\partial x_m}(L\psi) - L\frac{\partial\psi}{\partial x_m} = B_mL\psi - LB_m\psi. \tag{8}$$

Taking the derivatives of (7) with respect to x_m and of (8) with respect to x_n gives

$$\frac{\partial^2}{\partial x_m \partial x_n}(L\psi) - \frac{\partial}{\partial x_m}\left(L\frac{\partial\psi}{\partial x_n}\right) = \frac{\partial}{\partial x_m}(B_nL\psi) - \frac{\partial}{\partial x_m}(LB_n\psi)$$

$$\frac{\partial^2}{\partial x_n \partial x_m}(L\psi) - \frac{\partial}{\partial x_n}\left(L\frac{\partial\psi}{\partial x_m}\right) = \frac{\partial}{\partial x_n}(B_mL\psi) - \frac{\partial}{\partial x_n}(LB_m\psi).$$

Subtracting these two equations we obtain

$$-\frac{\partial}{\partial x_m}\left(L\frac{\partial\psi}{\partial x_n}\right) + \frac{\partial}{\partial x_n}\left(L\frac{\partial\psi}{\partial x_m}\right) = \frac{\partial}{\partial x_m}(B_nL\psi) - \frac{\partial}{\partial x_m}(LB_n\psi)$$

$$+ \frac{\partial}{\partial x_n}(LB_m\psi) - \frac{\partial}{\partial x_m}(LB_n\psi).$$

Inserting $\partial\psi/\partial x_n = B_n\psi$, $\partial\psi/\partial x_m = B_m\psi$ into this equation and taking into account (1) gives

$$\frac{\partial}{\partial x_m}(B_nL\psi) - \frac{\partial}{\partial x_n}(LB_m\psi) = 0.$$

From this equation we obtain

$$\frac{\partial B_n}{\partial x_m}\psi + B_n\frac{\partial\psi}{\partial x_m} - \frac{\partial B_m}{\partial x_n}\psi - B_m\frac{\partial\psi}{\partial x_n} = 0.$$

Inserting $\partial\psi/\partial x_n = B_n\psi$ and $\partial\psi/px_m = B_m\psi$ into this equation gives

$$\frac{\partial B_n}{\partial x_m}\psi + B_nB_m\psi - \frac{\partial B_m}{\partial x_n}\psi - B_mB_n\psi = 0.$$

Since ψ is arbitrary, equation (4) follows.

(ii) From L it is obvious that $B_1 = \partial$. Let f be a smooth function. Then

$$L^2 f = L(Lf)$$
$$= L(\partial f + u_2\partial^{-1}f + u_3\partial^{-2}f + u_4\partial^{-3}f + u_5\partial^{-4}f + \cdots)$$
$$= \partial^2 f + u_2 f + (\partial u_2)\partial^{-1}f + u_3(\partial^{-1}f) + (\partial u_3)\partial^{-2}f + u_4\partial^{-2}f$$
$$+ (\partial u_4)\partial^{-2}f + \cdots$$

where

$$\partial^{-2} f := \int^{x_1} \left(\int^{s_1} f(s_2) ds_2 \right) ds_1 \,.$$

Since f is an arbitrary smooth function we have $B_2 = \partial^2 + u_2$. In the same manner we find B_3 and B_4.

(iii) Since

$$\frac{\partial B_3}{\partial x_2} \psi = \frac{\partial}{\partial x_2} (\partial^3 \psi) + 3 \frac{\partial u_2}{\partial x_2} (\partial \psi) + 3 u_2 \partial \left(\frac{\partial \psi}{\partial x_2} \right) + 3 \frac{\partial u_3}{\partial x_2} \psi + 3 u_3 \frac{\partial \psi}{\partial x_2}$$

$$+ 3 \frac{\partial^2 u_2}{\partial x_1 \partial x_2} \psi + 3 \frac{\partial u_2}{\partial x_1} \frac{\partial \psi}{\partial x_2}$$

and

$$-\frac{\partial B_2}{\partial x_3} \psi = -\partial^2 \frac{\partial \psi}{\partial x_3} - 2 \frac{\partial u_2}{\partial x_3} \psi - 2 u_2 \frac{\partial \psi}{\partial x_3}$$

we find

$$\frac{\partial}{\partial x_1} \left(\frac{\partial u_2}{\partial x_3} - \frac{1}{4} \frac{\partial^3 u_2}{\partial x_1^3} - 3 u_2 \frac{\partial u_2}{\partial x_1} \right) - \frac{3}{4} \frac{\partial^2 u_2}{\partial x_2^2} = 0 \,.$$

This partial differential equation is the called the *Kadomtsev-Petviashvili equation*.

Problem 6. The nonlinear partial differential equation

$$\frac{\partial^2 u}{\partial x \partial y} + \alpha \frac{\partial u}{\partial x} + \beta \frac{\partial u}{\partial y} + \gamma \frac{\partial u}{\partial x} \frac{\partial u}{\partial y} = 0 \tag{1}$$

is called *Thomas equation*, where α, β and γ are real constants with $\gamma \neq 0$. Show that

$$\frac{\partial \phi}{\partial x} = -\frac{\partial u}{\partial x} - \left(2\beta + \gamma \frac{\partial u}{\partial x} \right) \phi \tag{2a}$$

$$\frac{\partial \phi}{\partial y} = \frac{\partial u}{\partial y} - \left(2\alpha + \gamma \frac{\partial u}{\partial y} \right) \phi \tag{2b}$$

provides a Lax representation for the Thomas equation.

Solution 6. Taking the derivative of equation (2b) with respect to x gives

$$\frac{\partial^2 \phi}{\partial x \partial y} = \frac{\partial^2 u}{\partial x \partial y} - 2\alpha \frac{\partial \phi}{\partial x} - \gamma \frac{\partial^2 u}{\partial x \partial y} \phi - \gamma \frac{\partial u}{\partial y} \frac{\partial \phi}{\partial x} \,. \tag{3}$$

Taking the derivative of (2a) with respect to y gives

$$\frac{\partial^2 \phi}{\partial y \partial x} = -\frac{\partial^2 u}{\partial y \partial x} - 2\beta \frac{\partial \phi}{\partial y} - \gamma \frac{\partial^2 u}{\partial y \partial x} \phi - \gamma \frac{\partial u}{\partial x} \frac{\partial \phi}{\partial y} \,. \tag{4}$$

Subtracting (3) and (4) leads to

$$0 = 2\frac{\partial^2 u}{\partial y \partial x} - \left(2\alpha + \gamma\frac{\partial u}{\partial y}\right)\frac{\partial\phi}{\partial x} + \left(2\beta + \gamma\frac{\partial u}{\partial x}\right)\frac{\partial\phi}{\partial y}.$$

Inserting equations (2a) and (2b) into this equation gives the Thomas equation (1). Equation (1) can be linearized by applying the transformation

$$u(x, y) = \frac{1}{\gamma}\ln v(x, y).$$

Problem 7. Consider the system of quasi-linear partial differential equations

$$\frac{\partial u}{\partial t} + \frac{\partial v}{\partial y} + u\frac{\partial v}{\partial x} - v\frac{\partial u}{\partial x} = 0, \qquad \frac{\partial u}{\partial y} + \frac{\partial v}{\partial x} = 0.$$

Show that this system arise as compatibility conditions $[L, M]\Psi = 0$ of an overdetermined system of linear equations

$$L\Psi = 0, \qquad M\Psi = 0$$

where $\Psi(x, y, t, \lambda)$ is a function, λ is a spectral parameter, and the *Lax pair* is given by

$$L = \frac{\partial}{\partial t} - v\frac{\partial}{\partial x} - \lambda\frac{\partial}{\partial y}, \qquad M = \frac{\partial}{\partial y} + u\frac{\partial}{\partial x} - \lambda\frac{\partial}{\partial x}.$$

Solution 7. Using the product rule we have

$$[L, M]\Psi = L(M\Psi) - M(L\Psi)$$

$$= \left(\frac{\partial}{\partial t} - v\frac{\partial}{\partial x} - \lambda\frac{\partial}{\partial y}\right)\left(\frac{\partial}{\partial y} + u\frac{\partial}{\partial x} - \lambda\frac{\partial}{\partial x}\right)\Psi$$

$$- \left(\frac{\partial}{\partial y} + u\frac{\partial}{\partial x} - \lambda\frac{\partial}{\partial x}\right)\left(\frac{\partial}{\partial t} - v\frac{\partial}{\partial x} - \lambda\frac{\partial}{\partial y}\right)\Psi$$

$$= \frac{\partial u}{\partial t}\frac{\partial\Psi}{\partial x} - v\frac{\partial u}{\partial x}\frac{\partial\Psi}{\partial x} - \lambda\frac{\partial u}{\partial y}\frac{\partial\Psi}{\partial x} + \frac{\partial v}{\partial y}\frac{\partial\Psi}{\partial x} + u\frac{\partial v}{\partial x}\frac{\partial\Psi}{\partial x} - \lambda\frac{\partial v}{\partial x}\frac{\partial\Psi}{\partial x}$$

$$= \left(\frac{\partial u}{\partial t} - v\frac{\partial u}{\partial x} - \lambda\frac{\partial u}{\partial y} + \frac{\partial v}{\partial y} + u\frac{\partial v}{\partial x} - \lambda\frac{\partial v}{\partial x}\right)\frac{\partial\Psi}{\partial x}$$

$$= \left(\frac{\partial u}{\partial t} - v\frac{\partial u}{\partial x} + \frac{\partial v}{\partial y} + u\frac{\partial v}{\partial x} - \lambda\left(\frac{\partial v}{\partial x} + \frac{\partial u}{\partial y}\right)\right)\frac{\partial\Psi}{\partial x}.$$

Thus from $[L, M]\Psi = 0$ the system of partial differential equations follows.

Chapter 16

Bäcklund Transformations

Problem 1. The one-dimensional *sine-Gordon equation* in light-cone coordinates ξ, η is given by

$$\frac{\partial^2 u}{\partial \xi \partial \eta} = \sin u. \tag{1}$$

Show that the transformation

$$\xi'(\xi, \eta) = \xi \tag{2a}$$

$$\eta'(\xi, \eta) = \eta \tag{2b}$$

$$\frac{\partial u'(\xi'(\xi,\eta), \eta'(\xi,\eta))}{\partial \xi'} = \frac{\partial u}{\partial \xi} - 2\lambda \sin\left(\frac{u(\xi,\eta) + u'(\xi'(\xi,\eta), \eta'(\xi,\eta))}{2}\right) \tag{2c}$$

$$\frac{\partial u'(\xi'(\xi,\eta), \eta'(\xi,\eta))}{\partial \eta'} = -\frac{\partial u}{\partial \eta} + \frac{2}{\lambda} \sin\left(\frac{u(\xi,\eta) - u'(\xi'(\xi,\eta), \eta'(\xi,\eta))}{2}\right) \tag{2d}$$

defines an *auto-Bäcklund transformation*, where λ is a nonzero real parameter. This means one has to show that

$$\frac{\partial^2 u'}{\partial \xi' \partial \eta'} = \sin u' \tag{3}$$

follows from (1) and (2). Vice versa one has to show that (1) follows from (3) and (2).

Solution 1. Differentiating (2c) with respect to η, taking into account (2a) and (2b) yields

$$\frac{\partial^2 u'}{\partial \eta' \partial \xi'} = \frac{\partial^2 u}{\partial \eta \partial \xi} - \lambda \cos\left(\frac{u + u'}{2}\right)\left(\frac{\partial u}{\partial \eta} + \frac{\partial u'}{\partial \eta'}\right). \tag{4}$$

181

Differentiating (2d) with respect to ξ, taking into account (2a) and (2b) yields

$$\frac{\partial^2 u'}{\partial \xi' \partial \eta'} = -\frac{\partial^2 u}{\partial \xi \partial \eta} + \frac{1}{\lambda} \cos\left(\frac{u - u'}{2}\right)\left(\frac{\partial u}{\partial \xi} - \frac{\partial u'}{\partial \xi'}\right).$$

For the sake of simplicity we have omitted the arguments. Adding the last two equations gives

$$2\frac{\partial^2 u'}{\partial \eta' \partial \xi'} = -\lambda \cos\left(\frac{u + u'}{2}\right)\left(\frac{\partial u}{\partial \eta} + \frac{\partial u'}{\partial \eta'}\right) + \frac{1}{\lambda} \cos\left(\frac{u - u'}{2}\right)\left(\frac{\partial u}{\partial \xi} - \frac{\partial u'}{\partial \xi'}\right).$$

Inserting $\partial u/\partial \xi$ from (2c) and $\partial u/\partial \eta$ from (2d) into this equation gives

$$2\frac{\partial^2 u'}{\partial \eta' \partial \xi'} = -\lambda \cos\left(\frac{u + u'}{2}\right)\frac{2}{\lambda} \sin\left(\frac{u - u'}{2}\right) + \frac{1}{\lambda} \cos\left(\frac{u - u'}{2}\right)2\lambda \sin\left(\frac{u + u'}{2}\right).$$

Using the identity

$$\sin \alpha \cos \beta \equiv \frac{1}{2}\sin(\alpha + \beta) + \frac{1}{2}\sin(\alpha - \beta)$$

we obtain

$$\frac{\partial^2 u'}{\partial \eta' \partial \xi'} = -\frac{1}{2}\sin u + \frac{1}{2}\sin u' + \frac{1}{2}\sin u + \frac{1}{2}\sin u'.$$

Consequently

$$\frac{\partial^2 u'}{\partial \eta' \partial \xi'} = \sin u'.$$

In the same manner we can show that (1) follows from (3).

Problem 2.   In problem 1 we showed that the one-dimensional sine-Gordon equation in light-cone coordinates ξ, η

$$\frac{\partial^2 u}{\partial \xi \partial \eta} = \sin u \tag{1}$$

admits an auto-Bäcklund transformation. Obviously

$$u(\xi, \eta) = 0 \tag{2}$$

is a trivial solution of (1). Apply the auto-Bäcklund transformation given in problem 1 to find a nontrivial solution of (1). This solution is also called the *vacuum solution* of (1).

Solution 2.   Inserting $u(\xi, \eta) = 0$ into the auto-Bäcklund transformation yields

$$\frac{\partial u'(\xi', \eta')}{\partial \xi'} = -2\lambda \sin\left(\frac{u'(\xi', \eta')}{2}\right)$$

$$\frac{\partial u'(\xi', \eta')}{\partial \eta'} = \frac{2}{\lambda} \sin\left(\frac{-u'(\xi', \eta')}{2}\right).$$

Since

$$\int \frac{du}{\sin\left(\dfrac{u}{2}\right)} = 2\ln\left(\tan\left(\frac{u}{4}\right)\right)$$

we obtain a solution of the one-dimensional sine-Gordon equation

$$u'(\xi', \eta') = 4 \tan^{-1} \exp(\lambda\xi' + \lambda^{-1}\eta' + C)$$

where C is a constant of integration.

Problem 3. Consider the two nonlinear ordinary differential equations

$$\frac{d^2 u}{dt^2} = \sin u, \qquad \frac{d^2 v}{dt^2} = \sinh v \tag{1}$$

where u and v are real-valued functions. Show that

$$\frac{du}{dt} - i\frac{dv}{dt} = 2e^{i\lambda} \sin\left(\frac{1}{2}(u + iv)\right) \tag{2}$$

defines a Bäcklund transformation, where λ is a real parameter.

Solution 3. Taking the time derivative of (2) gives

$$\frac{d^2 u}{dt^2} - i\frac{d^2 v}{dt^2} = e^{i\lambda} \cos\left(\frac{1}{2}(u + iv)\right)\left(\frac{du}{dt} + i\frac{dv}{dt}\right). \tag{3}$$

The complex conjugate of (2) is

$$\frac{du}{dt} + i\frac{dv}{dt} = 2e^{-i\lambda} \sin\left(\frac{1}{2}(u - iv)\right)$$

where we have used $\overline{\sin z} = \sin \bar{z}$. Inserting this equation into (3) yields

$$\frac{d^2 u}{dt^2} - i\frac{d^2 v}{dt^2} = 2\cos\left(\frac{1}{2}(u + iv)\right) \sin\left(\frac{1}{2}(u - iv)\right).$$

Using the identity

$$\cos\left(\frac{1}{2}(u + iv)\right) \sin\left(\frac{1}{2}(u - iv)\right) \equiv \frac{1}{2}\sin u + \frac{1}{2}\sin(-iv)$$

$$\equiv \frac{1}{2}\sin u - \frac{1}{2}i\sinh v$$

we find $d^2 u/dt^2 - i d^2 v/dt^2 = \sin u - i \sinh v$. Therefore (1) follows.

Problem 4.   The nonlinear partial differential equation

$$\frac{\partial u}{\partial t} + u\frac{\partial u}{\partial x} = D\frac{\partial^2 u}{\partial x^2} \tag{1}$$

is called *Burgers equation*. Let

$$u(x,t) = -2D\frac{\partial}{\partial x}\ln v(x,t). \tag{2}$$

Show that v satisfies the linear diffusion equation $\partial v/\partial t = D\partial^2 v/\partial x^2$.

Solution 4.   From (2) we find

$$u = -2D\frac{1}{v}\frac{\partial v}{\partial x}. \tag{3}$$

It follows that

$$\frac{\partial u}{\partial t} = 2D\frac{1}{v^2}\frac{\partial v}{\partial t}\frac{\partial v}{\partial x} - 2D\frac{1}{v}\frac{\partial^2 v}{\partial t\partial x}, \qquad \frac{\partial u}{\partial x} = 2D\frac{1}{v^2}\left(\frac{\partial v}{\partial x}\right)^2 - 2D\frac{1}{v}\frac{\partial^2 v}{\partial x^2}$$

$$\frac{\partial^2 u}{\partial x^2} = -4D\frac{1}{v^3}\left(\frac{\partial v}{\partial x}\right)^3 + 6D\frac{1}{v^2}\frac{\partial v}{\partial x}\frac{\partial^2 v}{\partial x^2} - 2D\frac{1}{v}\frac{\partial^3 v}{\partial x^3}.$$

Inserting these three equations into (1) yields

$$\frac{1}{v}\frac{\partial v}{\partial t}\frac{\partial v}{\partial x} - \frac{\partial^2 v}{\partial t\partial x} = \frac{D}{v}\frac{\partial v}{\partial x}\frac{\partial^2 v}{\partial x^2} - D\frac{\partial^3 v}{\partial x^3}$$

or

$$\frac{\partial v}{\partial x}\left(\frac{\partial v}{\partial t} - D\frac{\partial^2 v}{\partial x^2}\right) = v\frac{\partial}{\partial x}\left(\frac{\partial v}{\partial t} - D\frac{\partial^2 v}{\partial x^2}\right).$$

This partial differential equation is satisfied if v satisfies the linear diffusion equation $\partial v/\partial t = D\partial^2 v/\partial x^2$.

Problem 5.   The *Korteweg-de Vries equation* is given by

$$\frac{\partial u}{\partial t} + 6u\frac{\partial u}{\partial x} + \frac{\partial^3 u}{\partial x^3} = 0 \tag{1}$$

and the *modified Korteweg-de Vries equation* takes the form

$$\frac{\partial v}{\partial t} - 6v^2\frac{\partial v}{\partial x} + \frac{\partial^3 v}{\partial x^3} = 0. \tag{2}$$

Show that the Korteweg-de Vries equation (1) and the modified Korteweg-de Vries equation (2) are related by the Bäcklund transformation

$$\frac{\partial v}{\partial x} = u + v^2 \tag{3a}$$

$$\frac{\partial v}{\partial t} = -\frac{\partial^2 u}{\partial x^2} - 2\left(v\frac{\partial u}{\partial x} + u\frac{\partial v}{\partial x}\right). \tag{3b}$$

Equation (3a) is called the *Miura transformation*.

Solution 5. First we show that the Korteweg-de Vries equation follows from system (3). Taking the derivative of (3a) with respect to t yields

$$\frac{\partial^2 v}{\partial x \partial t} = \frac{\partial u}{\partial t} + 2v\frac{\partial v}{\partial t}. \tag{4}$$

Inserting (3b) gives

$$\frac{\partial^2 v}{\partial x \partial t} = \frac{\partial u}{\partial t} - 2v\frac{\partial^2 u}{\partial x^2} - 4v^2\frac{\partial u}{\partial x} - 4uv(u + v^2) \tag{5}$$

where we used $\partial v/\partial x = u + v^2$. Taking the derivative of (3b) with respect to x yields

$$\frac{\partial^2 v}{\partial x \partial t} = -\frac{\partial^3 u}{\partial x^3} - 2\left(2\frac{\partial u}{\partial x}\frac{\partial v}{\partial x} + v\frac{\partial^2 u}{\partial x^2} + u\frac{\partial^2 v}{\partial x^2}\right). \tag{6}$$

Taking the derivative of (3a) with respect to x yields

$$\frac{\partial^2 v}{\partial x^2} = \frac{\partial u}{\partial x} + 2v\frac{\partial v}{\partial x} = \frac{\partial u}{\partial x} + 2v(u + v^2). \tag{7}$$

Inserting (3a) and (7) into (6) leads to

$$\frac{\partial^2 v}{\partial x \partial t} = -\frac{\partial^3 u}{\partial x^3} - 2\left(2\frac{\partial u}{\partial x}(u + v^2) + v\frac{\partial^2 u}{\partial x^2} + u\left(\frac{\partial u}{\partial x} + 2uv + 2v^3\right)\right).$$

Subtracting this equation from (5) gives the Korteweg-de Vries equation. Now we have to show that the modified Korteweg-de Vries equation follows from system (3). Differentiating (3a) with respect to x yields

$$\frac{\partial^2 v}{\partial x^2} = \frac{\partial u}{\partial x} + 2v\frac{\partial v}{\partial x} = \frac{\partial u}{\partial x} + 2v(u + v^2)$$

and

$$\frac{\partial^3 v}{\partial x^3} = \frac{\partial^2 u}{\partial x^2} + 2u\frac{\partial v}{\partial x} + 2v\frac{\partial u}{\partial x} + 6v^2\frac{\partial v}{\partial x} = \frac{\partial^2 u}{\partial x^2} + 2v\frac{\partial u}{\partial x} + (2u + 6v^2)(u + v^2).$$

Adding this equation and (3b) gives

$$\frac{\partial v}{\partial t} + \frac{\partial^3 v}{\partial x^3} = -2u\frac{\partial v}{\partial x} + (2u + 6v^2)(u + v^2).$$

From (3) we obtain $u = \partial v/\partial x - v^2$. Inserting it into this partial differential equation we obtain the modified Korteweg-de Vries equation (2).

Problem 6.   Consider the partial differential equations

$$\frac{\partial^2 u}{\partial x \partial y} = \exp(u), \qquad \frac{\partial^2 v}{\partial x \partial y} = 0. \tag{1}$$

The first equation is called the *Liouville equation*. Show that

$$\frac{\partial v}{\partial x} = \frac{\partial u}{\partial x} + \lambda \exp\left(\frac{1}{2}(v+u)\right), \qquad \frac{\partial v}{\partial y} = -\frac{\partial u}{\partial y} - \frac{2}{\lambda} \exp\left(\frac{1}{2}(u-v)\right) \tag{2}$$

provides a Bäcklund transformation of these two equations, where λ is a nonzero parameter.

Solution 6.   Differentiating (2a) with respect to y yields

$$\frac{\partial^2 v}{\partial x \partial y} = \frac{\partial^2 u}{\partial x \partial y} + \lambda \exp\left(\frac{1}{2}(u+v)\right) \frac{1}{2}\left(\frac{\partial v}{\partial y} + \frac{\partial u}{\partial y}\right). \tag{3}$$

Inserting (2b) into (3) yields

$$\frac{\partial^2 v}{\partial x \partial y} = \frac{\partial^2 u}{\partial x \partial y} - \exp\left(\frac{1}{2}(u+v)\right)\exp\left(\frac{1}{2}(u-v)\right) = \frac{\partial^2 u}{\partial x \partial y} - \exp(u). \tag{4}$$

Differentiating (2b) with respect to x leads to

$$\frac{\partial^2 v}{\partial x \partial y} = -\frac{\partial^2 u}{\partial x \partial y} + \frac{1}{\lambda}\exp\left(\frac{1}{2}(u-v)\right)\left(\frac{\partial v}{\partial x} - \frac{\partial u}{\partial x}\right). \tag{5}$$

Inserting (2a) into (5) gives

$$\frac{\partial^2 v}{\partial x \partial y} = -\frac{\partial^2 u}{\partial x \partial y} + \exp\left(\frac{1}{2}(u-v)\right)\exp\left(\frac{1}{2}(u+v)\right)$$
$$= -\frac{\partial^2 u}{\partial x \partial y} + \exp(u).$$

Adding (4) and this equation leads to (1b). Subtracting (4) from this equation leads to (1a).

Chapter 17

Hirota Technique

Let f, g be smooth functions. The *Hirota bilinear operators* D_x and D_t are defined as

$$D_t^n D_x^m f \circ g := \left(\frac{\partial}{\partial t} - \frac{\partial}{\partial t'} \right)^n \left(\frac{\partial}{\partial x} - \frac{\partial}{\partial x'} \right)^m f(t,x)g(t',x') \bigg|_{x'=x, t'=t}$$

where $m, n = 0, 1, 2, \ldots$. The Hirota bilinear operator is linear. From this definition it follows that

$$D_x^m f \circ f = 0 \quad \text{for odd } m$$

$$D_x^m f \circ g = (-1)^m D_x^m g \circ f$$

$$D_x f \circ g = \frac{\partial f}{\partial x} g - f \frac{\partial g}{\partial x}$$

$$D_t f \circ g = \frac{\partial f}{\partial t} g - f \frac{\partial g}{\partial t}$$

$$\frac{\partial^2}{\partial x^2} \ln f = \frac{1}{2f^2} (D_x^2 f \circ f)$$

$$\frac{\partial^2}{\partial x \partial t} \ln f = \frac{1}{2f^2} D_x D_t f \circ f$$

$$\frac{\partial^4}{\partial x^4} \ln f = \frac{1}{2f^2} D_x^4 f \circ f - 6 \left(\frac{1}{2f^2} D_x^2 f \circ f \right)^2 .$$

Problem 1. The *Boussinesq equation* is given by

$$\frac{\partial^2 u}{\partial t^2} - \frac{\partial^2 u}{\partial x^2} - 3 \frac{\partial^2}{\partial x^2} u^2 - \frac{\partial^4 u}{\partial x^4} = 0. \tag{1}$$

Express the Boussinesq equation using Hirota bilinear operators setting

$$u = 2 \frac{\partial^2}{\partial x^2} \ln f \tag{2}$$

where $f > 0$.

Solution 1. Obviously

$$\frac{\partial^2}{\partial x^2} (\ln f) = \frac{1}{2f^2} D_x^2 f \circ f$$

and

$$\frac{\partial^4}{\partial x^4} (\ln f) = \frac{1}{2f^2} D_x^4 f \circ f - 6 \left(\frac{1}{2f^2} D_x^2 f \circ f \right)^2.$$

Then we obtain

$$u = \frac{1}{f^2} D_x^2 f \circ f$$

and

$$\frac{\partial^2 u}{\partial x^2} = \frac{1}{f^2} D_x^4 f \circ f - 3u^2.$$

Furthermore

$$2 \frac{\partial^2 \ln f}{\partial t^2} = \frac{1}{f^2} D_t^2 f \circ f.$$

It follows that

$$2 \frac{\partial^2 \ln \cdot f}{\partial t^2} - u - 3u^2 - \frac{\partial^2 u}{\partial x^2} = \frac{1}{f^2} ((D_t^2 - D_x^2 - D_x^4) f \circ f).$$

Taking the second derivative of this equation with respect to x we find

$$2 \frac{\partial^4 \ln f}{\partial t^2 \partial x^2} - \frac{\partial^2 u}{\partial x^2} - 3 \frac{\partial^2}{\partial x^2} u^2 - \frac{\partial^4 u}{\partial x^4} = \frac{\partial^2}{\partial x^2} \left(\frac{1}{f^2} (D_t^2 - D_x^2 - D_x^4) f \circ f \right).$$

Hence

$$\frac{\partial^2 u}{\partial t^2} - \frac{\partial^2 u}{\partial x^2} - 3 \frac{\partial^2}{\partial x^2} u^2 - \frac{\partial^4 u}{\partial x^4} = \frac{\partial^2}{\partial x^2} \left(\frac{1}{f^2} (D_t^2 - D_x^2 - D_x^4) f \circ f \right).$$

Thus, if f is a solution of

$$(D_t^2 - D_x^2 - D_x^4) f \circ f = 0$$

then u as given by (2) is a solution of the Boussinesq equation. The Boussinesq equation admits a Lax representation and can be solved by the inverse scattering method. Furthermore, one finds auto-Bäcklund transformations, infinite hierarchies of conservation laws and infinite hierarchies of Lie-Bäcklund vector fields.

Problem 2. The system of partial differential equations

$$\left(\frac{\partial}{\partial t} + c_1 \frac{\partial}{\partial x}\right) u_1 = -u_1 u_2 \tag{1a}$$

$$\left(\frac{\partial}{\partial t} + c_2 \frac{\partial}{\partial x}\right) u_2 = u_1 u_2 \tag{1b}$$

describes the *interaction of two waves* u_1 and u_2 propagating with constant velocities c_1 and c_2, respectively.
(i) Find a conservation law.
(ii) Let

$$u_1 := \frac{G_1}{F}, \qquad u_2 := \frac{G_2}{F}. \tag{2}$$

Show that system (1) can be written as

$$D_1 G_1 \circ F = -G_1 G_2 \tag{3a}$$

$$D_2 G_2 \circ F = G_1 G_2 \tag{3b}$$

where

$$D_j := D_t + c_j D_x, \qquad j = 1, 2.$$

(iii) Expand F, G_1 and G_2 in power series, i.e.

$$F = 1 + \epsilon f_1 + \epsilon^2 f_2 + \cdots$$
$$G_1 = \epsilon g_1 + \epsilon^2 g_2 + \cdots$$
$$G_2 = \epsilon h_1 + \epsilon^2 h_2 + \cdots$$

and construct a solution to system (1).

Solution 2. (i) Adding (1a) and (1b) leads to the conservation law

$$\frac{\partial}{\partial t}(u_1 + u_2) + \frac{\partial}{\partial x}(c_1 u_1 + c_2 u_2) = 0.$$

(ii) Since

$$\frac{\partial u_j}{\partial t} = \frac{1}{F^2}\left(\frac{\partial G_j}{\partial t} F - \frac{\partial F}{\partial t} G_j\right)$$

$$\frac{\partial u_j}{\partial x} = \frac{1}{F^2}\left(\frac{\partial G_j}{\partial x} F - \frac{\partial F}{\partial x} G_j\right)$$

where $j = 1, 2$, we find (3a) and (3b).

(iii) Substituting the power series for F, G_1, G_2 into system (3) and collecting terms with the same power in ϵ, we find

$$D_1 g_1 \circ 1 = 0$$

$$D_2 h_1 \circ 1 = 0$$

$$D_1 (g_1 \circ f_1 + g_2 \circ 1) = -g_1 h_1$$

$$D_2 (h_1 \circ f_1 + h_2 \circ 1) = g_1 h_1$$

(4)

and so on. From the first and second of these equations we have

$$g_1(x, t) = g_1(x - c_1 t) \tag{5a}$$

$$h_1(x, t) = h_1(x - c_2 t) \tag{5b}$$

where g_1 and h_1 are arbitrary functions. Inserting (5a) and (5b) into (4c) and (4d) we find that all higher terms can be chosen to be zero if f_1 satisfies the differential equations

$$\left(\frac{\partial}{\partial t} + c_1 \frac{\partial}{\partial x} \right) f_1 = h_1$$

$$\left(\frac{\partial}{\partial t} + c_2 \frac{\partial}{\partial x} \right) f_1 = -g_1.$$

The general solution to this system of partial differential equations is given by

$$f_1(x, t) = F_1(x - c_1 t) + F_2(x - c_2 t)$$

where F_1 and F_2 are related to g_1 and h_1 by

$$\left(\frac{\partial}{\partial t} + c_1 \frac{\partial}{\partial x} \right) F_2(x - c_2 t) = h_1(x - c_2 t)$$

$$\left(\frac{\partial}{\partial t} + c_2 \frac{\partial}{\partial x} \right) F_1(x - c_1 t) = -g_1(x - c_1 t).$$

Therefore we have an exact solution of system (1)

$$u_1(x, t) = \frac{-\left(\dfrac{\partial}{\partial t} + c_2 \dfrac{\partial}{\partial x} \right) F_1(x - c_1 t)}{F_1(x - c_1 t) + F_2(x - c_2 t)}$$

$$u_2(x, t) = \frac{\left(\dfrac{\partial}{\partial t} + c_1 \dfrac{\partial}{\partial x} \right) F_2(x - c_2 t)}{F_1(x - c_1 t) + F_2(x - c_2 t)}.$$

Problem 3. Show that Hirota's operators $D_x^n(f \circ g)$ and $D_x^m(f \circ g)$ given in the introduction can be written as

$$D_x^n(f \circ g) = \sum_{j=0}^n \frac{(-1)^{(n-j)} n!}{j!(n-j)!} \frac{\partial^j f}{\partial x^j} \frac{\partial^{n-j} g}{\partial x^{n-j}}, \tag{1}$$

$$D_x^m D_t^n(f \circ g) = \sum_{j=0}^m \sum_{i=0}^n \frac{(-1)^{(m+n-j-i)} m!}{j!(m-j)!} \frac{n!}{i!(n-i)!} \frac{\partial^{i+j} f}{\partial t^i \partial x^j} \frac{\partial^{n+m-i-j} g}{\partial t^{n-i} \partial x^{m-j}}. \tag{2}$$

Solution 3. We prove (1) by mathematical induction. The formula (2) can be proven in a similar way. We first try to show that

$$\left(\frac{\partial}{\partial x} - \frac{\partial}{\partial x'} \right)^n (f(x) \circ g(x')) = \sum_{j=0}^n \frac{(-1)^{(n-j)} n!}{j!(n-j)!} \frac{\partial^j f(x)}{\partial x^j} \frac{\partial^{n-j} g(x')}{\partial x'^{n-j}}. \tag{3}$$

For $n = 1$, we have

$$\left(\frac{\partial}{\partial x} - \frac{\partial}{\partial x'} \right) (f(x) \circ g(x')) = \frac{\partial f(x)}{\partial x} \circ g(x') - f(x) \circ \frac{\partial g(x')}{\partial x'}$$

$$= \sum_{j=0}^1 \frac{(-1)^{(1-j)} 1!}{j!(1-j)!} \frac{\partial^j f(x)}{\partial x^j} \frac{\partial^{1-j} g(x')}{\partial x'^{1-j}}.$$

Expression (3) obviously holds. If we assume (3) to be true for $n-1$, then

$$\left(\frac{\partial}{\partial x} - \frac{\partial}{\partial x'} \right)^{n-1} (f(x) \circ g(x')) = \sum_{j=0}^{n-1} \frac{(-1)^{(n-1-j)} (n-1)!}{j!(n-1-j)!} \frac{\partial^j f(x)}{\partial x^j} \frac{\partial^{n-1-j} g(x')}{\partial x'^{n-1-j}}. \tag{4}$$

Using (4), we have

$$\left(\frac{\partial}{\partial x} - \frac{\partial}{\partial x'} \right)^n (f(x) \circ g(x')) = \left(\frac{\partial}{\partial x} - \frac{\partial}{\partial x'} \right) \left(\frac{\partial}{\partial x} - \frac{\partial}{\partial x'} \right)^{n-1} (f(x) \circ g(x'))$$

$$= \sum_{j=0}^{n-1} \frac{(-1)^{(n-1-j)} (n-1)!}{j!(n-1-j)!} \frac{\partial^{j+1} f(x)}{\partial x^{j+1}} \frac{\partial^{n-1-j} g(x')}{\partial x'^{n-1-j}}$$

$$- \sum_{j=0}^{n-1} \frac{(-1)^{(n-1-j)} (n-1)!}{j!(n-1-j)!} \frac{\partial^j f(x)}{\partial x^j} \frac{\partial^{n-j} g(x')}{\partial x'^{n-j}}$$

$$= \sum_{j=1}^n \frac{(-1)^{(n-j)} (n-1)!}{(j-1)!(n-j)!} \frac{\partial^j f(x)}{\partial x^j} \frac{\partial^{n-j} g(x')}{\partial x'^{n-j}}$$

$$+ \sum_{j=0}^{n-1} \frac{(-1)^{(n-j)} (n-1)!}{j!(n-1-j)!} \frac{\partial^j f(x)}{\partial x^j} \frac{\partial^{n-j} g(x')}{\partial x'^{n-j}}$$

$$= \sum_{j=1}^{n-1} \frac{(-1)^{(n-j)} n!}{j!(n-j)!} \frac{\partial^j f(x)}{\partial x^j} \frac{\partial^{n-j} g(x')}{\partial x'^{m-j}}$$

$$+ \frac{\partial^n f(x)}{\partial x^n} + (-1)^n f(x) \frac{\partial^n g(x')}{\partial x'^n}$$

$$= \sum_{j=0}^{n} \frac{(-1)^{(n-j)} n!}{j!(n-j)!} \frac{\partial^j f(x)}{\partial x^j} \frac{\partial^{n-j} g(x')}{\partial x'^{n-j}}.$$

Thus, (4) is true for all $n = 1, 2, \ldots$. Setting $x' = x$ we have

$$D_x^n (f \circ g) = \sum_{j=0}^{n} \frac{(-1)^{(n-j)} n!}{j!(n-j)!} \frac{\partial^j f(x)}{\partial x^j} \frac{\partial^{n-j} g(x)}{\partial x^{n-j}}.$$

Chapter 18

Painlevé Test

A partial differential equation has the *Painlevé property* when the solution of the partial differential equation considered in the complex domain is single-valued about the movable, singularity manifold. One requires that the solution be a single-valued functional of the data, i.e., arbitrary functions. To prove that a partial differential equation has the Painlevé property one expands a solution about a movable singular manifold

$$\phi(z_1, \ldots, z_n) = 0.$$

Let $u = u(z_1, \ldots, z_n)$ be a solution of the partial differential equation and assume that

$$u = \phi^n \sum_{j=0}^{\infty} u_j \phi^j$$

where ϕ and u_j are analytic functions of $z_1, \ldots, z_n$ in a neighbourhood of the manifold $\phi = 0$.

Problem 1. (i) *Burgers' equation* is given by

$$\frac{\partial u}{\partial t} + u \frac{\partial u}{\partial x} = \frac{\partial^2 u}{\partial x^2}. \tag{1}$$

Show that Burgers' equation has the Painlevé property.
(ii) Show that one can set $u_j = 0$ for $j \geq 2$. Find the partial differential equation for u_1 in this case.

Solution 1. (i) Inserting the expansion

$$u = \phi^n \sum_{j=0}^{\infty} u_j \phi^j$$

into Burgers equation yields $n = -1$ and

$$(j-2)(j+1)\left(\frac{\partial \phi}{\partial x}\right)^2 = F_j(u_{j-1}, \ldots, u_0, \partial\phi/\partial t, \partial\phi/\partial x).$$

In order that the expansion is valid F_2 has to vanish identically. We find

$$j = 0, \qquad u_0 = -2\frac{\partial\phi}{\partial x}$$

$$j = 1, \qquad \frac{\partial\phi}{\partial t} + u_1\frac{\partial\phi}{\partial x} = \frac{\partial^2\phi}{\partial x^2}$$

$$j = 2, \qquad \frac{\partial}{\partial x}\left(\frac{\partial\phi}{\partial t} + u_1\frac{\partial\phi}{\partial x} - \frac{\partial^2\phi}{\partial x^2}\right) = 0$$

The relation at $j = 2$ (so-called *compatibility condition*) is satisfied identically owing to the relation for $j = 1$. One calls $j = 2$ a resonance. Therefore the expansion is valid for arbitrary functions ϕ and u_2.

(ii) If we set the arbitrary function u_2 equal to zero

$$u_2(x, t) = 0$$

and require that the function u_1 satisfies Burgers' equation, then $u_j = 0$ for $j \geq 2$. We obtain

$$u = -2\frac{1}{\phi}\frac{\partial\phi}{\partial x} + u_1$$

where u and u_1 satisfy Burgers' equation and

$$\frac{\partial\phi}{\partial t} + u_1\frac{\partial\phi}{\partial x} = \frac{\partial^2\phi}{\partial x^2}.$$

Since $u_1 = 0$ is a solution of Burgers' equation, we find that this partial differential equation takes the form (linear diffusion equation)

$$\frac{\partial\phi}{\partial t} = \frac{\partial^2\phi}{\partial x^2}.$$

With $u_1 = 0$ we obtain the transformation

$$u = -2\frac{1}{\phi}\frac{\partial\phi}{\partial x}.$$

This is the *Cole-Hopf transformation*.

Problem 2. Perform a Painlevé test for the *inviscid Burgers' equation*

$$\frac{\partial u}{\partial t} = u\frac{\partial u}{\partial x} \tag{1}$$

Solution 2. We find an expansion of the form

$$u = u_0 + u_1\phi^n + u_2\phi^{2n} + \cdots$$

where $n = 1/2$. For the first two expansion coefficients, i.e., u_0 and u_1 we find

$$\phi^{-1/2} : \quad u_0\frac{\partial\phi}{\partial x} = \frac{\partial\phi}{\partial t}, \qquad \phi^0 : \quad \frac{\partial u_0}{\partial t} + u_0\frac{\partial u_0}{\partial x} + \frac{1}{2}u_1^2\frac{\partial\phi}{\partial x} = 0.$$

If we require that u_0 satisfies the inviscid Burgers' equation, then we find from the infinite coupled system for the expansion coefficients that $u_1 = u_2 = \cdots = 0$. Then inserting the partial differential equation for $\phi^{-1/2}$ into the partial differential equation for ϕ^0 we obtain the partial differential equation

$$\left(\frac{\partial\phi}{\partial x}\right)^2\frac{\partial^2\phi}{\partial t^2} + \left(\frac{\partial\phi}{\partial t}\right)^2\frac{\partial^2\phi}{\partial x^2} - 2\frac{\partial\phi}{\partial x}\frac{\partial\phi}{\partial t}\frac{\partial^2\phi}{\partial x\partial t} = 0.$$

This equation can be linearized by applying a *Legendre transformation*.

Problem 3. The $SO(2,1)$ invariant *nonlinear σ-model* can be written as

$$(w + \bar{w})\frac{\partial^2 w}{\partial x\partial t} - 2\frac{\partial w}{\partial x}\frac{\partial w}{\partial t} = 0 \tag{1}$$

where $\bar{w}$ denotes the complex conjugate of w. Show that the partial differential equation (1) has the Painlevé property. Introduce the real fields u and v, i.e. $w = u + iv$.

Solution 3. Introducing the real fields u and v with $w + \bar{w} = 2u$, we find that (1) can be written as a coupled system of partial differential equations

$$u\frac{\partial^2 u}{\partial x\partial t} - \frac{\partial u}{\partial x}\frac{\partial u}{\partial t} + \frac{\partial v}{\partial x}\frac{\partial v}{\partial t} = 0 \tag{2a}$$

$$u\frac{\partial^2 v}{\partial x\partial t} - \frac{\partial u}{\partial t}\frac{\partial v}{\partial x} - \frac{\partial u}{\partial x}\frac{\partial v}{\partial t} = 0. \tag{2b}$$

Inserting the ansatz

$$u = \phi^m\sum_{j=0}^{\infty}u_j\phi^j, \qquad v = \phi^n\sum_{j=0}^{\infty}v_j\phi^j \tag{3}$$

into system (2) yields, for the first terms in the expansion $n = m = -1$ and

$$u_0^2 + v_0^2 = 0, \qquad 0 \cdot u_0 v_0 \frac{\partial \phi}{\partial x} \frac{\partial \phi}{\partial t} = 0.$$

Thus u_0 or v_0 is arbitrary. Inserting the expansions (3) into (2) with the values given above we find that at $j = 1$ u_1 or v_1 can be chosen arbitrarily. For $j = 2$ we find that u_2 or v_2 can be chosen arbitrarily. Thus (1) has the Painlevé property.

Problem 4.   (i) Show that the *Kadomtsev-Petviashvili equation*

$$\frac{\partial^2 u}{\partial y^2} + \frac{\partial}{\partial x} \left(\frac{\partial u}{\partial t} + u \frac{\partial u}{\partial x} + \frac{\partial^3 u}{\partial x^3} \right) = 0$$

possesses the Painlevé property.
(ii) Find an auto-Bäcklund transformation.

Solution 4.   (i) The expansion about the singular manifold $\phi = 0$ is

$$u = \phi^{-2} \sum_{j=0}^{\infty} u_j \phi^j$$

with resonances at $j = -1, 4, 5, 6$. Therefore, owing to the noncharacteristic condition $\partial \phi / \partial x \neq 0$ when $\phi = 0$ we obtain that ϕ, u_4, u_5, u_6 are arbitrary functions of x, y, t in the expansion.
(ii) The Bäcklund transformation is

$$u = u_0 \phi^{-2} + u_1 \phi^{-1} + u_2$$

where

$$u_0 = -12 \left(\frac{\partial \phi}{\partial x} \right)^2, \qquad u_1 = 12 \frac{\partial^2 \phi}{\partial x^2}$$

$$u_2 + \frac{\partial \phi / \partial t}{\partial \phi / \partial x} + 4 \frac{\partial^3 \phi / \partial x^3}{\partial \phi / \partial x} - 3 \left(\frac{\partial^2 \phi / \partial x^2}{\partial \phi / \partial x} \right)^2 + \left(\frac{\partial \phi / \partial y}{\partial \phi / \partial x} \right)^2 = 0$$

and

$$\frac{\partial^2 \phi}{\partial x \partial t} + \frac{\partial^4 \phi}{\partial x^4} + \frac{\partial^2 \phi}{\partial y^2} + \frac{\partial^2 \phi}{\partial x^2} u_2 = 0.$$

Thus the auto Bäcklund transformation is

$$u = 12 \frac{\partial^2}{\partial x^2} \ln \phi + u_2.$$

Chapter 19

Lie Algebras

A *Lie algebra* is defined as follows: A vector space L over a field F, with an operation $L \times L \to L$ denoted by $(x, y) \to [x, y]$ and called the commutator of x and y, is called a Lie algebra over F if the following axioms are satisfied:

    (L1)   The bracket operation is bilinear.

    (L2) $[x, x] = 0$ for all $x \in L$

    (L3) $[x, [y, z]] + [y, [z, x]] + [z, [x, y]] = 0$ $(x, y, z \in L)$.

Axiom (L3) is called the *Jacobi identity*.

Bilinearity means that

$$[\alpha x + \beta y, z] = \alpha[x, z] + \beta[y, z]$$
$$[x, \alpha y + \beta z] = \alpha[x, y] + \beta[x, z]$$

where $\alpha, \beta \in F$. A Lie algebra is called real when $F = \mathbb{R}$. A Lie algebra is called complex when $F = \mathbb{C}$. If $F = \mathbb{R}$ or $F = \mathbb{C}$, then $[x, x] = 0$ and $[x, y] = -[y, x]$ are equivalent.

A Lie algebra is called abelian or commutative if $[x, y] = 0$ for all $x, y \in L$. Let L_1 and L_2 be Lie algebras and $\Phi : L_1 \to L_2$ a bijection such that for all $\alpha, \beta \in F$ and all $x, y \in L_1$

$$\Phi(\alpha x + \beta y) = \alpha\Phi(x) + \beta\Phi(y), \qquad \Phi([x, y]) = [\Phi(x), \Phi(y)]$$

then Φ is called an isomorphism and the Lie algebras L_1 and L_2 are isomorphic.

Problem 1.   (i) Classify all real two-dimensional Lie algebras, where $F = \mathbb{R}$. Denote the basis elements by X and Y.
(ii) Find a representation of the non-abelian two-dimensional Lie algebra using differential operators.
(iii) Find a representation of the non-abelian two-dimensional Lie algebra using 2×2 matrices.

Solution 1.   (i) There exists only two non-isomorphic two-dimensional Lie algebras. Let X and Y span a two-dimensional Lie algebra. Then their commutator is either 0 or $[X, Y] = Z \neq 0$. In the first case we have a two-dimensional abelian Lie algebra. In the second case we have

$$[X, Y] = Z = \alpha X + \beta Y, \qquad \alpha, \beta \in F$$

where α and β are not both zero. Let us suppose $\alpha \neq 0$, then one has

$$[Z, \alpha^{-1} Y] = Z.$$

Hence, for all non-abelian two-dimensional Lie algebras one can choose a basis $\{ Z, U \}$ such that
$$[Z, U] = Z.$$
Consequently, all non-abelian two-dimensional Lie algebras are isomorphic.
(ii) An example is the set of differential operators

$$\left\{ \frac{d}{dx}, \, x\frac{d}{dx} \right\}$$

where $Z \to d/dx$ and $U \to xd/dx$.
(iii) Since $\text{tr}([Z, U]) = 0$ we have $\text{tr} Z = 0$. A representation using 2×2 matrices is

$$Z = \begin{pmatrix} 0 & 1 \\ 0 & 0 \end{pmatrix}, \qquad U = \begin{pmatrix} 0 & 0 \\ 0 & 1 \end{pmatrix}.$$

Problem 2.   Let L be the real vector space $\mathbb{R}^3$. Let $\mathbf{a}, \mathbf{b} \in L$. Define

$$\mathbf{a} \times \mathbf{b} := \begin{pmatrix} a_2 b_3 - a_3 b_2 \\ a_3 b_1 - a_1 b_3 \\ a_1 b_2 - a_2 b_1 \end{pmatrix}. \tag{1}$$

Equation (1) is the *cross* or *vector product*.
(i) Show that L is a Lie algebra.
(ii) Show that in general

$$\mathbf{a} \times (\mathbf{b} \times \mathbf{c}) \neq (\mathbf{a} \times \mathbf{b}) \times \mathbf{c}.$$

Solution 2. (i) Obviously, we have

$$(\mathbf{a} + \mathbf{b}) \times (\mathbf{c} + \mathbf{d}) = \mathbf{a} \times \mathbf{c} + \mathbf{a} \times \mathbf{d} + \mathbf{b} \times \mathbf{c} + \mathbf{b} \times \mathbf{d}.$$

From (1) we see that $\mathbf{a} \times \mathbf{b} = -\mathbf{b} \times \mathbf{a}$. Using definition (1) we obtain, after a lengthy calculation, that

$$\mathbf{a} \times (\mathbf{b} \times \mathbf{c}) + \mathbf{c} \times (\mathbf{a} \times \mathbf{b}) + \mathbf{b} \times (\mathbf{c} \times \mathbf{a}) = \mathbf{0}.$$

This equation is called the *Jacobi identity*.
(ii) Let

$$\mathbf{a} := \begin{pmatrix} 1 \\ 0 \\ 0 \end{pmatrix}, \qquad \mathbf{b} := \begin{pmatrix} 0 \\ 1 \\ 0 \end{pmatrix}, \qquad \mathbf{c} := \begin{pmatrix} 0 \\ 1 \\ 0 \end{pmatrix}$$

be the standard basis in $\mathbb{R}^3$. Then

$$\mathbf{a} \times (\mathbf{b} \times \mathbf{c}) \neq (\mathbf{a} \times \mathbf{b}) \times \mathbf{c}.$$

Problem 3. Let v, w be two elements of a Lie algebra L. Then

$$\mathrm{ad}v(w) := [v, w]. \tag{1}$$

Let

$$x := \begin{pmatrix} 0 & 1 \\ 0 & 0 \end{pmatrix}, \qquad h := \begin{pmatrix} 1 & 0 \\ 0 & -1 \end{pmatrix}, \qquad y := \begin{pmatrix} 0 & 0 \\ 1 & 0 \end{pmatrix} \tag{2}$$

be an ordered basis for the semi-simple Lie algebra $sl(2, \mathbb{R})$.
(i) Compute the matrices $\mathrm{ad}x$, $\mathrm{ad}h$ and $\mathrm{ad}y$ relative to this basis. This is the so-called *adjoint representation*.
(ii) Calculate the eigenvalues and eigenvectors of $\mathrm{ad}x$, $\mathrm{ad}h$ and $\mathrm{ad}y$.
(iii) Do the matrices

$$\mathrm{ad}x, \qquad \mathrm{ad}h, \qquad \mathrm{ad}y$$

form basis of a Lie algebra under the commutator? If so, is this Lie algebra isomorphic to $sl(2, \mathbb{R})$?

Solution 3. Since

$$[x, h] = -2x, \qquad [x, y] = h, \qquad [y, h] = 2y$$

we find

$$\mathrm{ad}x(x) = 0, \qquad \mathrm{ad}x(h) = -2x, \qquad \mathrm{ad}x(y) = h$$
$$\mathrm{ad}h(x) = 2x, \qquad \mathrm{ad}h(h) = 0, \qquad \mathrm{ad}h(y) = -2y$$
$$\mathrm{ad}y(x) = -h \qquad \mathrm{ad}y(h) = 2y, \qquad \mathrm{ad}y(y) = 0.$$

From
$$(x, h, y) \, \mathrm{ad}x = (0, -2x, h)$$

we find
$$\mathrm{ad}x = \begin{pmatrix} 0 & -2 & 0 \\ 0 & 0 & 1 \\ 0 & 0 & 0 \end{pmatrix}.$$

Analogously,

$$\mathrm{ad}h = \begin{pmatrix} 2 & 0 & 0 \\ 0 & 0 & 0 \\ 0 & 0 & -2 \end{pmatrix}, \qquad \mathrm{ad}y = \begin{pmatrix} 0 & 0 & 0 \\ -1 & 0 & 0 \\ 0 & 2 & 0 \end{pmatrix}.$$

(iii) Since

$$[\mathrm{ad}x, \mathrm{ad}h] = -2\mathrm{ad}x, \quad [\mathrm{ad}x, \mathrm{ad}y] = \mathrm{ad}h, \quad [\mathrm{ad}y, \mathrm{ad}h] = 2\mathrm{ad}y$$

we obtain that the set $\{\,\mathrm{ad}x, \,\mathrm{ad}h, \,\mathrm{ad}y\,\}$ is closed under the commutator. Thus this set forms a basis of a Lie algebra L. The vector space isomorphism $\phi : sl(2, \mathbb{R}) \to L$ defined on the basis elements by

$$\phi(x) \to \mathrm{ad}x \qquad \phi(h) \to \mathrm{ad}h \qquad \phi(y) \to \mathrm{ad}y$$

is a Lie algebra isomorphism. Thus the Lie algebra $sl(2, \mathbb{R})$ is isomorphic to L.

Problem 4.   (i) Show that the vector fields

$$\left\{ \frac{d}{dx}, \ x\frac{d}{dx}, \ x^2\frac{d}{dx} \right\} \tag{1}$$

form a Lie algebra L under the *commutator*. Recall that the commutator is given by

$$\left[f_1(x)\frac{d}{dx}, \ f_2(x)\frac{d}{dx} \right] g(x) := \left(f_1\frac{df_2}{dx} - f_2\frac{df_1}{dx} \right) \frac{dg}{dx} \tag{2}$$

where g is an arbitrary smooth function. Thus (2) can also be written as

$$\left[f_1(x)\frac{d}{dx}, \ f_2(x)\frac{d}{dx} \right] := \left(f_1\frac{df_2}{dx} - f_2\frac{df_1}{dx} \right) \frac{d}{dx}. \tag{3}$$

(ii) Determine the *centre* Z of the Lie algebra L. The centre of the Lie algebra L is defined as

$$Z(L) := \{z \in L \mid [z, x] = 0 \quad \text{for all } x \in L\}.$$

(iii) Determine

$$\text{ad}\left(\frac{d}{dx}\right), \qquad \text{ad}\left(x\frac{d}{dx}\right), \qquad \text{ad}\left(x^2\frac{d}{dx}\right).$$

(iv) Do the vector fields

$$\left\{\frac{d}{dx}, \ x\frac{d}{dx}, \ x^2\frac{d}{dx}, \ x^3\frac{d}{dx}\right\}$$

form a basis of a Lie algebra? If not can the set be extended to form a Lie algebra.

Solution 4. (i) Straightforward calculation yields

$$\left[\frac{d}{dx}, x\frac{d}{dx}\right] = \frac{d}{dx}, \qquad \left[\frac{d}{dx}, x^2\frac{d}{dx}\right] = 2x\frac{d}{dx}, \qquad \left[x\frac{d}{dx}, x^2\frac{d}{dx}\right] = x^2\frac{d}{dx}.$$

Since the Jacobi identity is satisfied for vector fields, we find that the vector fields given by (1) form a basis of a Lie algebra.

(ii) Now an arbitrary element of L can be written as

$$f(x)\frac{d}{dx}$$

where $f(x) = a + bx + cx^2$ with $a, b, c \in \mathbb{R}$. Then we find from the condition

$$\left[f(x)\frac{d}{dx}, \frac{d}{dx}\right] = 0$$

that $df/dx = 0$. From the condition

$$\left[f(x)\frac{d}{dx}, x\frac{d}{dx}\right] = 0$$

together with condition $df/dx = 0$, we obtain $f(x) = 0$. Thus the centre Z of the Lie algebra L is given by $Z = \{0\}$.

(iii) Let

$$A := \frac{d}{dx}, \qquad B := x\frac{d}{dx}, \qquad C := x^2\frac{d}{dx}.$$

We introduce the order A, B, C. Then we have

$$\text{ad}A(A) = 0, \qquad \text{ad}A(B) = A, \qquad \text{ad}A(C) = 2B.$$

Thus we find

$$\text{ad}A = \begin{pmatrix} 0 & 1 & 0 \\ 0 & 0 & 2 \\ 0 & 0 & 0 \end{pmatrix}.$$

Analogously

$$\text{ad}B = \begin{pmatrix} -1 & 0 & 0 \\ 0 & 0 & 0 \\ 0 & 0 & 1 \end{pmatrix}, \quad \text{ad}C = \begin{pmatrix} 0 & 0 & 0 \\ -2 & 0 & 0 \\ 0 & -1 & 0 \end{pmatrix}.$$

(iv) Let

$$A_k := x^k \frac{d}{dx}$$

where $k = 0, 1, 2, \dots$. Since

$$[A_2, A_3] = x^4 \frac{d}{dx} \equiv A_4$$

we find that the set $\{\, A_0, A_1, A_2, A_3 \,\}$ does not form a basis of a Lie algebra. Since

$$[A_j, A_k] = (k - j)A_{j+k-1}$$

we find that the vector fields given by the set

$$S := \{\, A_k \; : \; k = 0, 1, 2, \dots \,\}$$

form an infinite-dimensional Lie algebra under the commutator.

Problem 5.   Let L be any Lie algebra. If $x, y \in L$, define

$$\kappa(x, y) := \text{tr}(\text{ad}x \, \text{ad}y) \tag{1}$$

where κ is called the *Killing form* and tr(.) denotes the trace.
(i) Show that

$$\kappa([x, y], z) = \kappa(x, [y, z]). \tag{2}$$

(ii) If B_1 and B_2 are subspaces of a Lie algebra L, we define $[B_1, B_2]$ to be the subspace spanned by all products $[b_1, b_2]$ with $b_1 \in B_1$ and $b_2 \in B_2$. This is the set of sums

$$\sum_j [b_{1j}, b_{2j}] \tag{3}$$

where $b_{1j} \in B_1$ and $b_{2j} \in B_2$. Let

$$L^2 := [L, L], \qquad L^3 := [L^2, L], \quad \dots \quad , L^k := [L^{k-1}, L]. \tag{4}$$

A Lie algebra L is called *nilpotent* if

$$L^n = \{0\}$$

for some positive integer n. Prove that if L is nilpotent, the Killing form of L is identically zero.

Solution 5.   (i) For $m \times m$ matrices we have the properties for the trace

$$\operatorname{tr}(A + B) = \operatorname{tr}A + \operatorname{tr}B, \qquad \operatorname{tr}(AB) = \operatorname{tr}(BA).$$

Moreover we apply $\operatorname{ad}[x, y] \equiv [\operatorname{ad}x, \operatorname{ad}y]$. Using these properties we obtain

$$
\begin{aligned}
\kappa([x, y], z) &= \operatorname{tr}(\operatorname{ad}[x, y] \operatorname{ad}z) = \operatorname{tr}([\operatorname{ad}x, \operatorname{ad}y]\operatorname{ad}z) \\
&= \operatorname{tr}(\operatorname{ad}x\,\operatorname{ad}y\,\operatorname{ad}z - \operatorname{ad}y\,\operatorname{ad}x\,\operatorname{ad}z) \\
&= \operatorname{tr}(\operatorname{ad}x\,\operatorname{ad}y\,\operatorname{ad}z) - \operatorname{tr}(\operatorname{ad}y\,\operatorname{ad}x\,\operatorname{ad}z) \\
&= \operatorname{tr}(\operatorname{ad}x\,\operatorname{ad}y\,\operatorname{ad}z) - \operatorname{tr}(\operatorname{ad}x\,\operatorname{ad}z\,\operatorname{ad}y) \\
&= \operatorname{tr}(\operatorname{ad}x\,[\operatorname{ad}y, \operatorname{ad}z]) = \operatorname{tr}(\operatorname{ad}x\,\operatorname{ad}[y, z]) \\
&= \kappa(x, [y, z]).
\end{aligned}
$$

(ii) For any $u \in L$ we have

$$(\operatorname{ad}x\,\operatorname{ad}y)^1(u) = [x, [y, u]] \in L^3$$

$$(\operatorname{ad}x\,\operatorname{ad}y)^2(u) = [x, [y, [x, [y, u]]]] \in L^5$$

In general

$$(\operatorname{ad}x\,\operatorname{ad}y)^k(u) \in L^{2k+1}.$$

Since L is nilpotent, say

$$L^n = \{0\}$$

we find that for $2k + 1 \geq n$

$$(\operatorname{ad}x\,\operatorname{ad}y)^k(u) = 0 \quad \text{for all} \quad u \in L$$

or equivalently

$$(\operatorname{ad}x\,\operatorname{ad}y)^k = 0$$

where 0 is the zero matrix. Thus the matrix $(\operatorname{ad}x\,\operatorname{ad}y)$ is nilpotent. If λ is an eigenvalue of $(\operatorname{ad}x\,\operatorname{ad}y)$ with eigenvector $\mathbf{v}$, i.e.

$$(\operatorname{ad}x\operatorname{ad}y)\mathbf{v} = \lambda\mathbf{v}$$

then

$$\lambda^k\mathbf{v} = (\operatorname{ad}x\,\operatorname{ad}y)^k\mathbf{v} = 0.$$

Therefore $\lambda^k = 0$ and thus $\lambda = 0$. Thus the eigenvalues $\lambda_1, \dots, \lambda_m$ of $(\operatorname{ad}x\,\operatorname{ad}y)$ are all zero, and we find

$$\kappa(x, y) = \operatorname{tr}(\operatorname{ad}x\,\operatorname{ad}y) = \sum_{j=1}^{m} \lambda_j = 0$$

since the trace of a square matrix is the sum of the eigenvalues. This is true for any $x, y \in L$ and the result follows.

Problem 6. Let $gl(n, \mathbb{R})$ be the Lie algebra of all $n \times n$ matrices over $\mathbb{R}$. Let

$$x \in gl(n, \mathbb{R}) \tag{1}$$

have n distinct eigenvalues $\lambda_1, \ldots, \lambda_n$ in $\mathbb{R}$. Prove that the eigenvalues of the $n^2 \times n^2$ matrix

$$\text{ad}x \tag{2}$$

are the n^2 scalars $\lambda_i - \lambda_j$, where $i, j = 1, 2, \ldots, n$.

Solution 6. Since the eigenvalues $\lambda_1, \ldots, \lambda_n$ of x are real and distinct, the corresponding eigenvectors $\mathbf{v}_1, \ldots, \mathbf{v}_n$ are linearly independent (not necessarily orthogonal), and thus form a basis for $\mathbb{R}^n$. The existence of a basis of eigenvectors of x means that x can be diagonalized, i.e., there exists a matrix $R \in GL(n, \mathbb{R})$ such that

$$R^{-1}xR = \text{diag}(\lambda_1, \ldots, \lambda_n)$$

where $GL(n, \mathbb{R})$ denotes the general linear group $(GL(n, \mathbb{R}) \subset gl(n, \mathbb{R}))$. Now

$$\text{ad}(R^{-1}xR)(E_{ij}) \equiv [R^{-1}xR, E_{ij}] = (\lambda_i - \lambda_j)E_{ij}$$

where E_{ij} are the matrices having 1 in the (i, j) position and 0 elsewhere. Thus

$$\lambda_i - \lambda_j, \qquad i, j = 1, 2, \ldots, n$$

are eigenvalues of

$$\text{ad}(R^{-1}xR).$$

Since the corresponding set of eigenvectors E_{ij} $(i, j = 1, 2, \ldots, n)$ is a basis of $gl(n, \mathbb{R})$ we conclude that these are the only eigenvalues of $\text{ad}(R^{-1}xR)$. Now if λ is an eigenvalue of $\text{ad}x$ with eigenvector $\mathbf{v}$, then

$$\begin{aligned}
\text{ad}(R^{-1}xR)(R^{-1}\mathbf{v}R) &= R^{-1}xRR^{-1}\mathbf{v}R - R^{-1}\mathbf{v}RR^{-1}xR \\
&= R^{-1}x\mathbf{v}R - R^{-1}\mathbf{v}xR \\
&= R^{-1}(\text{ad}x(\mathbf{v}))R.
\end{aligned}$$

Thus

$$\text{ad}(R^{-1}xR)(R^{-1}\mathbf{v}R) = \lambda(R^{-1}\mathbf{v}R)$$

so that λ is an eigenvalue of $\text{ad}(R^{-1}xR)$. Thus we find $\lambda = \lambda_i - \lambda_j$ for some $i, j \in \{1, \ldots, n\}$. Conversely we have

$$\text{ad}(R^{-1}xR)E_{ij} = (\lambda_i - \lambda_j)E_{ij}.$$

It follows that

$$R^{-1}xRE_{ij} - E_{ij}R^{-1}xR = (\lambda_i - \lambda_j)E_{ij}.$$

Multiplying this equation from the left with R and from the right with R^{-1} yields

$$xRE_{ij}R^{-1} - RE_{ij}R^{-1}x = (\lambda_i - \lambda_j)RE_{ij}R^{-1}$$

or

$$\mathrm{ad}x(RE_{ij}R^{-1}) = (\lambda_i - \lambda_j)(RE_{ij}R^{-1}).$$

Thus $\lambda_i - \lambda_j$ is indeed an eigenvalue of $\mathrm{ad}x$ for all $i, j \in \{1, \ldots, n\}$. This proves the theorem.

Problem 7. Let

$$H(\mathbf{p}, \mathbf{q}) = \frac{1}{2}(p_1^2 + p_2^2 + p_3^2) + U(\|\mathbf{q}\|) \tag{1}$$

be a Hamilton function in $\mathbb{R}^6$, where $\|.\|$ denotes the Euclidean norm, i.e.

$$\|\mathbf{q}\| = \sqrt{q_1^2 + q_2^2 + q_3^2}. \tag{2}$$

(i) Show that

$$h_1(\mathbf{p}, \mathbf{q}) = q_2 p_3 - q_3 p_2, \quad h_2(\mathbf{p}, \mathbf{q}) = q_3 p_1 - q_1 p_3, \quad h_3(\mathbf{p}, \mathbf{q}) = q_1 p_2 - q_2 p_1$$

are first integrals.

(ii) Show that the functions h_1, h_2, h_3 form a Lie algebra under the Poisson bracket. The *Poisson bracket* of two smooth functions f and g is defined as

$$\{f(\mathbf{p}, \mathbf{q}), g(\mathbf{p}, \mathbf{q})\} := \sum_{j=1}^{N} \left(\frac{\partial f}{\partial q_j}\frac{\partial g}{\partial p_j} - \frac{\partial f}{\partial p_j}\frac{\partial g}{\partial q_j}\right)$$

where $N = 3$ in the present case.

(iii) Is the Lie algebra simple? A Lie algebra is called *simple* when it contains no proper ideals.

Solution 7. (i) A function $I(\mathbf{p}, \mathbf{q})$ is called a *first integral* with respect to H if

$$\{I, H\} = 0.$$

This equation is equivalent to $dH/dt = 0$. Since

$$\{h_1, H\} = 0, \quad \{h_2, H\} = 0, \quad \{h_3, H\} = 0$$

we find that h_1, h_2 and h_3 are first integrals of H.

(ii) We obtain

$$\{h_1, h_2\} = h_3, \quad \{h_2, h_3\} = h_1, \quad \{h_3, h_1\} = h_2.$$

We conclude that h_1, h_2, and h_3 form a basis of a Lie algebra.

(iii) A subspace I of a Lie algebra L is called an *ideal* of L if $x \in L$, $y \in I$ together imply $[x, y] \in I$. If L has no ideals except itself and $\{0\}$, and if moreover

$$[L, L] \neq 0$$

we call L simple. Obviously, if L is simple it implies that $Z(L) = \{0\}$, where $Z(L)$ denotes the center of the Lie algebra L. Moreover $[L, L] = L$. From the commutation relations for h_1, h_2, h_3, we find that the Lie algebra is simple.

Problem 8.   Let L be a Lie algebra. If $H \in L$, then we call $A \in L$ a recursion element with respect to H with value $\mu \in \mathbb{C}$ if $A \neq 0$ and

$$[H, A] = \mu A.$$

If $\mu > 0$, then A is called a raising element with value μ and if $\mu < 0$, then A is called a lowering element with value μ. Assume that the elements of L are linear operators in a Hilbert space. Consider the eigenvalue equation

$$H\phi = \lambda \phi \tag{1}$$

and

$$[H, A] = \mu A. \tag{2}$$

Find $H(A\phi)$.

Solution 8.   From (2) we obtain

$$HA = AH + \mu A.$$

Thus

$$H(A\phi) = (\mu A + AH)\phi = \mu(A\phi) + A(H\phi).$$

Using (1) we find

$$H(A\phi) = (\mu + \lambda)(A\phi).$$

This is an eigenvalue equation. Consequently, if $A\phi \neq 0$, then $A\phi$ is an eigenfunction of H with eigenvalue $\lambda + \mu$. Thus, if $\mu \neq 0$ and we have one eigenfunction ϕ for H and a recursion element A for H, then we can generate infinitely many eigenfunctions. If A and B are recursion elements with respect to H with values μ_1 and μ_2 and the product AB is well defined, then

$$[H, AB] = [H, A]B + A[H, B] = (\mu_1 + \mu_2)AB.$$

Thus AB is a recursion element with respect to H with value $\mu_1 + \mu_2$.

Problem 9. Let A, B be $m \times m$ matrices. The *Baker-Campbell-Hausdorff formula* is given by

$$\exp(A)B\exp(-A) = \sum_{n=0}^{\infty} \frac{[A,B]_n}{n!}$$

where the repeated commutator is defined by

$$[A,B]_n := [A,[A,B]_{n-1}]$$

with $[A,B]_0 = B$. Consider the nontrivial Lie algebra with two generators X and Y and the commutation relation

$$[X,Y] = Y.$$

Let

$$U = \exp(\alpha X + \beta Y), \qquad V = \exp(aX)\exp(bY).$$

Find UXU^{-1}, UYU^{-1}, VXV^{-1}, and VYV^{-1}.

Solution 9. Since

$$[\alpha X + \beta Y, X]_n = -\frac{\beta}{\alpha}\alpha^n Y, \qquad n > 0$$

and

$$[\alpha X + \beta Y, Y]_n = \alpha^n Y$$

we obtain

$$UXU^{-1} = X - \frac{\beta}{\alpha}(\exp(\alpha) - 1)Y, \qquad UYU^{-1} = \exp(\alpha)Y.$$

On the other hand, V transforms the generators to

$$VXV^{-1} = X - b\exp(a)Y, \qquad VYV^{-1} = \exp(a)Y.$$

Comparing the two sets of similarity transformations, one obtains

$$a = \alpha, \qquad b = \frac{\beta}{\alpha}(1 - \exp(-\alpha)).$$

Problem 10. Let G be a Lie group with Lie algebra g. For complex z near 1 we consider the function

$$f(z) := \frac{\ln(z)}{z-1} = \sum_{n=0}^{\infty} \frac{(-1)^n}{n+1}(z-1)^n.$$

Then for X, Y near 0 in g we have

$$\exp X \exp Y = \exp C(X, Y)$$

where

$$C(X, Y) = X + Y + \sum_{n=1}^{\infty} \frac{(-1)^n}{n+1} \int_0^1 \left(\sum_{\substack{k, \ell \geq 0 \\ k + \ell \geq 1}} \frac{t^k}{k! \ell!} (\operatorname{ad}X)^k (\operatorname{ad}Y)^\ell \right)^n X \, dt$$

$$= X + Y + \sum_{n=1}^{\infty} \frac{(-1)^n}{n+1} \sum_{\substack{k_1, \ldots, k_n \geq 0 \\ \ell_1, \ldots, \ell_n \geq 0 \\ k_i + \ell_i \geq 1}} \frac{(\operatorname{ad}X)^{k_1} (\operatorname{ad}Y)^{\ell_1} \cdots (\operatorname{ad}X)^{k_n} (\operatorname{ad}Y)^{\ell_n}}{(k_1 + \cdots + k_n + 1) k_1! \cdots k_n! \ell_1! \cdots \ell_n!} X$$

where $(\operatorname{ad}X)Y := [X, Y]$. Thus

$$C(X, Y) = X + Y + \frac{1}{2}[X, Y] + \frac{1}{12}([X, [X, Y]] - [Y, [Y, X]]) + \cdots .$$

Apply this equation to the Bose operators $X = b$ and $Y = b^\dagger$.

Solution 10.   The commutator of b and $b^\dagger$ is given by

$$[b, b^\dagger] = I$$

where I is the identity operator. Since $[b, I] = 0$, $[b^\dagger, I] = 0$ we find

$$C(b, b^\dagger) = b + b^\dagger + I .$$

Problem 11.   The *Nahm equations* is a system of non-linear ordinary differential equations given by

$$\frac{dT_j}{dt} = \frac{1}{2} \sum_{k=1}^{3} \sum_{l=1}^{3} \epsilon_{jkl} [T_k, T_l](t) \tag{1}$$

for three $n \times n$ matrices T_j of complex-valued functions of the variable t. The indices j, k, l range over $1, 2, 3$. The tensor ϵ_{jkl} is the totally anti-symmetric tensor with $\epsilon_{123} = 1$. Let $\{ H_\alpha, \alpha = 1, \ldots, n-1 \}$ be generators of the Cartan-Lie subalgebra $s\ell(n)$, and let $\{ E_\alpha, E_{-\alpha} \}$ be step operators satisfying

$$[H_\alpha, E_{\pm \beta}] = \pm K_{\beta \alpha} E_{\pm \beta} \tag{2a}$$

$$[E_\alpha, E_{-\beta}] = \delta_{\alpha \beta} H_\beta \tag{2b}$$

where $\delta_{\alpha\beta}$ is the Kronecker delta. The $(n-1) \times (n-1)$ matrix $K_{\beta\alpha}$ is the *Cartan matrix* of the Lie algebra $s\ell(n,\mathbb{R})$. The *Cartan matrix* is given by

$$(K_{\beta\alpha}) = \begin{pmatrix} 2 & -1 & 0 & \cdots & 0 & 0 \\ -1 & 2 & -1 & \cdots & 0 & 0 \\ 0 & -1 & 2 & \cdots & 0 & 0 \\ \vdots & \vdots & \vdots & \ddots & \vdots & \vdots \\ 0 & 0 & 0 & \cdots & 2 & -1 \\ 0 & 0 & 0 & \cdots & -1 & 2 \end{pmatrix}.$$

Assume that

$$T_1(t) = \frac{1}{2}i \sum_{\alpha=1}^{n-1} q_\alpha(t)(E_\alpha + E_{-\alpha}), \quad T_2(t) = -\frac{1}{2} \sum_{\alpha=1}^{n-1} q_\alpha(t)(E_\alpha - E_{-\alpha})$$

$$T_3(t) = \frac{1}{2}i \sum_{\alpha=1}^{n-1} p_\alpha(t)H_\alpha$$

where q_α and p_α are smooth functions of t. Find the equations of motion for q_α and p_α.

Solution 11. Substituting T_1, T_2 and T_3 into the Nahm equation (1) and using the commutation relations (2a) and (2b) yields

$$\frac{dp_\alpha}{dt} = q_\alpha^2, \qquad \frac{dq_\alpha}{dt} = \frac{1}{2} \sum_\beta p_\beta K_{\alpha\beta} q_\alpha.$$

If we set

$$\phi_\alpha = 2\ln q_\alpha$$

we obtain

$$\frac{d^2\phi_\alpha}{dt^2} = \sum_{\beta=1}^{n-1} K_{\alpha\beta} \exp(\phi_\beta)$$

which are the *Toda-molecule equations*.

Problem 12. Let A and B be $n \times n$ matrices over $\mathbb{C}$. It is known that if $[A, B] = 0$, then $\exp(A + B) = \exp(A)\exp(B)$. Let C and D be $n \times n$ matrices over $\mathbb{C}$. Assume that $\exp(C + D) = \exp(C)\exp(D)$. Can we conclude that $[C, D] = 0$?

Solution 12. The answer is no. A counterexample is as follows. Let $\boldsymbol{\sigma} = (\sigma_1, \sigma_2, \sigma_3)$ be the Pauli spin matrices. Consider the two unit vectors **m** and **n** in $\mathbb{C}^3$

$$\mathbf{m} = (\cos(\alpha), \sin(\alpha), 0)^T, \qquad \mathbf{n} = (\cos(\beta), \sin(\beta), 0)^T.$$

Then the sum $\mathbf{p} = \mathbf{m} + \mathbf{n}$ is a unit vector if and only if $\cos(\alpha - \beta) = -1/2$. Thus $\alpha - \beta = 2\pi/3$. Moreover, for the vector product of $\mathbf{m}$ and $\mathbf{n}$ we find

$$\mathbf{m} \times \mathbf{n} = (0, 0, \cos(\alpha)\sin(\beta) - \sin(\alpha)\cos(\beta))^T$$

where

$$\cos(\alpha)\sin(\beta) - \sin(\alpha)\cos(\beta) = -\sin(\alpha - \beta).$$

If $\alpha - \beta = 2\pi/3$ we have

$$-\sin(\alpha - \beta) = -\frac{1}{2}\sqrt{3}.$$

Now let

$$C = 2i\pi\sigma \cdot \mathbf{m}, \qquad D = 2i\pi\sigma \cdot \mathbf{n}.$$

Thus $C + D = 2i\pi\sigma \cdot \mathbf{p}$. $\exp(C)$, $\exp(D)$ and $\exp(C + D)$ now represent rotations by 4π around $\mathbf{m}$, $\mathbf{n}$ and $\mathbf{p}$, respectively. Thus

$$\exp(C) = \exp(D) = \exp(C + D) = I$$

although

$$[C, D] = -i4\pi^2\sigma \cdot (\mathbf{m} \times \mathbf{n}) \neq 0.$$

Problem 13.   The semi-simple Lie algebra $s\ell(2, \mathbb{R})$ is spanned by the 2×2 matrices

$$h = \begin{pmatrix} 1 & 0 \\ 0 & -1 \end{pmatrix}, \qquad e = \begin{pmatrix} 0 & 1 \\ 0 & 0 \end{pmatrix}, \qquad f = \begin{pmatrix} 0 & 0 \\ 1 & 0 \end{pmatrix}$$

with

$$[e, f] = h, \qquad [h, e] = 2e, \qquad [h, f] = -2f.$$

(i) Show that the bracket relations are also satisfied by the 3×3 matrices

$$\rho_2(h) = \begin{pmatrix} 2 & 0 & 0 \\ 0 & 0 & 0 \\ 0 & 0 & -2 \end{pmatrix}, \quad \rho_2(e) = \begin{pmatrix} 0 & 2 & 0 \\ 0 & 0 & 1 \\ 0 & 0 & 0 \end{pmatrix}, \quad \rho_2(f) = \begin{pmatrix} 0 & 0 & 0 \\ 1 & 0 & 0 \\ 0 & 2 & 0 \end{pmatrix}.$$

(ii) Extend the result to general n.

Solution 13.   (i) Straightforward calculation yields

$$[\rho_2(e), \rho_2(f)] = \rho_2(h), \quad [\rho_2(h), \rho_2(e)] = 2\rho_2(e), \quad [\rho_2(h), \rho_2(f)] = -2\rho_2(f).$$

For general n we have the $(n + 1) \times (n + 1)$ matrices

$$\rho_n(h) = \begin{pmatrix} n & 0 & \cdots & \cdots & 0 \\ 0 & n-2 & \cdots & \cdots & 0 \\ \vdots & \vdots & \ddots & \vdots & \vdots \\ 0 & 0 & \cdots & -n+2 & 0 \\ 0 & 0 & \cdots & 0 & -n \end{pmatrix}$$

$$\rho_n(e) = \begin{pmatrix} 0 & n & \cdots & \cdots & 0 \\ 0 & 0 & n-1 & \cdots & 0 \\ \vdots & \vdots & \ddots & \vdots & \vdots \\ 0 & 0 & \cdots & \cdots & 1 \\ 0 & 0 & \cdots & \cdots & 0 \end{pmatrix}, \qquad \rho_n(f) = \begin{pmatrix} 0 & 0 & \cdots & \cdots & 0 \\ 1 & 0 & \cdots & \cdots & 0 \\ \vdots & \vdots & \ddots & \vdots & \vdots \\ 0 & 0 & \cdots & 0 & 0 \\ 0 & 0 & \cdots & n & 0 \end{pmatrix}$$

as representations.

Problem 14. The Lie algebra $su(n)$ are the $n \times n$ matrices X with the conditions

$$X^* = -X, \qquad \mathrm{tr}X = 0.$$

Find a basis for $su(3)$.

Solution 14. Let σ_1, σ_2, σ_3 be the Pauli spin matrices. Consider the eight 3×3 matrices

$$S_j = \begin{pmatrix} 0 & 0 \\ 0 & -i\sigma_j \end{pmatrix}, \qquad j = 1, 2, 3$$

$$S_4 = \begin{pmatrix} -2i & 0 & 0 \\ 0 & i & 0 \\ 0 & 0 & i \end{pmatrix}, \quad S_5 = \begin{pmatrix} 0 & -1 & 0 \\ 1 & 0 & 0 \\ 0 & 0 & 0 \end{pmatrix}, \quad S_6 = \begin{pmatrix} 0 & 0 & -1 \\ 0 & 0 & 0 \\ 1 & 0 & 0 \end{pmatrix}$$

$$S_7 = \begin{pmatrix} 0 & i & 0 \\ i & 0 & 0 \\ 0 & 0 & 0 \end{pmatrix}, \quad S_8 = \begin{pmatrix} 0 & 0 & i \\ 0 & 0 & 0 \\ i & 0 & 0 \end{pmatrix}.$$

Obviously, the matrices $S_1, \ldots, S_8$ satisfy the conditions given above. From the equation

$$\sum_{j=1}^{8} c_j S_j = 0_3$$

we find that $c_1 = c_2 = \cdots = c_8 = 0$, where 0_3 is the 3×3 zero matrix. Thus the matrices are linearly independent. Therefore the matrices S_j ($j = 1, 2, \ldots, 8$) form a basis for the Lie algebra $su(3)$. Furthermore, a 9 element basis cannot exist since that would be a basis for the 3×3 matrices and would not satisfy the conditions above.

Problem 15. The Lie algebra $su(4)$ consists of all 4×4 matrices over $\mathbb{C}$ such that

$$A^* = -A, \qquad \mathrm{tr}A = 0.$$

The first condition is that the matrices are skew-hermitian. Give a basis for the Lie algebra under the condition that the elements of the basis are orthogonal to each other with respect to the scalar product

$$\langle A, B \rangle := \mathrm{tr}(AB^*).$$

Solution 15. Since $\text{tr}A = 0$ the dimension of the Lie algebra is 16-1=15. We can built the basis on three diagonal matrices with trace 0 and 12 non-diagonal matrices. We have the three diagonal matrices

$$X_1 = \begin{pmatrix} i & 0 & 0 & 0 \\ 0 & -i & 0 & 0 \\ 0 & 0 & 0 & 0 \\ 0 & 0 & 0 & 0 \end{pmatrix}, \quad X_2 = \frac{1}{\sqrt{3}} \begin{pmatrix} i & 0 & 0 & 0 \\ 0 & i & 0 & 0 \\ 0 & 0 & -2i & 0 \\ 0 & 0 & 0 & 0 \end{pmatrix},$$

$$X_3 = \frac{1}{\sqrt{6}} \begin{pmatrix} i & 0 & 0 & 0 \\ 0 & i & 0 & 0 \\ 0 & 0 & i & 0 \\ 0 & 0 & 0 & -3i \end{pmatrix}$$

and the 12 nondiagonal matrices

$$X_4 = \begin{pmatrix} 0 & i & 0 & 0 \\ i & 0 & 0 & 0 \\ 0 & 0 & 0 & 0 \\ 0 & 0 & 0 & 0 \end{pmatrix}, \quad X_5 = \begin{pmatrix} 0 & 1 & 0 & 0 \\ -1 & 0 & 0 & 0 \\ 0 & 0 & 0 & 0 \\ 0 & 0 & 0 & 0 \end{pmatrix}, \quad X_6 = \begin{pmatrix} 0 & 0 & i & 0 \\ 0 & 0 & 0 & 0 \\ i & 0 & 0 & 0 \\ 0 & 0 & 0 & 0 \end{pmatrix},$$

$$X_7 = \begin{pmatrix} 0 & 0 & 1 & 0 \\ 0 & 0 & 0 & 0 \\ -1 & 0 & 0 & 0 \\ 0 & 0 & 0 & 0 \end{pmatrix}, \quad X_8 = \begin{pmatrix} 0 & 0 & 0 & 0 \\ 0 & 0 & i & 0 \\ 0 & i & 0 & 0 \\ 0 & 0 & 0 & 0 \end{pmatrix}, \quad X_9 = \begin{pmatrix} 0 & 0 & 0 & 0 \\ 0 & 0 & 1 & 0 \\ 0 & -1 & 0 & 0 \\ 0 & 0 & 0 & 0 \end{pmatrix},$$

$$X_{10} = \begin{pmatrix} 0 & 0 & 0 & i \\ 0 & 0 & 0 & 0 \\ 0 & 0 & 0 & 0 \\ i & 0 & 0 & 0 \end{pmatrix}, \quad X_{11} = \begin{pmatrix} 0 & 0 & 0 & 1 \\ 0 & 0 & 0 & 0 \\ 0 & 0 & 0 & 0 \\ -1 & 0 & 0 & 0 \end{pmatrix}, \quad X_{12} = \begin{pmatrix} 0 & 0 & 0 & 0 \\ 0 & 0 & 0 & i \\ 0 & 0 & 0 & 0 \\ 0 & i & 0 & 0 \end{pmatrix},$$

$$X_{13} = \begin{pmatrix} 0 & 0 & 0 & 0 \\ 0 & 0 & 0 & 1 \\ 0 & 0 & 0 & 0 \\ 0 & -1 & 0 & 0 \end{pmatrix}, \quad X_{14} = \begin{pmatrix} 0 & 0 & 0 & 0 \\ 0 & 0 & 0 & 0 \\ 0 & 0 & 0 & i \\ 0 & 0 & i & 0 \end{pmatrix}, \quad X_{15} = \begin{pmatrix} 0 & 0 & 0 & 0 \\ 0 & 0 & 0 & 0 \\ 0 & 0 & 0 & 1 \\ 0 & 0 & -1 & 0 \end{pmatrix}.$$

Problem 16. Let $X_1, X_2, \ldots, X_r$ be the basis of a Lie algebra with the commutator

$$[X_i, X_j] = \sum_{k=1}^{r} C_{ij}^k X_k$$

where the C_{ij}^k are the structure constants. The structure constants satisfy (third fundamental theorem)

$$C_{ij}^k = -C_{ji}^k$$

$$\sum_{m=1}^{r} \left(C_{ij}^m C_{mk}^\ell + C_{jk}^m C_{mi}^\ell + C_{ki}^m C_{mj}^\ell \right) = 0.$$

We replace the X_i's by c-number differential operators (vector fields)

$$X_i \mapsto V_i = \sum_{\ell=1}^{r} \sum_{k=1}^{r} x_k C_{i\ell}^k \frac{\partial}{\partial x_\ell}, \qquad i = 1, 2, \ldots, r.$$

Let

$$V_j = \sum_{n=1}^{r} \sum_{m=1}^{r} x_m C_{jn}^m \frac{\partial}{\partial x_n}.$$

Show that

$$[V_i, V_j] = \sum_{k=1}^{n} C_{ij}^k V_k$$

where

$$V_k = \sum_{n=1}^{r} \sum_{m=1}^{r} x_m C_{kn}^m \frac{\partial}{\partial x_n}.$$

Solution 16.   Since $\partial x_i / \partial x_j = \delta_{ij}$ where δ_{ij} is the Kronecker delta we have

$$[V_i, V_j] = \sum_{k=1}^{r} \sum_{n=1}^{r} \sum_{m=1}^{r} x_k C_{im}^k C_{jn}^m \frac{\partial}{\partial x_n} - \sum_{k=1}^{r} \sum_{m=1}^{r} \sum_{\ell=1}^{r} x_m C_{jk}^m C_{i\ell}^k \frac{\partial}{\partial x_\ell}.$$

Using the first part of the third fundamental theorem ($C_{kj}^m = -C_{jk}^m$) and renaming the index summations we find

$$[V_i, V_j] = \sum_{n=1}^{r} \sum_{m=1}^{r} \sum_{k=1}^{r} x_m C_{nj}^k C_{ki}^m \frac{\partial}{\partial x_n} + \sum_{n=1}^{r} \sum_{m=1}^{r} \sum_{k=1}^{r} x_m C_{in}^k C_{kj}^m \frac{\partial}{\partial x_n}$$

$$= \sum_{n=1}^{r} \sum_{m=1}^{r} x_m \left(\sum_{k=1}^{r} (C_{nj}^k C_{ki}^m + C_{in}^k C_{kj}^m) \right) \frac{\partial}{\partial x_n}.$$

Using the second part of the third fundamental theorem we finally obtain

$$[V_i, V_j] = \sum_{k=1}^{r} C_{ij}^k \left(\sum_{n=1}^{r} \sum_{m=1}^{r} x_m C_{kn}^m \frac{\partial}{\partial x_n} \right) = \sum_{k=1}^{r} C_{ij}^k V_k.$$

Problem 17.   Consider the Lie group $G = O(2,1)$ and its algebra $o(2,1) = \{K_1, K_2, L_3\}$, where K_1, K_2 are Lorentz boosts and L_3 an infinitesimal rotation. The maximal subalgebras of $o(2,1)$ are represented by $\{K_1, K_2 + L_3\}$ and $\{L_3\}$, nonmaximal subalgebras by $\{K_1\}$ and $\{K_2 + L_3\}$. The two-dimensional subalgebra corresponds to the projective group of a real line. The one-dimensional subalgebras correspond to the groups $O(2)$, $O(1,1)$ and the translations $T(1)$, respectively. Find the $o(2,1)$ infinitesimal generators.

Solution 17.   Consider the manifold $O(2,1)/T(1)$, isomorphic to the upper sheet of the cone

$$x_0^2 - x_1^2 - x_2^2 = 0, \qquad x_0 > 0.$$

We introduce *horospheric coordinates* (ξ, η)

$$x_0(\eta, \xi) = \eta(\xi^2 + 1)/2, \qquad x_1(\eta, \xi) = \eta(\xi^2 - 1)/2, \qquad x_2(\eta, \xi) = \xi\eta$$

where $-\infty < \xi < \infty$, $\eta > 0$. The $o(2,1)$ infinitesimal operators are

$$L_3 = \frac{1+\xi^2}{2}\frac{\partial}{\partial \xi} - \xi\eta\frac{\partial}{\partial \eta}, \quad K_1 = \xi\frac{\partial}{\partial \xi} - \eta\frac{\partial}{\partial \eta}, \quad K_2 = \frac{1-\xi^2}{2}\frac{\partial}{\partial \xi} + \xi\eta\frac{\partial}{\partial \eta}.$$

Problem 18.   Show that the operators

$$L_+ = \bar{z}z, \qquad L_- = -\frac{\partial}{\partial z}\frac{\partial}{\partial \bar{z}}$$

$$L_3 = -\frac{1}{2}\left(z\frac{\partial}{\partial z} + \bar{z}\frac{\partial}{\partial \bar{z}} + 1\right), \qquad L_0 = -\frac{1}{2}\left(z\frac{\partial}{\partial z} - \bar{z}\frac{\partial}{\partial \bar{z}} + 1\right)$$

form a basis for the Lie algebra $u(1,1)$ under the commutator.

Solution 18.   We have $[L_0, L_+] = [L_0, L_-] = [L_0, L_+] = 0$ and for the nonzero commutators

$$[L_+, L_-] = -2L_3, \quad [L_+, L_3] = -L_+, \quad [L_-, L_3] = L_-.$$

The Lie algebra $u(1,1)$ can be decomposed into the direct sum $su(1,1) \oplus u(1)$.

Problem 19.   Let $z \in \mathbb{C}$. Consider the vector field

$$L_n := z^{n+1}\frac{d}{dz}, \qquad n \in \mathbb{Z}$$

Calculate the commutator $[L_m, L_n]$.

Solution 19.   We obtain

$$[L_m, L_n] = z^{m+1}\frac{d}{dz}z^{n+1}\frac{d}{dz} - z^{n+1}\frac{d}{dz}z^{m+1}\frac{d}{dz}$$

$$= (n+1)z^{m+n+1}\frac{d}{dz} - (m+1)z^{m+n+1}\frac{d}{dz}$$

$$= (n-m)z^{m+n+1}\frac{d}{dz}$$

$$= (n-m)L_{m+n}.$$

Problem 20. Let $b_1^\dagger$, $b_2^\dagger$ be Bose creation operators. The semisimple Lie algebra $su(1,1)$ is generated by

$$K_+ := b_1^\dagger b_2^\dagger, \quad K_- := b_1 b_2, \quad K_0 := \frac{1}{2}(b_1^\dagger b_1 + b_2^\dagger b_2 + I)$$

with the commutation relations

$$[K_0, K_+] = K_+, \quad [K_0, K_-] = -K_-, \quad [K_-, K_+] = 2K_0$$

where I is the identity operator. We use the ordering K_+, K_-, K_0 for the basis.
(i) Find the adjoint representation for the Lie algebra.
(ii) Find the *Killing form* and metric tensor (g_{ij}) for this Lie algebra.
(iii) Find the *Casimir invariant* C using

$$C = \sum_{i=1}^{3}\sum_{j=1}^{3} g^{ij} X_i X_j$$

where we set $X_1 = K_+$, $X_2 = K_-$, $X_3 = K_0$ and (g^{ij}) is the inverse matrix of (g_{ij}).

Solution 20. (i) Using the ordering given above for the basis we have

$$(\mathrm{ad}K_+)K_+ = [K_+, K_+] = 0$$
$$(\mathrm{ad}K_+)K_- = [K_+, K_-] = -2K_0$$
$$(\mathrm{ad}K_+)K_0 = [K_+, K_0] = -K_+.$$

Thus

$$(K_+ \quad K_- \quad K_0)\begin{pmatrix} 0 & 0 & -1 \\ 0 & 0 & 0 \\ 0 & -2 & 0 \end{pmatrix} = (0 \quad -2K_0 \quad -K_+).$$

Since K_+, K_-, K_0 is a basis of the Lie algebra it follows that

$$\mathrm{ad}K_+ = \begin{pmatrix} 0 & 0 & -1 \\ 0 & 0 & 0 \\ 0 & -2 & 0 \end{pmatrix}.$$

Analogously we find

$$\mathrm{ad}K_- = \begin{pmatrix} 0 & 0 & 0 \\ 0 & 0 & 1 \\ 2 & 0 & 0 \end{pmatrix}, \qquad \mathrm{ad}K_0 = \begin{pmatrix} 1 & 0 & 0 \\ 0 & -1 & 0 \\ 0 & 0 & 0 \end{pmatrix}.$$

With $X_1 = K_+$, $X_2 = K_-$, $X_3 = K_0$ we have

$$g_{ij} = K(X_i, X_j) = \mathrm{tr}(\mathrm{ad}X_i \mathrm{ad}X_j)$$

where tr denotes the trace. We obtain

$$\mathrm{tr}(\mathrm{ad}K_+\mathrm{ad}K_+) = 0, \quad \mathrm{tr}(\mathrm{ad}K_+\mathrm{ad}K_-) = -4, \quad \mathrm{tr}(\mathrm{ad}K_+\mathrm{ad}K_0) = 0$$

$$\mathrm{tr}(\mathrm{ad}K_-\mathrm{ad}K_+) = 0, \quad \mathrm{tr}(\mathrm{ad}K_-\mathrm{ad}K_-) = 0, \quad \mathrm{tr}(\mathrm{ad}K_-\mathrm{ad}K_0) = 0$$

$$\mathrm{tr}(\mathrm{ad}K_0\mathrm{ad}K_+) = 0, \quad \mathrm{tr}(\mathrm{ad}K_0\mathrm{ad}K_-) = 0, \quad \mathrm{tr}(\mathrm{ad}K_0\mathrm{ad}K_0) = 2.$$

Thus we obtain the metric tensor

$$g = \begin{pmatrix} 0 & -4 & 0 \\ -4 & 0 & 0 \\ 0 & 0 & 2 \end{pmatrix}$$

with the inverse matrix

$$g^{-1} = \begin{pmatrix} 0 & -1/4 & 0 \\ -1/4 & 0 & 0 \\ 0 & 0 & 1/2 \end{pmatrix}.$$

Thus

$$C = \sum_{i=1}^{3} \sum_{j=1}^{3} g^{ij} X_i X_j = \frac{1}{2}\left(K_0^2 - \frac{1}{2}(K_+K_- + K_-K_+) \right).$$

Problem 21. Let b, $b^\dagger$ be Bose annihilation and creation operators, respectively. Let I be the identity operator. The *harmonic oscillator Lie algebra* $ho(1)$ is spanned by $\{ I, b, b^\dagger, b^\dagger b \}$. The Lie subalgebra spanned by $\{ I, b, b^\dagger \}$ is known as the *Heisenberg-Weyl Lie algebra* $h(1)$.
(i) Find the commutators for the harmonic oscillator algebra.
(ii) Is the harmonic oscillator Lie algebra semisimple?

Solution 21. (i) We have $[b^\dagger b, b] = -b$, $[b^\dagger b, b^\dagger] = b^\dagger$ and

$$[b, b^\dagger] = I, \quad [b, I] = [b^\dagger, I] = [b^\dagger b, I] = 0.$$

(ii) Owing to the element I the harmonic oscillator Lie algebra is not semisimple.

Chapter 20

Lie Groups

A *Lie group* is a group G which is also an analytic manifold such that the mapping $(a, b) \to ab^{-1}$ $(a, b \in G)$ of the product manifold $G \times G$ into G is analytic. Thus a Lie group is set endowed with compatible structures of a group and an analytic manifold. A subgroup H of a Lie group G is said to be a Lie subgroup if it is a submanifold of the underlying manifold of G. A group G is called a Lie transformation group of a differentiable manifold M if there is a differentiable map $\varphi : G \times M \to M$, $\varphi(g, \mathbf{x}) = g\mathbf{x}$ $(\mathbf{x} \in M)$ such that (i) $(g_1 \cdot g_2)\mathbf{x} = g_1 \cdot (g_2\mathbf{x})$ for $\mathbf{x} \in M$ and (ii) $e\mathbf{x} = \mathbf{x}$ for the identity element e of G and $\mathbf{x} \in M$.

Problem 1. A (global) group of $n \times n$ matrices is compact if it is a bounded, closed subset of the set of all $n \times n$ matrices. A set U of $n \times n$ matrices is bounded if there exists a constant $K > 0$ such that $|A_{ik}| \leq K$ for $1 \leq i, k \leq n$ and all $A \in U$. The set U is closed provided every Cauchy sequence in U converges to a matrix in U. A sequence of $n \times n$ matrices $\{A^{(p)}\}$ is a *Cauchy sequence* if each of the sequences of matrix elements $\{A^{(p)}_{ik}\}$, $1 \leq i, k \leq n$, is Cauchy. Show that the orthogonal group $O(3, \mathbb{R})$ is compact.

Solution 1. If $A \in O(3, \mathbb{R})$ then $A^T A = I_3$, i.e.

$$\sum_{j=1}^{3} A_{j\ell} A_{jk} = \delta_{\ell k} .$$

Setting $\ell = k$ we obtain

$$\sum_{j=1}^{3} (A_{jk})^2 = 1.$$

Thus $|A_{jk}| \leq 1$ for all j, k. Thus the matrix elements are bounded. Let $\{ A^{(p)} \}$ be a Cauchy sequence in $O(3, \mathbb{R})$ with limit A. Then

$$I_3 = \lim_{p \to \infty} (A^{(p)})^T A^{(p)} = A^T A$$

so $A \in O(3, \mathbb{R})$ and the Lie group $O(3, \mathbb{R})$ is compact.

Problem 2.   The Lie group $SU(1,1)$ consists of the set of all 2×2 pseudo-unitary matrices (with determinant 1) preserving the quadratic form $|z_1|^2 - |z_2|^2$ ($z_1, z_2 \in \mathbb{C}$).
(i) Show that the 2×2 matrix

$$U(\tau, \alpha, \beta) = \begin{pmatrix} \cosh(\tau/2) e^{-i\alpha} & \sinh(\tau/2) e^{-i\beta} \\ \sinh(\tau/2) e^{i\beta} & \cosh(\tau/2) e^{i\alpha} \end{pmatrix} \tag{1}$$

preserves the quadratic form $|z_1|^2 - |z_2|^2$ ($z_1, z_2 \in \mathbb{C}$), where $\tau, \alpha, \beta \in \mathbb{R}$.
(ii) Show that $\det U = 1$.
(iii) Give the inverse of U.
(iv) Show that $U_1(\tau_1, \alpha_1, \beta_1) U_2(\tau_2, \alpha_2, \beta_2)$ with $\alpha_1 + \alpha_2 = \beta_1 - \beta_2$ is again a matrix of the form (1).

Solution 2.   (i) We have

$$\begin{pmatrix} \tilde{z}_1 \\ \tilde{z}_2 \end{pmatrix} = U \begin{pmatrix} z_1 \\ z_2 \end{pmatrix} = \begin{pmatrix} z_1 \cosh(\tau/2) e^{-i\alpha} + z_2 \sinh(\tau/2) e^{-i\beta} \\ z_1 \sinh(\tau/2) e^{i\beta} + z_2 \cosh(\tau/2) e^{i\alpha} \end{pmatrix}.$$

Using that $\cosh^2(\tau/2) - \sinh^2(\tau/2) = 1$ we obtain

$$\begin{aligned}
|\tilde{z}_1|^2 - |\tilde{z}_2|^2 &= z_1 z_1^* \cosh^2(\tau/2) + z_2 z_2^* \sinh^2(\tau/2) \\
&\quad - z_1 z_1^* \sinh^2(\tau/2) - z_2 z_2^* \cosh^2(\tau/2) \\
&= z_1 z_1^* - z_2 z_2^* \\
&= |z_1|^2 - |z_2|^2.
\end{aligned}$$

(ii) Using that $\cosh^2(\tau/2) - \sinh^2(\tau/2) = 1$ we obtain $\det U = 1$.
(iii) The inverse of U is given by the replacements $\tau \to -\tau$, $\alpha \to -\alpha$, $\beta \to \beta$. Thus

$$U^{-1} = \begin{pmatrix} \cosh(\tau/2) e^{i\alpha} & -\sinh(\tau/2) e^{-i\beta} \\ -\sinh(\tau/2) e^{i\beta} & \cosh(\tau/2) e^{-i\alpha} \end{pmatrix}.$$

(iv) Using the identities

$$\cosh(x + y) \equiv \cosh(x) \cosh(y) + \sinh(x) \sinh(y)$$

$$\sinh(x+y) \equiv \sinh(x)\cosh(y) + \cosh(x)\sinh(y)$$

and $\alpha_1 + \alpha_2 = \beta_1 - \beta_2$ we obtain

$$U_1 U_2 = \begin{pmatrix} \cosh((\tau_1+\tau_2)/2)e^{-i(\alpha_1+\alpha_2)} & \sinh((\tau_1+\tau_2)/2)e^{-i(\beta_1-\alpha_2)} \\ \sinh((\tau_1+\tau_2)/2)e^{i(\beta_1-\alpha_2)} & \cosh((\tau_1+\tau_2)/2)e^{i(\alpha_1+\alpha_2)} \end{pmatrix}.$$

Problem 3. Let $A \in SL(2,\mathbb{C})$. Show that

$$A^2 = \text{tr}(A)A - I_2$$
$$A + A^{-1} = \text{tr}(A)I_2$$
$$A^n = U_{n-1}(x)A - U_{n-2}I_2$$

where $x := \frac{1}{2}\text{tr}A$ and $U_n(x)$ are Chebyshev's polynomial of the second kind defined by

$$U_{-1}(x) = 0, \qquad U_0(x) = 1, \qquad U_{n+1}(x) = 2xU_n(x) - U_{n-1}(x).$$

Solution 3. These identities are a consequence of the Cayley-Hamilton theorem.

Problem 4. The group $Sp(2n,\mathbb{R})$ consists of all real $2n \times 2n$ matrices S which obey the condition

$$S^T J S = J$$

where J is the $2n \times 2n$ skew-symmetric matrix

$$J := \begin{pmatrix} 0_n & I_n \\ -I_n & 0_n \end{pmatrix}$$

with I_n the $n \times n$ identity matrix and 0_n the $n \times n$ zero matrix. Let V be a $2n \times 2n$ real symmetric positive definite matrix. Show that there exists an $S \in Sp(2n,\mathbb{R})$ such that

$$S^T V S = D^2 > 0, \qquad D^2 = \text{diag}(\kappa_1, \kappa_2, \ldots, \kappa_n, \kappa_1, \kappa_2, \ldots, \kappa_n).$$

Solution 4. Since $J^T = -J$, it follows that $V^{-1/2}JV^{-1/2}$ is antisymmetric. Hence there exists a $2n \times 2n$ matrix $R \in SO(2n)$ such that

$$R^T V^{-1/2}JV^{-1/2}R = \begin{pmatrix} 0_n & \Omega \\ -\Omega & 0_n \end{pmatrix}, \qquad \Omega = \text{diagonal} > 0.$$

We define a diagonal positive definite matrix

$$D = \begin{pmatrix} \Omega^{-1/2} & 0_n \\ 0_n & \Omega^{-1/2} \end{pmatrix}.$$

Then we have
$$DR^T V^{-1/2} JV^{-1/2} RD = J.$$
Now we define $S := V^{-1/2} RD$. Then $S^T JS = J$ and $S^T VS = D^2$, where D is a diagonal matrix with $d_{jj} > 0$ for $j = 1, 2, \ldots, 2n$.

Problem 5.   The group of complex rotations $O(n, \mathbb{C})$ is defined as the group of all $n \times n$ complex matrices O, such that $OO^T = O^T O = I_n$, where T means transpose. These transformations preserve the real scalar product

$$\mathbf{x} \cdot \mathbf{y} = \sum_{j=1}^{n} x_j y_j$$

so that $(O\mathbf{x}) \cdot O\mathbf{y} = \mathbf{x} \cdot \mathbf{y}$, where $\mathbf{x}$ and $\mathbf{y}$ are complex vectors in general, i.e. $x_j, y_j \in \mathbb{C}$.
(i) Show that the matrix ($\alpha \in \mathbb{R}$)

$$O = \begin{pmatrix} \cosh \alpha & i \sinh \alpha \\ -i \sinh \alpha & \cosh \alpha \end{pmatrix}$$

is an element of $O(n, \mathbb{C})$.
(ii) Find the partial derivatives under complex orthogonal transformations $O \in O(n, \mathbb{C})$

$$w_j(\mathbf{x}) := (O\mathbf{x})_j = \sum_{k=1}^{n} O_{jk} x_k, \qquad j = 1, 2, \ldots, n$$

i.e. $\partial/\partial w_j$ with $j = 1, 2, \ldots, n$.

Solution 5.   (i) We have

$$O^T = \begin{pmatrix} \cosh \alpha & -i \sinh \alpha \\ i \sinh \alpha & \cosh \alpha \end{pmatrix}.$$

Thus $O^T O = OO^T = I_2$.
(ii) Using the chain rule we have

$$\frac{\partial}{\partial w_j} = \sum_{k=1}^{n} \frac{\partial x_k}{\partial w_j} \frac{\partial}{\partial x_k} = \sum_{k=1}^{n} O_{kj}^{-1} \frac{\partial}{\partial x_k} = \sum_{k=1}^{n} O_{jk} \frac{\partial}{\partial x_k}.$$

Thus the partial derivatives transform exactly as the coordinates, since $O^{-1} = O^T$.

Problem 6.   Consider the Lie group $SL(2, \mathbb{R})$, i.e. the set of all real 2×2 matrices with determinant equal to 1. A dynamical system in $SL(2, \mathbb{R})$ can be defined by
$$M_{k+2} = M_k M_{k+1} \qquad k = 0, 1, 2, \ldots \tag{1}$$

with the initial matrices $M_0, M_1 \in SL(2, \mathbb{R})$. Let $F_k := \mathrm{tr} M_k$. Show that

$$F_{k+3} = F_{k+2} F_{k+1} - F_k \qquad k = 0, 1, 2, \ldots. \tag{2}$$

Hint. Use that property that for any 2×2 matrix A we have

$$A^2 - A \mathrm{tr}(A) + I_2 \det(A) = 0. \tag{3}$$

Solution 6. From (1) it follows that

$$M_{k+3} = M_{k+1} M_{k+2} = M_{k+1} M_k M_{k+1}.$$

Taking the trace of this equation and cyclic invariance provides

$$\mathrm{tr}(M_{k+3}) = \mathrm{tr}(M_{k+1} M_k M_{k+1}) = \mathrm{tr}(M_{k+1}^2 M_k).$$

Using (3) with $\det M_k = 1$ we arrive at

$$\begin{aligned}
\mathrm{tr}(M_{k+3}) &= \mathrm{tr}((M_{k+1} \mathrm{tr}(M_{k+1}) - I_2) M_k) \\
&= \mathrm{tr}(M_{k+1}) \mathrm{tr}(M_{k+1} M_k) - \mathrm{tr}(M_k) \\
&= \mathrm{tr}(M_{k+1}) \mathrm{tr}(M_{k+2}) - \mathrm{tr}(M_k).
\end{aligned}$$

Thus (2) follows.

Problem 7. The Lie group $SO(m, n)$ consists of all real matrices S that satisfy

$$S^T g S = g$$

where $\det S = 1$ and $g = \mathrm{diag}(+1, +1, \ldots, +1, -1, \ldots, -1)$ with n $+1$'s and m -1's. Let V be a real symmetric positive definite matrix of dimension N. Show that for any choice of partition $N = m + n$, there exists an $S \in SO(m, n)$ such that

$$S^T V S = D^2 = \mathrm{diagonal} \ (\text{and} \ > 0).$$

Solution 7. Consider the matrix $V^{-1/2} g V^{-1/2}$ constructed from the given matrix V. Since $V^{-1/2} g V^{-1/2}$ is real symmetric, there exists a rotation matrix $R \in SO(N)$ which diagonalizes $V^{-1/2} g V^{-1/2}$

$$R^T V^{-1/2} g V^{-1/2} R = \mathrm{diagonal} \equiv \Lambda.$$

This may be viewed also as a congruence of g using $V^{-1/2} R$, and signatures are preserved under congruence. As a consequence, the diagonal matrix Λ can be expressed as the product of a positive diagonal matrix and g

$$R^T V^{-1/2} g V^{-1/2} R = D^{-2} g = D^{-1} g D^{-1}.$$

Here D is diagonal and positive definite. Taking the inverse of the matrices on both sides of this equation we find that the diagonal entries of $gD^2 = D^2g$ are the eigenvalues of $V^{1/2}gV^{1/2}$ and that the columns of R are the eigenvectors of $V^{1/2}gV^{1/2}$. Since $V^{1/2}gV^{1/2}$, gV, and Vg are conjugate to one another, we conclude that D^2 is determined by the eigenvalues of $gV \sim Vg$. We define $S := V^{-1/2}RD$. Then S satisfies the following two equations

$$S^T gS = g, \qquad S^T VS = D^2 = \text{diagonal}.$$

The first equation tells us that $S \in SO(m, n)$ and the second tells us that V is diagonalized through congruence by S.

Problem 8.   Show that the matrices

$$A(\phi, \theta, \psi) = \begin{pmatrix} e^{i\frac{\phi+\psi}{2}} \cos\frac{\theta}{2} & ie^{i\frac{\phi-\psi}{2}} \sin\frac{\theta}{2} \\ ie^{i\frac{\psi-\phi}{2}} \sin\frac{\theta}{2} & e^{-i\frac{\phi+\psi}{2}} \cos\frac{\theta}{2} \end{pmatrix}$$

form a group under matrix multiplication, where ϕ, θ, ψ are the *Euler angles*. The Euler angles are coordinates on the sphere S^3.

Solution 8.   Since

$$A_1(\phi_1, \theta_1, \psi_1)A_2(\phi_2, \theta_2, \psi_2) = A(\phi, \theta, \psi) \tag{2}$$

with

$$\cos\theta = \cos\theta_1 \cos\theta_2 - \sin\theta_1 \sin\theta_2 \cos(\phi_2 - \psi_1)$$

$$e^{i\phi} = \frac{e^{i\phi_1}}{\sin\theta}(\sin\theta_1 \cos\theta_2 + \cos\theta_1 \sin\theta_2 \cos(\phi_2 + \psi_1) + i\sin\theta_2 \sin(\phi_2 + \psi_1))$$

$$e^{i(\phi+\psi)/2} = \frac{e^{i(\phi_1+\psi_1)/2}}{\cos\frac{\theta}{2}}(\cos\frac{\theta_1}{2}\cos\frac{\theta_2}{2}e^{i(\phi_2+\psi_1)/2} - \sin\frac{\theta_1}{2}\sin\frac{\theta_2}{2}e^{-i(\phi_2+\psi_2)/2})$$

we find that the matrices $A(\phi, \theta, \psi)$ are closed under matrix multiplication. The neutral element is given by

$$A(\phi = 0, \theta = 0, \psi = 0) = \begin{pmatrix} 1 & 0 \\ 0 & 1 \end{pmatrix}.$$

Since

$$\det A(\phi, \theta, \psi) = \cos^2\left(\frac{\theta}{2}\right) + \sin^2\left(\frac{\theta}{2}\right) = 1$$

we find that the inverse exists. The matrix A is unitary, i.e.

$$A^{-1}(\phi, \theta, \psi) = A^*(\phi, \theta, \psi) \equiv \bar{A}^T(\phi, \theta, \psi) \tag{1}$$

where A^* denotes the hermitian conjugate matrix of A. Since for arbitrary $n \times n$ matrices the associative law holds we have proved that the matrices $A(\phi, \theta, \psi)$ form a group under matrix multiplication.

Problem 9. The *commutator* $[A, B]$ of two elements A and B of a group G is defined by

$$[A, B] := A^{-1}B^{-1}AB.$$

Consider the Lie group $U(2)$ and $A, B \in U(2)$

$$A = \begin{pmatrix} 0 & 1 \\ 1 & 0 \end{pmatrix}, \qquad B = \frac{1}{\sqrt{2}} \begin{pmatrix} 1 & 1 \\ 1 & -1 \end{pmatrix}.$$

Calculate $[A, B]$. Is $[A, B]$ an element of the Lie group $U(2)$?

Solution 9. Since $A^{-1} = A$, $B = B^{-1}$ we have

$$[A, B] = A^{-1}B^{-1}AB = ABAB = (AB)^2.$$

Thus

$$[A, B] = \begin{pmatrix} 0 & -1 \\ 1 & 0 \end{pmatrix}.$$

Since

$$([A, B])([A, B])^* = I_2$$

the commutator is an element of $U(2)$.

Problem 10. We consider the following subgroups of the Lie group $SL(2, \mathbb{R})$. Let

$$K := \left\{ \begin{pmatrix} \cos\theta & -\sin\theta \\ \sin\theta & \cos\theta \end{pmatrix} : \theta \in [0, 2\pi) \right\}$$

$$A := \left\{ \begin{pmatrix} r^{1/2} & 0 \\ 0 & r^{-1/2} \end{pmatrix} : r > 0 \right\}$$

$$N := \left\{ \begin{pmatrix} 1 & t \\ 0 & 1 \end{pmatrix} : t \in \mathbb{R} \right\}.$$

It can be shown that any matrix $m \in SL(2, \mathbb{R})$ can be written in a unique way as the product $m = kan$ with $k \in K$, $a \in A$ and $n \in N$. This decomposition is called *Iwasawa decomposition* and has a natural generalization to $SL(n, \mathbb{R})$, $n \geq 3$. The notation of the subgroups comes from the fact that K is a compact subgroup, A is an abelian subgroup and N is a nilpotent subgroup of $SL(2, \mathbb{R})$. Find the Iwasawa decomposition of the matrix

$$\begin{pmatrix} 1 & 1 \\ 1 & 2 \end{pmatrix}.$$

Solution 10. From

$$\begin{pmatrix} 1 & 1 \\ 1 & 2 \end{pmatrix} = \begin{pmatrix} \cos\theta & -\sin\theta \\ \sin\theta & \cos\theta \end{pmatrix} \begin{pmatrix} r^{1/2} & 0 \\ 0 & r^{-1/2} \end{pmatrix} \begin{pmatrix} 1 & t \\ 0 & 1 \end{pmatrix}$$

we obtain

$$\begin{pmatrix} 1 & 1 \\ 1 & 2 \end{pmatrix} = \begin{pmatrix} r^{1/2}\cos\theta & tr^{1/2}\cos\theta - r^{-1/2}\sin\theta \\ r^{1/2}\sin\theta & tr^{1/2}\sin\theta + r^{-1/2}\cos\theta \end{pmatrix}.$$

Thus we have the four conditions

$$r^{1/2}\cos\theta = 1$$
$$r^{1/2}t\cos\theta - r^{-1/2}\sin\theta = 1$$
$$r^{1/2}\sin\theta = 1$$
$$r^{1/2}t\sin\theta + r^{-1/2}\cos\theta = 2$$

for the three unknowns r, t, θ. We obtain the solution

$$\cos\theta = \sin\theta = \frac{1}{\sqrt{2}}, \qquad r = 2, \qquad t = \frac{3}{2}.$$

Problem 11.   Consider the vector fields

$$x\frac{d}{dx}, \qquad \frac{d}{dx}.$$

The manifold is $M = \mathbb{R}$, i.e. $x \in \mathbb{R}$. Calculate

$$\exp\left(tx\frac{d}{dx}\right)\exp\left(t\frac{d}{dx}\right)x, \qquad \exp\left(t\frac{d}{dx}\right)\exp\left(tx\frac{d}{dx}\right)x.$$

Solution 11.   For the first case we have

$$\exp\left(tx\frac{d}{dx}\right)\exp\left(t\frac{d}{dx}\right)x = \exp\left(tx\frac{d}{dx}\right)\left(1 + \frac{t}{1!}\frac{d}{dx} + \frac{t^2}{2!}\frac{d^2}{dx^2} + \cdots\right)x$$

$$= \exp\left(tx\frac{d}{dx}\right)(x + t)$$

$$= \exp\left(tx\frac{d}{dx}\right)x + \exp\left(tx\frac{d}{dx}\right)t$$

$$= xe^t + t.$$

Analogously for the second case we find

$$\exp\left(t\frac{d}{dx}\right)\exp\left(tx\frac{d}{dx}\right)x = e^t(x + t).$$

Chapter 21

Differential Forms

We define differential p-forms of class C^∞ on an open set Ω of $\mathbb{R}^n$ to be the expressions

$$\omega := \sum_{j_1 < j_2 \ldots < j_p}^{n} c_{j_1 j_2 \ldots j_p}(\mathbf{x}) dx_{j_1} \wedge dx_{j_2} \wedge \cdots \wedge dx_{j_p}$$

where the functions $c_{j_1 j_2 \ldots j_p} \in C^\infty(\Omega)$ and the integers j_1, ..., j_p lie between 1 and n. Two such differential forms may be added componentwise. One defines the *Graßmann product* (also called *exterior product* or *wedge product*) of a p-form and a q-form as follows: For any permutation σ of the indices $j_1, \ldots, j_p$,

$$dx_{\sigma(j_1)} \wedge dx_{\sigma(j_2)} \wedge \cdots \wedge dx_{\sigma(j_p)} = \operatorname{sgn}(\sigma) dx_{j_1} \wedge dx_{j_2} \wedge \cdots \wedge dx_{j_p}$$

where $\operatorname{sgn}(\sigma)$ denotes the sign of the permutation σ. Let

$$\omega' = \sum_{k_1 < k_2 \ldots < k_q}^{n} b_{k_1 k_2 \ldots k_q}(\mathbf{x}) dx_{k_1} \wedge dx_{k_2} \wedge \cdots \wedge dx_{k_q}.$$

Then

$$\omega \wedge \omega' := \sum_{\substack{j_1 < j_2 < \ldots < j_p \\ k_1 < k_2 < \ldots < k_q}}^{n} c_{j_1 j_2 \ldots j_p}(\mathbf{x}) b_{k_1 k_2 \ldots k_q}(\mathbf{x}) dx_{j_1} \wedge \cdots \wedge dx_{j_p} \wedge dx_{k_1} \wedge \cdots \wedge dx_{k_q}.$$

The *exterior derivative* of a p-form is defined by

$$d\omega = \sum_{j_1 < j_2 \ldots < j_p}^{n} dc_{j_1 j_2 \ldots j_p}(\mathbf{x}) \wedge dx_{j_1} \wedge dx_{j_2} \wedge \cdots \wedge dx_{j_p}.$$

For a smooth function f (a zero form) we have

$$df = \sum_{j=1}^{n} \frac{\partial f}{\partial x_j} dx_j.$$

Problem 1.   Let $\Omega = \mathbb{R}^3$ and consider the one-form

$$\alpha := f_1(\mathbf{x})dx_1 + f_2(\mathbf{x})dx_2 + f_3(\mathbf{x})dx_3. \tag{1}$$

Calculate $d\alpha$. Let $d\alpha = 0$. Find the condition on the functions f_1, f_2 and f_3.

Solution 1.   Applying $dx_j \wedge dx_k = -dx_k \wedge dx_j$ yields

$$d\alpha = \left(\frac{\partial f_2}{\partial x_1} - \frac{\partial f_1}{\partial x_2} \right) dx_1 \wedge dx_2 + \left(\frac{\partial f_3}{\partial x_2} - \frac{\partial f_2}{\partial x_3} \right) dx_2 \wedge dx_3$$
$$+ \left(\frac{\partial f_1}{\partial x_3} - \frac{\partial f_3}{\partial x_1} \right) dx_3 \wedge dx_1.$$

Thus the condition $d\alpha = 0$ leads to

$$\frac{\partial f_2}{\partial x_1} - \frac{\partial f_1}{\partial x_2} = 0, \qquad \frac{\partial f_3}{\partial x_2} - \frac{\partial f_2}{\partial x_3} = 0, \qquad \frac{\partial f_1}{\partial x_3} - \frac{\partial f_3}{\partial x_1} = 0$$

since $dx_1 \wedge dx_2$, $dx_2 \wedge dx_3$ and $dx_3 \wedge dx_1$ are linearly independent.

Problem 2.   Let $f : \mathbb{R}^n \to \mathbb{R}$ be a C^1 function with $f(\mathbf{x}) \neq 0$ for all $\mathbf{x} \in \mathbb{R}^n$. Let ω be a differential one-form defined on $\mathbb{R}^n$. Assume that

$$d(f\omega) = 0. \tag{1}$$

Show that $\omega \wedge d\omega = 0$.

Solution 2.   From (1) it follows that

$$d(f\omega) = (df) \wedge \omega + fd\omega = 0.$$

Taking the exterior product with ω yields

$$\omega \wedge (df) \wedge \omega + \omega \wedge (fd\omega) = 0.$$

Applying the associative law for differential forms and that $\omega \wedge \omega = 0$ for differential one-forms gives

$$\omega \wedge (fd\omega) = 0.$$

Thus
$$f\omega \wedge d\omega = 0.$$
Since $f(\mathbf{x}) \neq 0$ for all $\mathbf{x} \in \mathbb{R}^n$, equation $\omega \wedge d\omega = 0$ follows.

Problem 3. Consider a differential p-form ω defined on $\mathbb{R}^n$. If there is a $(p-1)$-form ψ defined on $\mathbb{R}^n$ such that

$$\omega = d\psi \tag{1}$$

then ω is called an *exact differential form*.
(i) Show that if ω_1 and ω_2 are exact p-differential forms then $\omega_1 \wedge \omega_2$ is also exact.
(ii) Give an example.

Solution 3. (i) Since ω_1 and ω_2 are exact we have

$$\omega_1 = d\psi_1, \qquad \omega_2 = d\psi_2.$$

Since ω_1 and ω_2 are differential p-forms we find that $\omega_1 \wedge \omega_2$ is a $2p$ differential form. We have to find a $2p - 1$ differential form α defined on $\mathbb{R}^n$ such that

$$d\alpha = \omega_1 \wedge \omega_2.$$

Thus $d\alpha = d\psi_1 \wedge d\psi_2$. Since

$$d(\psi_1 \wedge d\psi_2) = d\psi_1 \wedge d\psi_2$$

where we used that $dd\psi_2 = 0$ we find $\alpha = \psi_1 \wedge d\psi_2$.
(ii) Consider
$$\omega_1 = dx_1 \wedge dx_2, \qquad \omega_2 = dx_3 \wedge dx_4$$
defined on $\mathbb{R}^4$. Then we can choose ψ_1 and ψ_2 as

$$\psi_1 = x_1 dx_2, \qquad \psi_2 = x_3 dx_4.$$

Therefore
$$\psi_1 \wedge d\psi_2 = \psi_1 \wedge \omega_2 = x_1 dx_2 \wedge dx_3 \wedge dx_4.$$

ψ_1 and ψ_2 are not unique, for example we could also choose

$$\psi_1 = \frac{1}{2}(x_1 dx_2 - x_2 dx_1), \qquad \psi_2 = \frac{1}{2}(x_3 dx_4 - x_4 dx_3).$$

Problem 4. Consider the complex number $z = re^{i\phi}$. Calculate

$$\frac{dz \wedge d\bar{z}}{z}.$$

Solution 4.   Note that $\bar{z} = re^{-i\phi}$. Then

$$dz = e^{i\phi}dr + ie^{i\phi}rd\phi, \qquad d\bar{z} = e^{-i\phi}dr - ie^{-i\phi}rd\phi.$$

Thus

$$dz \wedge d\bar{z} = (e^{i\phi}dr + ie^{i\phi}rd\phi) \wedge (e^{-i\phi}dr - ie^{-i\phi}rd\phi)$$
$$= -irdr \wedge d\phi + ird\phi \wedge dr = -2irdr \wedge d\phi.$$

It follows that
$$\frac{dz \wedge d\bar{z}}{z} = -2ie^{-i\phi}dr \wedge d\phi.$$

Problem 5.   The differential form

$$\alpha = \frac{dz_1 \wedge dz_2}{z_1 z_2}$$

where $z_j = x_j + iy_j$ $(x_j, y_j \in \mathbb{R})$ is defined on

$$\mathbb{C}^2 \setminus \{(z_1, z_2) : z_1 = 0 \vee z_2 = 0\}.$$

Find the real and imaginary part of α.

Solution 5.   Since $dz_j = dx_j + idy_j$, $j = 1, 2$ and

$$z_1 z_2 = x_1 x_2 - y_1 y_2 + i(x_1 y_2 + x_2 y_1)$$

we obtain written as a sum of real and imaginary part

$$\frac{dz_1 \wedge dz_2}{z_1 z_2} = \frac{(x_1 x_2 - y_1 y_2)\omega + (x_1 y_2 + x_2 y_1)(dx_1 \wedge dy_2 + dy_1 \wedge dx_2)}{x_1^2 x_2^2 + y_1^2 y_2^2 + x_1^2 y_2^2 + x_2^2 y_1^2}$$
$$+ i\frac{(x_1 x_2 - y_1 y_2)(dx_1 \wedge dy_2 + dy_1 \wedge dx_2) - (x_1 y_2 + x_2 y_1)\omega}{x_1^2 x_2^2 + y_1^2 y_2^2 + x_1^2 y_2^2 + x_2^2 y_1^2}$$

where $\omega := dx_1 \wedge dx_2 - dy_1 \wedge dy_2$.

Problem 6.   Let

$$g := \sum_{j,k=1}^{m} g_{jk}(\mathbf{x})dx_j \otimes dx_k$$

be the *metric tensor field* of a Riemannian or pseudo-Riemannian real C^∞ manifold M (dim $M = m < \infty$). Let $(x_1, x_2, \ldots, x_m)$ be the local coordinate system in a local coordinate neighbourhood (U, ψ). Since $(dx_j)_p$ $(j = 1, \ldots, m)$ is a basis of T_p^* (dual space of the tangent vector space T_p)

at each point p of U, we can express an r-form ω_p ($p \in U$) uniquely in the form

$$\omega_p = \sum_{j_1 < \cdots < j_r}^{m} a_{j_1 \cdots j_r}(p)(dx_{j_1})_p \wedge \cdots \wedge (dx_{j_r})_p.$$

The *Hodge star operator* is an f-linear mapping which transforms an r-form into its dual $(m - r)$-form. The *-operator which is applied to an r-form defined on an arbitrary Riemannian or pseudo-Riemannian manifold M with metric tensor field g, is defined by

$$*(dx_{j_1} \wedge dx_{j_2} \wedge \cdots \wedge dx_{j_r}) :=$$

$$\sum_{k_1, \cdots, k_m = 1}^{m} g^{j_1 k_1} \cdots g^{j_r k_r} \frac{1}{(m - r)!} \cdot \frac{g}{\sqrt{|g|}} \varepsilon_{k_1 \ldots k_m} dx_{k_{r+1}} \wedge \cdots \wedge dx_{k_m}$$

where $\varepsilon_{k_1 \ldots k_m}$ is the totally antisymmetric tensor with $\varepsilon_{12 \ldots m} = +1$ and

$$g \equiv \det(g_{ij}), \qquad \sum_{j=1}^{m} g^{ij} g_{jk} = \delta_k^i$$

where δ_k^i denotes the Kronecker delta.

(i) Let $M = \mathbb{R}^2$ and $g = dx_1 \otimes dx_1 + dx_2 \otimes dx_2$. Calculate $(*dx_1)$ and $(*dx_2)$.

(ii) Let $M = \mathbb{R}^3$ and $g = dx_1 \otimes dx_1 + dx_2 \otimes dx_2 + dx_3 \otimes dx_3$. Calculate $*dx_1, *dx_2, *dx_3, *(dx_1 \wedge dx_2), *(dx_2 \wedge dx_3), *(dx_3 \wedge dx_1)$.

(iii) Let $M = \mathbb{R}^4$ and

$$g = dx_1 \otimes dx_1 + dx_2 \otimes dx_2 + dx_3 \otimes dx_3 - dx_4 \otimes dx_4.$$

This is the *Minkowski metric*. Calculate $*dx_1, *dx_2, *dx_3, *dx_4$ and

$$*(dx_1 \wedge dx_2), \qquad *(dx_2 \wedge dx_3), \qquad *(dx_3 \wedge dx_1)$$

$$*(dx_1 \wedge dx_4), \qquad *(dx_2 \wedge dx_4), \qquad *(dx_3 \wedge dx_4)$$

Solution 6. (i) Since $g_{11} = g_{22} = 1$, $g_{12} = g_{21} = 0$ we find

$$g^{11} = g^{22} = 1, \qquad g^{12} = g^{21} = 0.$$

Therefore

$$*dx_1 = \sum_{k_1, k_2 = 1}^{2} g^{1 k_1} \frac{1}{(2 - 1)!} \frac{g}{\sqrt{|g|}} \varepsilon_{k_1 k_2} dx_{k_2}.$$

We have $g = \det(g_{ij}) = 1$ and $\varepsilon_{12} = 1$, $\varepsilon_{21} = -1$, $\varepsilon_{11} = \varepsilon_{22} = 0$. It follows that

$$*dx_1 = \sum_{j_2 = 1}^{2} g^{11} \varepsilon_{1 j_2} dx_{j_2} = dx_2.$$

Similarly, $*dx_2 = -dx_1$.

(ii) Since $g_{11} = g_{22} = g_{33} = 1$ and $g_{jk} = 0$ for $j \neq k$ we obtain

$$g^{11} = g^{22} = g^{33} = 1 \quad \text{and} \quad g^{jk} = 0 \quad \text{for} \quad j \neq k.$$

Therefore we find for $*dx_1$

$$*dx_1 = \sum_{k_1,k_2,k_3=1}^{3} g^{1k_1} \frac{1}{(3-1)!} \varepsilon_{k_1 k_2 k_3} dx_{k_2} \wedge dx_{k_3} = \frac{1}{2} \sum_{k_2,k_3=1}^{3} \varepsilon_{1k_2 k_3} dx_{k_2} \wedge dx_{k_3}.$$

Thus $*dx_1 = dx_2 \wedge dx_3$. Analogously $*dx_2 = dx_3 \wedge dx_1$, $*dx_3 = dx_1 \wedge dx_2$. Furthermore

$$*(dx_1 \wedge dx_2) = \sum_{k_1,k_2,k_3=1}^{3} g^{1k_1} g^{2k_2} \varepsilon_{k_1 k_2 k_3} dx_{k_3} = \sum_{k_3=1}^{3} \varepsilon_{12k_3} dx_{k_3} = dx_3.$$

Analogously $*(dx_2 \wedge dx_3) = dx_1$, $*(dx_3 \wedge dx_1) = dx_2$.

(iii) Here we have $g_{11} = g_{22} = g_{33} = 1$, $g_{44} = -1$. Therefore

$$g^{11} = g^{22} = g^{33} = 1 \qquad g^{44} = -1.$$

Moreover $g = \det(g_{jk}) = -1$. Consequently, we find

$$*dx_1 = -dx_2 \wedge dx_3 \wedge dx_4, \qquad *dx_2 = -dx_3 \wedge dx_1 \wedge dx_4$$

$$*dx_3 = -dx_1 \wedge dx_2 \wedge dx_4, \qquad *dx_4 = -dx_1 \wedge dx_2 \wedge dx_3$$

and

$$*(dx_1 \wedge dx_2) = -dx_3 \wedge dx_4, \qquad *(dx_2 \wedge dx_3) = -dx_1 \wedge dx_4$$

$$*(dx_3 \wedge dx_1) = -dx_2 \wedge dx_4, \qquad *(dx_1 \wedge dx_4) = dx_2 \wedge dx_3$$

$$*(dx_2 \wedge dx_4) = dx_3 \wedge dx_1, \qquad *(dx_3 \wedge dx_4) = dx_1 \wedge dx_2.$$

Problem 7. In electrodynamics we have the differential two-forms

$$\beta = E_1 dx_1 \wedge dt + E_2 dx_2 \wedge dt + E_3 dx_3 \wedge dt$$
$$+ B_3 dx_1 \wedge dx_2 + B_1 dx_2 \wedge dx_3 + B_2 dx_3 \wedge dx_1$$

and

$$*\beta = \frac{E_1}{c} dx_2 \wedge dx_3 + \frac{E_2}{c} dx_3 \wedge dx_1 + \frac{E_3}{c} dx_1 \wedge dx_2$$
$$- B_3 c\, dx_3 \wedge dt - B_1 c\, dx_1 \wedge dt - B_2 c\, dx_2 \wedge dt$$

where c is a positive constant (speed of light).

(i) Calculate $d\beta$, $d(*\beta)$ and give an interpretation of $d\beta = 0$ and $d(*\beta) = 0$.

(ii) Calculate $\beta \wedge \beta$ and $\beta \wedge (*\beta)$ and give an interpretation.

Solution 7. (i) We set $x_4 \equiv ct$. Since

$$dE_k = \sum_{j=1}^{4} \frac{\partial E_k}{\partial x_j} dx_j, \qquad dB_k = \sum_{j=1}^{4} \frac{\partial B_k}{\partial x_j} dx_j$$

for $k = 1, 2, 3$ and $dx_j \wedge dx_k = -dx_k \wedge dx_j$ we find

$$\begin{aligned}
d\beta = &\left(\frac{\partial B_1}{\partial x_1} + \frac{\partial B_2}{\partial x_2} + \frac{\partial B_3}{\partial x_3} \right) dx_1 \wedge dx_2 \wedge dx_3 \\
&+ \left(\frac{\partial B_3}{\partial t} - \frac{\partial E_1}{\partial x_2} + \frac{\partial E_2}{\partial x_1} \right) dx_1 \wedge dx_2 \wedge dt \\
&+ \left(\frac{\partial B_2}{\partial t} - \frac{\partial E_3}{\partial x_1} + \frac{\partial E_1}{\partial x_3} \right) dx_3 \wedge dx_1 \wedge dt \\
&+ \left(\frac{\partial B_1}{\partial t} - \frac{\partial E_2}{\partial x_3} + \frac{\partial E_3}{\partial x_2} \right) dx_2 \wedge dx_3 \wedge dt.
\end{aligned}$$

From the condition $d\beta = 0$ we obtain

$$0 = \left(\frac{\partial B_1}{\partial x_1} + \frac{\partial B_2}{\partial x_2} + \frac{\partial B_3}{\partial x_3} \right)$$

$$0 = \left(\frac{\partial B_3}{\partial t} - \frac{\partial E_1}{\partial x_2} + \frac{\partial E_2}{\partial x_1} \right)$$

$$0 = \left(\frac{\partial B_2}{\partial t} - \frac{\partial E_3}{\partial x_1} + \frac{\partial E_1}{\partial x_3} \right)$$

$$0 = \left(\frac{\partial B_1}{\partial t} - \frac{\partial E_2}{\partial x_3} + \frac{\partial E_3}{\partial x_2} \right).$$

Calculating $d(*\beta)$ yields

$$\begin{aligned}
d(*\beta) = &\frac{1}{c} \left(\frac{\partial E_1}{\partial x_1} + \frac{\partial E_2}{\partial x_2} + \frac{\partial E_3}{\partial x_3} \right) dx_1 \wedge dx_2 \wedge dx_3 \\
&+ \left(\frac{1}{c} \frac{\partial E_3}{\partial t} - c \frac{\partial B_2}{\partial x_1} + c \frac{\partial B_1}{\partial x_2} \right) dx_1 \wedge dx_2 \wedge dt \\
&+ \left(\frac{1}{c} \frac{\partial E_2}{\partial t} - c \frac{\partial B_1}{\partial x_3} + c \frac{\partial B_3}{\partial x_1} \right) dx_3 \wedge dx_1 \wedge dt \\
&+ \left(\frac{1}{c} \frac{\partial E_1}{\partial t} - c \frac{\partial B_3}{\partial x_2} + c \frac{\partial B_2}{\partial x_3} \right) dx_2 \wedge dx_3 \wedge dt.
\end{aligned}$$

From the condition $d(*\beta) = 0$ we obtain

$$0 = \left(\frac{\partial E_1}{\partial x_1} + \frac{\partial E_2}{\partial x_2} + \frac{\partial E_3}{\partial x_3} \right)$$

$$0 = \left(\frac{1}{c} \frac{\partial E_3}{\partial t} - c \frac{\partial B_2}{\partial x_1} + c \frac{\partial B_1}{\partial x_2} \right)$$

$$0 = \left(\frac{1}{c} \frac{\partial E_2}{\partial t} - c \frac{\partial B_1}{\partial x_3} + c \frac{\partial B_3}{\partial x_1} \right)$$

$$0 = \left(\frac{1}{c} \frac{\partial E_1}{\partial t} - c \frac{\partial B_3}{\partial x_2} + c \frac{\partial B_2}{\partial x_3} \right).$$

These systems of partial differential equations are *Maxwell's equations* in free space, i.e.

$$\operatorname{div} \mathbf{B} = 0, \qquad -\frac{\partial \mathbf{B}}{\partial t} = \nabla \times \mathbf{E}$$

$$\operatorname{div} \mathbf{E} = 0, \qquad \frac{1}{c^2} \frac{\partial \mathbf{E}}{\partial t} = \nabla \times \mathbf{B}.$$

(ii) Straightforward calculation yields

$$\beta \wedge \beta = 2(B_1 E_1 + B_2 E_2 + B_3 E_3) dx_1 \wedge dx_2 \wedge dx_3 \wedge dt$$

$$\beta \wedge (*\beta) = \left(\frac{E_1^2}{c} + \frac{E_2^2}{c} + \frac{E_3^2}{c} - B_1^2 c - B_2^2 c - B_3^2 c \right) dx_1 \wedge dx_2 \wedge dx_3 \wedge dt.$$

This equation describes the energy density of the electromagnetic field. Both equations are invariant under the Lorentz transformation. Note that $** \beta = -\beta$.

Problem 8. Within the techniques of differential forms the *vector potential* and *scalar potential* is given by the one-form

$$\alpha = A_1 dx_1 + A_2 dx_2 + A_3 dx_3 - U dt$$

and the electromagnetic field by the two-form

$$\beta = E_1 dx_1 \wedge dt + E_2 dx_2 \wedge dt + E_3 dx_3 \wedge dt$$
$$+ B_3 dx_1 \wedge dx_2 + B_1 dx_2 \wedge dx_3 + B_2 dx_3 \wedge dx_1.$$

Find the relations which follow from $d\alpha = \beta$.

Solution 8. We set $ct \equiv x_4$. Since $dx_j \wedge dx_k = -dx_k \wedge dx_j$ and

$$dA_k = \frac{\partial A_k}{\partial x_1} dx_1 + \frac{\partial A_k}{\partial x_2} dx_2 + \frac{\partial A_k}{\partial x_3} dx_3 + \frac{\partial A_k}{\partial t} dt$$

$$dU = \frac{\partial U}{\partial x_1} dx_1 + \frac{\partial U}{\partial x_2} dx_2 + \frac{\partial U}{\partial x_3} dx_3 + \frac{\partial U}{\partial t} dt$$

we find

$$d\alpha = \left(\frac{\partial A_2}{\partial x_1} - \frac{\partial A_1}{\partial x_2} \right) dx_1 \wedge dx_2 + \left(\frac{\partial A_3}{\partial x_2} - \frac{\partial A_2}{\partial x_3} \right) dx_2 \wedge dx_3$$

$$+\left(\frac{\partial A_1}{\partial x_3} - \frac{\partial A_3}{\partial x_1}\right)dx_3 \wedge dx_1 + \left(-\frac{\partial A_1}{\partial t} - \frac{\partial U}{\partial x_1}\right)dx_1 \wedge dt$$

$$+\left(-\frac{\partial A_2}{\partial t} - \frac{\partial U}{\partial x_2}\right)dx_2 \wedge dt + \left(-\frac{\partial A_3}{\partial t} - \frac{\partial U}{\partial x_3}\right)dx_3 \wedge dt .$$

Comparing the six basis elements $dx_1 \wedge dx_2$, $dx_2 \wedge dx_3$, $dx_3 \wedge dx_1$, $dx_1 \wedge dt$, $dx_2 \wedge dt$, $dx_3 \wedge dt$ we find

$$B_3 = \frac{\partial A_2}{\partial x_1} - \frac{\partial A_1}{\partial x_2}, \quad B_1 = \frac{\partial A_3}{\partial x_2} - \frac{\partial A_2}{\partial x_3}, \quad B_2 = \frac{\partial A_1}{\partial x_3} - \frac{\partial A_3}{\partial x_1}$$

$$E_1 = -\frac{\partial A_1}{\partial t} - \frac{\partial U}{\partial x_1}, \quad E_2 = -\frac{\partial A_2}{\partial t} - \frac{\partial U}{\partial x_2}, \quad E_3 = -\frac{\partial A_3}{\partial t} - \frac{\partial U}{\partial x_3} .$$

It follows that

$$\mathbf{B} = \nabla \times \mathbf{A}, \qquad \mathbf{E} = -\nabla U - \frac{\partial \mathbf{A}}{\partial t} .$$

Problem 9. The basic quantity in electromagnetism is the differential two-form

$$\beta = E_1(\mathbf{x},t)dx_1 \wedge dt + E_2(\mathbf{x},t)dx_2 \wedge dt + E_3(\mathbf{x},t)dx_3 \wedge dt$$

$$+ B_3(\mathbf{x},t)dx_1 \wedge dx_2 + B_1(\mathbf{x},t)dx_2 \wedge dx_3 + B_2(\mathbf{x},t)dx_3 \wedge dx_1 .$$

The system $'$ (prime) and without prime are connected by the *Lorentz transformation*

$$x_1' = \gamma(x_1 - vt), \qquad t' = \gamma(t - v\frac{x_1}{c^2}), \qquad x_2' = x_2, \qquad x_3' = x_3 \qquad (1)$$

where $\gamma := 1/\sqrt{1 - v^2/c^2}$ and v with $0 \le v < c$ is a constant. If β is a "physical quantity", then

$$\beta' = \beta \qquad (2)$$

where

$$\beta = E_1(\mathbf{x},t)dx_1 \wedge dt + \cdots + B_3(\mathbf{x},t)dx_1 \wedge dx_2$$

and

$$\beta' = E_1'(\mathbf{x}'(\mathbf{x},t),t'(\mathbf{x},t))dx_1'(\mathbf{x},t) \wedge dt'(\mathbf{x},t) + \cdots$$

$$+ B_3'(\mathbf{x}'(\mathbf{x},t),t'(\mathbf{x},t))dx_1'(\mathbf{x},t) \wedge dx_2'(\mathbf{x},t) .$$

Find the transformation law for $\mathbf{B}$ and $\mathbf{E}$ from the condition (2).

Solution 9. From the Lorentz transformation it follows that

$$dx_1'(\mathbf{x},t) = \gamma(dx_1 - vdt), \qquad dt'(\mathbf{x},t) = \gamma(dt - (v/c^2)dx_1)$$

$$dx_2'(\mathbf{x}, t) = dx_2, \qquad dx_3'(\mathbf{x}, t) = dx_3.$$

Consequently,

$$dx_1'(\mathbf{x}, t) \wedge dt'(\mathbf{x}, t) = \gamma^2 dx_1 \wedge dt - \gamma^2 (v^2/c^2) dx_1 \wedge dt = dx_1 \wedge dt$$

$$dx_2'(\mathbf{x}, t) \wedge dt'(\mathbf{x}, t) = dx_2 \wedge \gamma(dt - (v/c^2)dx_1) = \gamma dx_2 \wedge dt + \gamma(v/c^2)dx_1 \wedge dx_2$$

$$dx_3'(\mathbf{x}, t) \wedge dt'(\mathbf{x}, t) = dx_3 \wedge \gamma(dt - (v/c^2)dx_1) = \gamma dx_3 \wedge dt - \gamma(v/c^2)dx_3 \wedge dx_1$$

$$dx_1'(\mathbf{x}, t) \wedge dx_2'(\mathbf{x}, t) = \gamma(dx_1 - vdt) \wedge dx_2 = \gamma dx_1 \wedge dx_2 + \gamma v dx_2 \wedge dt$$

$$dx_2'(\mathbf{x}, t) \wedge dx_3'(\mathbf{x}, t) = dx_2 \wedge dx_3$$

$$dx_3'(\mathbf{x}, t) \wedge dx_1'(\mathbf{x}, t) = dx_3 \wedge \gamma(dx_1 - vdt) = \gamma dx_3 \wedge dx_1 - \gamma v dx_3 \wedge dt.$$

Inserting these equations into the two-form β' yields

$$\begin{aligned}
\beta = \ & E_1'(\mathbf{x}'(\mathbf{x}, t), t'(\mathbf{x}, t))dx_1 \wedge dt \\
& + E_2'(\mathbf{x}'(\mathbf{x}, t), t'(\mathbf{x}, t))(\gamma dx_2 \wedge dt + \gamma(v/c^2)dx_1 \wedge dx_2) \\
& + E_3'(\mathbf{x}'(\mathbf{x}, t), t'(\mathbf{x}, t))(\gamma dx_3 \wedge dt - \gamma(v/c^2)dx_3 \wedge dx_1) \\
& + B_3'(\mathbf{x}'(\mathbf{x}, t), t'(\mathbf{x}, t))(\gamma dx_1 \wedge dx_2 + \gamma v dx_2 \wedge dt) \\
& + B_1'(\mathbf{x}'(\mathbf{x}, t), t'(c, t))dx_2 \wedge dx_3 \\
& + B_2'(\mathbf{x}'(\mathbf{x}, t), t'(\mathbf{x}, t))(\gamma dx_3 \wedge dx_1 - \gamma v dx_3 \wedge dt).
\end{aligned}$$

Comparing the six basis elements of the two-forms

$$dx_1 \wedge dt, \qquad dx_2 \wedge dt, \qquad dx_3 \wedge dt, \qquad dx_1 \wedge dx_2, \qquad dx_2 \wedge dx_3, \qquad dx_3 \wedge dx_1$$

we find the transformation law for $\mathbf{E}$ and $\mathbf{B}$

$$\begin{aligned}
E_1(\mathbf{x}, t) &= E_1'(\mathbf{x}'(\mathbf{x}, t), t'(\mathbf{x}, t)) \\
B_1(\mathbf{x}, t) &= B_1'(\mathbf{x}'(\mathbf{x}, t), t'(\mathbf{x}, t)) \\
E_2(\mathbf{x}, t) &= \gamma E_2'(\mathbf{x}'(\mathbf{x}, t), t'(\mathbf{x}, t)) + \gamma v B_3'(\mathbf{x}'(\mathbf{x}, t), t'(\mathbf{x}, t)) \\
E_3(\mathbf{x}, t) &= \gamma E_3'(\mathbf{x}'(\mathbf{x}, t), t'(\mathbf{x}, t)) - \gamma v B_2'(\mathbf{x}'(\mathbf{x}, t), t'(\mathbf{x}, t)) \\
B_3(\mathbf{x}, t) &= \gamma B_3'(\mathbf{x}'(\mathbf{x}, t), t'(\mathbf{x}, t)) + \gamma(v/c^2) E_2'(\mathbf{x}'(\mathbf{x}, t), t'(\mathbf{x}, t)) \\
B_2(\mathbf{x}, t) &= \gamma B_2'(\mathbf{x}'(\mathbf{x}, t), t'(\mathbf{x}, t)) - \gamma(v/c^2) E_3'(\mathbf{x}'(\mathbf{x}, t), t'(\mathbf{x}, t)).
\end{aligned}$$

Problem 10.   Let the volume V of the system and the temperature T of the system be the independent variables of the given thermodynamic system. Furthermore, let the external pressure P and the internal energy U of the system be the dependent variables. All the objects under consideration are smooth. Prove the following:

Theorem. Let $\omega := dU + PdV$ be a one-form in a two-dimensional space

(V, T). Assume that $\partial P / \partial T \neq 0$ and that P and U are related by the equation

$$\frac{\partial U}{\partial V} = T \frac{\partial P}{\partial T} - P \tag{1}$$

which is the so-called *thermodynamical equation of state*. Then
(i) The one-form ω is not closed.
(ii) There exists a one-form $\delta \equiv f(V, T)\omega$ such that $d\delta = 0$.
(iii) The function $f(V, T)$ is given by

$$f(V, T) = \frac{g(K(V, T))}{T} \tag{2}$$

where g is a smooth function of $K(V, T)$ and $K(V, T)$ must satisfy the partial differential equations

$$\frac{\partial K}{\partial V} - \frac{1}{T} \frac{\partial U}{\partial V} - \frac{P}{T} = 0 \tag{3}$$

$$\frac{\partial K}{\partial T} - \frac{1}{T} \frac{\partial U}{\partial T} = 0. \tag{4}$$

In particular we can choose $g(K) = 1$.

Solution 10. (i) The exterior derivative of the one-form

$$\omega = dU(V, T) + P(V, T)dV = \frac{\partial U}{\partial T} dT + \left(\frac{\partial U}{\partial V} + P \right) dV$$

yields

$$d\omega = dP \wedge dV = \frac{\partial P}{\partial T} dT \wedge dV \neq 0$$

since $\partial P / \partial T \neq 0$. $\partial P / \partial T \neq 0$ is a reasonable assumption to make.
(ii) Owing to $\omega \wedge d\omega = 0$, the *theorem of Frobenius* tells us that there is a function $f(V, T)$ such that $d(f\omega) = 0$. The theorem of Frobenius can be given in this special form because there are only two independent variables. This is locally trivially true (though not necessarily globally, the Frobenius theorem itself being a local result) whether or not (1) holds, since ω is a one-form in only two variables.
(iii) The condition $d(f\omega) = 0$ yields

$$0 = d(f\omega) = (df) \wedge \omega + f d\omega = \left(\left(\frac{\partial U}{\partial V} + P \right) \frac{\partial f}{\partial T} - \frac{\partial U}{\partial T} \frac{\partial f}{\partial V} + \frac{\partial P}{\partial T} f \right) dT \wedge dV.$$

Thus we obtain a linear partial differential equation of first order

$$\left(\frac{\partial U}{\partial V} + P \right) \frac{\partial f}{\partial T} - \frac{\partial U}{\partial T} \frac{\partial f}{\partial V} = -\frac{\partial P}{\partial T} f. \tag{5}$$

Inserting (2) through (4) into (5), we find that (5) is satisfied identically. When we interpret that ω represents an "infinitesimal quantity of heat" the equation $\omega = dU + PdV$ is the *first law of thermodynamics*. If we assume that the internal energy U and pressure P (external pressure) are related by (1) (the so-called thermodynamical equation of state, where the right-hand side of the equation can be computed from the equation of state), then there is a function $f(V,T) = g(K(V,T))/T$ such that $g\omega/T$ is a closed form. $K(V,T)$ must satisfy (3) and (4). Let $g = 1$. The quantity ω/T is usually called the "infinitesimal entropy". We write $\omega = TdS$. This means that

$$dS = \frac{dU + PdV}{T}. \tag{6}$$

S is the first integral of the exterior differential equation $dU + PdV = 0$. Since

$$dS = \frac{\partial S}{\partial V}dV + \frac{\partial S}{\partial T}dT, \qquad dU = \frac{\partial U}{\partial V}dV + \frac{\partial U}{\partial T}dT$$

we can write (6) as

$$\frac{\partial S}{\partial V} - \frac{1}{T}\frac{\partial U}{\partial V} - \frac{P}{T} = 0 \tag{7}$$

and

$$\frac{\partial S}{\partial T} - \frac{1}{T}\frac{\partial U}{\partial T} = 0. \tag{8}$$

Comparing (3) and (4) with (7) and (8) we must put $K = S$. Consequently, the most general local integrating factor for ω is

$$f(V,T) = \frac{1}{T} \times \text{arbitrary smooth function of } S.$$

To prove that physically fT is actually a constant, one must invoke the zeroth law and assume that T is the temperature. The given derivation can also be considered from a converse point of view. From dS we are able to derive the thermodynamical equation of state. Since $ddS = 0$, it follows that

$$0 = d\left(\frac{dU + PdV}{T}\right).$$

Consequently,

$$0 = \frac{1}{T^2}\left(\frac{\partial U}{\partial V} + P - T\frac{\partial P}{\partial T}\right)dV \wedge dT.$$

Therefore

$$\frac{\partial U}{\partial V} + P - T\frac{\partial P}{\partial T} = 0.$$

Thus we obtain the thermodynamical equation of state.

Problem 11. The differential form

$$\alpha = \frac{dz}{z}$$

with $z = x + iy$ $(x, y \in \mathbb{R})$ is defined on $\mathbb{C} \setminus \{0\}$.
(i) Find $d\alpha$.
(ii) Calculate

$$\oint_C \alpha$$

where C is the unit circle around the origin in the complex plane $\mathbb{C}$.

Solution 11. (i) We have

$$d\alpha = d\left(\frac{1}{z}\right) \wedge dz = -\frac{1}{z^2} dz \wedge dz = 0$$

since $dz \wedge dz = 0$.
(ii) From $z = x + iy$ it follows that $dz = dx + i dy$. Thus the differential form α takes the form

$$\alpha = \frac{dx + i dy}{x + iy} = \frac{x dx + y dy + i(x dy - y dx)}{x^2 + y^2}.$$

Introducing $z = r \exp(i\phi)$ with $r = 1$ we have $dz = ri \exp(i\phi) d\phi$. Therefore

$$\oint_C \frac{dz}{z} = i \int_0^{2\pi} d\phi = 2\pi i.$$

Problem 12. Let $\Omega = \mathbb{R}^3 \setminus \{(0,0,0)\}$. Consider

$$\alpha = \frac{\lambda}{r}(x_1 dx_2 \wedge dx_3 + x_2 dx_3 \wedge dx_1 + x_3 dx_1 \wedge dx_2)$$

where $\lambda > 0$ and $r^2 := x_1^2 + x_2^2 + x_3^2$. Consider the map f given by

$$x_1(u, v) = \sin u \cos v, \qquad x_2(u, v) = \sin u \sin v, \qquad x_3(u, v) = \cos u$$

where $0 \leq u < \pi$ and $0 \leq v < 2\pi$. Calculate

$$\int_{S^2} f^* \alpha$$

where $S^2 := \{ (x_1, x_2, x_3) : x_1^2 + x_2^2 + x_3^2 = 1 \}$.

Solution 12.   Since

$$f^*(dx_1 \wedge dx_2) = d(\sin u \cos v) \wedge d(\sin u \sin v) = \sin u \cos u du \wedge dv$$
$$f^*(dx_2 \wedge dx_3) = d(\sin u \sin v) \wedge d(\cos u) = \sin^2 u \cos v du \wedge dv$$
$$f^*(dx_3 \wedge dx_1) = d(\cos u) \wedge d(\sin u \cos v) = \sin^2 u \sin v du \wedge dv$$

we find

$$f^*\alpha = \lambda \sin u du \wedge dv$$

with $r = 1$. Consequently

$$\int_{S^2} f^*\alpha = \lambda \int_{u=0}^{\pi} \int_{v=0}^{2\pi} \sin u du dv = 4\pi\lambda.$$

Problem 13.   Let M be an oriented n-manifold and let D be a regular domain in M. Let ω be a differential form of degree $(n - 1)$ of compact support. Then

$$\int_D d\omega = \int_{\partial D} i^*\omega \tag{1}$$

where the boundary ∂D of D is considered as an oriented submanifold with the orientation induced by that of M. $i : \partial D \to M$ is the canonical injection map. Equation (1) is called *Stokes theorem*.

Let $M = \mathbb{R}^3$ and

$$K := \left\{ (x, y, z) : \frac{x^2}{a^2} + \frac{y^2}{b^2} + \frac{z^2}{c^2} \leq 1, \quad a, b, c > 0 \right\}$$

and

$$\omega := \frac{x^3}{a^2} dy \wedge dz + \frac{y^3}{b^2} dz \wedge dx + \frac{z^3}{c^2} dx \wedge dy.$$

Calculate ∂K and $d\omega$. The orientation is x, y, z. Show that

$$\int_{\partial K} \omega = \int_K d\omega.$$

Solution 13.   We find that the boundary of K is given by

$$\partial K = \left\{ (x, y, z) : \frac{x^2}{a^2} + \frac{y^2}{b^2} + \frac{z^2}{c^2} = 1, \quad a, b, c > 0 \right\}.$$

The exterior derivative of ω leads to the differential three-form

$$d\omega = \frac{3x^2}{a^2} dx \wedge dy \wedge dz + \frac{3y^2}{b^2} dy \wedge dz \wedge dx + \frac{3z^2}{c^2} dz \wedge dx \wedge dy$$

$$= 3 \left(\frac{x^2}{a^2} + \frac{y^2}{b^2} + \frac{z^2}{c^2} \right) dx \wedge dy \wedge dx.$$

To calculate the right-hand side of (1) we introduce *spherical coordinates*

$$x(r, \phi, \theta) = ar \cos\phi \cos\theta, \quad y(r, \phi, \theta) = br \sin\phi \cos\theta, \quad z(r, \phi, \theta) = cr \sin\theta$$

where $0 \le \phi < 2\pi$, $-\pi/2 \le \theta < \pi/2$ and $0 \le r \le 1$. Then

$$dx \wedge dy \wedge dz = abcr^2 \cos\theta dr \wedge d\phi \wedge d\theta$$

and

$$\frac{x^2}{a^2} + \frac{y^2}{b^2} + \frac{z^2}{c^2} = r^2(\cos^2\phi\cos^2\theta + \sin^2\phi\cos^2\theta + \sin^2\theta) = r^2.$$

Consequently

$$\int_K d\omega = \int_0^1 \int_{-\pi/2}^{\pi/2} \int_0^{2\pi} 3abcr^4 \cos\theta dr d\theta d\phi = \frac{4\pi}{5} 3abc.$$

Now we calculate the left-hand side of (1). We set

$$x(\phi, \theta) = a \cos\phi \cos\theta, \quad y(\phi, \theta) = b \sin\phi \cos\theta, \quad z(\phi, \theta) = c \sin\theta$$

where $0 \le \phi < 2\pi$ and $-\pi/2 \le \theta < \pi/2$. It follows that

$$dx \wedge dy = ab \cos\theta \sin\theta d\phi \wedge d\theta$$
$$dy \wedge dz = cb \cos\phi \cos^2\theta d\phi \wedge d\theta$$
$$dz \wedge dx = ac \sin\phi \cos^2\theta d\phi \wedge d\theta.$$

Therefore

$$\int_{\partial K} \omega = I_1 + I_2 + I_3$$

where

$$I_1 = \int_{-\pi/2}^{\pi/2} \int_0^{2\pi} abc \cos^4\phi \cos^5\theta d\phi d\theta = \frac{4}{5} abc\pi$$

$$I_2 = \int_{-\pi/2}^{\pi/2} \int_0^{2\pi} abc \sin^4\phi \cos^5\theta d\phi d\theta = \frac{4}{5} abc\pi$$

$$I_3 = \int_{-\pi/2}^{\pi/2} \int_0^{2\pi} abc \sin^4\theta \cos\theta d\phi d\theta = \frac{4}{5} abc\pi.$$

Thus

$$\int_{\partial K} \omega = \frac{4\pi}{5} 3abc.$$

Problem 14. *Poincaré's lemma* tells us that if ω is a p-differential form on M ($\dim M = n$) for which there exists a $(p-1)$-differential form α such that $d\alpha = \omega$, then $d\omega = 0$. The converse of Poincaré's lemma tells us that if ω is a p-differential form on an open set $U \subset M$ (which is contractible to a point) such that $d\omega = 0$, then there exists a $(p-1)$ differential form α such that $\omega = d\alpha$. The exception is $p = 0$. Then $\omega = f$ and the vanishing of df simply means f is constant.

Let

$$\omega = a(\mathbf{x})dx_{i_1} \wedge dx_{i_2} \wedge \cdots \wedge dx_{i_p}$$

and let λ be a real parameter with $\lambda \in [0,1]$. We introduce the linear operator T_λ which is defined as

$$T_\lambda \omega := \int_0^1 \lambda^{p-1}\left(x_1\frac{\partial}{\partial x_1} + \cdots + x_n\frac{\partial}{\partial x_n}\right)\rfloor a(\lambda\mathbf{x})dx_{i_1} \wedge dx_{i_2} \wedge \cdots \wedge dx_{i_p}d\lambda$$

where $\rfloor$ denotes the *contraction* (also called the *interior product* of a differential form and a vector field). We have

$$\frac{\partial}{\partial x_j}\rfloor dx_k = \delta_{jk}.$$

Since the operator T_λ is linear it must only be defined for a monom. Note that

$$d(T_\lambda\omega) = \omega.$$

(i) Apply the converse of Poincaré's lemma to $\omega = dx_1 \wedge dx_2$ defined on $\mathbb{R}^2$.
(ii) Apply the converse of Poincaré's lemma to $\omega = dx_1 \wedge dx_2 \wedge dx_3$ defined on $\mathbb{R}^3$.

Solution 14.   (i) We have $d\omega = 0$ and $\mathbb{R}^2$ is contractible to $0 \in \mathbb{R}^2$. From the definition for the contraction we obtain

$$T_\lambda\omega = \left(x_1\frac{\partial}{\partial x_1} + \cdots + x_n\frac{\partial}{\partial x_n}\right)\rfloor(dx_{i_1} \wedge dx_{i_2} \wedge \cdots \wedge dx_{i_p})\int_0^1 \lambda^{p-1}a(\lambda\mathbf{x})d\lambda.$$

We have $n = 2$ and $p = 2$. Since

$$\left(x_1\frac{\partial}{\partial x_1} + x_2\frac{\partial}{\partial x_2}\right)\rfloor(dx_1 \wedge dx_2) = -x_2dx_1 + x_1dx_2$$

we find

$$T_\lambda\omega = (x_1dx_2 - x_2dx_1)\int_0^1 \lambda d\lambda = \frac{1}{2}(x_1dx_2 - x_2dx_1).$$

(ii) We have $d\omega = 0$ and $\mathbb{R}^3$ is contractible to $0 \in \mathbb{R}^3$. Since

$$\left(\sum_{j=1}^3 x_j\frac{\partial}{\partial x_j}\right)\rfloor(dx_1 \wedge dx_2 \wedge dx_3) = x_1dx_2 \wedge dx_3 - x_2dx_1 \wedge dx_3 + x_3dx_1 \wedge dx_2$$

and

$$\int_0^1 \lambda^2 d\lambda = \frac{1}{3}$$

we find

$$T_\lambda \omega = \frac{1}{3}(x_1 dx_2 \wedge dx_3 + x_2 dx_3 \wedge dx_1 + x_3 dx_1 \wedge dx_2)$$

where we used $dx_1 \wedge dx_3 = -dx_3 \wedge dx_1$.

Problem 15. Find the closed plane curve of a given length L which encloses a maximum area.

Solution 15. Given the curve $(x(t), y(t))$, where $t \in [t_0, t_1]$. We assume that $x(t)$ and $y(t)$ are continuously differentiable and $x(t_0) = x(t_1)$, $y(t_0) = y(t_1)$. Let A be the area enclosed and L be the given length. Then we have

$$L = \int_{t_0}^{t_1} \left(\left(\frac{dx}{dt} \right)^2 + \left(\frac{dy}{dt} \right)^2 \right)^{1/2} dt.$$

Since the exterior derivative d of $(xdy - ydx)/2$ is given by the two form

$$\frac{1}{2}d(xdy - ydx) = dx \wedge dy$$

we can apply *Stokes theorem* and find the area

$$A = \int_{t_0}^{t_1} \left(x\frac{dy}{dt} - y\frac{dx}{dt} \right) dt.$$

To find the maximum area we apply the *Lagrange multiplier method* and consider

$$H = \int_{t_0}^{t_1} \left(\left(x\frac{dy}{dt} - y\frac{dx}{dt} \right) + \lambda \left(\left(\frac{dx}{dt} \right)^2 + \left(\frac{dy}{dt} \right)^2 \right)^{1/2} \right) dt$$

where λ is the Lagrange multiplier. Thus we consider

$$\phi(x, y, \dot{x}, \dot{y}) = \frac{1}{2}(x\dot{y} - y\dot{x}) + \lambda(\dot{x}^2 + \dot{y}^2)^{1/2}$$

where ϕ satisfies the Euler-Lagrange equation. Thus, by partial differentiation

$$\frac{\partial \phi}{\partial x} = \frac{\dot{y}}{2}, \qquad \frac{\partial \phi}{\partial y} = -\frac{\dot{x}}{2}$$

$$\frac{\partial \phi}{\partial \dot{x}} = -\frac{y}{2} + \frac{\lambda \dot{x}}{(\dot{x}^2 + \dot{y}^2)^{1/2}}, \qquad \frac{\partial \phi}{\partial \dot{y}} = \frac{x}{2} + \frac{\lambda \dot{y}}{(\dot{x}^2 + \dot{y}^2)^{1/2}}.$$

The *Euler-Lagrange equations* are

$$\frac{\partial \phi}{\partial x} - \frac{d}{dt}\frac{\partial \phi}{\partial \dot{x}} = 0, \qquad \frac{\partial \phi}{\partial y} - \frac{d}{dt}\frac{\partial \phi}{\partial \dot{y}} = 0.$$

It follows that

$$\frac{\dot{y}}{2} - \frac{d}{dt}\left(-\frac{y}{2} + \frac{\lambda\dot{x}}{(\dot{x}^2 + \dot{y}^2)^{1/2}}\right) = 0, \qquad -\frac{\dot{x}}{2} - \frac{d}{dt}\left(\frac{x}{2} + \frac{\lambda\dot{y}}{(\dot{x}^2 + \dot{y}^2)^{1/2}}\right) = 0.$$

If we choose a special parameter s, the arc length along the competing curves, then

$$(\dot{x}^2 + \dot{y}^2)^{1/2} = 1$$

and

$$ds = (\dot{x}^2 + \dot{y}^2)^{1/2}dt$$

so that Euler-Lagrange equations reduce to

$$\frac{dy}{ds} - \lambda\frac{d^2x}{ds^2} = 0, \qquad \frac{dx}{ds} + \lambda\frac{d^2y}{ds^2} = 0.$$

Integration of these linear system of differential equations yields

$$y - \lambda\frac{dx}{ds} = C_1, \qquad x + \lambda\frac{dy}{ds} = C_2$$

where C_1 and C_2 are constants of integration. Elimination of y yields

$$\lambda^2\frac{d^2x}{ds^2} + x = C_2$$

with the solution

$$x(s) = a\sin\left(\frac{s}{\lambda}\right) + b\cos\left(\frac{s}{\lambda}\right) + C_2$$

where a and b are constants of integration. For y we find

$$y(s) = a\cos\left(\frac{s}{\lambda}\right) - b\sin\left(\frac{s}{\lambda}\right) + C_1.$$

Thus $(x(s), y(s))$ describe a circle.

Problem 16. Consider the manifold $M = \mathbb{R}^2$ and the metric tensor field $g = dx_1 \otimes dx_1 + dx_2 \otimes dx_2$. Let

$$\omega = \omega_1(\mathbf{x})dx_1 + \omega_2(\mathbf{x})dx_2$$

be a differential one-form in M with $\omega_1, \omega_2 \in C^\infty(\mathbb{R}^2)$. Show that ω can be written as

$$\omega = d\alpha + \delta\beta + \gamma$$

where α is a $C^\infty(\mathbb{R}^2)$ function, β is a two-form given by $\beta = b(\mathbf{x})dx_1 \wedge dx_2$ ($b(\mathbf{x}) \in C^\infty(\mathbb{R}^2)$) and $\gamma = \gamma_1(\mathbf{x})dx_1 + \gamma_2(\mathbf{x})dx_2$ is a harmonic one-form, i.e. $(d\delta + \delta d)\gamma = 0$. We define

$$\delta\beta := (-1) * d * \beta .$$

Solution 16. We have

$$d\alpha = \frac{\partial\alpha}{\partial x_1}dx_1 + \frac{\partial\alpha}{\partial x_2}dx_2 .$$

Since $\delta\beta = (-1) * d * \beta$ and $*(dx_1 \wedge dx_2) = 1$, $*dx_1 = dx_2$, $*dx_2 = -dx_1$ we obtain

$$\begin{aligned}
\delta\beta &= (-1) * d * b(\mathbf{x})dx_1 \wedge dx_2 \\
&= (-1) * db(\mathbf{x})(*(dx_1 \wedge dx_2)) \\
&= (-1) * \left(\frac{\partial b}{\partial x_1}dx_1 + \frac{\partial b}{\partial x_2}dx_2 \right) \\
&= (-1) \left(\frac{\partial b}{\partial x_1}dx_2 - \frac{\partial b}{\partial x_2}dx_1 \right) \\
&= \frac{\partial b}{\partial x_2}dx_1 - \frac{\partial b}{\partial x_1}dx_2 .
\end{aligned}$$

From the condition that γ is a harmonic one-form we obtain

$$\begin{aligned}
(d\delta + \delta d)\gamma &= d(\delta\gamma) + \delta(d\gamma) \\
&= d(\delta\gamma_1(\mathbf{x})dx_1 + \delta\gamma_2(\mathbf{x})dx_2) + \delta\left(-\frac{\partial\gamma_1}{\partial x_2} + \frac{\partial\gamma_2}{\partial x_1} \right)dx_1 \wedge dx_2 \\
&= d((-1) * d(\gamma_1(\mathbf{x})dx_2) - (-1) * d(\gamma_2(\mathbf{x})dx_1) \\
&\quad + \delta\left(\frac{\partial\gamma_2}{\partial x_1}dx_1 \wedge dx_2 \right) - \delta\left(\frac{\partial\gamma_1}{\partial x_2}dx_1 \wedge dx_2 \right) \\
&= d\left((-1) * \frac{\partial\gamma_1}{\partial x_1}dx_1 \wedge dx_2 - *\frac{\partial\gamma_2}{\partial x_2}dx_1 \wedge dx_2 \right) \\
&\quad + (-1) * d\left(\frac{\partial\gamma_2}{\partial x_1} \right) - (-1) * d\left(\frac{\partial\gamma_1}{\partial x_2} \right) \\
&= d\left(-\frac{\partial\gamma_1}{\partial x_1} - \frac{\partial\gamma_2}{\partial x_2} \right) + (-1) * \left(\frac{\partial^2\gamma_2}{\partial x_1^2}dx_1 + \frac{\partial^2\gamma_2}{\partial x_1\partial x_2}dx_2 \right) \\
&\quad + *\frac{\partial^2\gamma_1}{\partial x_1\partial x_2}dx_1 + *\frac{\partial^2\gamma_1}{\partial x_2^2}dx_2 \\
&= -\left(\frac{\partial^2\gamma_1}{\partial x_1^2} + \frac{\partial^2\gamma_1}{\partial x_2^2} \right)dx_1 - \left(\frac{\partial^2\gamma_2}{\partial x_1^2} + \frac{\partial^2\gamma_2}{\partial x_2^2} \right)dx_2 .
\end{aligned}$$

Thus from $\omega = d\alpha + \delta\beta + \gamma$ we obtain

$$\omega_1(\mathbf{x}) = \frac{\partial a}{\partial x_1} + \frac{\partial b}{\partial x_2} + \gamma_1(\mathbf{x}), \qquad \omega_2(\mathbf{x}) = \frac{\partial a}{\partial x_2} - \frac{\partial b}{\partial x_1} + \gamma_2(\mathbf{x})$$

and that γ is a harmonic one-form, i.e.

$$\frac{\partial^2 \gamma_1}{\partial x_1^2} + \frac{\partial^2 \gamma_1}{\partial x_2^2} = 0, \qquad \frac{\partial^2 \gamma_2}{\partial x_1^2} + \frac{\partial^2 \gamma_2}{\partial x_2^2} = 0.$$

Problem 17. Let $f : \mathbb{R}^2 \to \mathbb{R}^2$ be a smooth planar mapping with constant Jacobian determinant $J = 1$, written as

$$Q = Q(p, q), \qquad P = P(p, q).$$

For coordinates in $\mathbb{R}^2$ the (area) differential two-form is given as

$$\omega = dp \wedge dq.$$

(i) Find $f^*\omega$.
(ii) Show that $pdq - f^*(pdq) = dF$ for some smooth function $F : \mathbb{R}^2 \to \mathbb{R}$.

Solution 17.   (i) Since

$$J \equiv \left(\frac{\partial P}{\partial p} \frac{\partial Q}{\partial q} - \frac{\partial P}{\partial q} \frac{\partial Q}{\partial p} \right) = 1$$

we have

$$
\begin{aligned}
f^*(dp \wedge dq) &= (f^* dp) \wedge (f^* dq) \\
&= \left(\frac{\partial P}{\partial q} dq + \frac{\partial P}{\partial p} dp \right) \wedge \left(\frac{\partial Q}{\partial q} dq + \frac{\partial Q}{\partial p} dp \right) \\
&= \left(\frac{\partial P}{\partial p} \frac{\partial Q}{\partial q} - \frac{\partial P}{\partial q} \frac{\partial Q}{\partial p} \right) dp \wedge dq \\
&= dp \wedge dq.
\end{aligned}
$$

(ii) Since $dp \wedge dq = d(pdq)$ and the exterior derivative commutes with the pull-back operator, i.e. $d(f^*pdq) = f^*d(pdq)$ we can write

$$d(pdq - f^*(pdq)) = 0.$$

This implies that $pdq - f^*(pdq) = dF$ for some smooth function $F : \mathbb{R}^2 \to \mathbb{R}$.

Chapter 22

Matrix-valued Differential Forms

Let

$$\widetilde{\alpha} := \sum_{j=1}^{n} (a_j(x,t)dx + A_j(x,t)dt) \otimes X_j$$

be a *Lie algebra-valued differential form* where $\{X_1, X_2, \ldots, X_n\}$ forms a basis of a semi-simple Lie algebra. The exterior derivative is defined as

$$d\widetilde{\alpha} := \sum_{j=1}^{n} \left(-\frac{\partial a_j}{\partial t} + \frac{\partial A_j}{\partial x} \right) dx \wedge dt \otimes X_j.$$

The commutator is defined as

$$[\widetilde{\alpha}, \widetilde{\alpha}] := \sum_{k=1}^{n} \sum_{j=1}^{n} (a_k A_j - a_j A_k) dx \wedge dt \otimes [X_k, X_j].$$

The *covariant exterior derivative* is defined as

$$D_{\widetilde{\alpha}} \widetilde{\alpha} := d\widetilde{\alpha} + \frac{1}{2}[\widetilde{\alpha}, \widetilde{\alpha}].$$

Problem 1. (i) Calculate the covariant derivative of $\widetilde{\alpha}$ and find the equation which follows from the condition

$$D_{\widetilde{\alpha}} \widetilde{\alpha} = 0. \tag{1}$$

245

(ii) Study the case where the basis of the semi-simple Lie algebra satisfies the commutation relations

$$[X_1, X_2] = 2X_2, \qquad [X_3, X_1] = 2X_3, \qquad [X_2, X_3] = X_1. \qquad (2)$$

(iii) Let

$$a_1 = -\eta, \qquad a_2 = \frac{1}{2}\frac{\partial u}{\partial x}, \qquad a_3 = -\frac{1}{2}\frac{\partial u}{\partial x} \qquad (3a)$$

$$A_1 = -\frac{1}{4\eta}\cos u, \qquad A_2 = A_3 = -\frac{1}{4\eta}\sin u \qquad (3b)$$

where η is an arbitrary constant with $\eta \neq 0$. Find the equation of motion.

Solution 1.   (i) Since

$$[X_k, X_j] = \sum_{i=1}^{n} C_{kj}^i X_i$$

it follows that

$$D_{\tilde{\alpha}}\tilde{\alpha} = \sum_{i=1}^{n} \left(\left(-\frac{\partial a_i}{\partial t} + \frac{\partial A_i}{\partial x} \right) + \frac{1}{2}\sum_{k=1}^{n}\sum_{j=1}^{n}(a_k A_j - a_j A_k)C_{kj}^i \right) dx \wedge dt \otimes X_i.$$

Since the X_j, $j = 1, \ldots, n$ form a basis of the semi-simple Lie algebra, the condition $D_{\tilde{\alpha}}\tilde{\alpha} = 0$ yields

$$\left(-\frac{\partial a_i}{\partial t} + \frac{\partial A_i}{\partial x} \right) + \frac{1}{2}\sum_{k=1}^{n}\sum_{j=1}^{n}(a_k A_j - a_j A_k)C_{kj}^i = 0$$

for $i = 1, \ldots, n$. Since $C_{kj}^i = -C_{jk}^i$, it follows that

$$\left(-\frac{\partial a_i}{\partial t} + \frac{\partial A_i}{\partial x} \right) + \sum_{k<j}^{n}(a_k A_j - a_j A_k)C_{kj}^i = 0.$$

(ii) Consider now a special case where $n = 3$ and X_1, X_2 and X_3 satisfy the commutation relations (2). We find

$$-\frac{\partial a_1}{\partial t} + \frac{\partial A_1}{\partial x} + a_2 A_3 - a_3 A_2 = 0$$

$$-\frac{\partial a_2}{\partial t} + \frac{\partial A_2}{\partial x} + 2(a_1 A_2 - a_2 A_1) = 0$$

$$-\frac{\partial a_3}{\partial t} + \frac{\partial A_3}{\partial x} - 2(a_1 A_3 - a_3 A_1) = 0.$$

A convenient choice of a basis $\{X_1, X_2, X_3\}$ is given by

$$X_1 = \begin{pmatrix} 1 & 0 \\ 0 & -1 \end{pmatrix}, \qquad X_2 = \begin{pmatrix} 0 & 1 \\ 0 & 0 \end{pmatrix}, \qquad X_3 = \begin{pmatrix} 0 & 0 \\ 1 & 0 \end{pmatrix}.$$

Consequently, the Lie algebra under consideration is $sl(2, \mathbb{R})$.

(iii) Inserting (3) into this system yields the sine-Gordon equation

$$\frac{\partial^2 u}{\partial x \partial t} = \sin u.$$

Other one-dimensional soliton equations (such as the Korteweg-de Vries equation, the nonlinear Schrödinger equation, the Liouville equation) can be derived with this method.

Problem 2. Let

$$\tilde{\alpha} := \sum_{k=1}^{3} \alpha_k \otimes T_k \tag{1}$$

be a Lie algebra-valued differential one-form. Here $\otimes$ denotes the tensor product and T_k ($k = 1, 2, 3$) are the generators given by

$$T_1 := \frac{1}{2} \begin{pmatrix} 0 & -i \\ -i & 0 \end{pmatrix}, \qquad T_2 := \frac{1}{2} \begin{pmatrix} 0 & -1 \\ 1 & 0 \end{pmatrix}, \qquad T_3 := \frac{1}{2} \begin{pmatrix} -i & 0 \\ 0 & i \end{pmatrix}. \tag{2}$$

T_1, T_2 and T_3 form a basis of the semi-simple Lie algebra $su(2)$. The quantities α_k are differential one-forms

$$\alpha_k := \sum_{j=1}^{4} A_{kj}(\mathbf{x}) dx_j \tag{3}$$

where $\mathbf{x} = (x_1, x_2, x_3, x_4)$. The quantity $\tilde{\alpha}$ is usually called the *connection* or *vector potential*. The actions of the Hodge operator $*$ for a given metric tensor field, and the exterior derivative d, may be consistently defined by

$$*\tilde{\alpha} := \sum_{k=1}^{3} (*\alpha_k) \otimes T_k \tag{4}$$

and

$$d\tilde{\alpha} := \sum_{k=1}^{3} (d\alpha_k) \otimes T_k. \tag{5}$$

The bracket $[\,,\,]$ of Lie algebra-valued differential forms, say $\tilde{\beta}$ and $\tilde{\gamma}$, is defined as

$$\left[\tilde{\beta}, \tilde{\gamma}\right] := \sum_{k=1}^{n} \sum_{b=1}^{n} (\beta_k \wedge \gamma_b) \otimes [T_k, T_b].$$

The covariant exterior derivative of a Lie algebra-valued differential p-form $\widetilde{\gamma}$ with respect to a Lie algebra-valued one-form $\widetilde{\beta}$ is defined as

$$D_{\widetilde{\beta}}\widetilde{\gamma} := d\widetilde{\gamma} - g\left[\widetilde{\beta}, \widetilde{\gamma}\right]$$

where

$$g = \begin{cases} -1 & p \text{ even} \\ -\frac{1}{2} & p \text{ odd.} \end{cases}$$

Therefore

$$D_{\widetilde{\beta}}\widetilde{\beta} = d\widetilde{\beta} + \frac{1}{2}\left[\widetilde{\beta}, \widetilde{\beta}\right], \qquad D_{\widetilde{\beta}}(D_{\widetilde{\beta}}\widetilde{\beta}) = 0.$$

The last equation is called the *Bianchi identity*. Let $\widetilde{\alpha}$ be the Lie algebra valued-differential form given by equation (1). The *Yang-Mills equations* are given by

$$D_{\widetilde{\alpha}}(*D_{\widetilde{\alpha}}\widetilde{\alpha}) = 0.$$

The quantity $D_{\widetilde{\alpha}}\widetilde{\alpha}$ is called the *curvature form* or *field strength tensor*. This equation is a coupled system of nonlinear partial differential equations of second order. In addition we have to impose gauge conditions. Assume the metric tensor field is given by

$$g = dx_1 \otimes dx_1 + dx_2 \otimes dx_2 + dx_3 \otimes dx_3 - dx_4 \otimes dx_4.$$

(i) Write down the Yang-Mills equation explicitly.
(ii) Impose the gauge conditions

$$A_{k4} = 0$$

$$\sum_{i=1}^{3} \frac{\partial A_{ki}}{\partial x_i} = 0$$

$$\frac{\partial A_{ki}}{\partial x_j} = 0, \qquad i, j = 1, 2, 3$$

where $k = 1, 2, 3$.

Solution 2.   (i) The commutation relations of T_1, T_2 and T_3 are given by

$$[T_1, T_2] = T_3, \qquad [T_2, T_3] = T_1, \qquad [T_3, T_1] = T_2.$$

Since

$$D_{\widetilde{\alpha}}\widetilde{\alpha} = d\widetilde{\alpha} + \frac{1}{2}[\widetilde{\alpha}, \widetilde{\alpha}]$$

$$\frac{1}{2}[\widetilde{\alpha}, \widetilde{\alpha}] = (\alpha_1 \wedge \alpha_2) \otimes T_3 + (\alpha_2 \wedge \alpha_3) \otimes T_1 + (\alpha_3 \wedge \alpha_1) \otimes T_2$$

$$d\widetilde{\alpha} = \sum_{a=1}^{3} d\alpha_a \otimes T_a$$

and

$$*(D_{\widetilde{\alpha}}\widetilde{\alpha}) = (*d\alpha_1 + *(\alpha_2 \wedge \alpha_3)) \otimes T_1 + (*d\alpha_2 + *(\alpha_3 \wedge \alpha_1)) \otimes T_2$$
$$+ (*d\alpha_3 + *(\alpha_1 \wedge \alpha_2)) \otimes T_3$$

we obtain

$$D_{\widetilde{\alpha}} * (D_{\widetilde{\alpha}}\widetilde{\alpha}) = d(*d\alpha_1 + *(\alpha_2 \wedge \alpha_3)) \otimes T_1$$
$$+ d(*d\alpha_2 + *(\alpha_3 \wedge \alpha_1)) \otimes T_2 + d(*d\alpha_3 + *(\alpha_1 \wedge \alpha_2)) \otimes T_3$$
$$+ (\alpha_2 \wedge (*d\alpha_3 + *(\alpha_1 \wedge \alpha_2)) - \alpha_3 \wedge (*d\alpha_2 + *(\alpha_3 \wedge \alpha_1))) \otimes T_1$$
$$+ (\alpha_3 \wedge (*d\alpha_1 + *(\alpha_2 \wedge \alpha_3)) - \alpha_1 \wedge (*d\alpha_3 + *(\alpha_1 \wedge \alpha_2))) \otimes T_2$$
$$+ (\alpha_1 \wedge (*d\alpha_2 + *(\alpha_3 \wedge \alpha_1)) - \alpha_2 \wedge (*d\alpha_1 + *(\alpha_2 \wedge \alpha_3))) \otimes T_3 .$$

Then from the condition $D_{\widetilde{\alpha}}(*D_{\widetilde{\alpha}}\widetilde{\alpha}) = 0$ it follows that

$$d(*d\alpha_1 + *(\alpha_2 \wedge \alpha_3)) + (\alpha_2 \wedge (*d\alpha_3 + *(\alpha_1 \wedge \alpha_2)) - \alpha_3 \wedge (*d\alpha_2 + *(\alpha_3 \wedge \alpha_1))) = 0$$
$$d(*d\alpha_2 + *(\alpha_3 \wedge \alpha_1)) + (\alpha_3 \wedge (*d\alpha_1 + *(\alpha_2 \wedge \alpha_3)) - \alpha_1 \wedge (*d\alpha_3 + *(\alpha_1 \wedge \alpha_2))) = 0$$
$$d(*d\alpha_3 + *(\alpha_1 \wedge \alpha_2)) + (\alpha_1 \wedge (*d\alpha_2 + *(\alpha_3 \wedge \alpha_1)) - \alpha_2 \wedge (*d\alpha_1 + *(\alpha_2 \wedge \alpha_3))) = 0 .$$

Now $*(\alpha_k \wedge \alpha_b)$ is a two-form and therefore $d * (\alpha_k \wedge \alpha_b)$ is a three-form. We find 12 coupled partial differential equations.

(ii) Let us now impose the gauge conditions. Thus we have eliminated the space dependence of the fields. Taking into account the gauge equations we arrive at the autonomous system of ordinary differential equations of second order

$$\frac{d^2 A_{ki}}{dt^2} + \sum_{b=1}^{3}\sum_{j=1}^{3}(A_{bj}A_{bj}A_{ki} - A_{kj}A_{bj}A_{bi}) = 0$$

together with

$$\sum_{j=1}^{3}\sum_{b=1}^{3}\sum_{c=1}^{3}\varepsilon_{kbc}A_{bj}\frac{dA_{cj}}{dt} = 0.$$

This system can be viewed as a Hamilton system with

$$H(A_{ki}, dA_{ki}/dt) = \frac{1}{2}\sum_{k=1}^{3}\sum_{i=1}^{3}\left(\frac{dA_{ki}}{dt}\right)^2 + \frac{1}{4}\sum_{k=1}^{3}\sum_{i=1}^{3}(A_{ki}^2)^2$$
$$- \frac{1}{4}\sum_{i=1}^{3}\sum_{j=1}^{3}\left(\sum_{k=1}^{3}A_{ki}A_{kj}\right)^2 .$$

To simplify further we put $A_{ki}(t) = O_{ki}f_k(t)$ (no summation), where $O = (O_{ki})$ is a time-independent 3×3 orthogonal matrix. We put

$$\frac{df_k}{dt} \equiv p_k, \qquad f_k \equiv q_k$$

where $k = 1, 2, 3$. Then we obtain

$$\frac{d^2q_1}{dt^2} + q_1(q_2^2 + q_3^2) = 0, \qquad \frac{d^2q_2}{dt^2} + q_2(q_1^2 + q_3^2) = 0, \qquad \frac{d^2q_3}{dt^2} + q_3(q_1^2 + q_2^2) = 0$$

with the Hamilton function

$$H(\mathbf{p}, \mathbf{q}) = \frac{1}{2}(p_1^2 + p_2^2 + p_3^2) + \frac{1}{2}(q_1^2 q_2^2 + q_1^2 q_3^2 + q_2^2 q_3^2).$$

Problem 3. Consider the Lie group

$$G := \left\{ \begin{pmatrix} e^\alpha & \beta \\ 0 & 1 \end{pmatrix} : \alpha \in \mathbb{R}, \, \beta \in \mathbb{R} \right\}. \tag{1}$$

Let

$$X := \begin{pmatrix} e^\alpha & \beta \\ 0 & 1 \end{pmatrix} \tag{2}$$

and

$$\Omega := X^{-1}dX. \tag{3}$$

Show that

$$d\Omega + \Omega \wedge \Omega = 0. \tag{4}$$

Solution 3. From (2) we obtain the inverse matrix

$$X^{-1} = \begin{pmatrix} e^{-\alpha} & -\beta e^{-\alpha} \\ 0 & 1 \end{pmatrix}.$$

From (2) we also obtain

$$dX = \begin{pmatrix} e^\alpha d\alpha & d\beta \\ 0 & 0 \end{pmatrix}.$$

Thus

$$\Omega = X^{-1}dX = \begin{pmatrix} d\alpha & e^{-\alpha}d\beta \\ 0 & 0 \end{pmatrix}.$$

Therefore

$$d\Omega = \begin{pmatrix} 0 & -e^{-\alpha}d\alpha \wedge d\beta \\ 0 & 0 \end{pmatrix}$$

and

$$\Omega \wedge \Omega = \begin{pmatrix} 0 & e^{-\alpha}d\alpha \wedge d\beta \\ 0 & 0 \end{pmatrix}.$$

Thus (4) follows.

Problem 4. Let

$$SL(2, \mathbb{R}) := \left\{ X = \begin{pmatrix} a & b \\ c & d \end{pmatrix} \middle| ad - bc = 1 \right\} \tag{1}$$

be the semi-simple Lie group of all (2×2)-real unimodular matrices. Its right-invariant Maurer-Cartan form is

$$\omega = dX X^{-1} = \begin{pmatrix} \omega_{11} & \omega_{12} \\ \omega_{21} & \omega_{22} \end{pmatrix} \tag{2}$$

where

$$\omega_{11} + \omega_{22} = 0). \tag{3}$$

(i) Show that ω satisfies the *structure equation* of $SL(2, \mathbb{R})$, (also called the Maurer-Cartan equation)

$$d\omega = \omega \wedge \omega. \tag{4}$$

(ii) Let U be a neighbourhood in the (x, t)-plane and consider the smooth mapping

$$f : U \to SL(2, \mathbb{R}). \tag{5}$$

The pull-backs of the Maurer-Cartan forms can be written

$$\omega_{11} = \eta(x, t)dx + A(x, t)dt$$
$$\omega_{12} = q(x, t)dx + B(x, t)dt$$
$$\omega_{21} = r(x, t)dx + C(x, t)dt$$

where the coefficients are functions of x, t. Find the differential equations for η, q, r, A, B and C.

(iii) Consider the special case that $r = +1$ and η is a parameter independent of x, t. Writing

$$q = u(x, t).$$

Find $A(x, t)$, $B(x, t)$ as functions of u and C. Let

$$C = \eta^2 - \frac{1}{2}u.$$

Find the differential equations for u.

Solution 4. (i) Since $\omega_{11} \wedge \omega_{11} = 0$, $\omega_{22} \wedge \omega_{22} = 0$, $\omega_{11} = -\omega_{22}$ we have

$$\omega \wedge \omega = \begin{pmatrix} \omega_{11} & \omega_{12} \\ \omega_{21} & \omega_{22} \end{pmatrix} \wedge \begin{pmatrix} \omega_{11} & \omega_{12} \\ \omega_{21} & \omega_{22} \end{pmatrix}$$

$$= \begin{pmatrix} \omega_{12} \wedge \omega_{21} & \omega_{11} \wedge \omega_{12} + \omega_{12} \wedge \omega_{22} \\ \omega_{21} \wedge \omega_{11} + \omega_{22} \wedge \omega_{21} & \omega_{21} \wedge \omega_{12} \end{pmatrix}$$

$$= \begin{pmatrix} \omega_{12} \wedge \omega_{21} & 2\omega_{11} \wedge \omega_{12} \\ -2\omega_{11} \wedge \omega_{21} & -\omega_{12} \wedge \omega_{21} \end{pmatrix}.$$

Or, written explicitly,

$$d\omega_{11} = \omega_{12} \wedge \omega_{21}, \qquad d\omega_{12} = 2\omega_{11} \wedge \omega_{12}, \qquad d\omega_{21} = 2\omega_{21} \wedge \omega_{11} \,.$$

(ii) We find

$$-\frac{\partial \eta}{\partial t} + \frac{\partial A}{\partial x} - qC + rB = 0$$

$$-\frac{\partial q}{\partial t} + \frac{\partial B}{\partial x} - 2\eta B + 2qA = 0$$

$$-\frac{\partial r}{\partial t} + \frac{\partial C}{\partial x} - 2rA + 2\eta C = 0 \,.$$

(iii) From the third equation in (ii) with $r = 1$ we obtain

$$A(x,t) = \eta C(x,t) + \frac{1}{2}\frac{\partial C}{\partial x} \,.$$

From the first equation in (ii) we obtain

$$B(x,t) = u(x,t)C(x,t) - \eta\frac{\partial C}{\partial x} - \frac{1}{2}\frac{\partial^2 C}{\partial x^2} \,.$$

Inserting this expressions into the second equation in (ii) with $q = u$ yields

$$\frac{\partial u}{\partial t} = \frac{\partial u}{\partial x}C + 2u\frac{\partial C}{\partial x} + 2\eta^2\frac{\partial C}{\partial x} - \frac{1}{2}\frac{\partial^3 C}{\partial x^3} \,.$$

With $C = \eta^2 - u/2$ we finally obtain

$$\frac{\partial u}{\partial t} = \frac{1}{4}\frac{\partial^3 u}{\partial x^3} - \frac{3}{2}u\frac{\partial u}{\partial x}$$

which is the well-known *Korteweg-de Vries equation*.

Chapter 23

Lie Derivative

Let M be an n-dimensional C^∞ differentiable manifold with local coordinates x_j, $j = 1, \ldots, n$. Real valued C^∞ vector fields and real valued C^∞ differential forms on M can be considered. The components of the vector field V are denoted by $V_j \partial/\partial x_j$. This means

$$V := V_1(\mathbf{x}) \frac{\partial}{\partial x_1} + V_2(\mathbf{x}) \frac{\partial}{\partial x_2} + \cdots + V_n(\mathbf{x}) \frac{\partial}{\partial x_n}$$

in local coordinates. The *Lie derivative* of a differential form α with respect to V is defined by the derivative of α along the integral curve $t \mapsto \Phi_t$ of V, i.e.,

$$L_V \alpha := \lim_{t \to 0} \frac{\Phi_t^* \alpha - \alpha}{t}.$$

It can be shown that this can be written as

$$L_V \alpha := d(V \rfloor \alpha) + V \rfloor (d\alpha)$$

where $d\alpha$ is the exterior derivative of the differential form α and $V \rfloor \alpha$ is the contraction of α by V. In local coordinates we have

$$\frac{\partial}{\partial x_j} \rfloor dx_k = \delta_{jk}$$

where δ_{jk} denotes the Kronecker delta. Furthermore, we have the *product rule*

$$V \rfloor (\alpha \wedge \beta) = (V \rfloor \alpha) \wedge \beta + (-1)^r \alpha \wedge (V \rfloor \beta)$$

where α is an r-form. The linear operators $d(.)$, $V\rfloor$ and $L_V(.)$ are coordinate-free operators.

Problem 1.   Consider the differential one form $\omega = x_1 dx_2 - x_2 dx_1$ on $\mathbb{R}^2$. Show that ω is invariant under the transformation

$$\begin{pmatrix} x_1' \\ x_2' \end{pmatrix} = \begin{pmatrix} \cos\alpha & -\sin\alpha \\ \sin\alpha & \cos\alpha \end{pmatrix} \begin{pmatrix} x_1 \\ x_2 \end{pmatrix}.$$

Solution 1.   (i) We have

$$dx_1' = \cos\alpha \, dx_1 - \sin\alpha \, dx_2, \qquad dx_2' = \sin\alpha \, dx_1 + \cos\alpha \, dx_2.$$

Thus

$$x_1' dx_2' - x_2' dx_1' = x_1 dx_2 - x_2 dx_1.$$

Let ω be the $(n-1)$ differential form on $\mathbb{R}^n$ given by

$$\omega = \sum_{j=1}^{n} (-1)^{j-1} x_j dx_1 \wedge \cdots \wedge \widehat{dx_j} \wedge \cdots \wedge dx_n$$

where $\widehat{}$ indicates omission. Then ω is invariant under the orthogonal group of $\mathbb{R}^n$.

Problem 2.   Let $M = \mathbb{R}^n$. Consider the volume form

$$\omega := dx_1 \wedge dx_2 \wedge \cdots \wedge dx_n.$$

Find

$$V\rfloor\omega$$

where the vector field X is given by

$$V := V_1(\mathbf{x})\frac{\partial}{\partial x_1} + V_2(\mathbf{x})\frac{\partial}{\partial x_2} + \cdots + V_n(\mathbf{x})\frac{\partial}{\partial x_n}$$

Find

$$d(V\rfloor\omega).$$

Give an interpretation of the result.

Solution 2.   Applying the product rule we obtain

$$V\rfloor\omega = \sum_{j=1}^{n} (-1)^{j+1} V_j dx_1 \wedge \cdots \wedge \widehat{dx_j} \wedge \cdots \wedge dx_n$$

where the circumflex indicates omission. Taking the exterior derivative of this equation we obtain

$$d(V \rfloor \omega) = \left(\sum_{j=1}^{n} \frac{\partial V_j}{\partial x_j} \right) dx_1 \wedge dx_2 \wedge \cdots \wedge dx_n .$$

This is the divergence of the vector field V. Since $d\omega = 0$ we have

$$L_V \omega = d(V \rfloor \omega) .$$

Problem 3. Consider the autonomous nonlinear system of ordinary differential equations

$$\frac{dx_1}{dt} = x_1 - x_1 x_2, \qquad \frac{dx_2}{dt} = -x_2 + x_1 x_2$$

where $x_1 > 0$ and $x_2 > 0$. This equation is a *Lotka-Volterra model*. The corresponding vector field is

$$V = (x_1 - x_1 x_2) \frac{\partial}{\partial x_1} + (-x_2 + x_1 x_2) \frac{\partial}{\partial x_2} .$$

Let

$$\Omega = \frac{dx_1 \wedge dx_2}{x_1 x_2} .$$

Calculate the Lie derivative $L_V \Omega$.

Solution 3. Since $d\Omega = 0$ we have

$$L_V \Omega = d(V \rfloor \Omega) .$$

Thus

$$L_V \Omega = d \left[\left(\frac{x_1 - x_1 x_2}{x_1 x_2} \right) dx_2 - \left(\frac{-x_2 + x_1 x_2}{x_1 x_2} \right) dx_1 \right]$$

$$= d \left(\frac{1}{x_2} dx_2 + \frac{1}{x_1} dx_1 \right) = 0$$

where we have used $ddx_1 = 0$, $ddx_2 = 0$. Since

$$L_V \Omega = 0$$

we say that Ω is *invariant* under V.

Problem 4. (i) Let $M = \mathbb{R}^n$. Let X_1, X_2 and Y be smooth vector fields defined on M. Assume that

$$[X_1, Y] = fY, \qquad [X_2, Y] = gY \tag{1}$$

where f and g are smooth functions defined on M. Let $\Omega := dx_1 \wedge \cdots \wedge dx_n$ be the volume form on M. Show that

$$L_Y(X_1 \rfloor X_2 \rfloor Y \rfloor \Omega) = (\text{div} Y)(X_1 \rfloor X_2 \rfloor Y \rfloor \Omega) \qquad (2)$$

where $\text{div} Y$ denotes the divergence of the vector field Y, i.e.,

$$\text{div} Y := \sum_{j=1}^{n} \frac{\partial Y_j}{\partial x_j}. \qquad (3)$$

(ii) Let X, Y be smooth vector fields and let f and g be smooth functions. Assume that

$$L_Y g = 0, \qquad [X, Y] = fY. \qquad (4)$$

Calculate

$$L_Y(L_X g). \qquad (5)$$

Apply the identity

$$L_{[X,Y]} \equiv [L_X, L_Y]. \qquad (6)$$

Solution 4.   (i) Straightforward calculation shows that

$$
\begin{aligned}
L_Y(X_1 \rfloor X_2 \rfloor Y \rfloor \Omega) &= [Y, X_1] \rfloor X_2 \rfloor Y \rfloor \Omega + X_1 \rfloor L_Y(X_2 \rfloor Y \rfloor \Omega) \\
&= -fY \rfloor X_2 \rfloor Y \rfloor \Omega + X_1 \rfloor L_Y(X_2 \rfloor Y \rfloor \Omega) \\
&= X_1 \rfloor ([Y, X_2] \rfloor Y \rfloor \Omega + X_2 \rfloor L_Y(Y \rfloor \Omega)) \\
&= -X_1 \rfloor (gY) \rfloor Y \rfloor \Omega + X_1 \rfloor X_2 \rfloor L_Y(Y \rfloor \Omega) \\
&= X_1 \rfloor X_2 \rfloor (\text{div} Y) Y \rfloor \Omega \\
&= (\text{div} Y)(X_1 \rfloor X_2 \rfloor Y \rfloor \Omega)
\end{aligned}
$$

where we have used that $Y \rfloor Y \rfloor \Omega = 0$.
(ii) Applying identity (6) and equation (4) we find

$$L_Y(L_X g) = L_X(L_Y g) - L_{[X,Y]} g = -L_{[X,Y]} g = -L_{fY} g = -fL_Y g = 0.$$

Therefore the function $L_X g$ is invariant under the vector field Y.

Problem 5.   (i) Let

$$T := \sum_{i,j=1}^{n} a_{ij}(\mathbf{x}) \frac{\partial}{\partial x_i} \otimes dx_j \qquad (1)$$

be a smooth (1,1) tensor field. Let

$$V = \sum_{k=1}^{n} V_k(\mathbf{x}) \frac{\partial}{\partial x_k} \qquad (2)$$

be a smooth vector field. Find $L_V T = 0$, where $L_V(\cdot)$ denotes the Lie derivative.

(ii) Assume that $L_V T = 0$. Show that

$$L_V \sum_{j=1}^{n} a_{jj} = 0. \tag{3}$$

Recall that the Lie derivative is linear and obeys the product rule, i.e.,

$$L_V \left(a_{ij} \frac{\partial}{\partial x_i} \otimes dx_j \right) \equiv (L_V a_{ij}) \frac{\partial}{\partial x_i} \otimes dx_j + a_{ij} \left(L_V \frac{\partial}{\partial x_i} \right) \otimes dx_j$$

$$+ {}_{ij} \frac{\partial}{\partial x_i} \otimes (L_V dx_j).$$

Solution 5. (i) First we calculate $L_V T$. Applying the linearity of the Lie derivative we find

$$L_V T = L_V \sum_{i,j=1}^{n} \left(a_{ij} \frac{\partial}{\partial x_i} \otimes dx_j \right) = \sum_{i,j=1}^{n} L_V \left(a_{ij} \frac{\partial}{\partial x_i} \otimes dx_j \right).$$

Applying the product rule for the Lie derivative and

$$\left[V_k \frac{\partial}{\partial x_k}, \frac{\partial}{\partial x_i} \right] = \frac{\partial V_k}{\partial x_i} \frac{\partial}{\partial x_k}$$

we find

$$L_V T = \sum_{i,j=1}^{n} \left(\sum_{k=1}^{n} \left(V_k \frac{\partial a_{ij}}{\partial x_k} - a_{kj} \frac{\partial V_i}{\partial x_k} + a_{ik} \frac{\partial V_k}{\partial x_j} \right) \right) \frac{\partial}{\partial x_i} \otimes dx_j.$$

Thus from $L_V T = 0$ we obtain

$$\sum_{k=1}^{n} \left(V_k \frac{\partial a_{ij}}{\partial x_k} - a_{kj} \frac{\partial V_i}{\partial x_k} + a_{ik} \frac{\partial V_k}{\partial x_j} \right) = 0 \qquad \text{for all} \quad i, j = 1, \ldots, n.$$

(ii) For $i = j$ we obtain

$$\sum_{k=1}^{n} \left(V_k \frac{\partial a_{jj}}{\partial x_k} - a_{kj} \frac{\partial V_j}{\partial x_k} + a_{jk} \frac{\partial V_k}{\partial x_j} \right) = 0$$

for $j = 1, \ldots, n$. Since

$$L_V \sum_{j=1}^{n} a_{jj} = \sum_{k=1}^{n} \sum_{j=1}^{n} V_k \frac{\partial a_{jj}}{\partial x_k}$$

we find (3).

Problem 6. Let $M = \mathbb{R}^n$ (or any open subset of $\mathbb{R}^n$). X is a vector field on M and α an r-form on M ($r < n$). We call α a *conformal invariant r-form* of X if

$$L_X \alpha = g \alpha.$$

Here $L_X \alpha$ is the Lie derivative of α with respect to X and g an arbitrary smooth function. Prove the following

Theorem. Let $M = \mathbb{R}^n$. The volume form is given by

$$\omega := dx_1 \wedge \cdots \wedge dx_n.$$

Let X and Y be two vector fields such that

$$[X, Y] = fY$$

and

$$\alpha := Y \rfloor \omega.$$

Then

$$L_X \alpha = (f + \operatorname{div} X) \alpha.$$

Solution 6. Applying the rules for the Lie derivative we find

$$
\begin{aligned}
L_X(Y \rfloor \omega) &= [X, Y] \rfloor \omega + Y \rfloor (L_X \omega) \\
&= (fY) \rfloor \omega + Y \rfloor (X \rfloor d\omega + d(X \rfloor \omega)) \\
&= f(Y \rfloor \omega) + Y \rfloor (d(X \rfloor \omega)) \\
&= f(Y \rfloor \omega) + Y \rfloor ((\operatorname{div} X) \omega) \\
&= (f + \operatorname{div} X)(Y \rfloor \omega).
\end{aligned}
$$

Hence we have

$$g = f + \operatorname{div} X.$$

Problem 7. The nonlinear partial differential equation

$$\frac{\partial^2 u}{\partial x \partial t} = \sin u \qquad (1)$$

is the so-called one-dimensional *sine-Gordon equation*. Show that

$$V = (4u_{xxx} + 2(u_x)^3) \frac{\partial}{\partial u} \qquad (2)$$

is a Lie-Bäcklund vector field of (1).

Solution 7. Taking the derivative of (1) with respect to x we find that (1) defines the manifolds

$$u_{xt} = \sin u$$
$$u_{xxt} = u_x \cos u$$
$$u_{xxxt} = u_{xx} \cos u - (u_x)^2 \sin u \tag{3}$$
$$u_{xxxxt} = u_{xxx} \cos u - 3u_x u_{xx} \sin u - (u_x)^3 \cos u .$$

The *prolongated vector field* $\bar{V}$ of V is given by

$$\bar{V} = V + (4u_{xxxt} + 6(u_x)^2 u_{xt})\frac{\partial}{\partial u_t} + (4u_{xxxxt} + 12u_x u_{xx} u_{xt} + 6(u_x)^2 u_{xxt})\frac{\partial}{\partial u_{xt}} .$$

It follows that the Lie derivative of $u_{xt} - \sin u$ with respect to $\bar{V}$ is given by

$$L_{\bar{V}}(u_{xt} - \sin u) = 4u_{xxxxt} + 12u_x u_{xx} u_{xt} + 6(u_x)^2 u_{xxt} - (4u_{xxx} + 2(u_x)^3) \cos u .$$

Inserting the equations given by (3) into this equation yields

$$L_{\bar{V}}(u_{xt} - \sin u) = 0 .$$

Consequently, the vector field V is a Lie-Bäcklund symmetry vector field of the sine-Gordon equation.

Problem 8. Let V be a smooth vector field defined on $\mathbb{R}^n$

$$V = \sum_{i=1}^{n} V_i(\mathbf{x})\frac{\partial}{\partial x_i} .$$

Let T be a $(1,1)$ smooth tensor field defined on $\mathbb{R}^n$

$$T = \sum_{i=1}^{n} \sum_{j=1}^{n} a_{ij}(\mathbf{x})\frac{\partial}{\partial x_i} \otimes dx_j .$$

Let $L_V T$ be the Lie derivative of T with respect to the vector field V. Show that if $L_V T = 0$ then

$$L_V \text{tr}(a(\mathbf{x})) = 0$$

where $a(\mathbf{x})$ is the $n \times n$ matrix $(a_{ij}(\mathbf{x}))$ and tr denotes the trace.

Solution 8. The Lie derivative is linear and obeys the product rule. Thus applying the product rule we have

$$L_V T = \sum_{i=1}^{n} \sum_{j=1}^{n} \left(L_V a_{ij}\frac{\partial}{\partial x_i} \otimes dx_j + a_{ij}\left(L_V \frac{\partial}{\partial x_i}\right) \otimes dx_j + a_{ij}\frac{\partial}{\partial x_i} \otimes L_V dx_i \right) .$$

Since

$$L_V(a_{ij}(\mathbf{x})) = V a_{ij}(\mathbf{x}) = \sum_{k=1}^n V_k \frac{\partial a_{ij}}{\partial x_k}$$

$$L_V \frac{\partial}{\partial x_i} = \left[\sum_{k=1}^n V_k \frac{\partial}{\partial x_k}, \frac{\partial}{\partial x_i}\right] = -\sum_{k=1}^n \frac{\partial V_k}{\partial x_i} \frac{\partial}{\partial x_k}$$

$$L_V dx_j = \sum_{k=1}^n \frac{\partial V_j}{\partial x_k} dx_k$$

we obtain for the Lie derivative

$$L_V T = \sum_{i=1}^n \sum_{j=1}^n \sum_{k=1}^n \left(V_k \frac{\partial a_{ij}}{\partial x_k} \frac{\partial}{\partial x_i} \otimes dx_j - a_{ij} \frac{\partial V_k}{\partial x_i} \frac{\partial}{\partial x_k} \otimes dx_j + a_{ij} \frac{\partial V_j}{\partial x_k} \frac{\partial}{\partial x_i} \otimes dx_k\right).$$

Using the *contraction operator* C which is f-linear and

$$C\left(\frac{\partial}{\partial x_i} \otimes dx_j\right) = \delta_{ij}$$

where δ_{ij} denotes the Kronecker delta we find $CL_V T = L_V(CT)$. Since $L_V T = 0$ by assumption we have $L_V(CT) = 0$. Therfore

$$L_V(CT) = \sum_{j=1}^n \sum_{k=1}^n \frac{\partial a_{jj}}{\partial x_k} = \sum_{k=1}^n \frac{\partial}{\partial x_k} \mathrm{tr}(a(\mathbf{x})) = 0.$$

Thus $\mathrm{tr}(a(\mathbf{x}))$ is a first integral with respect to V.

Problem 9. Let V, W be vector fields. Let f, g be C^∞ functions and α be a differential form. Assume that

$$L_V \alpha = f\alpha, \qquad L_W \alpha = g\alpha.$$

Show that

$$L_{[V,W]}\alpha = (L_V f - L_W g)\alpha. \qquad (1)$$

Solution 9. We have

$$L_{[V,W]}\alpha = [L_V, L_W]\alpha = L_V(L_W\alpha) - L_W(L_V\alpha) = L_V(g\alpha) - L_W(f\alpha).$$

Applying the product rule we have

$$L_V(g\alpha) = (L_V g)\alpha + g L_V\alpha = (L_V g)\alpha + gf\alpha$$
$$L_W(f\alpha) = (L_W f)\alpha + f L_W\alpha = (L_W f)\alpha + fg\alpha.$$

Thus (1) follows.

Problem 10. Consider the vector fields

$$V = x\frac{\partial}{\partial x} + y\frac{\partial}{\partial y}, \qquad W = x\frac{\partial}{\partial y} - y\frac{\partial}{\partial x}$$

defined on $\mathbb{R}^2$.
(i) Do the vector fields V, W form a basis of a Lie algebra? If so, what type of Lie algebra do we have.
(ii) Express the two vector fields in polar coordinates $x(r, \theta) = r\cos\theta$, $y(r, \theta) = r\sin\theta$.
(iii) Calculate the commutator of the two vector fields expressed in polar coordinates. Compare with the result of (i).

Solution 10. (i) Calculating the commutator of V and W we find

$$[V, W] = 0.$$

Thus we have a basis of a commutative Lie algebra.
(ii) Applying the chain rule to $f(x(r, \theta), y(r, \theta))$ we obtain

$$\frac{\partial f}{\partial r} = \frac{\partial f}{\partial x}\frac{\partial x}{\partial r} + \frac{\partial f}{\partial y}\frac{\partial y}{\partial r} = \frac{\partial f}{\partial x}\cos\theta + \frac{\partial f}{\partial y}\sin\theta$$

$$\frac{\partial f}{\partial \theta} = \frac{\partial f}{\partial x}\frac{\partial x}{\partial \theta} + \frac{\partial f}{\partial y}\frac{\partial y}{\partial \theta} = -\frac{\partial f}{\partial x}r\sin\theta + \frac{\partial f}{\partial y}r\cos\theta.$$

Thus

$$\frac{\partial f}{\partial r}r\sin\theta = \frac{\partial f}{\partial x}r\cos\theta\sin\theta + \frac{\partial f}{\partial y}r\sin^2\theta$$

$$\frac{\partial f}{\partial \theta}\cos\theta = -\frac{\partial f}{\partial x}r\sin\theta\cos\theta + \frac{\partial f}{\partial y}r\cos^2\theta.$$

Therefore

$$\frac{\partial}{\partial y} = \sin\theta\frac{\partial}{\partial r} + \frac{\cos\theta}{r}\frac{\partial}{\partial \theta}.$$

Analogously

$$-\frac{\partial f}{\partial r}r\cos\theta = -\frac{\partial f}{\partial x}r\cos^2\theta - \frac{\partial f}{\partial y}r\sin\theta\cos\theta$$

$$\frac{\partial f}{\partial \theta}\sin\theta = -\frac{\partial f}{\partial x}r\sin^2\theta + \frac{\partial f}{\partial y}r\cos\theta\sin\theta.$$

Therefore

$$\frac{\partial}{\partial x} = \cos\theta\frac{\partial}{\partial r} - \frac{\sin\theta}{r}\frac{\partial}{\partial \theta}.$$

Thus the vector field V takes the form

$$r\frac{\partial}{\partial r}$$

and the vector field W takes the form

$$\frac{\partial}{\partial \theta}.$$

(iii) For the commutator we find

$$\left[r\frac{\partial}{\partial r},\frac{\partial}{\partial \theta}\right]=0.$$

Thus under the transformation to polar coordinates the commutator is preserved.

Problem 11.   Let V, W be two smooth vector fields defined on $\mathbb{R}^3$. We write

$$V = V_1(\mathbf{x})\frac{\partial}{\partial x_1} + V_2(\mathbf{x})\frac{\partial}{\partial x_2} + V_3(\mathbf{x})\frac{\partial}{\partial x_3}$$

$$W = W_1(\mathbf{x})\frac{\partial}{\partial x_1} + W_2(\mathbf{x})\frac{\partial}{\partial x_2} + W_3(\mathbf{x})\frac{\partial}{\partial x_3}.$$

Let

$$\omega = dx_1 \wedge dx_2 \wedge dx_3$$

be the volume form in $\mathbb{R}^3$. Then $L_V\omega = (\operatorname{div}(V))\omega$, where $L_V(.)$ denotes the Lie derivative and $\operatorname{div}V$ denotes the divergence of the vector field V. Find the divergence of the vector field given by the commutator $[V, W]$. Apply it to the vector fields asscociated with the autonomous systems of first order differential equations

$$\frac{dx_1}{dt}=\sigma(x_2-x_1),\quad \frac{dx_2}{dt}=\alpha x_1-x_2-x_1x_3,\quad \frac{dx_3}{dt}=-\beta x_3+x_1x_2$$

and

$$\frac{dx_1}{dt}=a(x_2-x_1),\quad \frac{dx_2}{dt}=(c-a)x_1+cx_2-x_1x_3,\quad \frac{dx_3}{dt}=-bx_3+x_1x_2.$$

The first system is the *Lorenz model* and the second system is *Chen's model*.

Solution 11.   Using the properties of the Lie derivative such as linearity and the product rule we find

$$L_{[V,W]}\omega = L_V(L_W\omega) - L_W(L_V\omega)$$
$$= L_V((\operatorname{div}W)\omega) - L_W((\operatorname{div}V)\omega)$$
$$= (L_V(\operatorname{div}W))\omega + (\operatorname{div}W)L_V\omega - (L_W(\operatorname{div}V))\omega - (\operatorname{div}V)L_W\omega$$
$$= (L_V(\operatorname{div}W) + (\operatorname{div}W)(\operatorname{div}V) - L_W(\operatorname{div}V) - (\operatorname{div}V)(\operatorname{div}W))\omega$$
$$= (L_V(\operatorname{div}W) - L_W(\operatorname{div}V))\omega.$$

Since divV for the first system is constant and divW for the second system is also constant, we find

$$L_{[V,W]}\omega = 0\,.$$

Thus the vector field given by the commutator is divergenceless.

Problem 12. Consider the smooth vector field

$$V = \sum_{j=1}^{n} V_j(\mathbf{u})\frac{\partial}{\partial u_j}$$

defined on $\mathbb{R}^n$. Consider the smooth differential one-form

$$\alpha = \sum_{k=1}^{n} f_k(\mathbf{u})du_k\,.$$

Find the Lie derivative $L_V\alpha$. What is the condition such that $L_V\alpha = 0$?

Solution 12. Using the linearity and the product rule of the Lie derivative we have

$$L_V\alpha = \sum_{k=1}^{n}(L_V f_k du_k)$$

$$= \sum_{k=1}^{n}((L_V f_k)du_k + f_k(\mathbf{u})(L_V du_k))$$

$$= \sum_{k=1}^{n}\left(\sum_{j=1}^{n} V_j\frac{\partial f_k}{\partial u_j}\right)du_k + \sum_{k=1}^{n} f_k dV_k$$

$$= \sum_{k=1}^{n}\left(\sum_{j=1}^{n} V_j\frac{\partial f_k}{\partial u_j}\right)du_k + \sum_{k=1}^{n} f_k\sum_{j=1}^{n}\frac{\partial V_k}{\partial u_j}du_j$$

$$= \sum_{k=1}^{n}\left(\sum_{j=1}^{n}\left(V_j\frac{\partial f_k}{\partial u_j} + f_j\frac{\partial V_J}{\partial u_k}\right)\right)du_k\,.$$

Thus the condition is

$$\sum_{j=1}^{n}\left(V_j\frac{\partial f_k}{\partial u_j} + f_j\frac{\partial V_j}{\partial u_k}\right) = 0,\qquad k = 1,2,\ldots,n\,.$$

For $n = 1$ we obtain $V\partial f/\partial u + f\partial V/\partial u = 0$.

Problem 13. Some quantities in physics, owing to the transformation laws have to be considered as *currents* instead of differential forms. Let

M be an orientable n-dimensional differentiable manifold of class C^∞. We denote by $\Phi_k(M)$ the set of all differential forms of degree k with compact support. Let $\phi \in \Phi_k(M)$ and let α be an exterior differential form of degree $n - k$ with locally integrable coefficients. Then, as an example of a current, we have

$$T_\alpha(\phi) \equiv \alpha(\phi) := \int_M \alpha \wedge \phi.$$

Define the Lie derivative for this current.

Solution 13.   The Lie derivative of this current can be defined as

$$L_V T(\phi) = -T(L_V \phi).$$

This definition can be motivated as follows. Assume that $L_V \alpha$ exists. Since

$$(L_V \alpha) \wedge \phi \equiv L_V(\alpha \wedge \phi) - \alpha \wedge (L_V \phi)$$

we obtain

$$\begin{aligned}
T_{L_V \alpha} \phi &= \int_M (L_V \alpha) \wedge \phi \\
&= \int_M (L_V(\alpha \wedge \phi) - \alpha \wedge (L_V \phi)) \\
&= - \int_M \alpha \wedge (L_V \phi)
\end{aligned}$$

where we used that

$$\int_M L_V(\alpha \wedge \phi) = \int_M d(V \rfloor (\alpha \wedge \phi)) = 0$$

noting that the $(n - 1)$ differential form $V \rfloor (\alpha \wedge \phi)$ has compact support.

Chapter 24

Metric Tensor Fields

Problem 1. Let

$$g = dx \otimes dx + \cos(u(x,t))dx \otimes dt + \cos(u(x,t))dt \otimes dx + dt \otimes dt \quad (1)$$

be a metric tensor field, where u is a smooth function of x and t. The *Christoffel symbols* are defined as

$$\Gamma^a_{mn} := \frac{1}{2}g^{ab}(g_{bm,n} + g_{bn,m} - g_{mn,b})$$

where the summation convention is used and $g_{bm,1} := \partial g_{bm}/\partial x$, $g_{bm,2} := \partial g_{bm}/\partial t$. The *Riemann curvature tensor* is defined by

$$R^r_{msq} := \Gamma^r_{mq,s} - \Gamma^r_{ms,q} + \Gamma^r_{ns}\Gamma^n_{mq} - \Gamma^r_{nq}\Gamma^n_{ms}.$$

The *Ricci tensor* R_{mq} is defined by

$$R_{mq} := R^a_{maq} = -R^a_{mqa}$$

i.e. the Ricci tensor is constructed by contraction. Finally, the *curvature scalar* R is given by

$$R = R^m_m.$$

(i) Calculate the *Riemann curvature scalar* R.
(ii) Find the equation which follows from the condition $R = -2$.

Solution 1. We set $1 \equiv x$ and $2 \equiv t$. Then $g_{11} = g_{22} = 1$ and

$$g_{12} = g_{21} = \cos(u(x,t)).$$

265

We write g as a 2×2 matrix

$$g = \begin{pmatrix} g_{11} & g_{12} \\ g_{21} & g_{22} \end{pmatrix}.$$

The inverse of g is given by

$$g^{-1} = \begin{pmatrix} g^{11} & g^{12} \\ g^{21} & g^{22} \end{pmatrix}$$

where

$$g^{11} = g^{22} = \frac{1}{\sin^2 u}, \qquad g^{12} = g^{21} = -\frac{\cos u}{\sin^2 u}.$$

Obviously $g_{11,1} = g_{11,2} = g_{22,1} = g_{22,2} = 0$ and

$$g_{12,1} = g_{21,1} = -\frac{\partial u}{\partial x} \sin u, \quad g_{12,2} = g_{21,2} = -\frac{\partial u}{\partial t} \sin u.$$

Therefore the Christoffel symbols are given by

$$\Gamma_{11}^1 = \frac{\partial u}{\partial x} \frac{\cos u}{\sin u}, \quad \Gamma_{12}^1 = \Gamma_{21}^1 = 0, \quad \Gamma_{22}^1 = -\frac{\partial u}{\partial t} \frac{1}{\sin u}$$

$$\Gamma_{11}^2 = -\frac{\partial u}{\partial x} \frac{1}{\sin u}, \quad \Gamma_{12}^2 = \Gamma_{21}^2 = 0, \quad \Gamma_{22}^2 = \frac{\partial u}{\partial t} \frac{\cos u}{\sin u}.$$

Obviously $R_{mss}^b = 0$, i.e.

$$R_{111}^1 = R_{122}^1 = R_{211}^1 = R_{222}^1 = R_{111}^2 = R_{122}^2 = R_{211}^2 = R_{222}^2 = 0.$$

Moreover

$$R_{112}^1 = -R_{121}^1 = -\frac{\partial^2 u}{\partial x \partial t} \frac{\cos u}{\sin u}, \qquad R_{212}^1 = -R_{221}^1 = -\frac{\partial^2 u}{\partial x \partial t} \frac{1}{\sin u}$$

$$R_{112}^2 = -R_{121}^2 = \frac{\partial^2 u}{\partial x \partial t} \frac{1}{\sin u}, \qquad R_{212}^2 = -R_{221}^2 = \frac{\partial^2 u}{\partial x \partial t} \frac{\cos u}{\sin u}.$$

For the Ricci tensor we find

$$R_{11} = R_{111}^1 + R_{121}^2 = -\frac{\partial^2 u}{\partial x \partial t} \frac{1}{\sin u}, \quad R_{12} = R_{112}^1 + R_{122}^2 = -\frac{\partial^2 u}{\partial x \partial t} \frac{\cos u}{\sin u}$$

$$R_{21} = R_{211}^1 + R_{221}^2 = -\frac{\partial^2 u}{\partial x \partial t} \frac{\cos u}{\sin u}, \quad R_{22} = R_{212}^1 + R_{222}^2 = -\frac{\partial^2 u}{\partial x \partial t} \frac{1}{\sin u}.$$

From R_{nq} we obtain R_q^m via $R_q^m = g^{mn} R_{nq}$. We find

$$R_1^1 = R_2^2 = -\frac{1}{\sin u} \frac{\partial^2 u}{\partial x \partial t}$$

and $R_1^2 = R_2^1 = 0$. Thus for the curvature scalar we find

$$R = -\frac{2}{\sin u}\frac{\partial^2 u}{\partial x \partial t}.$$

If $R = -2$, then

$$\frac{\partial^2 u}{\partial x \partial t} = \sin u.$$

This is the so-called *sine-Gordon equation*.

Problem 2. Given the metric tensor field

$$g = g_{11}(q_1, q_2)dq_1 \otimes dq_1 + g_{22}(q_1, q_2)dq_2 \otimes dq_2. \tag{1}$$

(i) Calculate the *Riemann curvature scalar R*.
(ii) Simplify g to the special case $g_{11}(q_1, q_2) = g_{22}(q_1, q_2) = E - V(q_1, q_2)$.

Solution 2. (i) First we have to calculate the *Christoffel symbols*. Since

$$\Gamma_{11}^1 := \frac{1}{2}\sum_{k=1}^{2} g^{1k}\left(\frac{\partial g_{k1}}{\partial q_1} + \frac{\partial g_{1k}}{\partial q_1} - \frac{\partial g_{11}}{\partial q_k}\right)$$

and $g_{12} = g_{21} = 0$ we have

$$\Gamma_{11}^1 = \frac{1}{2}g^{11}\left(\frac{\partial g_{11}}{\partial q_1} + \frac{\partial g_{11}}{\partial q_1} - \frac{\partial g_{11}}{\partial q_1}\right) = \frac{1}{2}g^{11}\frac{\partial g_{11}}{\partial q_1}.$$

Analogously

$$\Gamma_{12}^1 = \frac{1}{2}\sum_{k=1}^{2} g^{1k}\left(\frac{\partial g_{k2}}{\partial q_1} + \frac{\partial g_{1k}}{\partial q_2} - \frac{\partial g_{12}}{\partial q_k}\right) = \frac{1}{2}g^{11}\frac{\partial g_{11}}{\partial q_2}$$

$$\Gamma_{21}^1 = \frac{1}{2}\sum_{k=1}^{2} g^{1k}\left(\frac{\partial g_{k1}}{\partial q_2} + \frac{\partial g_{2k}}{\partial q_1} - \frac{\partial g_{21}}{\partial q_k}\right) = \frac{1}{2}g^{11}\frac{\partial g_{11}}{\partial q_2}$$

$$\Gamma_{22}^1 = \frac{1}{2}\sum_{k=1}^{2} g^{1k}\left(\frac{\partial g_{k2}}{\partial q_2} + \frac{\partial g_{2k}}{\partial q_2} - \frac{\partial g_{22}}{\partial q_k}\right) = -\frac{1}{2}g^{11}\frac{\partial g_{22}}{\partial q_1}$$

$$\Gamma_{11}^2 = \frac{1}{2}\sum_{k=1}^{2} g^{2k}\left(\frac{\partial g_{k1}}{\partial q_1} + \frac{\partial g_{1k}}{\partial q_1} - \frac{\partial g_{11}}{\partial q_k}\right) = -\frac{1}{2}g^{22}\frac{\partial g_{11}}{\partial q_2}$$

$$\Gamma_{12}^2 = \frac{1}{2}\sum_{k=1}^{2} g^{2k}\left(\frac{\partial g_{k2}}{\partial q_1} + \frac{\partial g_{1k}}{\partial q_2} - \frac{\partial g_{12}}{\partial q_k}\right) = \frac{1}{2}g^{22}\frac{\partial g_{22}}{\partial q_1}$$

$$\Gamma_{21}^2 = \frac{1}{2}\sum_{k=1}^{2} g^{2k}\left(\frac{\partial g_{k1}}{\partial q_2} + \frac{\partial g_{2k}}{\partial q_1} - \frac{\partial g_{21}}{\partial q_k}\right) = \frac{1}{2}g^{22}\frac{\partial g_{22}}{\partial q_1}$$

$$\Gamma_{22}^2 = \frac{1}{2} \sum_{k=1}^{2} g^{2k} \left(\frac{\partial g_{k2}}{\partial q_2} + \frac{\partial g_{2k}}{\partial q_2} - \frac{\partial g_{22}}{\partial q_k} \right) = \frac{1}{2} g^{22} \frac{\partial g_{22}}{\partial q_2}$$

where $g^{11} = 1/g_{11}$, $g^{22} = 1/g_{22}$. Next we have to calculate the curvature. We find

$$R_{12} = \sum_{h=1}^{2} \left(\frac{\partial \Gamma_{12}^h}{\partial q_h} - \frac{\partial \Gamma_{h1}^h}{\partial q_2} + \sum_{\ell=1}^{2} \left(\Gamma_{h\ell}^h \Gamma_{12}^\ell - \Gamma_{1h}^\ell \Gamma_{2\ell}^h \right) \right) = 0$$

$$R_{21} = 0 \qquad \text{for symmetry reasons}$$

$$R_{11} = \sum_{h=1}^{2} \left(\frac{\partial \Gamma_{11}^h}{\partial q_h} - \frac{\partial \Gamma_{h1}^h}{\partial q_1} + \sum_{\ell=1}^{2} \left(\Gamma_{h\ell}^h \Gamma_{11}^\ell - \Gamma_{1h}^\ell \Gamma_{1\ell}^h \right) \right)$$

$$= \frac{\partial \Gamma_{11}^2}{\partial q_2} - \frac{\partial \Gamma_{21}^2}{\partial q_1} + \left(\Gamma_{21}^2 \Gamma_{11}^1 + \Gamma_{22}^2 \Gamma_{11}^2 - \Gamma_{11}^2 \Gamma_{12}^1 - \Gamma_{12}^2 \Gamma_{12}^2 \right)$$

$$R_{22} = \sum_{h=1}^{2} \left(\frac{\partial \Gamma_{22}^h}{\partial q_h} - \frac{\partial \Gamma_{h2}^h}{\partial q_2} + \sum_{\ell=1}^{2} \left(\Gamma_{h\ell}^h \Gamma_{22}^\ell - \Gamma_{2h}^\ell \Gamma_{2\ell}^h \right) \right)$$

$$= \frac{\partial \Gamma_{22}^1}{\partial q_1} - \frac{\Gamma_{12}^1}{\partial q_2} + \Gamma_{11}^1 \Gamma_{22}^1 + \Gamma_{12}^1 \Gamma_{22}^2 - \Gamma_{21}^1 \Gamma_{21}^1 - \Gamma_{21}^2 \Gamma_{22}^1.$$

Now the Riemann curvature scalar R is given by $R = g^{11} R_{11} + g^{22} R_{22}$. Consequently,

$$R = g^{11} \left(\frac{\partial \Gamma_{11}^2}{\partial q_2} - \frac{\partial \Gamma_{21}^2}{\partial q_1} \right) + g^{11} \left(\Gamma_{11}^1 \Gamma_{21}^2 + \Gamma_{22}^2 \Gamma_{11}^2 - \Gamma_{11}^2 \Gamma_{12}^1 - \Gamma_{12}^2 \Gamma_{12}^2 \right)$$

$$+ g^{22} \left(\frac{\partial \Gamma_{22}^1}{\partial q_1} - \frac{\partial \Gamma_{12}^1}{\partial q_2} \right) + g^{22} \left(\Gamma_{22}^2 \Gamma_{12}^1 + \Gamma_{11}^1 \Gamma_{22}^1 - \Gamma_{22}^1 \Gamma_{21}^2 - \Gamma_{21}^1 \Gamma_{21}^1 \right)$$

where

$$\frac{\partial \Gamma_{11}^2}{\partial q_2} = -\frac{1}{2} \frac{\partial}{\partial q_2} \left(g^{22} \frac{\partial g_{11}}{\partial q_2} \right) = -\frac{1}{2} \left(\frac{\partial g^{22}}{\partial q_2} \right) \left(\frac{\partial g_{11}}{\partial q_2} \right) - \frac{1}{2} g^{22} \frac{\partial^2 g_{11}}{\partial q_2^2}$$

$$\frac{\partial \Gamma_{21}^2}{\partial q_1} = \frac{1}{2} \frac{\partial}{\partial q_1} \left(g^{22} \frac{\partial g_{22}}{\partial q_1} \right) = \frac{1}{2} \left(\frac{\partial g^{22}}{\partial q_1} \right) \left(\frac{\partial g_{22}}{\partial q_1} \right) + \frac{1}{2} g^{22} \frac{\partial^2 g_{22}}{\partial q_1^2}$$

$$\frac{\partial \Gamma_{22}^1}{\partial q_1} = -\frac{1}{2} \frac{\partial}{\partial q_1} \left(g^{11} \frac{\partial g_{22}}{\partial q_1} \right) = -\frac{1}{2} \left(\frac{\partial g^{11}}{\partial q_1} \right) \left(\frac{\partial g_{22}}{\partial q_1} \right) - \frac{1}{2} g^{11} \frac{\partial^2 g_{22}}{\partial q_1^2}$$

$$\frac{\partial \Gamma_{12}^1}{\partial q_2} = \frac{1}{2} \frac{\partial}{\partial q_2} \left(g^{11} \frac{\partial g_{11}}{\partial q_2} \right) = \frac{1}{2} \left(\frac{\partial g^{11}}{\partial q_2} \right) \left(\frac{\partial g_{11}}{\partial q_2} \right) + \frac{1}{2} g^{11} \frac{\partial^2 g_{11}}{\partial q_2^2} .$$

Therefore

$$R = \frac{g^{11} g^{22}}{2} \left(-\frac{\partial^2 g_{11}}{\partial q_2^2} - \frac{\partial^2 g_{22}}{\partial q_1^2} - \frac{\partial^2 g_{22}}{\partial q_1^2} - \frac{\partial^2 g_{11}}{\partial q_2^2} \right)$$

$$- \frac{1}{2} \left(g^{11} \frac{\partial g^{22}}{\partial q_2} \frac{\partial g_{11}}{\partial q_2} + g^{11} \frac{\partial g^{22}}{\partial q_1} \frac{\partial g_{22}}{\partial q_1} + g^{22} \frac{\partial g^{11}}{\partial q_1} \frac{\partial g_{22}}{\partial q_1} + g^{22} \frac{\partial g^{11}}{\partial q_2} \frac{\partial g_{11}}{\partial q_2} \right).$$

(ii) If $g_{11} = g_{22} = E - V(q_1, q_2)$ we obtain $g^{11} = 1/(E-V)$, $g^{22} = 1/(E-V)$. Therefore

$$R = \frac{1}{(E-V)^3} \left(\left(\frac{\partial^2 V}{\partial q_1^2} + \frac{\partial^2 V}{\partial q_2^2} \right) (E - V) + \frac{\partial V}{\partial q_1} \frac{\partial V}{\partial q_1} + \frac{\partial V}{\partial q_2} \frac{\partial V}{\partial q_2} \right).$$

The motion of a classical mechanical system with Hamilton function

$$H(\mathbf{p}, \mathbf{q}) = \frac{1}{2} \alpha^{jk} p_j p_k + V(q_1, q_2, \dots, q_N)$$

(summation convention; $j, k = 1, 2, \dots, N$) can be represented as a *geodesic flow* on a Riemannian manifold with metric $g_{jk} = (E - V(\mathbf{q})) \alpha_{jk}$. The matrix (α^{jk}) is symmetric and the matrix (α_{kl}) is its inverse

$$\alpha^{jk} \alpha_{kl} = \delta_l^j$$

where δ_l^j is the Kronecker symbol. In our case we have $N = 2$ and $\alpha_{jk} = \delta_{jk}$. The geodesic flow on a closed Riemannian manifold of negative curvature is a so-called *C-flow* and is therefore ergodic.

Problem 3. Let $a > b > 0$ and define $f : \mathbb{R}^2 \to \mathbb{R}^3$ by

$$f(\theta, \phi) = ((a + b \cos \phi) \cos \theta, (a + b \cos \phi) \sin \theta, b \sin \phi)$$

where $0 \le \phi < 2\pi$ and $0 \le \theta < 2\pi$. The function f is a *parametrized torus* T^2 on $\mathbb{R}^3$. Let

$$g = dx_1 \otimes dx_1 + dx_2 \otimes dx_2 + dx_3 \otimes dx_3.$$

The parametrized torus is doubly-periodic, i.e.

$$f(\theta + 2k\pi, \phi) = f(\theta, \phi), \qquad f(\theta, \phi + 2k\pi) = f(\theta, \phi), \qquad k \in \mathbb{Z}.$$

(i) Calculate $g|_{T^2}$.
(ii) Calculate the Christoffel symbols Γ_{ab}^m from $g|_{T^2}$.
(iii) Give the differential equations of the geodesics.

Solution 3. (i) Since

$$x_1(\phi, \theta) = (a + b \cos \phi) \cos \theta$$
$$x_2(\phi, \theta) = (a + b \cos \phi) \sin \theta$$
$$x_3(\phi, \theta) = b \sin \phi$$

and $dx_j = (\partial x_j / \partial \theta) d\theta + (\partial x_j / \partial \phi) d\phi$ $(j = 1, 2, 3)$ we find

$$dx_1 \otimes dx_1 + dx_2 \otimes dx_2 + dx_3 \otimes dx_3 = (a + b \cos \phi)^2 d\theta \otimes d\theta + b^2 d\phi \otimes d\phi$$

where we used that $\sin^2\theta + \cos^2\theta = 1$ and $\sin^2\phi + \cos^2\phi = 1$. We identify θ with the first coordinate and ϕ with the second coordinate. Thus

$$g_{11} = (a + b\cos(\phi))^2, \quad g_{12} = 0, \quad g_{21} = 0, \quad g_{22} = b^2$$

and therefore

$$g^{11} = \frac{1}{(a + b\cos(\phi))^2}, \quad g^{12} = 0, \quad g^{21} = 0, \quad g^{22} = \frac{1}{b^2}.$$

(ii) We set $y^1 = \theta$ and $y^2 = \phi$. The *Christoffel symbols* are given by

$$\Gamma^\alpha_{\beta\gamma} = \frac{1}{2}g^{\alpha\delta}\left(\frac{\partial g_{\beta\delta}}{\partial y^\gamma} + \frac{\partial g_{\gamma\delta}}{\partial y^\beta} - \frac{\partial g_{\beta\gamma}}{\partial y^\delta}\right)$$

where we make use of the summation convention (summation over δ, $\delta = 1, 2, 3$). Thus we find

$$\Gamma^1_{11} = 0, \quad \Gamma^1_{12} = -\frac{b\sin\phi}{a + b\cos\phi}, \quad \Gamma^1_{21} = -\frac{b\sin\phi}{a + b\cos\phi}, \quad \Gamma^1_{22} = 0$$

$$\Gamma^2_{11} = \frac{(a + b\cos\phi)\sin\phi}{b}, \quad \Gamma^2_{12} = 0, \quad \Gamma^2_{21} = 0, \quad \Gamma^2_{22} = 0.$$

(iii) The differential equations for the *geodesics* are determined by

$$\frac{d^2 y^\alpha}{ds^2} + \Gamma^\alpha_{\beta\gamma}\frac{dy^\beta}{ds}\frac{dy^\gamma}{ds} = 0$$

where we used the summation convention (summation over β and γ, $\beta = 1, 2, 3$, $\gamma = 1, 2, 3$). Thus we find

$$\frac{d^2\theta}{ds^2} - \frac{2b\sin\phi}{a + b\cos\phi}\frac{d\theta}{ds}\frac{d\phi}{ds} = 0, \quad \frac{d^2\phi}{ds^2} + \frac{(a + b\cos\phi)\sin\phi}{b}\left(\frac{d\theta}{ds}\right)^2 = 0.$$

Problem 4. The two-dimensional *de Sitter space* $\mathbb{V}$ with the topology $\mathbb{R}\times\mathbb{S}$ may be visualized as a one-sheet hyperboloid $\mathbb{H}_{r_0}$ embedded in 3-dimensional Minkowski space $\mathbb{M}$, i.e.

$$\mathbb{H}_{r_0} = \{(y^0, y^1, y^2) \in \mathbb{M} \,|\, (y^2)^2 + (y^1)^2 - (y^0)^2 = r_0^2, \; r_0 > 0\}$$

where r_0 is the parameter of the one-sheet hyperboloid $\mathbb{H}_{r_0}$. The induced metric, $g_{\mu\nu}$ ($\mu, \nu = 0, 1$), on $\mathbb{H}_{r_0}$ is the de Sitter metric.

(i) Show that we can parametrize (parameters ρ and θ) the *hyperboloid* as follows

$$y^0(\rho, \theta) = -\frac{r_0\cos(\rho/r_0)}{\sin(\rho/r_0)}, \quad y^1(\rho, \theta) = \frac{r_0\cos(\theta/r_0)}{\sin(\rho/r_0)}, \quad y^2(\rho, \theta) = \frac{r_0\sin(\theta/r_0)}{\sin(\rho/r_0)}$$

where $0 < \rho < \pi r_0$ and $0 \le \theta < 2\pi r_0$.

(ii) Using this parametrization find the metric tensor field induced on $\mathbb{H}_{r_0}$.

Solution 4. (i) Inserting y^0, y^1, y^2 into $(y^2)^2 + (y^1)^2 - (y^0)^2$ and using $\sin^2(\rho/r_0) + \cos^2(\rho/r_0) = 1$ we obtain r_0^2.

(ii) We start from the metric tensor field

$$g = dy^0 \otimes dy^0 - dy^1 \otimes dy^1 - dy^2 \otimes dy^2 .$$

Since $\partial y^0/\partial\rho = \csc^2(\rho/r_0)$, $\partial y^0/\partial\theta = 0$ and

$$\frac{\partial y^1}{\partial\rho} = -\frac{\cos(\theta/r_0)\cos(\rho/r_0)}{\sin^2(\rho/r_0)}, \quad \frac{\partial y^1}{\partial\theta} = -\frac{\sin(\theta/r_0)}{\sin(\rho/r_0)}$$

$$\frac{\partial y^2}{\partial\rho} = -\frac{\sin(\theta/r_0)\cos(\rho/r_0)}{\sin^2(\rho/r_0)}, \quad \frac{\partial y^2}{\partial\theta} = \frac{\cos(\theta/r_0)}{\sin(\rho/r_0)}$$

we find

$$g = \left(\frac{1}{\sin^4(\rho/r_0)} - \frac{\cos^2(\theta/r_0)\cos^2(\rho/r_0)}{\sin^4(\rho/r_0)} - \frac{\sin^2(\theta/r_0)\cos^2(\rho/r_0)}{\sin^4(\rho/r_0)} \right) d\rho \otimes d\rho$$

$$+ \left(-\frac{\sin^2(\theta/r_0)}{\sin^2(\rho/r_0)} - \frac{\cos^2(\theta/r_0)}{\sin^2(\rho/r_0)} \right) d\theta \otimes d\theta .$$

Using $\sin^2(\theta/r_0) + \cos^2(\theta/r_0) = 1$ we arrive at

$$g = \frac{1}{\sin^2(\rho/r_0)} d\rho \otimes d\rho - \frac{1}{\sin^2(\rho/r_0)} d\theta \otimes d\theta .$$

Problem 5. The *anti-de Sitter space* is defined as the surface

$$X^2 + Y^2 + Z^2 - U^2 - V^2 = -1$$

embedded in a five-dimensional flat space with the metric tensor field

$$g = dX \otimes dX + dY \otimes dY + dZ \otimes dZ - dU \otimes dU - dV \otimes dV .$$

This is a solution of Einstein's equations with the cosmological constant $\Lambda = -3$. Its intrinsic curvature is constant and negative. Find the metric tensor field in terms of the intrinsic coordinates (ρ, θ, ϕ, t) where

$$X(\rho, \theta, \phi, t) = \frac{2\rho}{1 - \rho^2} \sin\theta \cos\phi$$

$$Y(\rho, \theta, \phi, t) = \frac{2\rho}{1 - \rho^2} \sin\theta \sin\phi$$

$$Z(\rho, \theta, \phi, t) = \frac{2\rho}{1 - \rho^2} \cos\theta$$

and

$$U(\rho, \theta, \phi, t) = \frac{1 + \rho^2}{1 - \rho^2} \cos t, \qquad V(\rho, \theta, \phi, t) = \frac{1 + \rho^2}{1 - \rho^2} \sin t$$

where $0 \le \rho < 1$, $0 \le \phi < 2\pi$, $0 \le \theta < \pi$, $-\pi \le t < \pi$.

Solution 5.   We find

$$g = -\left(\frac{1 + \rho^2}{1 - \rho^2}\right)^2 dt \otimes dt + \frac{4}{(1 - \rho^2)^2}(d\rho \otimes d\rho + \rho^2 d\theta \otimes d\theta + \rho^2 \sin^2 \theta d\phi \otimes d\phi).$$

What are the Killing vector fields of g?

Problem 6.   Consider the *Poincaré upper half-plane*

$$H_+^2 := \{\, (x, y) \in \mathbb{R}^2 \, : \, y > 0 \,\}$$

with metric tensor field

$$g = \frac{1}{y} dx \otimes \frac{1}{y} dx + \frac{1}{y} dy \otimes \frac{1}{y} dy$$

which is conformal with the standard inner product. Find the curvature forms.

Solution 6.   The orthonormal coframe and frame are

$$\sigma^1 = \frac{1}{y} dx, \quad \sigma^2 = \frac{1}{y} dy, \quad s_1 = y \frac{\partial}{\partial x}, \quad s_2 = y \frac{\partial}{\partial y}.$$

It follows that

$$d\sigma^1 = d\left(\frac{1}{y} dx\right) = \frac{1}{y^2} dx \wedge dy = \sigma^1 \wedge \sigma^2$$

and obviously $d\sigma^2 = 0$. The connection forms we compute from $d\sigma + \omega \wedge \sigma = 0$. We find

$$-d\sigma^1 = 0 + \omega_2^1 \wedge \sigma^2 = -\sigma^1 \wedge \sigma^2, \qquad -d\sigma^2 = \omega_1^2 \wedge \sigma^1 + 0 = 0.$$

Therefore $\omega_2^1 = -\sigma^1 = -y^{-1} dx$. Then

$$\omega = \begin{pmatrix} 0 & -\sigma^1 \\ \sigma^1 & 0 \end{pmatrix}.$$

For the curvature forms we obtain

$$\Omega_2^1 = d\omega_2^1 + \omega_1^1 \wedge \omega_2^1 + \omega_2^1 \wedge \omega_2^2 = d(-y^{-1} dx) = -\sigma^1 \wedge \sigma^2$$

and therefore

$$\Omega = \begin{pmatrix} 0 & -\sigma^1 \wedge \sigma^2 \\ \sigma^1 \wedge \sigma^2 & 0 \end{pmatrix}.$$

Problem 7. Given a smooth surface in the Euclidean space $\mathbb{R}^3$ described by

$$\mathbf{x}(u,v) = \begin{pmatrix} x_1(u,v) \\ x_2(u,v) \\ x_3(u,v) \end{pmatrix}.$$

The *Gaussian curvature* is calculated as follows. First we calculate $E(u,v)$, $F(u,v)$, $G(u,v)$ of the first fundamental form

$$E(u,v) = \frac{\partial \mathbf{x}}{\partial u} \cdot \frac{\partial \mathbf{x}}{\partial u}, \qquad F(u,v) = \frac{\partial \mathbf{x}}{\partial u} \cdot \frac{\partial \mathbf{x}}{\partial v}, \qquad G(u,v) = \frac{\partial \mathbf{x}}{\partial v} \cdot \frac{\partial \mathbf{x}}{\partial v}$$

where $\cdot$ denotes the scalar product. Next we calculate the normal vector field

$$\mathbf{n}^+(u,v) := \frac{\frac{\partial \mathbf{x}}{\partial u} \times \frac{\partial \mathbf{x}}{\partial v}}{\left| \frac{\partial \mathbf{x}}{\partial u} \times \frac{\partial \mathbf{x}}{\partial v} \right|}$$

where $\times$ denotes the vector product. Using $\mathbf{n}^+$ we calculate $L(u,v)$, $M(u,v)$, $N(u,v)$ of the second fundamental form

$$L(u,v) = \mathbf{n}^+ \cdot \frac{\partial^2 \mathbf{x}}{\partial u^2}, \qquad M(u,v) = \mathbf{n}^+ \cdot \frac{\partial^2 \mathbf{x}}{\partial u \partial v}, \qquad N(u,v) = \mathbf{n}^+ \cdot \frac{\partial^2 \mathbf{x}}{\partial v^2}.$$

Then the Gaussian curvature $K(u,v)$ is given by

$$K := \frac{LN - M^2}{EG - F^2}.$$

The *Möbius band* embedded in $\mathbb{R}^3$ can be parametrized as

$$\mathbf{x}(u,v) = \begin{pmatrix} (2 - v\sin(u/2))\sin(u) \\ (2 - v\sin(u/2))\cos(u) \\ v\cos(u/2) \end{pmatrix}$$

where $-1/2 < v < 1/2$ and $u \in [0, 2\pi)$. Find the Gaussian curvature for the Möbius band.

Solution 7. This is a typical application for computer algebra. A SymbolicC++ program that calculates K is given by

```
// curvature.cpp

#include <iostream>
```

```
#include "symbolicc++.h"
using namespace std;

Symbolic curvature(const Symbolic &x,const Symbolic &u,
                   const Symbolic &v,const Equations &rules)
{
 Symbolic dxdu = df(x,u);
 Symbolic dxdv = df(x,v);
 Symbolic d2xdu2  = df(dxdu,u);
 Symbolic d2xdudv = df(dxdu,v);
 Symbolic d2xdv2  = df(dxdv,v);
 Symbolic E = (dxdu | dxdu).subst(rules);
 Symbolic F = (dxdu | dxdv).subst(rules);
 Symbolic G = (dxdv | dxdv).subst(rules);
 Symbolic n = (dxdu % dxdv).subst(rules);
 n = n/sqrt((n | n).subst(rules));
 Symbolic L = (n | d2xdu2).subst(rules);
 Symbolic M = (n | d2xdudv).subst(rules);
 Symbolic N = (n | d2xdv2).subst(rules);
 return (L*N - M*M)/(E*G - F*F);
}

int main(void)
{
 Symbolic x("x", 3), u("u"), v("v");
 Equations rules = (sin(u/2)*sin(u/2)==1-cos(u/2)*cos(u/2),
                    cos(u/2)*cos(u/2)==(cos(u)+1)/2,
                    sin(u)*sin(u)==1-cos(u)*cos(u));
 x(0) = (2-v*sin(u/2))*sin(u);
 x(1) = (2-v*sin(u/2))*cos(u);
 x(2) = v*cos(u/2);
 cout << "x = " << x << endl;
 cout << "K = " << curvature(x,u,v,rules) << endl;
 return 0;
}
```

where % denotes the vector product and | denotes the scalar product. The Gaussian curvature follows as

$$K(u,v) = \frac{1}{(v^2/4 + (2 - v\sin(u/2))^2)^2}.$$

Chapter 25

Killing Vector Fields

Let M be a smooth manifold and g be a metric tensor field. Then V is called a *Killing vector field* with respect to g if

$$L_V g = 0$$

where $L_V(.)$ denotes the Lie derivative. If V and W are Killing vector fields of g, then the commutator $[V, W]$ is also a Killing vector field.

Problem 1. Let g be a metric tensor field. Assume that V and W are Killing vector fields. Show that $[V, W]$ is also a Killing vector field.

Solution 1. By assumption we have $L_V g = 0$ and $L_W g = 0$. Using the identity

$$L_{[V,W]}g \equiv [L_V, L_W]g \equiv L_V(L_W g) - L_W(L_V g)$$

it follows that

$$L_{[V,W]}g = 0.$$

Problem 2. Let

$$g = \frac{1}{2}\sum_{j,k=1}^{3} g_{jk}(\mathbf{x})dx_j \otimes dx_k$$

be a metric tensor field in $\mathbb{R}^3$ and let

$$V(\mathbf{x}) = V_1(\mathbf{x})\frac{\partial}{\partial x_1} + V_2(\mathbf{x})\frac{\partial}{\partial x_2} + V_3(\mathbf{x})\frac{\partial}{\partial x_3}.$$

Assume that $g_{jk} = \delta_{jk}$ where δ_{jk} is the Kronecker symbol, i.e.

$$\delta_{jk} := \begin{cases} 1 & j = k \\ 0 & j \neq k \end{cases}.$$

(i) Calculate $L_V g$.
(ii) Give an interpretation of $L_V g = 0$.

Solution 2.   (i) The Lie derivative is linear and satisfies the product rule

$$L_V g = L_V(dx_1 \otimes dx_1) + L_V(dx_2 \otimes dx_2) + L_V(dx_3 \otimes dx_3)$$

$$L_V(dx_j \otimes dx_k) = (L_V dx_j) \otimes dx_k + dx_j \otimes (L_V dx_k).$$

Furthermore the Lie derivative satisfies

$$L_V dx_j = d(V \rfloor dx_j) = dV_j = \sum_{k=1}^{3} \frac{\partial V_j}{\partial x_k} dx_k.$$

Here $\rfloor$ denotes the contraction, i.e.

$$\frac{\partial}{\partial x_j} \rfloor dx_k = \delta_{jk}.$$

It follows that

$$L_V g = \frac{1}{2} \sum_{j,k=1}^{3} \left(\frac{\partial V_j}{\partial x_k} + \frac{\partial V_k}{\partial x_j} \right) dx_j \otimes dx_k.$$

Since $dx_j \otimes dx_k$ $(j, k = 1, 2, 3)$ are basic elements, the right-hand side of (11) can also be written as a 3×3 matrix, namely

$$\frac{1}{2} \begin{pmatrix} 2\dfrac{\partial V_1}{\partial x_1} & \dfrac{\partial V_1}{\partial x_2} + \dfrac{\partial V_2}{\partial x_1} & \dfrac{\partial V_1}{\partial x_3} + \dfrac{\partial V_3}{\partial x_1} \\[2ex] \dfrac{\partial V_1}{\partial x_2} + \dfrac{\partial V_2}{\partial x_1} & 2\dfrac{\partial V_2}{\partial x_2} & \dfrac{\partial V_2}{\partial x_3} + \dfrac{\partial V_3}{\partial x_2} \\[2ex] \dfrac{\partial V_1}{\partial x_3} + \dfrac{\partial V_3}{\partial x_1} & \dfrac{\partial V_2}{\partial x_3} + \dfrac{\partial V_3}{\partial x_2} & 2\dfrac{\partial V_3}{\partial x_3} \end{pmatrix}.$$

(ii) The physical meaning is as follows: Consider a deformable body B in $\mathbb{R}^3$, $B \subset \mathbb{R}^3$. A displacement with or without deformation of the body B is a diffeomorphism of $\mathbb{R}^3$ defined in a neighbourhood of B. All such diffeomorphisms form a local group generated by the so-called displacement vector field $V(\mathbf{x})$. Consequently, the strain tensor field can be considered as the Lie derivative of the metric tensor field g in $\mathbb{R}^3$ with respect to the vector field $V(\mathbf{x})$. The metric tensor field gives rise to the distance between two points, namely

$$(ds)^2 = \sum_{j,k=1}^{3} g_{jk}(\mathbf{x})dx_j dx_k \,.$$

Then the strain tensor field measures the variation of the distance between two points under a displacement generated by $V(\mathbf{x})$. The equation (invariance condition)

$$L_V g = 0$$

tells us that two points of the body B do not change during the displacement. To summarize. The strain tensor is the Lie derivative of the metric tensor field with respect to the deformation (or exactly the displacement vector field).

Problem 3.   Let

$$g = dx_1 \otimes dx_1 + dx_2 \otimes dx_2$$

be the metric tensor field of the two-dimensional Euclidean space. Assume that

$$L_V g = 0$$

where

$$V = V_1(\mathbf{x})\frac{\partial}{\partial x_1} + V_2(\mathbf{x})\frac{\partial}{\partial x_2}$$

is a smooth vector field defined on the two-dimensional Euclidean space. Show that

$$L_V(dx_1 \wedge dx_2) = 0 \,.$$

Solution 3.   From $L_V g = 0$ it follows that

$$\frac{\partial V_1}{\partial x_1} = 0, \quad \frac{\partial V_2}{\partial x_2} = 0, \quad \frac{\partial V_1}{\partial x_2} + \frac{\partial V_2}{\partial x_1} = 0 \,.$$

Now

$$L_V(dx_1 \wedge dx_2) = (L_V dx_1) \wedge dx_2 + dx_1 \wedge (L_V dx_2)$$

$$= dV_1 \wedge dx_2 + dx_1 \wedge dV_2$$

$$= \frac{\partial V_1}{\partial x_1}dx_1 \wedge dx_2 + \frac{\partial V_2}{\partial x_2}dx_1 \wedge dx_2$$

$$= \left(\frac{\partial V_1}{\partial x_1} + \frac{\partial V_2}{\partial x_2}\right) dx_1 \wedge dx_2 \,.$$

Inserting $\partial V_1/\partial x_1 = \partial V_2/\partial x_2 = 0$ into this equation yields $L_V(dx_1 \wedge dx_2) = 0$.

Problem 4.   Let E^3 be the three-dimensional Euclidean space with metric tensor field

$$g = dx_1 \otimes dx_1 + dx_2 \otimes dx_2 + dx_3 \otimes dx_3 \,. \qquad (1)$$

Let

$$M \equiv S^2 = \{(x_1, x_2, x_3) : x_1^2 + x_2^2 + x_3^2 = 1\} \,.$$

This means the manifold M is the unit sphere.
(i) Calculate the metric tensor field $\widetilde{g}$ for M. The parametrization is given by

$$x_1(u_1, u_2) = \cos u_1 \sin u_2, \quad x_2(u_1, u_2) = \sin u_1 \sin u_2, \quad x_3(u_1, u_2) = \cos u_2$$

where $u_1 \in [0, 2\pi)$ and $u_2 \in [0, \pi)$.
(ii) Find the vector fields V such that

$$L_V \widetilde{g} = 0 \,.$$

(iii) Show that these vector fields form a Lie algebra under the commutator.

Solution 4.   (i) Since ·

$$dx_1 = -\sin u_1 \sin u_2 du_1 + \cos u_1 \cos u_2 du_2$$

$$dx_2 = \cos u_1 \sin u_2 du_1 + \sin u_1 \cos u_2 du_2$$

$$dx_3 = -\sin u_2 du_2$$

we obtain from (1) that

$$\widetilde{g} = \sin^2 u_2 du_1 \otimes du_1 + du_2 \otimes du_2 \,.$$

(ii) From the condition

$$L_V \widetilde{g} = 0$$

with

$$V = V_1(u_1, u_2) \frac{\partial}{\partial u_1} + V_2(u_1, u_2) \frac{\partial}{\partial u_2}$$

we find

$$2 \left(V_2 \sin u_2 \cos u_2 + \frac{\partial V_1}{\partial u_1} \sin^2 u_2 \right) du_1 \otimes du_1 + \left(\sin^2 u_2 \frac{\partial V_1}{\partial u_2} + \frac{\partial V_2}{\partial u_1} \right) du_1 \otimes du_2$$

$$+ \left(\sin^2 u_2 \frac{\partial V_1}{\partial u_2} + \frac{\partial V_2}{\partial u_1} \right) du_2 \otimes du_1 + 2 \frac{\partial V_2}{\partial u_2} du_2 \otimes du_2 = 0 \,.$$

This leads to the system of linear partial differential equations

$$V_2 \sin u_2 \cos u_2 + \frac{\partial V_1}{\partial u_1} \sin^2 u_2 = 0$$

$$\sin^2 u_2 \frac{\partial V_1}{\partial u_2} + \frac{\partial V_2}{\partial u_1} = 0$$

$$\frac{\partial V_2}{\partial u_2} = 0 \,.$$

From these equations we conclude that V_2 depends only on u_1. Thus the solution of this system of partial differential equations leads to the three independent vector fields

$$V^I = \frac{\partial}{\partial u_1}$$

$$V^{II} = \cos u_1 \cot u_2 \frac{\partial}{\partial u_1} + \sin u_1 \frac{\partial}{\partial u_2}$$

$$V^{III} = -\sin u_1 \cot u_2 \frac{\partial}{\partial u_1} + \cos u_1 \frac{\partial}{\partial u_2}.$$

(iii) We find

$$[V^I, V^{II}] = -\sin u_1 \cot u_2 \frac{\partial}{\partial u_1} + \cos u_1 \frac{\partial}{\partial u_2} = V^{III}$$

$$[V^I, V^{III}] = -\cos u_1 \cot u_2 \frac{\partial}{\partial u_1} - \sin u_1 \frac{\partial}{\partial u_2} = -V^{II}$$

$$[V^{II}, V^{III}] = \frac{\partial}{\partial u_1} = V^I.$$

Thus the vector fields V^I, V^{II} and V^{III} form a basis of a Lie algebra under the commutator.

Problem 5. The *Gödel metric tensor field* is given by

$$g = a^2[dx \otimes dx - \frac{1}{2}e^{2x}dy \otimes dy + dz \otimes dz - c^2 dt \otimes dt - ce^x dy \otimes dt - ce^x dt \otimes dy] \quad (1)$$

where a and c are constants.
(i) Show that the vector fields

$$Y = \frac{\partial}{\partial y}, \quad Z = \frac{\partial}{\partial z}, \quad T = \frac{\partial}{\partial t}, \quad A = \frac{\partial}{\partial x} - y\frac{\partial}{\partial y}$$

$$B = y\frac{\partial}{\partial x} + (ce^{-2x} - \frac{1}{2}y^2)\frac{\partial}{\partial y} - 2e^{-x}\frac{\partial}{\partial t}$$

are Killing vector fields of the Gödel metric tensor .
(ii) Calculate the commutators.

Solution 5. (i) First we recall that the Lie derivative is a linear operation. Since the functions g_{jk} do not depend on y, z and t it is obvious that Y, Z and T are Killing vector fields. Applying the product rule and

$$L_{\partial/\partial x}e^{2x}dt = 2e^{2x}dt, \quad L_{\partial/\partial x}e^{2x}dy = 2e^{2x}dy$$

$$L_{y\partial/\partial y}dy = dy, \quad L_{y\partial/\partial y}e^x dy = e^x dy$$

we find that
$$L_A g = 0.$$

Now we evaluate the Lie derivative of g with respect to B. Applying the product rule and the rules

$$L_{fV}\alpha = f(L_V\alpha) + df \wedge (V \rfloor \alpha), \qquad L_V(f\alpha) = (Vf)\alpha + f(L_V\alpha)$$

we obtain
$$L_B g = 0.$$

The Gödel metric tensor field contains *closed timelike lines*; that is, an observer can influence his own past.

(ii) For the zero commutators are

$$[Y, Z] = 0, \qquad [Y, T] = 0, \qquad [Z, T] = 0, \qquad [Z, A] = 0$$

and

$$[Z, B] = 0, \qquad [T, A] = 0, \qquad [T, B] = 0.$$

The nonzero commutators are

$$[Y, A] = -\frac{\partial}{\partial y} = -Y, \qquad [Y, B] = \frac{\partial}{\partial x} - y\frac{\partial}{\partial y} = A$$

and

$$[A, B] = -y\frac{\partial}{\partial x} + \left(\frac{1}{2}y^2 - ce^{-2x}\right)\frac{\partial}{\partial y} + 2e^{-x}\frac{\partial}{\partial t} = -B.$$

Problem 6.   Consider the *Poincaré upper half-plane*

$$H_+^2 := \{ (x, y) \in \mathbb{R}^2 : y > 0 \}$$

with metric tensor field

$$g = \frac{1}{y}dx \otimes \frac{1}{y}dx + \frac{1}{y}dy \otimes \frac{1}{y}dy$$

which is conformal with the standard inner product. Find the Killing vector fields.

Solution 6.   We obtain

$$\frac{\partial}{\partial x}, \qquad x\frac{\partial}{\partial x} + y\frac{\partial}{\partial y}, \qquad \frac{x^2 - y^2}{2}\frac{\partial}{\partial x} + xy\frac{\partial}{\partial y}.$$

Chapter 26

Inequalities

Problem 1. Let f and g be two integrable functions. Assume that

$$\int_I f(x)\,dx = \int_I g(x)\,dx \tag{1}$$

and $f(x) > 0$, $g(x) > 0$ for $x \in I$, where $I \subset \mathbb{R}$. Show that

$$\int_I f(x)\ln f(x)\,dx \geq \int_I f(x)\ln g(x)\,dx. \tag{2}$$

Solution 1. We have

$$\int_I f\ln f\,dx - \int_I f\ln g\,dx \equiv \int_I f(\ln f - \ln g)dx \equiv \int_I f\ln\frac{f}{g}dx$$

$$\equiv \int_I g\left(\frac{f}{g}\ln\left(\frac{f}{g}\right)\right)dx.$$

Thus

$$\int_I f\ln f\,dx - \int_I f\ln g\,dx \equiv \int_I g\left(\frac{f}{g}\ln\left(\frac{f}{g}\right) - \frac{f}{g} + 1\right)dx \geq 0$$

where we have used (1) and the fact that

$$\frac{f}{g}\ln\left(\frac{f}{g}\right) - \frac{f}{g} + 1 \geq 0.$$

Problem 2.   (i) Let $0 < \mu < 1$ and $a \geq 0$, $b \geq 0$. Show that

$$a^\mu b^{1-\mu} \leq \mu a + (1 - \mu)b. \tag{1}$$

(ii) Let p and q be two positive numbers which satisfy the condition

$$\frac{1}{p} + \frac{1}{q} = 1. \tag{2}$$

Let $c, d \in \mathbb{C}$. Show that

$$|cd| \leq \frac{|c|^p}{p} + \frac{|d|^q}{q}. \tag{3}$$

Solution 2.   (i) For $a = b$ the inequality (1) is obviously satisfied. Without loss of generality we can assume that $b > a > 0$. Applying the *theorem of the mean*, we obtain

$$b^{1-\mu} - a^{1-\mu} = (1 - \mu)(b - a)\xi^{-\mu}$$

with $a < \xi < b$. Since

$$\xi^{-\mu} < a^{-\mu}$$

we obtain

$$b^{1-\mu} - a^{1-\mu} \leq (1 - \mu)(b - a)a^{-\mu}.$$

Multiplying the left and right-hand sides of this inequality by a^μ leads to inequality (1).

(ii) If we set

$$\mu = \frac{1}{p}, \qquad 1 - \mu = \frac{1}{q}, \qquad a = |c|^p, \qquad b = |d|^q$$

in inequality (1) we obtain (3).

Problem 3.   Let $a_1, \ldots, a_n$ and $b_1, \ldots, b_n$ be arbitrary complex numbers. Let p and q be two real positive numbers which satisfy

$$\frac{1}{p} + \frac{1}{q} = 1. \tag{1}$$

Show that

$$\sum_{k=1}^{n} |a_k b_k| \leq \left[\sum_{k=1}^{n} |a_k|^p \right]^{1/p} \left[\sum_{k=1}^{n} |b_k|^q \right]^{1/q}. \tag{2}$$

Solution 3. Let

$$A^p := \sum_{k=1}^n |a_k|^p, \qquad B^q := \sum_{k=1}^n |b_k|^q.$$

We can assume that both A and B are positive. Let

$$\bar{a}_k := \frac{a_k}{A}, \qquad \bar{b}_k := \frac{b_k}{B}.$$

Using the result from problem 2 we obtain

$$|\bar{a}_k \bar{b}_k| \le \frac{|\bar{a}_k|^p}{p} + \frac{|\bar{b}_k|^q}{q}.$$

Taking the sum over k on both sides yields

$$\sum_{k=1}^n |\bar{a}_k \bar{b}_k| \le \sum_{k=1}^n \left(\frac{|\bar{a}_k|^p}{p} + \frac{|\bar{b}_k|^q}{q} \right) = \frac{1}{p} + \frac{1}{q} = 1.$$

Consequently,

$$\sum_{k=1}^n |a_k b_k| \le AB.$$

Problem 4. Let B be an $n \times n$ matrix over $\mathbb{R}$. Let B^T be the transpose. Show that

$$\operatorname{tr}(B^{2m}) \le \operatorname{tr}(B^m B^{mT}) \tag{1}$$

where tr denotes the trace.

Solution 4. Let A be an $n \times n$ matrix over $\mathbb{R}$. Then

$$(A^T A)_{jl} = \sum_{k=1}^n a_{kj} a_{kl} \tag{2}$$

and therefore

$$(A^T A)_{jj} = \sum_{k=1}^n a_{kj} a_{kj} \ge 0. \tag{3}$$

It follows that

$$\operatorname{tr}(A^T A) \ge 0.$$

We set

$$A^T := B^m - B^{mT}.$$

Hence $A = B^{mT} - B^m$. Using (3) we obtain

$$\operatorname{tr}((B^m - B^{mT})(B^{mT} - B^m)) \ge 0.$$

It follows that

$$\mathrm{tr}(B^m B^{mT}) + \mathrm{tr}(B^{mT} B^m) - \mathrm{tr}(B^{mT} B^{mT}) - \mathrm{tr}(B^m B^m) \geq 0 \,.$$

Since

$$\mathrm{tr}(B^m B^{mT}) \equiv \mathrm{tr}(B^{mT} B^m), \qquad \mathrm{tr}(B^m B^m) \equiv \mathrm{tr}(B^{mT} B^{mT})$$

inequality (1) follows.

Problem 5. Show that for two $n \times n$ positive-semidefinite real matrices A and B and $0 \leq \alpha \leq 1$ we have

$$\mathrm{tr}(A^\alpha B^{1-\alpha}) \leq (\mathrm{tr}(A))^\alpha (\mathrm{tr}(B))^{1-\alpha} \,. \tag{1}$$

Solution 5. Since A and B are positive-semidefinite their eigenvalues are real and nonnegative. We order the eigenvalues of A and B in decreasing order

$$\lambda_1 \geq \lambda_2 \geq \cdots \geq \lambda_n \geq 0, \qquad \mu_1 \geq \mu_2 \geq \cdots \geq \mu_n \geq 0 \,. \tag{2}$$

For an arbitrary orthonormal set of vectors $\mathbf{x}_i$ we have

$$\sum_{i=1}^{k} (\mathbf{x}_i, B^{1-\alpha} \mathbf{x}_i) \leq \sum_{i=1}^{k} \mu_i^{1-\alpha}, \qquad k = 1, 2, \ldots, n \tag{3}$$

where $(\,,\,)$ denotes the scalar product. Choosing $\mathbf{x}_i$ to be the eigenvectors of A and summing by parts gives

$$\mathrm{tr}(A^\alpha B^{1-\alpha}) = \sum_{i=1}^{n} \lambda_i^\alpha (\mathbf{x}_i, B^{1-\alpha} \mathbf{x}_i) \,. \tag{4}$$

Thus

$$\mathrm{tr}(A^\alpha B^{1-\alpha}) \leq \sum_{i=1}^{n} \lambda_i^\alpha \mu_i^{1-\alpha} \leq \left(\sum_{i=1}^{n} \lambda_i \right)^\alpha \left(\sum_{i=1}^{n} \mu_i \right)^{1-\alpha} = (\mathrm{tr}(A))^\alpha (\mathrm{tr}(B))^{1-\alpha}$$

where the last inequality is *Hölder's inequality* for positive real numbers. We also used that

$$\mathrm{tr}(A) = \sum_{i=1}^{n} \lambda_i, \qquad \mathrm{tr}(B) = \sum_{i=1}^{n} \mu_i \,.$$

Problem 6. Let A be a bounded linear operator in a Banach space. Let

$$f_{n,m}(A) := \left(\sum_{k=0}^{m} \frac{1}{k!} \left(\frac{A}{n} \right)^k \right)^n \,. \tag{1}$$

Show that

$$\|e^A - f_{n,m}(A)\| \le \frac{1}{n^m(m+1)!}\|A\|^{m+1}e^{\|A\|}. \tag{2}$$

Solution 6. Using the properties of the norm we have the inequality

$$\|e^A - f_{n,m}(A)\| \le \|e^{A/n} - h\| \cdot \|(e^{A/n})^{n-1} + (e^{A/n})^{n-2}h + \cdots + h^{n-1}\|$$
$$\le n\|e^{A/n} - h\| \cdot e^{(n-1)\|A\|/n}$$

where

$$h := \sum_{k=0}^{m} \frac{1}{k!}\left(\frac{A}{n}\right)^k.$$

Using Taylor's theorem we find

$$\|e^{A/n} - h\| = \|\sum_{k=m+1}^{\infty} \frac{1}{k!}\left(\frac{A}{n}\right)^k\|$$

$$\le \sum_{k=m+1}^{\infty} \frac{1}{k!}\left(\frac{\|A\|}{n}\right)^k$$

$$= e^{\|A\|/n} - \sum_{k=0}^{m} \frac{1}{k!}\left(\frac{\|A\|}{n}\right)^k$$

$$= \frac{1}{(m+1)!}\left(\frac{\|A\|}{n}\right)^{m+1} e^{\theta\|A\|/n}, \qquad 0 < \theta < 1.$$

Inserting this equation into the inequality above we obtain (1). $f_{n,m}$ converges to e^A if $n \to \infty$ or if $m \to \infty$.

Problem 7. Let A and B be two symmetric $n \times n$ matrices over $\mathbb{R}$. It can be shown that

$$\mathrm{tr}e^{A+B} \le \mathrm{tr}(e^A e^B) \le \frac{1}{2}\mathrm{tr}(e^{2A} + e^{2B})$$

$$\mathrm{tr}e^{A+B} \le \mathrm{tr}(e^A e^B) \le (\mathrm{tr}e^{pA})^{1/p}(\mathrm{tr}e^{qB})^{1/q}$$

where $p > 1$, $q > 1$ with

$$\frac{1}{p} + \frac{1}{q} = 1.$$

Is

$$(\mathrm{tr}e^{pA})^{1/p}(\mathrm{tr}e^{qB})^{1/q} \le \frac{1}{2}\mathrm{tr}(e^{2A} + e^{2B})?$$

Prove or disprove.

Solution 7.   Since A and B are symmetric the eigenvalues are real numbers. Let $\ln \lambda_1, \ldots, \ln \lambda_n$ and $\ln \mu_1, \ldots, \ln \mu_n$ be the (real) eigenvalues of A and B, respectively. Then the conjectured inequality takes the form

$$(\lambda_1^p + \cdots + \lambda_n^p)^{1/p} (\mu_1^q + \cdots + \mu_n^q)^{1/q} \le \frac{1}{2} \left(\lambda_1^2 + \cdots + \lambda_n^2 + \mu_1^2 + \cdots + \mu_n^2 \right).$$

If $n = 1$ or $p = q = 2$, this inequality is trivially true. It is generally false in all other cases. Let $n > 1$ and take $q < 2$. Choose $\lambda_1 = n^{1/4}$, $\lambda_2 = \cdots = \lambda_n = \epsilon$ and $\mu_1 = \mu_2 = \cdots = \mu_n = n^{-1/4}$. As $\epsilon \to 0$, the inequality tends to

$$n^{1/q} \le n^{1/2}$$

which is clearly false. Thus the desired counterexample is obtained for all sufficiently small ϵ.

Problem 8.   Let A, B be hermitian matrices. Then

$$\mathrm{tr}(e^{A+B}) \le \mathrm{tr}(e^A e^B).$$

Assume that

$$A = \begin{pmatrix} 0 & 0 \\ 0 & 1 \end{pmatrix}, \qquad B = \begin{pmatrix} 0 & 1 \\ 1 & 0 \end{pmatrix}.$$

Calculate the left and right-hand side of the inequality. Does equality hold?

Solution 8.   We note that the commutator of A and B is given by

$$[A, B] = \begin{pmatrix} 0 & -1 \\ 1 & 0 \end{pmatrix}.$$

Thus equality does not hold. We find since $A^2 = A$ and $B^2 = I_2$

$$e^A = \begin{pmatrix} 1 & 0 \\ 0 & e \end{pmatrix}, \qquad e^B = \begin{pmatrix} \cosh(1) & \sinh(1) \\ \sinh(1) & \cosh(1) \end{pmatrix}.$$

Thus

$$\mathrm{tr}(e^A e^B) = \cosh(1) + e \sinh(1).$$

The eigenvalues of $A + B$ are $\lambda_1 = (1 + \sqrt{5})/2$ and $\lambda_2 = (1 - \sqrt{5})/2$. Thus

$$\mathrm{tr}(e^{A+B}) = e^{\lambda_1} + e^{\lambda_2} = e^{1/2}(e^{\sqrt{5}/2} + e^{-\sqrt{5}/2}).$$

Chapter 27

Ising Model and Heisenberg Model

Problem 1. (i) Find the spectrum for the four-point *Ising model*

$$\hat{H} = J \sum_{j=1}^{4} \sigma_j \sigma_{j+1} \tag{1}$$

with cyclic boundary condition $\sigma_5 = \sigma_1$ and $J < 0$. The quantity J is called the exchange constant.

(ii) Calculate the *partition function*

$$Z(\beta) := \mathrm{tr} e^{-\beta \hat{H}}$$

and the *Helmholtz free energy*

$$F(\beta) := -\frac{1}{\beta} \ln Z(\beta) = -\frac{1}{\beta} \ln \mathrm{tr} e^{-\beta \hat{H}}$$

where $\beta := \frac{1}{kT}$ for the four point Ising model.

Solution 1. (i) Let I_2 be the 2×2 identity matrix and σ_z the Pauli spin matrix. Then

$$\sigma_1 = \sigma_z \otimes I_2 \otimes I_2 \otimes I_2, \quad \sigma_2 = I_2 \otimes \sigma_z \otimes I_2 \otimes I_2,$$

$$\sigma_3 = I_2 \otimes I_2 \otimes \sigma_z \otimes I_2, \quad \sigma_4 = I_2 \otimes I_2 \otimes I_2 \otimes \sigma_z$$

287

where

$$\sigma_z := \begin{pmatrix} 1 & 0 \\ 0 & -1 \end{pmatrix}$$

Thus we find the diagonal matrices

$$\sigma_1\sigma_2 = \text{diag}(1,1,1,1,-1,-1,-1,-1,-1,-1,-1,-1,1,1,1,1)$$
$$\sigma_2\sigma_3 = \text{diag}(1,1,-1,-1,-1,-1,1,1,1,1,-1,-1,-1,-1,1,1)$$
$$\sigma_3\sigma_4 = \text{diag}(1,-1,-1,1,1,-1,-1,1,1,-1,-1,1,1,-1,-1,1)$$
$$\sigma_4\sigma_1 = \text{diag}(1,-1,1,-1,1,-1,1,-1,-1,1,-1,1,-1,1,-1,1)\,.$$

It follows that

$$\sum_{j=1}^{4} \sigma_j\sigma_{j+1} = \text{diag}(4,0,0,0,0,-4,0,0,0,0,-4,0,0,0,0,4).$$

Consequently the spectrum of $\hat{H}$ is given by

$$
\begin{array}{cl}
4J & \text{twofold degenerate} \\
0 & \text{12 times degenerate} \\
-4J & \text{twofold degenerate}\,.
\end{array}
$$

(ii) From the result of (i) (eigenvalues) we find that the partition function is given by

$$Z(\beta) := \text{tr} e^{-\beta\hat{H}} = 2e^{4\beta J} + 12 + 2e^{-4\beta J}.$$

Thus the Helmholtz free energy is given by

$$F(\beta) = -\frac{1}{\beta}\ln(2e^{4\beta J} + 12 + 2e^{-4\beta J}).$$

Problem 2.   Calculate the eigenvalues and eigenvectors for the two-point *Heisenberg model*

$$\hat{H} = J\sum_{j=1}^{2} \mathbf{S}_j \cdot \mathbf{S}_{j+1} \tag{1}$$

with cyclic boundary conditions, i.e. $\mathbf{S}_3 \equiv \mathbf{S}_1$ where J is the so-called exchange constant ($J > 0$ or $J < 0$).

Solution 2.   It follows that

$$\hat{H} = J(\mathbf{S}_1 \cdot \mathbf{S}_2 + \mathbf{S}_2 \cdot \mathbf{S}_3) \equiv J(\mathbf{S}_1 \cdot \mathbf{S}_2 + \mathbf{S}_2 \cdot \mathbf{S}_1).$$

Therefore

$$\hat{H} = J(S_{1x}S_{2x} + S_{1y}S_{2y} + S_{1z}S_{2z} + S_{2x}S_{1x} + S_{2y}S_{1y} + S_{2z}S_{1z}).$$

Since $S_{1x} := S_x \otimes I$, $S_{2x} = I \otimes S_x$ etc., it follows that

$$\hat{H} = J[(S_x \otimes I_2)(I_2 \otimes S_x) + (S_y \otimes I_2)(I_2 \otimes S_y) + (S_z \otimes I_2)(I_2 \otimes S_z) \\ + (I_2 \otimes S_x)(S_x \otimes I_2) + (I_2 \otimes S_y)(S_y \otimes I_2) + (I_2 \otimes S_z)(S_z \otimes I_2)].$$

We find

$$\hat{H} = J[(S_x \otimes S_x) + (S_y \otimes S_y) + (S_z \otimes S_z) + (S_x \otimes S_x) + (S_y \otimes S_y) + (S_z \otimes S_z)].$$

Therefore

$$\hat{H} = 2J((S_x \otimes S_x) + (S_y \otimes S_y) + (S_z \otimes S_z)).$$

Since

$$S_x := \frac{1}{2}\sigma_x, \qquad S_y := \frac{1}{2}\sigma_y, \qquad S_z := \frac{1}{2}\sigma_z$$

we obtain

$$S_x \otimes S_x = \frac{1}{4}\begin{pmatrix} 0 & 1 \\ 1 & 0 \end{pmatrix} \otimes \begin{pmatrix} 0 & 1 \\ 1 & 0 \end{pmatrix} = \frac{1}{4}\begin{pmatrix} 0 & 0 & 0 & 1 \\ 0 & 0 & 1 & 0 \\ 0 & 1 & 0 & 0 \\ 1 & 0 & 0 & 0 \end{pmatrix}$$

etc. Then the Hamilton operator $\hat{H}$ is given by

$$\hat{H} = \frac{J}{2}\begin{pmatrix} 1 & 0 & 0 & 0 \\ 0 & -1 & 2 & 0 \\ 0 & 2 & -1 & 0 \\ 0 & 0 & 0 & 1 \end{pmatrix}.$$

The eigenvalues and eigenvectors can now be calculated. We set

$$|\uparrow\rangle := \begin{pmatrix} 1 \\ 0 \end{pmatrix} \quad \text{spin up,} \qquad |\downarrow\rangle := \begin{pmatrix} 0 \\ 1 \end{pmatrix} \quad \text{spin down.}$$

Then

$$|\uparrow\uparrow\rangle := |\uparrow\rangle \otimes |\uparrow\rangle, \quad |\uparrow\downarrow\rangle := |\uparrow\rangle \otimes |\downarrow\rangle, \quad |\downarrow\uparrow\rangle := |\downarrow\rangle \otimes |\uparrow\rangle, \quad |\downarrow\downarrow\rangle := |\downarrow\rangle \otimes |\downarrow\rangle.$$

Consequently,

$$|\uparrow\uparrow\rangle = \begin{pmatrix} 1 \\ 0 \\ 0 \\ 0 \end{pmatrix}, \quad |\uparrow\downarrow\rangle = \begin{pmatrix} 0 \\ 1 \\ 0 \\ 0 \end{pmatrix}, \quad |\downarrow\uparrow\rangle = \begin{pmatrix} 0 \\ 0 \\ 1 \\ 0 \end{pmatrix}, \quad |\downarrow\downarrow\rangle = \begin{pmatrix} 0 \\ 0 \\ 0 \\ 1 \end{pmatrix}.$$

Obviously $| \uparrow\uparrow\rangle$ and $| \downarrow\downarrow\rangle$ are eigenvectors of the Hamilton operator with eigenvalues $J/2$ and $J/2$, respectively. This means the eigenvalue $J/2$ is degenerate. The eigenvalues of the matrix

$$\frac{J}{2}\begin{pmatrix} -1 & 2 \\ 2 & -1 \end{pmatrix}$$

are given by $J/2$, $-3J/2$. The eigenvectors are linear combinations of $| \uparrow\downarrow\rangle$ and $| \downarrow\uparrow\rangle$, i.e.

$$\frac{1}{\sqrt{2}}(| \uparrow\downarrow\rangle + | \downarrow\uparrow\rangle), \qquad \frac{1}{\sqrt{2}}(| \uparrow\downarrow\rangle - | \downarrow\uparrow\rangle).$$

Problem 3. Consider the spin Hamilton operator

$$\hat{H} = a\sum_{j=1}^{4} \sigma_3(j)\sigma_3(j+1) + b\sum_{j=1}^{4} \sigma_1(j) \tag{1}$$

with cyclic boundary conditions, i.e. $\sigma_3(5) \equiv \sigma_3(1)$. Here a, b are real constants and σ_1, σ_2 and σ_3 are the Pauli matrices. Thus the underlying Hilbert space is $\mathbb{C}^{16}$.

(i) Calculate the matrix representation of $\hat{H}$. Recall that

$$\sigma_a(1) = \sigma_a \otimes I_2 \otimes I_2 \otimes I_2, \qquad \sigma_a(2) = I_2 \otimes \sigma_a \otimes I_2 \otimes I_2$$

$$\sigma_a(3) = I_2 \otimes I_2 \otimes \sigma_a \otimes I_2, \qquad \sigma_a(4) = I_2 \otimes I_2 \otimes I_2 \otimes \sigma_a.$$

(ii) Show that the Hamilton operator $\hat{H}$ admits the C_{4v} symmetry group.
(iii) Calculate the matrix representation of $\hat{H}$ for the subspace which belongs to the representation A_1.
(iv) Show that the time-evolution of any expectation value of an observable, Ω, can be expressed as the sum of a time-independent and a time-dependent term

$$\langle \Omega \rangle(t) = \sum_{n=1}^{16} |c_n(0)|^2 \langle n|\Omega|n\rangle + \sum_{\substack{m,n=1 \\ n\neq m}}^{16} c_m^*(0)c_n(0)e^{i(E_m - E_n)t/\hbar}\langle m|\Omega|n\rangle$$

where $c_n(0)$ are the coefficients of expansion of the initial state in terms of the energy eigenstates $|n\rangle$ of $\hat{H}$, E_n are the eigenvalues of $\hat{H}$.

Solution 3. (i) The Pauli matrices are given by

$$\sigma_1 := \begin{pmatrix} 0 & 1 \\ 1 & 0 \end{pmatrix}, \qquad \sigma_2 := \begin{pmatrix} 0 & -i \\ i & 0 \end{pmatrix}, \qquad \sigma_3 := \begin{pmatrix} 1 & 0 \\ 0 & -1 \end{pmatrix}.$$

The matrix representation of the first term of the right-hand side of (1) has been calculated in problem 1. We find the diagonal matrix

$$\sum_{j=1}^{4} \sigma_3(j)\sigma_3(j+1) = \text{diag}(4,0,0,0,0,-4,0,0,0,0,-4,0,0,0,0,4).$$

The second term leads to non-diagonal terms. Using (2) we find the symmetric 16×16 matrix for $\hat{H}$

$$\begin{pmatrix}
4a & b & b & 0 & b & 0 & 0 & 0 & b & 0 & 0 & 0 & 0 & 0 & 0 & 0 \\
b & 0 & 0 & b & 0 & b & 0 & 0 & 0 & b & 0 & 0 & 0 & 0 & 0 & 0 \\
b & 0 & 0 & b & 0 & 0 & b & 0 & 0 & 0 & b & 0 & 0 & 0 & 0 & 0 \\
0 & b & b & 0 & 0 & 0 & 0 & b & 0 & 0 & 0 & b & 0 & 0 & 0 & 0 \\
b & 0 & 0 & 0 & 0 & b & b & 0 & 0 & 0 & 0 & 0 & b & 0 & 0 & 0 \\
0 & b & 0 & 0 & b & -4a & 0 & b & 0 & 0 & 0 & 0 & 0 & b & 0 & 0 \\
0 & 0 & b & 0 & b & 0 & 0 & b & 0 & 0 & 0 & 0 & 0 & 0 & b & 0 \\
0 & 0 & 0 & b & 0 & b & b & 0 & 0 & 0 & 0 & 0 & 0 & 0 & 0 & b \\
b & 0 & 0 & 0 & 0 & 0 & 0 & 0 & 0 & b & b & 0 & b & 0 & 0 & 0 \\
0 & b & 0 & 0 & 0 & 0 & 0 & 0 & b & 0 & 0 & b & 0 & b & 0 & 0 \\
0 & 0 & b & 0 & 0 & 0 & 0 & 0 & b & 0 & -4a & b & 0 & 0 & b & 0 \\
0 & 0 & 0 & b & 0 & 0 & 0 & 0 & 0 & b & b & 0 & 0 & 0 & 0 & b \\
0 & 0 & 0 & 0 & b & 0 & 0 & 0 & b & 0 & 0 & 0 & 0 & b & b & 0 \\
0 & 0 & 0 & 0 & 0 & b & 0 & 0 & 0 & b & 0 & 0 & b & 0 & 0 & b \\
0 & 0 & 0 & 0 & 0 & 0 & b & 0 & 0 & 0 & b & 0 & b & 0 & 0 & b \\
0 & 0 & 0 & 0 & 0 & 0 & 0 & b & 0 & 0 & 0 & b & 0 & b & b & 4a
\end{pmatrix}.$$

(ii) To find the discrete symmetries of the Hamilton operator (1) we study the discrete coordinate transformations $\tau : \{1,2,3,4\} \rightarrow \{1,2,3,4\}$ i.e., permutations of the set $\{1,2,3,4\}$ which preserve $\hat{H}$. Any permutation preserves the second term of $\hat{H}$. The first term is preserved by cyclic permutations of 1234 or 4321. Therefore we find the following set of symmetries

$$E : (1,2,3,4) \rightarrow (1,2,3,4) \qquad C_2 : (1,2,3,4) \rightarrow (3,4,1,2)$$

$$C_4 : (1,2,3,4) \rightarrow (2,3,4,1) \qquad C_4^3 : (1,2,3,4) \rightarrow (4,1,2,3)$$

$$\sigma_v : (1,2,3,4) \rightarrow (2,1,4,3) \qquad \sigma_v' : (1,2,3,4) \rightarrow (4,3,2,1)$$

$$\sigma_d : (1,2,3,4) \rightarrow (1,4,3,2) \qquad \sigma_d' : (1,2,3,4) \rightarrow (3,2,1,4)$$

which form a group isomorphic to C_{4v}. The classes are given by

$$\{E\} \quad \{C_2\} \quad \{C_4, C_4^3\} \quad \{\sigma_v, \sigma_v'\} \quad \{\sigma_d, \sigma_d'\}.$$

The *character table* is as follows

	$\{E\}$	$\{C_2\}$	$\{2C_4\}$	$\{2\sigma_v\}$	$\{2\sigma_d\}$
A_1	1	1	1	1	1
A_2	1	1	1	-1	-1
B_1	1	1	-1	1	-1
B_2	1	1	-1	-1	1
E	2	-2	0	0	0

(iii) From the character table it follows that the projection operator of the representation A_1 is given by

$$\Pi_1 = \frac{1}{8}(O_E + O_{C_2} + O_{C_4} + O_{C_4^3} + O_{\sigma_v} + O_{\sigma_v'} + O_{\sigma_d} + O_{\sigma_d'}).$$

We recall that $O_p f(x) := f(P^{-1}x)$. We have $C_4^{-1} = C_4^3$, $(C_4^3)^{-1} = C_4$. The other elements of the groups are their own inverse. Thus for $\mathbf{e}_1, \cdots, \mathbf{e}_4 \in \mathbb{C}^2$ we have

$$8\Pi_1(\mathbf{e}_1 \otimes \mathbf{e}_2 \otimes \mathbf{e}_3 \otimes \mathbf{e}_4) = \mathbf{e}_1 \otimes \mathbf{e}_2 \otimes \mathbf{e}_3 \otimes \mathbf{e}_4 + \mathbf{e}_3 \otimes \mathbf{e}_4 \otimes \mathbf{e}_1 \otimes \mathbf{e}_2$$
$$+ \mathbf{e}_4 \otimes \mathbf{e}_1 \otimes \mathbf{e}_2 \otimes \mathbf{e}_3 + \mathbf{e}_2 \otimes \mathbf{e}_3 \otimes \mathbf{e}_4 \otimes \mathbf{e}_1$$
$$+ \mathbf{e}_2 \otimes \mathbf{e}_1 \otimes \mathbf{e}_4 \otimes \mathbf{e}_3 + \mathbf{e}_4 \otimes \mathbf{e}_3 \otimes \mathbf{e}_2 \otimes \mathbf{e}_1$$
$$+ \mathbf{e}_1 \otimes \mathbf{e}_4 \otimes \mathbf{e}_3 \otimes \mathbf{e}_2 + \mathbf{e}_3 \otimes \mathbf{e}_2 \otimes \mathbf{e}_1 \otimes \mathbf{e}_4.$$

We set

$$|\uparrow\rangle := \begin{pmatrix} 1 \\ 0 \end{pmatrix} \quad \text{spin up}, \qquad |\downarrow\rangle := \begin{pmatrix} 0 \\ 1 \end{pmatrix} \quad \text{spin down}.$$

These are the eigenfunctions of σ_z. As a basis for the total space we take all Kronecker products of the form $\mathbf{e}_1 \otimes \mathbf{e}_2 \otimes \mathbf{e}_3 \otimes \mathbf{e}_4$, where $\mathbf{e}_1, \mathbf{e}_2, \mathbf{e}_3, \mathbf{e}_4 \in \{|\uparrow\rangle, |\downarrow\rangle\}$. We define

$$|\uparrow\uparrow\uparrow\uparrow\rangle := |\uparrow\rangle \otimes |\uparrow\rangle \otimes |\uparrow\rangle \otimes |\uparrow\rangle, \qquad |\downarrow\uparrow\uparrow\uparrow\rangle := |\downarrow\rangle \otimes |\uparrow\rangle \otimes |\uparrow\rangle \otimes |\uparrow\rangle$$

and so on. From the symmetry of Π_1 it follows that we only have to consider the projections of basis elements which are independent under the symmetry operations. We find the following basis for the subspace A_1

$$\Pi_1|\uparrow\uparrow\uparrow\uparrow\rangle : |1\rangle = |\uparrow\uparrow\uparrow\uparrow\rangle$$
$$\Pi_1|\uparrow\uparrow\uparrow\downarrow\rangle : |2\rangle = \tfrac{1}{2}(|\uparrow\uparrow\uparrow\downarrow\rangle + |\uparrow\uparrow\downarrow\uparrow\rangle + |\uparrow\downarrow\uparrow\uparrow\rangle + |\downarrow\uparrow\uparrow\uparrow\rangle))$$
$$\Pi_1|\uparrow\uparrow\downarrow\downarrow\rangle : |3\rangle = \tfrac{1}{2}(|\uparrow\uparrow\downarrow\downarrow\rangle + |\uparrow\downarrow\downarrow\uparrow\rangle + |\downarrow\downarrow\uparrow\uparrow\rangle + |\downarrow\uparrow\uparrow\downarrow\rangle))$$
$$\Pi_1|\uparrow\downarrow\uparrow\downarrow\rangle : |4\rangle = \tfrac{1}{\sqrt{2}}(|\uparrow\downarrow\uparrow\downarrow\rangle + |\downarrow\uparrow\downarrow\uparrow\rangle))$$
$$\Pi_1|\uparrow\downarrow\downarrow\downarrow\rangle : |5\rangle = \tfrac{1}{2}(|\uparrow\downarrow\downarrow\downarrow\rangle + |\downarrow\downarrow\downarrow\uparrow\rangle + |\downarrow\downarrow\uparrow\downarrow\rangle + |\downarrow\uparrow\downarrow\downarrow\rangle))$$
$$\Pi_1|\downarrow\downarrow\downarrow\downarrow\rangle : |6\rangle = |\downarrow\downarrow\downarrow\downarrow\rangle.$$

Thus the subspace which belongs to A_1 is six-dimensional. Calculating $\langle j|\hat{H}|k\rangle$, where $j, k = 1, 2, \ldots, 6$ we find the 6×6 symmetric matrix

$$\begin{pmatrix} 4a & 2b & 0 & 0 & 0 & 0 \\ 2b & 0 & 2b & \sqrt{2}b & 0 & 0 \\ 0 & 2b & 0 & 0 & 2b & 0 \\ 0 & \sqrt{2}b & 0 & -4a & \sqrt{2}b & 0 \\ 0 & 0 & 2b & \sqrt{2}b & 0 & 2b \\ 0 & 0 & 0 & 0 & 2b & 4a \end{pmatrix}.$$

(iv) Let $|j\rangle$ be the eigenstates of $\hat{H}$ with eigenvalues E_j. Then we have

$$|\psi(0)\rangle = \sum_{j=1}^{16} c_j(0)|j\rangle.$$

From the *Schrödinger equation*

$$i\hbar \frac{\partial \psi}{\partial t} = \hat{H}\psi$$

we obtain

$$|\psi(t)\rangle = e^{-i\hat{H}t/\hbar}|\psi(0)\rangle = \sum_{j=1}^{16} c_j(0)e^{-i\hat{H}t/\hbar}|j\rangle = \sum_{j=1}^{16} c_j(0)e^{-iE_jt/\hbar}|j\rangle$$

as $|j\rangle$ is, by assumption, an eigenstate (eigenvector) of $\hat{H}$ with eigenvalue E_j. Since the dual state of $|\psi(t)\rangle$ is given by

$$\langle \psi(t)| = \sum_{k=1}^{16} c_k^*(0)e^{iE_kt/\hbar}\langle k|$$

we obtain $\langle \Omega \rangle(t)$.

Problem 4. (i) Consider the Hamilton operator

$$\hat{H}_2 = \hbar\omega(\sigma_z \otimes \sigma_z) + \Delta(\sigma_x \otimes \sigma_x).$$

Find the eigenvalues and eigenvectors. Discuss whether the eigenvectors can be written as product states. Find the unitary operator $U_2(t) = \exp(-i\hat{H}_2t/\hbar)$.
(ii) Consider the Hamilton operator

$$\hat{H}_3 = \hbar\omega(\sigma_z \otimes \sigma_z \otimes \sigma_z) + \Delta(\sigma_x \otimes \sigma_x \otimes \sigma_x).$$

Find the eigenvalues and eigenvectors. Discuss whether the eigenvectors can be written as product states. Find the unitary operator $U_3(t) = \exp(-i\hat{H}_3t/\hbar)$.

Solution 4.   (i) The Hamilton operator $\hat{H}_2$ acts in the Hilbert space $\mathcal{H} = \mathbb{C}^4$. Since $\mathrm{tr}\hat{H}_2 = 0$ we obtain

$$\sum_{j=1}^{2^2} E_j = 0.$$

where E_j are the eigenvalues of $\hat{H}_2$. Consider the operators

$$\Sigma_{z,2} := \sigma_z \otimes \sigma_z, \qquad \Sigma_{x,2} := \sigma_x \otimes \sigma_x.$$

We have $[\Sigma_{z,2}, \Sigma_{x,2}] = 0$. The four eigenvalues are given by

$$E_1 = \Delta + \hbar\omega, \quad E_2 = -(\Delta + \hbar\omega), \quad E_3 = \Delta - \hbar\omega, \quad E_4 = -(\Delta - \hbar\omega).$$

with the corresponding normalized eigenvectors

$$|\Phi^+\rangle = \frac{1}{\sqrt{2}} \begin{pmatrix} 1 \\ 0 \\ 0 \\ 1 \end{pmatrix}, \quad |\Psi^-\rangle = \frac{1}{\sqrt{2}} \begin{pmatrix} 0 \\ 1 \\ -1 \\ 0 \end{pmatrix}$$

$$|\Psi^+\rangle = \frac{1}{\sqrt{2}} \begin{pmatrix} 0 \\ 1 \\ 1 \\ 0 \end{pmatrix}, \quad |\Phi^-\rangle = \frac{1}{\sqrt{2}} \begin{pmatrix} 1 \\ 0 \\ 0 \\ -1 \end{pmatrix}.$$

Note that the states do not depend on the parameters ω and Δ. These states are the Bell states. The Bell states are fully entangled. The measure of entanglement for bipartite states are the von Neumann entropy, concurrence and the 2-tangle. As a measure of entanglement we apply the tangle which is the squared concurrence. The *concurrence* $\mathcal{C}$ for a pure state $|\psi\rangle$ in $\mathcal{H} = \mathbb{C}^4$ is given by

$$\mathcal{C} = 2 \left| \det \begin{pmatrix} c_{00} & c_{01} \\ c_{10} & c_{11} \end{pmatrix} \right|$$

with the state $|\psi\rangle$ written in the form

$$|\psi\rangle = \sum_{j,k=0}^{1} c_{jk} |j\rangle \otimes |k\rangle$$

and $|j\rangle$ $(j = 0, 1)$ denotes the standard basis in the Hilbert space $\mathbb{C}^2$. Next we calculate $\exp(-i\hat{H}_2 t/\hbar)$. The unitary operator $U_2(t) = \exp(-i\hat{H}_2 t/\hbar)$ can easily be calculated since

$$U_2(t) = \exp(-i\hat{H}_{20} t/\hbar) \exp(-i\hat{H}_{21} t/\hbar).$$

Thus

$$U_2(t) = \exp(-i\hat{H}_2 t/\hbar) = e^{-i\omega t(\sigma_z \otimes \sigma_z)} e^{-it\Delta(\sigma_x \otimes \sigma_x)/\hbar}.$$

and

$$e^{-i\omega t(\sigma_z \otimes \sigma_z)} = I_4 \cos(\omega t) - i(\sigma_z \otimes \sigma_z)\sin(\omega t)$$
$$e^{-i\Delta t(\sigma_x \otimes \sigma_x)/\hbar} = I_4 \cos(t\Delta/\hbar) - i(\sigma_x \otimes \sigma_x)\sin(t\Delta/\hbar)$$

we obtain

$$e^{-i\hat{H}_2 t/\hbar} = I_4 \cos(\omega t)\cos(t\Delta/\hbar) - i(\sigma_z \otimes \sigma_z)\sin(\omega t)\cos(t\Delta/\hbar)$$
$$-i(\sigma_x \otimes \sigma_x)\cos(\omega t)\sin(t\Delta/\hbar) - (\sigma_z\sigma_x) \otimes (\sigma_z\sigma_x)\sin(\omega t)\sin(t\Delta/\hbar).$$

The Hamilton operator $\hat{H}_2$ shows energy level crossing (when keeping $\hbar\omega$ fixed and varying Δ). The unitary operator $U_2(t) = \exp(-i\hat{H}_2 t/\hbar)$ can generate entangled states from unentangled states. However note that applying the unitary operator $U_2(t)$ to one of the Bell states given above cannot disentangle these states since they are eigenstates. For example, we have $U_2(t)|\Phi^+\rangle = e^{-iE_1 t/\hbar}|\Phi^+\rangle$.

If we start with the unentangled state (product state)

$$|\psi\rangle = (1\,0\,0\,0)^T$$

under the evolution $U_2(t)|\psi\rangle$ depending on t, $\hbar\omega$ and Δ we can find entangled states using the concurrence as measure. For the case $\hbar\omega = \Delta$ (level crossing) the state reduces to

$$\begin{pmatrix} \cos^2(\omega t) - i\sin(\omega t)\cos(\omega t) \\ 0 \\ 0 \\ -\sin^2(\omega t) - i\cos(\omega t)\sin(\omega t) \end{pmatrix}.$$

(ii) Let

$$\Sigma_{z,3} = \sigma_z \otimes \sigma_z \otimes \sigma_z, \quad \Sigma_{x,3} = \sigma_x \otimes \sigma_x \otimes \sigma_x.$$

We have for the anticommutator

$$[\Sigma_{z,3}, \Sigma_{x,3}]_+ = 0.$$

We find the eigenvalue $E = \sqrt{\hbar^2\omega^2 + \Delta^2}$ (four-times degenerate) with the normalized eigenvectors

$$\frac{1}{\sqrt{\Delta^2 + (E - \hbar\omega)^2}} \begin{pmatrix} \Delta \\ 0 \\ 0 \\ 0 \\ 0 \\ 0 \\ 0 \\ E - \hbar\omega \end{pmatrix}, \quad \frac{1}{\sqrt{\Delta^2 + (E - \hbar\omega)^2}} \begin{pmatrix} 0 \\ 0 \\ 0 \\ \Delta \\ E - \hbar\omega \\ 0 \\ 0 \\ 0 \end{pmatrix},$$

$$\frac{1}{\sqrt{\Delta^2 + (E + \hbar\omega)^2}} \begin{pmatrix} 0 \\ \Delta \\ 0 \\ 0 \\ 0 \\ 0 \\ E + \hbar\omega \\ 0 \end{pmatrix}, \quad \frac{1}{\sqrt{\Delta^2 + (E + \hbar\omega)^2}} \begin{pmatrix} 0 \\ 0 \\ \Delta \\ 0 \\ 0 \\ E + \hbar\omega \\ 0 \\ 0 \end{pmatrix}.$$

and the eigenvalue $-E$ (four times degenerate) with the normalized eigen-vectors

$$\frac{1}{\sqrt{\Delta^2 + (E + \hbar\omega)^2}} \begin{pmatrix} \Delta \\ 0 \\ 0 \\ 0 \\ 0 \\ 0 \\ 0 \\ -E - \hbar\omega \end{pmatrix}, \quad \frac{1}{\sqrt{\Delta^2 + (E + \hbar\omega)^2}} \begin{pmatrix} 0 \\ 0 \\ 0 \\ \Delta \\ -E - \hbar\omega \\ 0 \\ 0 \\ 0 \end{pmatrix},$$

$$\frac{1}{\sqrt{\Delta^2 + (E - \hbar\omega)^2}} \begin{pmatrix} 0 \\ \Delta \\ 0 \\ 0 \\ 0 \\ 0 \\ -E + \hbar\omega \\ 0 \end{pmatrix}, \quad \frac{1}{\sqrt{\Delta^2 + (E - \hbar\omega)^2}} \begin{pmatrix} 0 \\ 0 \\ \Delta \\ 0 \\ 0 \\ -E + \hbar\omega \\ 0 \\ 0 \end{pmatrix}.$$

If $\Delta = E - \hbar\omega$ then the first and second eigenstates are fully entangled. If $\Delta = E + \hbar$ the third and fourth states are fully entangled. As measure we can use the 3-tangle.

Since we have the eigenvalues and eigenvectors of $\hat{H}_3$ the unitary operator $U_3(t)$ can easily be calculated. We find

$$U_3(t) = I_8 \cos(Et/\hbar) + \frac{i}{E}(\hbar\omega\sigma_z \otimes \sigma_z \otimes \sigma_z + \Delta\sigma_x \otimes \sigma_x \otimes \sigma_x) \sin(Et/\hbar)$$

If we start with the unentangled state (product state)

$$|\psi\rangle = (1\,0\,0\,0\,0\,0\,0\,0)^T$$

under the evolution $U(t)|\psi\rangle$ depending on t, $\hbar\omega$ and Δ we can find entangled states.

Chapter 28

Number Theory

Problem 1. (i) Show that

$$\frac{\sqrt{5}-1}{2} = \cfrac{1}{1 + \cfrac{1}{1 + \cfrac{1}{1 + \cfrac{1}{\cdots}}}} \tag{1}$$

The irrational number $(\sqrt{5}-1)/2$ is called the *golden mean number*. This is the *worst irrational number* due to the representation (1).

(ii) The resistances of the resistors of the following network

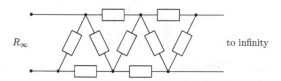

are all equal (say R). Calculate R_∞.

Solution 1. (i) We have the identity

$$\frac{\sqrt{5}-1}{2} \equiv \frac{(\sqrt{5}-1)(\sqrt{5}+1)}{2(\sqrt{5}+1)} \equiv \frac{4}{2(\sqrt{5}+1)} \equiv \frac{2}{\sqrt{5}+1} \equiv \frac{1}{(\sqrt{5}+1)/2}.$$

297

Now we also have the identity

$$\frac{\sqrt{5}+1}{2} \equiv \frac{2+\sqrt{5}-2+1}{2} \equiv 1 + \frac{\sqrt{5}-1}{2} \equiv 1 + \frac{(\sqrt{5}-1)(\sqrt{5}+1)}{2(\sqrt{5}+1)}$$

$$\equiv 1 + \frac{4}{2(\sqrt{5}+1)}.$$

Thus the continued fraction (1) follows.

(ii) Method 1. The network

has resistance $R_1 = R$. The network

has resistance

$$R_2 = \frac{1}{1/R + 1/(2R)} \equiv \frac{1}{1+1/2}R$$

The network

has the resistance

$$R_3 = \frac{1}{\dfrac{1}{R} + \dfrac{1}{R+R_2}}$$

Obviously, we find the recursion relation

$$R_n = \frac{1}{\dfrac{1}{R} + \dfrac{1}{R+R_{n-1}}}$$

where $(n = 2, 3, \ldots)$. Therefore

$$R_\infty = \frac{1}{1 + \dfrac{1}{1 + \dfrac{1}{1 + \dfrac{1}{\cdots}}}} R = \frac{(\sqrt{5}-1)}{2} R.$$

Method 2. Since

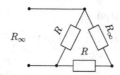

we obtain

$$R_\infty = \cfrac{1}{\cfrac{1}{R} + \cfrac{1}{R + R_\infty}}.$$

It follows that

$$R_\infty = \cfrac{1}{\cfrac{R_\infty + 2R}{R(R + R_\infty)}} = \frac{R(R + R_\infty)}{R_\infty + 2R}.$$

Therefore we obtain the quadratic equation

$$R_\infty^2 + R_\infty R - R^2 = 0.$$

The solution of the quadratic equation is given by

$$R_\infty = \frac{(\sqrt{5} - 1)}{2} R.$$

Problem 2. (i) Show that

$$\sqrt{3} - 1 \equiv \cfrac{1}{1 + \cfrac{1}{2 + \cfrac{1}{1 + \cfrac{1}{2 + \cfrac{1}{\cdots}}}}}$$

(ii) The resistances of the resistors of the following network

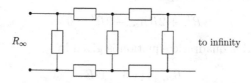

to infinity

are all equal (say R). Calculate R_∞.

Solution 2. (i) We have

$$\sqrt{3} - 1 \equiv \frac{(\sqrt{3} - 1)(\sqrt{3} + 1)}{\sqrt{3} + 1} \equiv \frac{2}{\sqrt{3} + 1} = \frac{1}{(\sqrt{3} + 1)/2}. \qquad (1)$$

The denominator can be written as

$$\frac{\sqrt{3}+1}{2} \equiv \frac{1+(\sqrt{3}-1)+1}{2} \equiv 1 + \frac{\sqrt{3}-1}{2}.$$

Now

$$\frac{\sqrt{3}-1}{2} \equiv \frac{(\sqrt{3}-1)(\sqrt{3}+1)}{2\sqrt{3}+1} \equiv \frac{1}{\sqrt{3}+1}$$

and

$$\sqrt{3}+1 \equiv 2 + (\sqrt{3}-1). \tag{2}$$

Combining (1) through (2) gives

$$\sqrt{3}-1 \equiv \cfrac{1}{1+\cfrac{\sqrt{3}-1}{2}} \equiv \cfrac{1}{1+\cfrac{1}{\sqrt{3}+1}} \equiv \cfrac{1}{1+\cfrac{1}{2+(\sqrt{3}-1)}}.$$

Thus

$$\sqrt{3}-1 = \cfrac{1}{1+\cfrac{1}{(2+\sqrt{3}-1)}}.$$

(ii) The same methods as in problem 1 can be used. We apply method 2. Since

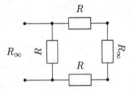

we have

$$R_\infty = \cfrac{1}{\cfrac{1}{R}+\cfrac{1}{2R+R_\infty}} \equiv \frac{R(2R+R_\infty)}{3R+R_\infty}.$$

Consequently, we obtain the quadratic equation

$$R_\infty^2 + 2RR_\infty - 2R^2 = 0.$$

The solution of this quadratic equation is given by

$$R_\infty = (\sqrt{3}-1)R.$$

Problem 3.   Let $x \in \mathbb{R}$. Let $N > 1$ be an integer. Show that there is an integer p and an integer q such that $1 \leq q \leq N$ and

$$\left| x - \frac{p}{q} \right| \leq \frac{1}{qN}.$$

Solution 3. The proof relies on the *pigeon hole principle* (also called the *Dirichlet box principle*). This principle says that if we put $N + 1$ objects into N boxes, then at least one of the boxes must contain more than one of the objects. We rewrite our statement as follows: There is an integer p and an integer q such that $1 \leq q \leq N$ and

$$|qx - p| \leq \frac{1}{N}.$$

We denote by $[x]$ the largest integer $\leq x$. For example $[\pi] = 3$. Consider the numbers

$$0 - 0 \cdot x, \quad x - [x], \quad 2x - [2x], \quad \ldots \quad Nx - [Nx].$$

This means we consider the numbers $kx - [kx]$ for $k = 0, 1, \ldots, N$. These $N + 1$ numbers will be our objects. They will lie between 0 and 1. We divide the interval $[0, 1]$ into N equal subintervals

$$[0, 1/N), \quad [1/N, 2/N), \quad [2/N, 3/N), \quad \ldots \quad [(N - 1)/N, 1].$$

These N subintervals are the boxes. There are N of them, and so two of the numbers $kx - [kx]$ $(0 \leq k \leq N)$ must lie in the same subinterval; that is, they can be no further apart than $1/N$. Say these two numbers are $nx - [nx]$ and $mx - [mx]$, where $0 \leq n < m \leq N$. Set $q = m - n$ and $p = [mx] - [nx]$. Then

$$|qx - p| = |(nx - [nx]) - (mx - [mx])| \leq \frac{1}{N}.$$

Moreover, $q = m - n \leq m \leq N$, and $q \geq 1$, since $m - n > 0$.

Problem 4. The *Farey sequence* F_N is the set of all fractions in lowest terms between 0 and 1 whose denominators do not exceed N, arranged in order of magnitude. For example, F_6 is given by

$$\frac{0}{1} \quad \frac{1}{6} \quad \frac{1}{5} \quad \frac{1}{4} \quad \frac{1}{3} \quad \frac{2}{5} \quad \frac{1}{2} \quad \frac{3}{5} \quad \frac{2}{3} \quad \frac{3}{4} \quad \frac{4}{5} \quad \frac{5}{6} \quad \frac{1}{1}.$$

N is known as the order of the series.
Let m_1/n_1 and m_2/n_2 be two successive terms in F_N. Then

$$m_2 n_1 - m_1 n_2 = 1. \tag{1}$$

For three successive terms m_1/n_1, m_2/n_2 and m_3/n_3, the middle term is the *mediant* of the other two

$$m_2/n_2 = (m_1 + m_3)/(n_1 + n_3). \tag{2}$$

(i) Find F_4.
(ii) Prove that (1) implies (2).
(iii) Prove that (2) implies (1).

Solution 4.   (i) We have

$$\frac{0}{1} \quad \frac{1}{4} \quad \frac{1}{3} \quad \frac{1}{2} \quad \frac{2}{3} \quad \frac{3}{4} \quad \frac{1}{1}.$$

(ii) To prove that (1) implies (2), assume we have two identities

$$m_2 n_1 - m_1 n_2 = 1, \qquad m_3 n_2 - m_2 n_3 = 1.$$

Subtract one from the other and recombine the terms yields

$$(m_3 + m_1)n_2 = m_2(n_3 + n_1)$$

which leads to (2).

(iii) To prove that (2) implies (1) we use mathematical induction. Assume that we are given a series of fractions m_i/n_i $(i = 1, 2, \ldots)$ with the condition that any three consecutive terms satisfy

$$\frac{m_i}{n_i} = \frac{m_{i-1} + m_{i+1}}{n_{i-1} + n_{i+1}}. \tag{3}$$

We show that for such a series (1) also holds. However (1) only contains two fractions. Therefore, let us assume that for $i = 2$, equation (1) indeed holds. This is clearly true for F_N which starts with $0/1$, $1/N$. Assume that for some $i > 2$

$$m_i n_{i-1} - m_{i-1} n_i = 1. \tag{4}$$

We rewrite (3) as

$$(m_{i+1} + m_{i-1})n_i = m_i(n_{i+1} + n_{i-1}). \tag{5}$$

Subtracting (5) from (3) gives

$$m_{i+1} n_i - m_i n_{i+1} = 1.$$

Equation (5) implies (3) only if $n_i \neq 0$.

Problem 5.   The *star-triangle transformation* is given by

$$R_u = \frac{R_2 R_3}{R_1 + R_2 + R_3}, \qquad R_1 = \frac{R_u R_v + R_v R_w + R_w R_u}{R_u}$$

$$R_v = \frac{R_1 R_3}{R_1 + R_2 + R_3}, \qquad R_2 = \frac{R_u R_v + R_v R_w + R_w R_u}{R_v}$$

$$R_w = \frac{R_1 R_2}{R_1 + R_2 + R_3}, \qquad R_3 = \frac{R_u R_v + R_v R_w + R_w R_u}{R_w}$$

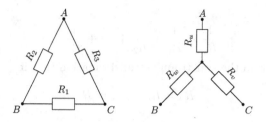

Figure 28.1. Star-triangle transformation.

Using the star-triangle transformation find the total resistance of the circuit given below between two opposite corners (say 1 and 7). This means the edges of a cube consist of equal resistors of resistance R, which are joined at the corners. Let a battery be connected to two opposite corners of a face of the cube (say 1 and 7). What is the resistance R_{17}?

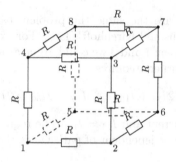

Solution 5. Drawing the circuit in the plane we have

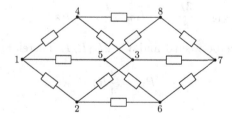

We solve the problem by successively applying the star-triangle transformation. With the star-triangle transformation we can simplify the circuit by using the standard rules for circuits in series

$$R_s = R_1 + R_2 + \cdots + R_n$$

and parallel

$$\frac{1}{R_s} = \frac{1}{R_1} + \frac{1}{R_2} + \cdots + \frac{1}{R_n}.$$

First we note from the star-triangle transformation that if

$$R_u = R_v = R_w = R$$

we have

$$R_1 = R_2 = R_3 = 3R.$$

In the first step, we convert the star 1, 5, 8, 6 to a triangle. Thus the node 5 disappears. In the second step we convert the star 1, 4, 8, 3 to a triangle. Thus the node 4 disappears. We repeat the step of eliminating the stars. At each step we apply the standard rules for circuits in series and parallel given above. Finally we arrive at

$$R_{17} = \frac{5}{6}R.$$

Since all the resistors are the same the problem can also be solved by symmetry considerations and Kirchhoff's laws. For example, if we want to find the resistance between 1 and 3 we could argue as follows. Conservation of the current at the corners requires

$$I = 2I_{14} + I_{15}, \qquad I_{15} = 2I_{56} = 2I_{58}$$

where I is the input current and $I_{14} = I_{12}$. The requirement that the voltage between 1 and 3 be independent of path yields the additional equation

$$2I_{14}R = 2(I_{15} + I_{58})R$$

where $I_{58} = I_{56}$. These three equations have the solution

$$I_{14} = \frac{3I}{8}, \qquad I_{15} = \frac{I}{4}, \qquad I_{58} = \frac{I}{8}.$$

Thus the resistance between 1 and 3 is $2I_{14}R/I$. Therefore

$$R_{13} = \frac{3}{4}R.$$

Problem 6.   The recursion relation

$$F_{n+2} = F_n + F_{n+1}$$

with the initial values $F_0 = F_1 = 1$ provides the *Fibonacci sequence*

$$1, 1, 2, 3, 5, 8, 13, 21, 34, \ldots.$$

A generalization of the Fibonacci sequence is the q-analogue of the sequence defined by the recursion relation

$$F_{n+2}(q) = F_n(q) + qF_{n+1}(q)$$

and the initial condition $F_0(q) = 1$ and $F_1(q) = q$. Here, q is a real or complex number.

(i) Give the first five terms of the sequence.
(ii) Find a generating function of $F_n(q)$.
(iii) Find an explicit expression for $F_n(q)$.

Solution 6. (i) The first five terms are

$$1, \; q, \; q^2 + 1, \; q^3 + 2q, \; q^4 + 3q^2 + 1, \ldots .$$

(ii) The generating function of $F_n(q)$ is

$$\frac{1}{1 - qs - s^2} = \sum_{n=0}^{\infty} F_n(q)s^n .$$

(iii) Writing $q = x - x^{-1}$ and partial fractioning the left-hand side of this equation yields

$$F_n(q) = \frac{x^{n+1} + (-1)^n x^{-(n+1)}}{x + x^{-1}}, \qquad n = 0, 1, 2, \ldots .$$

Problem 7. A symmetric $(r \times r)$ matrix $B = (B_{jk})$ with negative definite real part $\Re B = (\Re B_{jk})$ is called a *Riemann matrix*. The *Riemann theta function* is defined by

$$\theta(\mathbf{z}|B) := \sum_{\mathbf{N} \in \mathbb{Z}^r} \exp\left(\frac{1}{2} \langle B\mathbf{N}, \mathbf{N} \rangle + \langle \mathbf{N}, \mathbf{z} \rangle \right) . \tag{1}$$

Here $\mathbf{z} = (z_1, \ldots, z_r) \in \mathbb{C}^r$ is a complex vector. The triangular brackets denote the Euclidean scalar product

$$\langle \mathbf{N}, \mathbf{z} \rangle := \sum_{j=1}^{r} N_j z_j, \qquad \langle B\mathbf{N}, \mathbf{N} \rangle := \sum_{i,j=1}^{r} B_{ij} N_i N_j .$$

The summation in (1) is taken over the lattice of integer vectors $\mathbf{N} = (N_1, \ldots, N_r)$. The general term of this series depends only on the symmetric part of the $r \times r$ matrix B. From the estimate

$$\Re \langle B\mathbf{N}, \mathbf{N} \rangle \le -b \langle \mathbf{N}, \mathbf{N} \rangle, \quad b > 0$$

where $-b$ is the largest eigenvalue of the matrix $\Re B$ one finds that the series (1) is absolutely convergent, uniformly on compact sets. Thus the function $\theta(\mathbf{z}|B)$ is analytic in the whole space vector space $\mathbb{C}^r$. Let $\mathbf{e}_1, \ldots, \mathbf{e}_r$ be the basis vectors in $\mathbb{C}^r$ with the coordinates $(e_k)_j = \delta_{kj}$. We also introduce vectors $\mathbf{f}_1, \ldots, \mathbf{f}_r$, setting

$$(f_k)_j = B_{kj}, \qquad k, j = 1, \ldots, r.$$

The vector $\mathbf{f}_k$ can also be written in the form $\mathbf{f}_k = B\mathbf{e}_k$.
(i) Show that
$$\theta(\mathbf{z} + 2\pi i\mathbf{e}_k) = \theta(\mathbf{z}). \tag{2}$$

(ii) Show that
$$\theta(\mathbf{z} + \mathbf{f}_k) = \exp\left(-\frac{1}{2}B_{kk} - z_k\right)\theta(\mathbf{z}). \tag{3}$$

Solution 7.   (i) The periodicity of (2) is obvious. The general term of (1) does not change under the shift $\mathbf{z} \to \mathbf{z} + 2\pi i\mathbf{e}_k$.
(ii) We have

$$\theta(\mathbf{z} + \mathbf{f}_k) = \sum_{\mathbf{N} \in \mathbb{Z}^r} \exp\left(\frac{1}{2}\langle B\mathbf{N}, \mathbf{N}\rangle + \langle \mathbf{N}, \mathbf{z} + \mathbf{f}_k\rangle\right).$$

To prove (3) we change the summation index $\mathbf{N}$ in (1) setting $\mathbf{N} = \mathbf{M} - \mathbf{e}_k$, $\mathbf{M} \in \mathbb{Z}^r$. Thus

$$\theta(\mathbf{z} + \mathbf{f}_k) = \sum_{\mathbf{M} \in \mathbb{Z}^r} \exp\left(\frac{1}{2}\langle B\mathbf{M}, \mathbf{M}\rangle - \langle \mathbf{M}, B\mathbf{e}_k\rangle + \frac{1}{2}\langle B\mathbf{e}_k, \mathbf{e}_k\rangle \right.$$
$$\left. + \langle \mathbf{M}, \mathbf{z}\rangle + \langle \mathbf{M}, \mathbf{f}_k\rangle - \langle \mathbf{e}_k, \mathbf{z}\rangle - \langle \mathbf{e}_k, \mathbf{f}_k\rangle\right).$$

It follows that

$$\theta(\mathbf{z} + \mathbf{f}_k) = \exp\left(-\frac{1}{2}B_{kk} - z_k\right) \sum_{\mathbf{M} \in \mathbb{Z}^r} \exp\left(\frac{1}{2}\langle B\mathbf{M}, \mathbf{M}\rangle + \langle \mathbf{M}, \mathbf{z}\rangle\right).$$

Finally

$$\theta(\mathbf{z} + \mathbf{f}_k) = \exp\left(-\frac{1}{2}B_{kk} - z_k\right)\theta(\mathbf{z}).$$

Chapter 29

Combinatorial Problems

Problem 1. (i) Consider N boxes and n particles with $n \leq N$. In every box we can put a maximum of one particle (*Pauli principle, Fermi particles*). In how many ways can we put the n particles (which are identical) in the N boxes ?

(ii) Consider N boxes and an arbitrary number of particles n. We can put an arbitrary number of particles in every box (*Bose-particles*). Again the particles are identical. In how many ways can we put the n particles in the N boxes?

(iii) The same as case (ii) but now we have distinguishable particles.

Solution 1. (i) The problem is equivalent to that of determining the total number of ways in which N objects of which n are indistinguishable from type I (say particles) and $N - n$ are indistinguishable from type II (say empty spaces, holes) can be arranged in N possible places. The total number of permutations of N objects is $N!$. From each of these $N!$ arrangements we obtain $n!$ mutually indistinguishable arrangements by permutation of the $n!$ identical particles among themselves. From each of these $n!$ we obtain again $(N - n)!$ mutually indistinguishable arrangements by permutating the $(N - n)!$ identical holes among themselves. The $N!$ possible arrangements can therefore be partitioned into disjoint sets, each containing

$$n!(N - n)!$$

indistinguishable arrangements. The total number of distinguishable arrangements is therefore

$$\frac{N!}{n!(N-n)!} \equiv \binom{N}{n}.$$

We have the recursion relation

$$\binom{N+1}{n+1} = \binom{N}{n} + \binom{N}{n+1}, \qquad \binom{1}{1} := 1.$$

(ii) We select a box to begin with. The number of ways we can choose the $N-1$ boxes and the n-particles are $(N-1+n)!$. However the boxes and particles are indistinguishable. Thus we find

$$\binom{n+N-1}{n} \equiv \frac{(n+N-1)!}{n!(n+N-1-n)!} \equiv \frac{(n+N-1)!}{n!(N-1)!}.$$

(iii) For distinguishable particles the first particle can be put in any of the N boxes, the second one in any of N boxes etc.. Thus we find the desired value to be

$$N^n.$$

Problem 2.   A young man, who lives at location A of the city street plan shown in the figure, walks daily to the home of his fiancée, who lives m blocks east and n blocks north of A, at location B. He can only walk north or east. In how many different ways can he go from A to B? Consider first the case $n = m = 2$.

B

A

Solution 2.   We consider first the case $n = m = 2$. Each of the paths can be written as vector

$$(u,u,r,r) \quad (u,r,u,r) \quad (u,r,r,u) \quad (r,u,u,r) \quad (r,u,r,u) \quad (r,r,u,u)$$

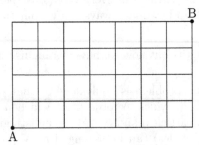

where u stands for one step north ("up") and r stands for one step east ("right"). Thus the number of paths for $n = m = 2$ is given by 6.
The general case is as follows: At each corner the man may walk either up or to the right. Consequently, a particular path is specified by a sequence of the form $(u, u, r, u, \ldots, r)$, where the total number of $u's$ and $r's$ are n and m, respectively. The number of different ways of writing such a sequence and therefore the total number of paths is given by

$$\frac{(m + n)!}{n! \, m!}.$$

Problem 3. (i) Let $\mathbb{Z}$ be the set of integers. Then $\mathbb{Z}^d$ $(d = 1, 2, 3, \ldots)$ denotes the d-tuple of the set of integers, i.e.

$$\mathbb{Z}^d := \mathbb{Z} \times \mathbb{Z} \times \cdots \times \mathbb{Z} \qquad d - \text{times}.$$

Find the number of all closed paths A_n with n-steps starting from $(0, \ldots, 0) \in \mathbb{Z}^d$ and returning to $(0, \ldots, 0) \in \mathbb{Z}^d$ (not necessarily the first time).
(ii) Calculate A_4 for $d = 2$.

Solution 3. (i) Obviously A_n is given by

$$A_n = \sum_{\{\boldsymbol{\Delta}_1, \boldsymbol{\Delta}_2, \cdots, \boldsymbol{\Delta}_n\}} \delta(\boldsymbol{\Delta}_1 + \boldsymbol{\Delta}_2 + \cdots + \boldsymbol{\Delta}_n)$$

where

$$\delta(\boldsymbol{\Delta}_1 + \boldsymbol{\Delta}_2 + \cdots + \boldsymbol{\Delta}_n) := \begin{cases} 1 & \text{for} \quad \boldsymbol{\Delta}_1 + \boldsymbol{\Delta}_2 + \cdots + \boldsymbol{\Delta}_n = 0 \\ 0 & \text{otherwise} \end{cases}$$

If n is odd we find $A_n = 0$. Now we apply the *completeness relation*

$$\delta(\boldsymbol{\Delta}_1 + \boldsymbol{\Delta}_2 + \cdots + \boldsymbol{\Delta}_n) := \frac{1}{N} \sum_{\mathbf{k} \in 1.BZ} e^{i\mathbf{k}(\boldsymbol{\Delta}_1 + \boldsymbol{\Delta}_2 + \cdots + \boldsymbol{\Delta}_n)}$$

where N denotes the lattice sites in the first Brillouin zone (1.BZ). It follows that

$$A_n = \sum_{\{\boldsymbol{\Delta}_1, \ldots, \boldsymbol{\Delta}_n\}} \frac{1}{N} \sum_{\mathbf{k} \in 1.BZ} e^{i\mathbf{k}(\boldsymbol{\Delta}_1 + \cdots + \boldsymbol{\Delta}_n)}$$

$$= \frac{1}{N} \sum_{\mathbf{k} \in 1.BZ} \sum_{\{\boldsymbol{\Delta}_1, \ldots, \boldsymbol{\Delta}_n\}} e^{i\mathbf{k}(\boldsymbol{\Delta}_1 + \cdots + \boldsymbol{\Delta}_n)}.$$

Thus we can write

$$A_n = \frac{1}{N} \sum_{\mathbf{k} \in 1.BZ} \sum_{\{\boldsymbol{\Delta}_1\}} e^{i\mathbf{k}\boldsymbol{\Delta}_1} \sum_{\{\boldsymbol{\Delta}_2\}} e^{i\mathbf{k}\boldsymbol{\Delta}_2} \cdots \sum_{\{\boldsymbol{\Delta}_n\}} e^{i\mathbf{k}\boldsymbol{\Delta}_n} = \frac{1}{N} \sum_{\mathbf{k} \in 1.BZ} (\epsilon(\mathbf{k}))^n$$

where

$$\epsilon(\mathbf{k}) := \sum_{\Delta} e^{i\mathbf{k}\Delta}.$$

(ii) In two dimensions ($d = 2$), we have $\epsilon(\mathbf{k}) = 2(\cos k_1 + \cos k_2)$ since $\Delta \in \{(1,0), (0,1), (-1,0), (0,-1)\}$. Thus we find

$$A_4 = \frac{1}{N} \sum_{\mathbf{k}\in 1.BZ} (\epsilon(\mathbf{k}))^4 = \frac{2^4}{4\pi^2} \int_{-\pi}^{\pi} \int_{-\pi}^{\pi} (\cos k_1 + \cos k_2)^4 dk_1 dk_2 = 36.$$

The *first Brillouin zone* in three dimensions is constructed as follows: Let $\mathbf{a}_1$, $\mathbf{a}_2$ and $\mathbf{a}_3$ be a set of primitive vectors for the direct lattice. Then the reciprocal lattice can be generated by the three primitive vectors

$$\mathbf{b}_1 = 2\pi \frac{\mathbf{a}_2 \times \mathbf{a}_3}{\mathbf{a}_1 \cdot (\mathbf{a}_2 \times \mathbf{a}_3)}, \quad \mathbf{b}_2 = 2\pi \frac{\mathbf{a}_3 \times \mathbf{a}_1}{\mathbf{a}_1 \cdot (\mathbf{a}_2 \times \mathbf{a}_3)}, \quad \mathbf{b}_3 = 2\pi \frac{\mathbf{a}_1 \times \mathbf{a}_2}{\mathbf{a}_1 \cdot (\mathbf{a}_2 \times \mathbf{a}_3)}.$$

The simple cubic Bravais lattice, with cubic primitive cell of side a, has as its reciprocal a simple cubic lattice with cubic primitive cell of side $2\pi/a$. The construction of the first Brillouin zone is as follows: choose any reciprocal lattice point as the origin. Draw the vectors connecting this point with (all) other lattice points. Next, construct a set of planes that are perpendicular bisectors of these vectors. The smallest solid figure containing the origin is the first Brillouin zone. Points $\mathbf{k}$ on the surface must satisfy the condition $\mathbf{k}^2 = (\mathbf{k} - \mathbf{K}_n)^2$ or $\mathbf{K}_n^2 - 2\mathbf{k} \cdot \mathbf{K}_n = 0$ for some reciprocal lattice vector $\mathbf{K}_n$. We need only consider those $\mathbf{k}$ values lying within the zone. Any $\mathbf{k}$ vector may be written as

$$\mathbf{k} = \left(\frac{h_1}{2N_1 + 1}\right)\mathbf{b}_1 + \left(\frac{h_2}{2N_2 + 1}\right)\mathbf{b}_2 + \left(\frac{h_3}{2N_3 + 1}\right)\mathbf{b}_3$$

where the h_i are integers in the range $-N_i \leq h_i \leq N_i$. The values of $\mathbf{k}$ that are given by the above formula may not all be within the first Brillouin zone. However, it is always possible to bring any such $\mathbf{k}$ outside the zone back into it by translation by a reciprocal lattice vector. The values of $\mathbf{k}$ form a uniform and dense distribution. In the limit in which the N_i are allowed to become infinite, we may convert sums over possible $\mathbf{k}$ values into integrals over the first Brillouin zone through the relation

$$\sum_{\mathbf{k}\in 1.BZ} \rightarrow \frac{N}{V} \int_{1.BZ} d\mathbf{k}$$

where $N = (2N_1 + 1)(2N_2 + 1)(2N_3 + 1)$ is the number of unit cells and V is the volume of the first Brillouin zone.

Problem 4. Find the probability $P_N(\mathbf{R})$ that a random walker reaches a lattice position $\mathbf{R}$,

$$\mathbf{R} = N_1 \mathbf{a}_1 + N_2 \mathbf{a}_2 + N_3 \mathbf{a}_3$$

after $N = N_1 + N_2 + N_3$ steps on a lattice with unit vectors $\mathbf{a}_i$. The lattice is the simple cubic lattice. The lattice is periodic with a periodicity L. The transition probability $W(\mathbf{R})$ for a unit displacement is given by

$$W(\mathbf{R}) := \begin{cases} \frac{1}{6} & \text{for } \mathbf{R} = (0,0,\pm1),(0,\pm1,0),(\pm1,0,0) \\ 0 & \text{otherwise.} \end{cases} \tag{1}$$

Solution 4. The probability satisfies the recurrence formula

$$P_{N+1}(\mathbf{R}) = \sum_{\mathbf{R}'} W(\mathbf{R} - \mathbf{R}') P_N(\mathbf{R}')$$

where $W(\mathbf{R} - \mathbf{R}')$ is the probability of transition from $\mathbf{R}'$ to $\mathbf{R}$. We introduce a *generating function*

$$\Phi(\mathbf{R}, z) := \sum_{N=0}^{\infty} z^N P_N(\mathbf{R}).$$

Obviously

$$\Phi(\mathbf{R}z) - z \sum_{\mathbf{R}'} W(\mathbf{R} - \mathbf{R}') \Phi(\mathbf{R}', z) = \delta_{\mathbf{R},\mathbf{0}} \tag{2}$$

which expresses the fact that the walker starts walking from the origin

$$P_0(\mathbf{R}) = \delta_{\mathbf{R},\mathbf{0}} = \begin{cases} 1, & \mathbf{R} = 0 \\ 0, & \text{otherwise} \end{cases}$$

where δ is the Kronecker delta. We solve (2) by using the *Fourier transform method*. Multiplying both sides by

$$\exp(2\pi i\mathbf{k} \cdot \mathbf{R}/L)$$

and summing over $\mathbf{R}$, we find

$$\widetilde{\Phi}(\mathbf{k}, z) - z\lambda(\mathbf{k})\widetilde{\Phi}(\mathbf{k}, z) = 1$$

with

$$\widetilde{\Phi}(\mathbf{k}, z) := \sum_{\mathbf{R}} \exp\left(\frac{2\pi i\mathbf{k} \cdot \mathbf{R}}{L}\right) \Phi(\mathbf{R}, z)$$

$$\lambda(\mathbf{k}) := \sum_{\mathbf{R}} W(\mathbf{R}) \exp\left(\frac{(2\pi i\mathbf{k} \cdot \mathbf{R})}{L}\right).$$

The Fourier inverse of $\widetilde{\Phi}(\mathbf{k}, z)$ is

$$\Phi(\mathbf{R}, z) = \frac{1}{L^3} \sum_{\mathbf{k}} \frac{\exp(-2\pi i\mathbf{k} \cdot \mathbf{R}/L)}{1 - z\lambda(\mathbf{k})}$$

which, in the limit of $L \to \infty$, gives

$$\lim_{L \to \infty} \Phi(\mathbf{R}, z) = \frac{1}{(2\pi)^3} \int_{-\pi}^{\pi} \int_{-\pi}^{\pi} \int_{-\pi}^{\pi} \frac{\exp(-i\boldsymbol{\Theta} \cdot \mathbf{R})}{1 - z\lambda(\boldsymbol{\Theta})} d\boldsymbol{\Theta}$$

where $d\boldsymbol{\Theta} \equiv d\Theta_1 d\Theta_2 d\Theta_3$ and $\boldsymbol{\Theta} \cdot \mathbf{R} := \Theta_1 R_1 + \Theta_2 R_2 + \Theta_3 R_3$. The probability $P_N(\mathbf{R})$ is then given by

$$P_N(\mathbf{R}) = \frac{1}{(2\pi)^3} \int_{-\pi}^{\pi} \int_{-\pi}^{\pi} \int_{-\pi}^{\pi} (\lambda(\boldsymbol{\Theta}))^N \exp(-i\boldsymbol{\Theta} \cdot \mathbf{R}) d\boldsymbol{\Theta}.$$

Owing to (1) we find

$$\lambda(\boldsymbol{\Theta}) = \frac{1}{6} \sum_{j=1}^{3} \left(e^{i\Theta_j} + e^{-i\Theta_j} \right) = \frac{1}{3} \sum_{j=1}^{3} \cos \Theta_j.$$

Thus

$$P_N(\mathbf{R}) = \frac{1}{(2\pi)^3} \int_{-\pi}^{\pi} \int_{-\pi}^{\pi} \int_{-\pi}^{\pi} \left(\frac{1}{3}(\cos \Theta_1 + \cos \Theta_2 + \cos \Theta_3) \right)^N e^{-i\boldsymbol{\Theta} \cdot \mathbf{R}} d\boldsymbol{\Theta}$$

and

$$\Phi(\mathbf{R}, z) = \frac{1}{(2\pi)^3} \int_{-\pi}^{\pi} \int_{-\pi}^{\pi} \int_{-\pi}^{\pi} \frac{\exp(-i\boldsymbol{\Theta} \cdot \mathbf{R})}{1 - z(\cos \Theta_1 + \cos \Theta_2 + \cos \Theta_3)/3} d\boldsymbol{\Theta}.$$

Problem 5. Consider a lattice with N lattice sites. Each lattice site can only be occupied by two electrons one with spin up and one with spin down (Pauli principle). Let N_e be the number of electrons we put in the lattice. Obviously, $0 \le N_e \le 2N$.
(i) Given N and N_e, find the number of ways to occupy the lattice with electrons.
(ii) Consider the case $N_e = N$.
(iii) Consider the case $N_e = N$ ($N =$ even) and $S_z = 0$, where S_z denotes the total spin.

Solution 5.   (i) Let a be the number of lattice sites occupied by one electron with spin up, b the number of lattice sites occupied by one electron with spin down, c the number of lattice sites occupied by two electrons (one spin up, one spin down) and d the unoccupied lattice sites. Obviously we have the conditions $a + b + 2c = N_e$, $a + b + c + d = N$. Thus the number of ways to occupy the lattice is

$$\frac{N!}{a!\, b!\, c!\, d!}$$

with the constraints given by (1).

(ii) If $N = N_e$ we have $d = c$ and thus

$$\frac{N!}{a!\,b!\,c!\,c!}.$$

(iii) If $N = N_e$ and if the total spin S_z is equal to 0 ($N =$ even) we find $a = b$ and therefore $a = N/2 - c$. Thus

$$\frac{N!}{\left(\frac{N}{2} - c\right)!\left(\frac{N}{2} - c\right)!\,c!\,c!}.$$

Consequently the total number of states with $N = N_e$ and $S_z = 0$ is given by

$$\sum_{c=0}^{N/2} \frac{N!}{\left(\frac{N}{2} - c\right)!\left(\frac{N}{2} - c\right)!\,c!\,c!}.$$

For example, for $N = N_e = 4$ and $S_z = 0$ we find 36.

Problem 6. Consider the problem of randomly placing r balls into n cells. The balls are indistinguishable. Let r_k be the number of balls in the k-th cell. Every n-tuple of integers satisfying

$$r_1 + r_2 + \cdots + r_n = r \tag{1}$$

describes a possible configuration of occupancy numbers. With indistinguishable balls two distributions are distinguishable only if the corresponding n-tuples are not identical.

(i) Show that the number of distinguishable distributions (i.e., the number of different solutions of (1)) is

$$A_{r,n} = \binom{n+r-1}{r} \equiv \frac{(n+r-1)!}{r!\,(n-1)!}.$$

(ii) Show that the number of distinguishable distributions, in which no cell remains empty, is

$$\binom{r-1}{n-1}.$$

Solution 6. (i) We represent the balls by stars and indicate the n cells by the n spaces between $n+1$ bars. Thus, for example,

$$|{\ast}{\ast}{\ast}|{\ast}|\,|\,|\,|{\ast}{\ast}{\ast}{\ast}|$$

is used as a symbol for a distribution of $r = 8$ balls in $n = 6$ cells with occupancy number $3, 1, 0, 0, 0, 4$. Such a symbol necessarily starts and ends with a bar, but the remaining $n - 1$ bars and r stars can appear in an arbitrary order. Thus the number of distinguishable distributions equals the number of ways of selecting r places out of $n + r - 1$, namely $A_{r,n}$.

(ii) The condition that no cell be empty imposes the restriction that no two bars are adjacent and $r \geq n$. The r stars leave $r - 1$ spaces of which $n - 1$ are to be occupied by bars. Thus we have

$$\binom{r-1}{n-1} \equiv \frac{(r-1)!}{(n-1)!(r-n)!}$$

choices.

Problem 7.   (i) A numerical partition of a positive integer n is a sequence

$$p_1 \geq p_2 \geq \cdots \geq p_k \geq 1$$

such that $p_1 + p_2 + \cdots + p_k = n$. Each p_j is called a part. For example, $18 = 7 + 4 + 4 + 1 + 1 + 1$ is a partition of 18 into 6 parts. The number of partitions of n into k parts is denoted by $p(n, k)$. Find $p(7, 3)$.

(ii) Show that the recurrence for $p(n, k)$ is given by

$$p(n, k) = p(n - 1, k - 1) + p(n - k, k)$$

with the initial conditions $p(n, 0) = 0$, $p(k, k) = 1$. Obviously, $p(n, 1) = 1$.

(iii) Every numerical partition of a positive integer n corresponds to a unique *Ferrer's diagram*. A Ferrer's diagram of a partition is an arrangement of n dots on a square grid, where a part j in the partition is represented by placing p_j dots in a row. This means we represent each term of the partition by a row of dots, the terms in descending order with the largest at the top. Sometimes it is more convenient to use squares instead of dots (in this case the diagram is called a *Young diagram*). The partition we obtain by reading the Ferrer's diagram by columns instead of rows is called the conjugate of the original partition. Find the conjugate of the partition $18 = 7 + 4 + 4 + 1 + 1 + 1$.

Solution 7.   (i) We have

$$7 = 5 + 1 + 1 = 4 + 2 + 1 = 3 + 3 + 1 = 3 + 2 + 2\,.$$

Thus $p(7, 3) = 4$.

(ii) Consider the cases $k = 1$ and $k > 1$. For $k = 1$ we have $p(n, 1) = 1$. For $k > 1$ we use induction.

(iii) We have $18 = 6 + 3 + 3 + 3 + 1 + 1 + 1$.

Chapter 30

Fermi Operators

For *Fermi operators* we have to take into account the *Pauli principle*. Let $c_j^\dagger$ be Fermi creation operators and let c_j be Fermi annihilation operators, where $j = 1, \dots, N$, with j denoting the quantum numbers (spin, wave vector, angular momentum, lattice site, etc.). Then we have

$$[c_i^\dagger, c_j]_+ \equiv c_i^\dagger c_j + c_j c_i^\dagger = \delta_{ij} I$$

$$[c_i, c_j]_+ = [c_i^\dagger, c_j^\dagger]_+ = 0$$

where δ_{ij} denotes the Kronecker delta and I is the identity operator. A consequence of these equations is

$$c_j^\dagger c_j^\dagger = 0$$

which describes the Pauli principle, i.e., two particles cannot be in the same state. The states are given by

$$|n_1, n_2, \dots, n_N\rangle := (c_1^\dagger)^{n_1} (c_2^\dagger)^{n_2} \cdots (c_N^\dagger)^{n_N} |0, \dots, 0, \dots, 0\rangle$$

where, because of the Pauli principle, $n_1, n_2, \dots, n_N \in \{0, 1\}$. We obtain

$$c_i |n_1, \dots, n_{i-1}, 0, n_{i+1}, \dots, n_N\rangle = 0$$

and

$$c_i^\dagger |n_1, \dots, n_{i-1}, 1, n_{i+1}, \dots, n_N\rangle = 0.$$

In the following we write $|0\rangle \equiv |0, \dots, 0\rangle \equiv |0\rangle \otimes \cdots \otimes |0\rangle$.

Problem 1. (i) Let $N = 1$. Find the matrix representation of the operators

$$c^\dagger, \quad c, \quad \hat{n}$$

and the states

$$|0\rangle, \quad c^\dagger|0\rangle$$

where $\hat{n} := c^\dagger c$ is the *number operator*.

(ii) Find the matrix representations for arbitrary N.

Solution 1. (i) A basis is given by

$$\{\, c^\dagger|0\rangle, \quad |0\rangle \,\}.$$

Then the dual basis is

$$\{\, \langle 0|c, \quad \langle 0| \,\}.$$

We obtain

$$c^\dagger \rightarrow \begin{pmatrix} \langle 0|cc^\dagger c^\dagger|0\rangle & \langle 0|cc^\dagger|0\rangle \\ \langle 0|c^\dagger c^\dagger|0\rangle & \langle 0|c^\dagger|0\rangle \end{pmatrix}.$$

Notice that

$$c|0\rangle = 0 \quad \Leftrightarrow \quad \langle 0|c^\dagger = 0.$$

Therefore

$$c^\dagger \rightarrow \begin{pmatrix} 0 & 1 \\ 0 & 0 \end{pmatrix} = \frac{1}{2}\sigma_+ = \frac{1}{2}(\sigma_x + i\sigma_y).$$

In an analogous manner we find

$$c \rightarrow \begin{pmatrix} 0 & 0 \\ 1 & 0 \end{pmatrix} = \frac{1}{2}\sigma_- \equiv \frac{1}{2}(\sigma_x - i\sigma_y)$$

and

$$c^\dagger c = \hat{n} \rightarrow \begin{pmatrix} 1 & 0 \\ 0 & 0 \end{pmatrix}.$$

The eigenvalues of $\hat{n}$ are $\{0, 1\}$. Next we give the matrix representations of $|0\rangle$ and $c^\dagger|0\rangle$. We find

$$|0\rangle \rightarrow \begin{pmatrix} \langle 0|c|0\rangle \\ \langle 0|0\rangle \end{pmatrix} = \begin{pmatrix} 0 \\ 1 \end{pmatrix}, \quad c^\dagger|0\rangle \rightarrow \begin{pmatrix} \langle 0|cc^\dagger|0\rangle \\ \langle 0|c^\dagger|0\rangle \end{pmatrix} = \begin{pmatrix} 1 \\ 0 \end{pmatrix}.$$

As an example we have

$$c|0\rangle = 0 \Leftrightarrow \begin{pmatrix} 0 & 0 \\ 1 & 0 \end{pmatrix}\begin{pmatrix} 0 \\ 1 \end{pmatrix} = \begin{pmatrix} 0 \\ 0 \end{pmatrix}, \quad c^\dagger|0\rangle \Leftrightarrow \begin{pmatrix} 0 & 1 \\ 0 & 0 \end{pmatrix}\begin{pmatrix} 0 \\ 1 \end{pmatrix} = \begin{pmatrix} 1 \\ 0 \end{pmatrix}.$$

(ii) For arbitrary N we have

$$c_k^\dagger \to \overbrace{\sigma_z \otimes \cdots \otimes \sigma_z \otimes \left(\frac{1}{2}\sigma_+\right) \otimes I_2 \otimes \cdots \otimes I_2}^{N\times}$$

where $\sigma_+/2$ is at the k-th place and

$$c_k \to \overbrace{\sigma_z \otimes \cdots \otimes \sigma_z \otimes \left(\frac{1}{2}\sigma_-\right) \otimes I_2 \otimes \cdots \otimes I_2}^{N\times}$$

where $\sigma_-/2$ is at the k-th place. Recall that

$$[c_k^\dagger, c_q]_+ = \delta_{kq} I, \qquad [c_k^\dagger, c_q^\dagger]_+ = [c_k, c_q]_+ = 0.$$

The commutation relations are satisfied by the faithful representations. For $N = 1$ the state $|0\rangle$ is given by

$$|0\rangle \to \begin{pmatrix} 0 \\ 1 \end{pmatrix}.$$

For arbitrary N the state $|0\rangle \equiv |0, \ldots, 0\rangle$ is given by

$$|\mathbf{0}\rangle = \underbrace{\begin{pmatrix} 0 \\ 1 \end{pmatrix} \otimes \begin{pmatrix} 0 \\ 1 \end{pmatrix} \otimes \begin{pmatrix} 0 \\ 1 \end{pmatrix} \otimes \cdots \otimes \begin{pmatrix} 0 \\ 1 \end{pmatrix}}_{N\times}$$

where $|0\rangle$ is called the *vacuum state*. The operator $\hat{n}_k$ takes the form

$$\hat{n}_k := c_k^\dagger c_k \to I_2 \otimes I_2 \otimes \cdots I_2 \otimes \left(\frac{1}{4}\sigma_+\sigma_-\right) \otimes I_2 \otimes \cdots \otimes I_2.$$

Problem 2. (i) Let $c^\dagger$ be a Fermi creation operator and c a Fermi annihilation operator. Let $\epsilon \in \mathbb{R}$. Show that

$$\exp(\epsilon c^\dagger c) c^\dagger \exp(-\epsilon c^\dagger c) = \exp(\epsilon) c^\dagger, \qquad \exp(\epsilon c^\dagger c) c \exp(-\epsilon c^\dagger c) = \exp(-\epsilon) c.$$

(ii) Let $\hat{n} := c^\dagger c$ be the number operator. Show that

$$\exp(-\epsilon \hat{n}) \equiv I + \hat{n}(e^{-\epsilon} - 1).$$

where I is the identity operator.

(iii) Let

$$\hat{n}_\uparrow \hat{n}_\downarrow := c_\uparrow^\dagger c_\uparrow c_\downarrow^\dagger c_\downarrow.$$

Show that

$$\exp(-\epsilon \hat{n}_\downarrow \hat{n}_\uparrow) \equiv I + \hat{n}_\uparrow \hat{n}_\downarrow (e^{-\epsilon} - 1).$$

Solution 2.   (i) We set

$$f(\epsilon) := \exp(\epsilon c^\dagger c) c^\dagger \exp(-\epsilon c^\dagger c)$$

and therefore $f(0) = c^\dagger$. Differentiating both sides with respect to ϵ yields

$$\frac{df}{d\epsilon} = \exp(\epsilon c^\dagger c) c^\dagger c c^\dagger \exp(-\epsilon c^\dagger c) - \exp(\epsilon c^\dagger c) c^\dagger c^\dagger c \exp(-\epsilon c^\dagger c).$$

Since $c^\dagger c^\dagger = 0$ and $cc^\dagger = I - c^\dagger c$ we obtain

$$\frac{df}{d\epsilon} = \exp(\epsilon c^\dagger c) c^\dagger \exp(-\epsilon c^\dagger c) = f$$

or $df/d\epsilon = f$. The solution of this linear differential equation with the initial condition $f(0) = c^\dagger$ gives

$$f(\epsilon) = c^\dagger e^\epsilon.$$

Analogously, we can prove the second identity.
(ii) We find the identity

$$\hat{n}^2 \equiv c^\dagger c c^\dagger c \equiv c^\dagger (I - c^\dagger c) c \equiv c^\dagger c \equiv \hat{n}$$

since $cc = 0$. Using this identity and

$$\exp(-\epsilon \hat{n}) := \sum_{k=0}^{\infty} \frac{(-\epsilon)^k \hat{n}^k}{k!}$$

we find the identity.
(iii) We have the identity

$$(\hat{n}_\uparrow \hat{n}_\downarrow)^2 \equiv \hat{n}_\uparrow \hat{n}_\downarrow \hat{n}_\uparrow \hat{n}_\downarrow \equiv \hat{n}_\uparrow \hat{n}_\uparrow \hat{n}_\downarrow \hat{n}_\downarrow \equiv \hat{n}_\uparrow \hat{n}_\downarrow.$$

Using this identity and

$$\exp(-\epsilon \hat{n}_\uparrow \hat{n}_\downarrow) := \sum_{k=0}^{\infty} \frac{(-\epsilon)^k (\hat{n}_\uparrow \hat{n}_\downarrow)^k}{k!}$$

we find the identity.

Problem 3.   (i) Let $c^\dagger$, c be Fermi creation and annihilation operators. Let $\gamma \in \mathbb{C}$. Calculate

$$(\gamma c^\dagger - \gamma^* c)^2$$

(ii) Find the operator

$$\exp(\gamma c^\dagger - \gamma^* c)$$

(iii) Find the state

$$\exp(\gamma c^\dagger - \gamma^* c)|0\rangle .$$

Solution 3. (i) Since $c^\dagger c^\dagger = 0$ and $cc = 0$ we have

$$(\gamma c^\dagger - \gamma^* c)(\gamma c^\dagger - \gamma^* c) = \gamma^2 c^\dagger c^\dagger - \gamma\gamma^* cc^\dagger - \gamma\gamma^* c^\dagger c + \gamma^* \gamma^* cc$$
$$= -\gamma\gamma^*(c^\dagger c + cc^\dagger) = -\gamma\gamma^* I .$$

(ii) Using the result from (i) we find

$$\exp(\gamma c^\dagger - \gamma^* c) = I \cos(\sqrt{\gamma\gamma^*}) + \frac{\gamma c^\dagger - \gamma^* c}{\sqrt{\gamma\gamma^*}} \sin(\sqrt{\gamma\gamma^*}) .$$

Note that if $\gamma\gamma^* = 0$, then $\exp(\gamma c^\dagger - \gamma^* c) = I$.
(iii) Using the result from (ii) and $c|0\rangle = 0$ we find

$$\exp(\gamma c^\dagger - \gamma^* c)|0\rangle = \cos(\sqrt{\gamma\gamma^*})|0\rangle + \frac{\gamma}{\sqrt{\gamma\gamma^*}} \sin(\sqrt{\gamma\gamma^*})c^\dagger|0\rangle .$$

Problem 4. Fermi creation and annihilation operators $c_j^\dagger$, c_j ($j = 0, 1, \ldots, N-1$) obey the anticommutation relations

$$[c_j, c_k]_+ = 0, \qquad [c_j^\dagger, c_k]_+ = \delta_{jk} I$$

where I is the identity operator. A basis for the Hilbert space is given by

$$\prod_{j=0}^{N-1} (c_j^\dagger)^{r_j}|0\rangle, \qquad r_j = 0, 1$$

and the vaccum state $|0\rangle$ is defined by $c_j|0\rangle = 0$, $j = 0, 1, \ldots, N - 1$. Consider the operators

$$\rho_j = \frac{1 - \nu_j}{2} c_j^\dagger c_j + \frac{1 + \nu_j}{2} c_j c_j^\dagger, \qquad \nu_j \in [-1, 1]$$

Show that the ρ_j's are density matrices. Use the matrix representation for $c_j^\dagger$ and c_j

$$c_j^\dagger = \sigma_z \otimes \sigma_z \otimes \cdots \otimes \sigma_z \otimes \left(\frac{1}{2}\sigma_+\right) \otimes I_2 \otimes I_2 \otimes \cdots \otimes I_2$$

$$c_j = \sigma_z \otimes \sigma_z \otimes \cdots \otimes \sigma_z \otimes \left(\frac{1}{2}\sigma_-\right) \otimes I_2 \otimes I_2 \otimes \cdots \otimes I_2$$

where σ_+ and σ_- are at the j-th position ($j = 0, 1, \ldots, N-1$) and

$$\sigma_+ := \sigma_x + i\sigma_y = \begin{pmatrix} 0 & 2 \\ 0 & 0 \end{pmatrix}, \qquad \sigma_- := \sigma_x - i\sigma_y = \begin{pmatrix} 0 & 0 \\ 2 & 0 \end{pmatrix}.$$

The vacuum state is $|0\rangle = |0\rangle \otimes |0\rangle \otimes \cdots \otimes |0\rangle$ with

$$|0\rangle = \begin{pmatrix} 0 \\ 1 \end{pmatrix}.$$

Solution 4.   Since $\sigma_z^2 = I_2$ we have

$$c_j^\dagger c_j = I_2 \otimes I_2 \otimes \cdots \otimes I_2 \otimes \begin{pmatrix} 1 & 0 \\ 0 & 0 \end{pmatrix} \otimes I_2 \otimes \cdots \otimes I_2$$

and

$$c_j c_j^\dagger = I_2 \otimes I_2 \otimes \cdots \otimes I_2 \otimes \begin{pmatrix} 0 & 0 \\ 0 & 1 \end{pmatrix} \otimes I_2 \otimes \cdots \otimes I_2.$$

Thus $c_j^\dagger c_j$ and $c_j c_j^\dagger$ are diagonal matrices and the eigenvalues of $c_j^\dagger c_j$ and $c_j c_j^\dagger$ are 1 and 0. Since $\nu_j \in [-1, 1]$ it follows that $(1-\nu_j)/2$ and $(1+\nu_j)/2$ cannot be negative. Thus the eigenvalues of ρ_j cannot be negative. Furthermore $\text{tr}\rho_j = 1$.

Problem 5.   Consider the Fermi creation and annihilation operators $c_j^\dagger$, $c_\ell^\dagger$, c_k, c_m, where $j, \ell, k, m = 1, 2, \ldots, N$. Find the commutator

$$[c_j^\dagger c_k, c_\ell^\dagger c_m].$$

Does the set $\{ c_j^\dagger c_k \}$ ($j, k = 1, 2, \ldots, N$) forms a Lie algebra under the commutator?

Solution 5.   We have

$$[c_j^\dagger c_k, c_\ell^\dagger c_m] = -\delta_{mj} c_\ell^\dagger c_k + \delta_{\ell k} c_j^\dagger c_m.$$

Thus the set $\{ c_j^\dagger c_k \}$ ($j, k = 1, 2, \ldots, N$) forms a Lie algebra under the commutator.

Problem 6.   Let $c_j^\dagger$, c_j ($j = 1, 2, 3$) be Fermi creation and annihilation operators. Consider the Hamilton operator

$$\hat{H} = t(c_1^\dagger c_2 + c_2^\dagger c_1 + c_2^\dagger c_3 + c_3^\dagger c_2 + c_1^\dagger c_3 + c_3^\dagger c_1) + k_1 c_1^\dagger c_1 + k_2 c_2^\dagger c_2 + k_3 c_3^\dagger c_3$$

and the *number operator*

$$\hat{N} = c_1^\dagger c_1 + c_2^\dagger c_2 + c_3^\dagger c_3 \, .$$

(i) Calculate the commutator $[\hat{H}, \hat{N}]$ and give an interpretation of the result.
(ii) Given a basis with two Fermi particles

$$c_1^\dagger c_2^\dagger |0\rangle, \quad c_1^\dagger c_3^\dagger |0\rangle, \quad c_2^\dagger c_3^\dagger |0\rangle \, .$$

Find the matrix representation of $\hat{H}$ and $\hat{N}$.
(iii) Given a basis with one Fermi particle

$$c_1^\dagger |0\rangle, \quad c_2^\dagger |0\rangle, \quad c_3^\dagger |0\rangle \, .$$

Find the matrix representation of $\hat{H}$.

Solution 6. (i) We have

$$[\hat{H}, c_1^\dagger c_1] = t(c_2^\dagger c_1 - c_1^\dagger c_2 + c_3^\dagger c_1 - c_1^\dagger c_3)$$
$$[\hat{H}, c_2^\dagger c_2] = t(c_3^\dagger c_2 - c_2^\dagger c_3 + c_1^\dagger c_2 - c_2^\dagger c_1)$$
$$[\hat{H}, c_3^\dagger c_3] = t(c_1^\dagger c_3 - c_3^\dagger c_1 + c_2^\dagger c_3 - c_3^\dagger c_2) \, .$$

It follows that

$$[\hat{H}, \hat{N}] = 0 \, .$$

$\hat{N}$ is the number operator. From the result that $[\hat{H}, \hat{N}] = 0$ we find $\hat{N}$ is a constant of motion, i.e. the total number of Fermi particles remains constant in the sense that if $|n\rangle$ is an eigenstate of the number operator $\hat{N}$ with eigenvalue n at time 0, then

$$|n\rangle(t) = e^{-i\hat{H}t/\hbar} |n\rangle$$

remains an eigenstate of $\hat{N}$ with eigenvalue n for all times.
(ii) Since

$$\hat{H} c_1^\dagger c_2^\dagger |0\rangle = t(c_1^\dagger c_3^\dagger |0\rangle - c_2^\dagger c_3^\dagger |0\rangle) + (k_1 + k_2) c_1^\dagger c_2^\dagger |0\rangle$$
$$\hat{H} c_1^\dagger c_3^\dagger |0\rangle = t(c_2^\dagger c_3^\dagger |0\rangle + c_1^\dagger c_2^\dagger |0\rangle) + (k_1 + k_3) c_1^\dagger c_3^\dagger |0\rangle$$
$$\hat{H} c_2^\dagger c_3^\dagger |0\rangle = t(c_1^\dagger c_3^\dagger |0\rangle - c_1^\dagger c_2^\dagger |0\rangle) + (k_2 + k_3) c_2^\dagger c_3^\dagger |0\rangle \, .$$

With the dual basis

$$\langle 0 | c_2 c_1, \quad \langle 0 | c_3 c_1, \quad \langle 0 | c_3 c_2 \qquad \bullet$$

the matrix representation of $\hat{H}$ is

$$\hat{H} = \begin{pmatrix} k_1 + k_2 & t & -t \\ t & k_1 + k_3 & t \\ -t & t & k_2 + k_3 \end{pmatrix} \, .$$

Obviously owing to the result of (i) the matrix representation of $\hat{N}$ is

$$\hat{N} = \begin{pmatrix} 2 & 0 & 0 \\ 0 & 2 & 0 \\ 0 & 0 & 2 \end{pmatrix} .$$

(iii) Since

$$\hat{H}c_1^\dagger|\mathbf{0}\rangle = t(c_2^\dagger|\mathbf{0}\rangle + c_3^\dagger|\mathbf{0}\rangle) + k_1 c_1^\dagger|\mathbf{0}\rangle$$
$$\hat{H}c_2^\dagger|\mathbf{0}\rangle = t(c_1^\dagger|\mathbf{0}\rangle + c_3^\dagger|\mathbf{0}\rangle) + k_2 c_2^\dagger|\mathbf{0}\rangle$$
$$\hat{H}c_3^\dagger|\mathbf{0}\rangle = t(c_1^\dagger|\mathbf{0}\rangle + c_2^\dagger|\mathbf{0}\rangle) + k_3 c_3^\dagger|\mathbf{0}\rangle$$

we obtain the matrix representation

$$\hat{H} = \begin{pmatrix} k_1 & t & t \\ t & k_2 & t \\ t & t & k_3 \end{pmatrix} .$$

Problem 7. In some cases it is convenient to consider the Fermi operator

$$\left\{ c_{k\sigma}^\dagger, \ c_{k\sigma}, \ k = 1, 2, \cdots, N; \ \sigma \in \{\uparrow, \downarrow\} \right\}$$

where k denotes the wave vector (or lattice site) and σ denotes the spin. Find the matrix representation of $c_{k\uparrow}^\dagger$ and $c_{k\downarrow}^\dagger$.

Solution 7. We have the commutation relation

$$[c_{k\sigma}^\dagger, c_{q\sigma'}]_+ = \delta_{kq}\delta_{\sigma\sigma'} I$$

$$[c_{k\sigma}, c_{q\sigma'}]_+ = [c_{k\sigma}^\dagger, c_{q\sigma'}^\dagger]_+ = 0.$$

The matrix representation is given by

$$c_{k\uparrow}^\dagger \rightarrow \overbrace{\sigma_z \otimes \sigma_z \otimes \cdots \otimes \sigma_z \otimes \left(\frac{1}{2}\sigma_+\right) \otimes I_2 \otimes \cdots \otimes I_2}^{2N\times}$$

where $\sigma_+/2$ is at the k-th place and

$$c_{k\downarrow}^\dagger \rightarrow \overbrace{\sigma_z \otimes \sigma_z \otimes \cdots \otimes \sigma_z \otimes \left(\frac{1}{2}\sigma_+\right) \otimes I_2 \otimes \cdots \otimes I_2}^{2N\times}$$

where $\sigma_+/2$ is at the $(k+N)$-th place.

Problem 8. Let $c_{j\uparrow}^{\dagger}$, $c_{j\uparrow}$, $c_{j\downarrow}^{\dagger}$, $c_{j\downarrow}$ be Fermi operators, i.e.

$$[c_{j\sigma}^{\dagger}, c_{k\sigma'}]_+ = \delta_{jk}\delta_{\sigma,\sigma'}I$$

$$[c_{j\sigma}^{\dagger}, c_{k\sigma'}^{\dagger}]_+ = 0, \qquad [c_{j\sigma}, c_{k\sigma'}]_+ = 0 \qquad (1)$$

for all $j, k = 1, 2, \ldots, N$. We define the linear operators

$$\xi_{j\uparrow} := i(c_{j\uparrow} - c_{j\uparrow}^{\dagger}), \qquad \xi_{j\downarrow} := i(c_{j\downarrow} - c_{j\downarrow}^{\dagger})(I - 2c_{j\uparrow}^{\dagger}c_{j\uparrow}) \qquad (2a)$$

$$\eta_{j\uparrow} := c_{j\uparrow} + c_{j\uparrow}^{\dagger}, \qquad \eta_{j\downarrow} := (c_{j\downarrow} + c_{j\downarrow}^{\dagger})(I - 2c_{j\uparrow}^{\dagger}c_{j\uparrow}). \qquad (2b)$$

(i) Find the properties of these operators.
(ii) Find the inverse transformation.

Solution 8. (i) By straightforward calculations using the anticommutation relations we find that the operators $\xi_{j\sigma}$ are unitary and self-adjoint and therefore involutions

$$\xi_{j\sigma} = \xi_{j\sigma}^{\dagger} = \xi_{j\sigma}^{-1}, \qquad \xi_{j\sigma}^2 = I$$

for all $j = 1, 2, \ldots, N$. The linear operators $\eta_{j\sigma}$ also satisfy

$$\eta_{j\sigma} = \eta_{j\sigma}^{\dagger} = \eta_{j\sigma}^{-1}, \qquad \eta_{j\sigma}^2 = I$$

for all $j = 1, 2, \ldots, N$.
(ii) The inverse transformation between $c_{j\sigma}^{\dagger}, c_{j\sigma}$ and $\xi_{j\sigma}, \eta_{j\sigma}$ is given by

$$c_{j\uparrow} = \frac{1}{2}(\eta_{j\uparrow} - i\xi_{j\uparrow}), \qquad c_{j\uparrow}^{\dagger} = \frac{1}{2}(\eta_{j\uparrow} + i\xi_{j\uparrow})$$

$$c_{j\downarrow} = \frac{1}{2}(i\eta_{j\downarrow} + \xi_{j\downarrow})\eta_{j\uparrow}\xi_{j\uparrow}, \qquad c_{j\downarrow}^{\dagger} = \frac{1}{2}(i\eta_{j\downarrow} - \xi_{j\downarrow})\eta_{j\uparrow}\xi_{j\uparrow}$$

where we have used that

$$i\eta_{j\uparrow}\xi_{j\uparrow} = I - 2c_{j\uparrow}^{\dagger}c_{j\uparrow}.$$

Problem 9. Let $c_{j\sigma}^{\dagger}$, $c_{j\sigma}$ ($j = 1, \ldots, N$) be Fermi creation and annihilation operators, where $\sigma \in \{\uparrow, \downarrow\}$. We define

$$\hat{n}_{j\uparrow} := c_{j\uparrow}^{\dagger}c_{j\uparrow}, \qquad \hat{n}_{j\downarrow} := c_{j\downarrow}^{\dagger}c_{j\downarrow}.$$

(i) Show that the operators

$$S_{jx} := \frac{1}{2}(c_{j\uparrow}^{\dagger}c_{j\downarrow} + c_{j\downarrow}^{\dagger}c_{j\uparrow}), \quad S_{jy} := \frac{1}{2i}(c_{j\uparrow}^{\dagger}c_{j\downarrow} - c_{j\downarrow}^{\dagger}c_{j\uparrow}), \quad S_{jz} := \frac{1}{2}(\hat{n}_{j\uparrow} - \hat{n}_{j\downarrow})$$

form a basis of a Lie algebra under the commutator.
(ii) Show that the operators

$$R_{jx} := \frac{1}{2}(c_{j\uparrow}^\dagger c_{j\downarrow}^\dagger + c_{j\downarrow} c_{j\uparrow}), \quad R_{jy} := \frac{1}{2i}(c_{j\uparrow}^\dagger c_{j\downarrow}^\dagger - c_{j\downarrow} c_{j\uparrow}), \quad R_{jz} := \frac{1}{2}(\hat{n}_{j\uparrow} + \hat{n}_{j\downarrow} - I)$$

form a Lie algebra under the commutator.
(iii) Prove the identity

$$\hat{n}_{j\uparrow}\hat{n}_{j\downarrow} \equiv \frac{1}{4}(1 - \alpha_j) + R_{jz} + \frac{1}{3}(\alpha_j - 1)\mathbf{S}_j^2 + \frac{1}{3}(\alpha_j + 1)\mathbf{R}_j^2$$

where

$$\mathbf{S}_j^2 := S_{jx}^2 + S_{jy}^2 + S_{jz}^2, \qquad \mathbf{R}_j^2 := R_{jx}^2 + R_{jy}^2 + R_{jz}^2.$$

Solution 9.   (i) Since

$$[c_{j\sigma}^\dagger, c_{j\sigma'}]_+ = \delta_{\sigma\sigma'} I, \quad [c_{j\sigma}^\dagger, c_{j\sigma'}^\dagger]_+ = 0, \quad [c_{j\sigma}, c_{j\sigma'}]_+ = 0$$

we obtain the commutators

$$[S_{jx}, S_{jy}] = iS_{jz}, \quad [S_{jy}, S_{jz}] = iS_{jx}, \quad [S_{jz}, S_{jx}] = iS_{jy}.$$

Thus the operators S_{jx}, S_{jy}, S_{jz} form a basis of a Lie algebra.
(ii) Analogously, we obtain

$$[R_{jx}, R_{jy}] = iR_{jz}, \quad [R_{jy}, R_{jz}] = iR_{jx}, \quad [R_{jz}, R_{jx}] = iR_{jy}.$$

Thus the operators R_{jx}, R_{jy}, R_{jz} form a Lie algebra. The two Lie algebras are isomorphic.
(iii) We find

$$\mathbf{S}_j^2 = \frac{3}{4}(\hat{n}_{j\uparrow} + \hat{n}_{j\downarrow} - 2\hat{n}_{j\uparrow}\hat{n}_{j\downarrow})$$

and

$$\mathbf{R}_j^2 = \frac{3}{2}\hat{n}_{j\uparrow}\hat{n}_{j\downarrow} - \frac{3}{4}(\hat{n}_{j\uparrow} + \hat{n}_{j\downarrow}) + \frac{3}{4}I.$$

Thus identity (1) follows.

Problem 10.   Let

$$n_{j\uparrow} := c_{j\uparrow}^\dagger c_{j\uparrow}, \quad n_{j\downarrow} := c_{j\downarrow}^\dagger c_{j\downarrow}.$$

Consider the Hamilton operator (*Hubbard model*)

$$\hat{H} = t\sum_{i,j}\sum_{\sigma\in\{\uparrow\downarrow\}} c_{i\sigma}^\dagger c_{j\sigma} + U\sum_{j=1}^N n_{j\uparrow}n_{j\downarrow}$$

where t and U is a real constant. Consider the operators

$$\hat{N}_e = \sum_j^N \sum_\sigma c_{j\sigma}^\dagger c_{j\sigma}, \qquad \hat{S}_z = \frac{1}{2} \sum_j^N (c_{j\uparrow}^\dagger c_{j\uparrow} - c_{j\downarrow}^\dagger c_{j\downarrow})$$

where $\hat{N}_e$ is the number operator and $\hat{S}_z$ is the total spin operator in z-direction. Calculate the commutators $[\hat{H}, \hat{N}_e]$ and $[\hat{H}, \hat{S}_z]$.

Solution 10. We find

$$[\hat{H}, \hat{N}_e] = 0, \qquad [\hat{H}, \hat{S}_z] = 0.$$

Thus $\hat{N}_e$ and $\hat{S}_z$ are constants of motion. Thus we can consider subspaces with fixed number of electrons and fixed number of total spin in z-direction.

Problem 11. Let $c_{k\sigma}^\dagger$ and $c_{k\sigma}$ be Fermi creation and annihilation operators, where $\sigma \in \{\uparrow, \downarrow\}$ and $j = 1, 2, \ldots, N$ and k refers to the momentum.. The Fermi operator obeys the anticommuation relations

$$[c_{k\sigma}, c_{p\sigma'}^\dagger]_+ = \delta_{kp}\delta_{\sigma\sigma'}I, \quad [c_{k\sigma}, c_{k\sigma'}]_+ = 0$$

where I is the identity operator and 0 the zero operator. From the second relation it follows that

$$[c_{k\sigma}^\dagger, c_{p\sigma'}^\dagger]_+ = 0.$$

Let

$$n_{j\uparrow} := c_{k\uparrow}^\dagger c_{k\uparrow}, \qquad n_{j\downarrow} := c_{k\downarrow}^\dagger c_{k\downarrow}.$$

Consider the four point Hamilton operator (*Hubbard model*) in *Bloch representation*

$$\hat{H} = \sum_k \sum_{\sigma \in \{\uparrow\downarrow\}} \epsilon(k) c_{k\sigma}^\dagger c_{k\sigma} + \frac{U}{4} \sum_{k_1, k_2, k_3, k_4} \delta(k_1 - k_2 + k_3 - k_4) c_{k_1\uparrow}^\dagger c_{k_2\uparrow} c_{k_3\downarrow}^\dagger c_{k_4\downarrow}$$

where $\epsilon(k) = 2t\cos(k)$. Here

$$k, k_1, k_2, k_3 \in S = \left\{ -\frac{\pi}{2},\ 0,\ \frac{\pi}{2},\ \pi \quad \mod 2\pi \right\}$$

and

$$\delta(k_1 - k_2 + k_3 - k_4) = \begin{cases} 1 & \text{if } k_1 - k_2 + k_3 - k_4 = 0 \mod 2\pi \\ 0 & \text{otherwise} \end{cases}$$

Consider the operators

$$\hat{N}_e = \sum_k \sum_\sigma c_{k\sigma}^\dagger c_{j\sigma}$$

$$\hat{S}_z = \frac{1}{2} \sum_k (c_{k\uparrow}^\dagger c_{k\uparrow} - c_{k\downarrow}^\dagger c_{k\downarrow})$$

$$\hat{P} = \sum_k k(c_{k\uparrow}^\dagger c_{k\uparrow} + c_{k\downarrow}^\dagger c_{k\downarrow})$$

where $\hat{N}_e$ is the number operator, $\hat{S}_z$ is the total spin operator in z-direction and $\hat{P}$ the total momentum operator and k runs over the set S given above.
(i) Calculate the commutators $[\hat{H}, \hat{N}_e]$, $[\hat{H}, \hat{S}_z]$, $[hatH, \hat{P}]$.
(ii) Find the eigenvalues of the total momentum operator $\hat{P}$.

Solution 11. (i) We find

$$[\hat{H}, \hat{N}_e] = 0, \quad [\hat{H}, \hat{S}_z] = 0, \quad [\hat{H}, \hat{P}] = 0.$$

Thus $\hat{N}_e$, $\hat{S}_z$ and $\hat{P}$ are constants of motion. Thus we can consider subspaces with fixed number of electrons, fixed number of total spin in z-direction and fixed total momentum.
(ii) Obviously we find $-\pi/2, 0, \pi/2, \pi$.

Problem 12. Consider the four point Hubbard model described in Bloch representation

$$\hat{H} = \sum_k \sum_{\sigma \in \{\uparrow\downarrow\}} \epsilon(k) c_{k\sigma}^\dagger c_{k\sigma} + \frac{U}{4} \sum_{k_1, k_2, k_3, k_4} \delta(k_1 - k_2 + k_3 - k_4) c_{k_1\uparrow}^\dagger c_{k_2\uparrow} c_{k_3\downarrow}^\dagger c_{k_4\downarrow}$$

where $\epsilon(k) = 2t \cos(k)$. Here

$$k, k_1, k_2, k_3 \in \{ -\frac{\pi}{2}, 0, \frac{\pi}{2}, \pi \mod 2\pi \}$$

Give a basis of the Hilbert space for $N_e = 4$, $S_z = 0$ and $P = 0$.

Solution 12. Owing to $N_e = 4$, $S_z = 0$ and the Pauli principle we have the states

$$c_{k_1\uparrow}^\dagger c_{k_2\uparrow}^\dagger c_{k_3\downarrow}^\dagger c_{k_4\downarrow}^\dagger |0\rangle, \quad k_1 < k_2, \quad k_3 < k_4, \quad k_1 + k_2 + k_3 + k_4 = 0 \mod 2\pi \}.$$

The dimensions of the subspaces with $P = -\pi/2, 0, \pi/2, \pi$ are $8, 10, 8, 10$. We obtain the ten states

$$c_{-\pi/2\uparrow}^\dagger c_{0\uparrow}^\dagger c_{-\pi/2\downarrow}^\dagger c_{\pi\downarrow}^\dagger |0\rangle, \quad c_{-\pi/2\uparrow}^\dagger c_{0\uparrow}^\dagger c_{0\downarrow}^\dagger c_{\pi/2\downarrow}^\dagger |0\rangle$$

$$c_{-\pi/2\uparrow}^\dagger c_{\pi/2\uparrow}^\dagger c_{-\pi/2\downarrow}^\dagger c_{\pi/2\downarrow}^\dagger |0\rangle, \quad c_{-\pi/2\uparrow}^\dagger c_{\pi\uparrow}^\dagger c_{-\pi/2\downarrow}^\dagger c_{0\downarrow}^\dagger |0\rangle$$

$$c_{-\pi/2\uparrow}^\dagger c_{\pi\uparrow}^\dagger c_{\pi/2\downarrow}^\dagger c_{\pi\downarrow}^\dagger |0\rangle, \quad c_{0\uparrow}^\dagger c_{\pi/2\uparrow}^\dagger c_{-\pi/2\downarrow}^\dagger c_{0\downarrow}^\dagger |0\rangle$$

$$c_{0\uparrow}^{\dagger} c_{\pi/2\uparrow}^{\dagger} c_{\pi/2\downarrow}^{\dagger} c_{\pi\downarrow}^{\dagger} |0\rangle, \quad c_{0\uparrow}^{\dagger} c_{\pi\uparrow}^{\dagger} c_{0\downarrow}^{\dagger} c_{\pi\downarrow}^{\dagger} |0\rangle$$

$$c_{\pi/2\uparrow}^{\dagger} c_{\pi\uparrow}^{\dagger} c_{-\pi/2\downarrow}^{\dagger} c_{\pi\downarrow}^{\dagger} |0\rangle, \quad c_{\pi/2\uparrow}^{\dagger} c_{\pi\uparrow}^{\dagger} c_{0\downarrow}^{\dagger} c_{\pi/2\downarrow}^{\dagger} |0\rangle.$$

Problem 13. Consider the Hamilton operator (two-point *Hubbard model*)

$$\hat{H} = t(c_{1\uparrow}^{\dagger} c_{2\uparrow} + c_{1\downarrow}^{\dagger} c_{2\downarrow} + c_{2\uparrow}^{\dagger} c_{1\uparrow} + c_{2\downarrow}^{\dagger} c_{1\downarrow}) + U(n_{1\uparrow}n_{1\downarrow} + n_{2\uparrow}n_{2\downarrow}) \quad (1)$$

where $n_{j\uparrow} := c_{j\uparrow}^{\dagger} c_{j\uparrow}$, $n_{j\downarrow} := c_{j\downarrow}^{\dagger} c_{j\downarrow}$. The operators $c_{j\uparrow}^{\dagger}, c_{j\downarrow}^{\dagger}, c_{j\uparrow}, c_{j\downarrow}$ are Fermi operators.

(i) Show that the Hubbard Hamilton operator (1) commutes with the total number operator $\hat{N}$ and the total spin operator $\hat{S}_z$ where

$$\hat{N} := \sum_{j=1}^{2} (c_{j\uparrow}^{\dagger} c_{j\uparrow} + c_{j\downarrow}^{\dagger} c_{j\downarrow}), \qquad \hat{S}_z := \frac{1}{2} \sum_{j=1}^{2} (c_{j\uparrow}^{\dagger} c_{j\uparrow} - c_{j\downarrow}^{\dagger} c_{j\downarrow}).$$

(ii) We consider the subspace with two particles $N = 2$ and total spin $S_z = 0$. A basis in this space is given by

$$c_{1\uparrow}^{\dagger} c_{1\downarrow}^{\dagger} |0\rangle, \quad c_{1\uparrow}^{\dagger} c_{2\downarrow}^{\dagger} |0\rangle, \quad c_{2\uparrow}^{\dagger} c_{1\downarrow}^{\dagger} |0\rangle, \quad c_{2\uparrow}^{\dagger} c_{2\downarrow}^{\dagger} |0\rangle.$$

Find the matrix representation of $\hat{H}$ for this basis.
(iii) Find the discrete symmetries of $\hat{H}$ and perform a group-theoretical reduction.

Solution 13. (i) Using the Fermi anti-commutation relations we obtain $[\hat{H}, \hat{N}] = 0$, $[\hat{H}, \hat{S}_z] = 0$. We also have $[\hat{N}, \hat{S}_z] = 0$.
(ii) Using the Fermi anti-commutation relations and $c_{j\uparrow}|0\rangle = 0$, $c_{j\downarrow}|0\rangle = 0$ we obtain the matrix representation of $\hat{H}$ with the given basis

$$\begin{pmatrix} U & t & t & 0 \\ t & 0 & 0 & t \\ t & 0 & 0 & t \\ 0 & t & t & U \end{pmatrix}.$$

(iii) The Hamilton operator (1) admits the symmetry $1 \to 2$, $2 \to 1$, i.e. swapping the sites 1 and 2 leaves the Hamilton operator invariant. Thus we have a finite group and two elements (identity and the swapping of the sites). There are two conjugacy classes and therefore two irreducible representations. We find the two invariant subspaces

$$\left\{ \frac{1}{\sqrt{2}}(c_{1\downarrow}^{\dagger} c_{1\uparrow}^{\dagger}|0\rangle + c_{2\downarrow}^{\dagger} c_{2\uparrow}^{\dagger}|0\rangle), \quad \frac{1}{\sqrt{2}}(c_{1\downarrow}^{\dagger} c_{2\uparrow}^{\dagger}|0\rangle + c_{2\downarrow}^{\dagger} c_{1\uparrow}^{\dagger}|0\rangle) \right\}$$

$$\left\{ \frac{1}{\sqrt{2}}(c_{1\downarrow}^{\dagger} c_{1\uparrow}^{\dagger}|0\rangle - c_{2\downarrow}^{\dagger} c_{2\uparrow}^{\dagger}|0\rangle), \quad \frac{1}{\sqrt{2}}(c_{1\downarrow}^{\dagger} c_{2\uparrow}^{\dagger}|0\rangle - c_{2\downarrow}^{\dagger} c_{1\uparrow}^{\dagger}|0\rangle) \right\}.$$

Problem 14.   Let

$$\hat{H} = \hat{H}_0 + \hat{H}_1$$

be a Hamilton operator describing a many-body Fermi or Bose system. The *grand thermodynamic potential* is given by

$$\Omega(\beta) := -\frac{1}{\beta} \ln \operatorname{tr} \exp(-\beta(\hat{H} - \mu\hat{N})) = -\frac{1}{\beta} \ln Z$$

where μ is the chemical potential, $\hat{N}$ is the number operator and Z is the grand thermodynamic partition function. Assume that the grand thermodynamic potential for the unperturbed Hamilton operator $\hat{H}_0$

$$\Omega_0(\beta) := -\frac{1}{\beta} \ln \operatorname{tr} \exp(-\beta(\hat{H}_0 - \mu\hat{N}))$$

can be calculated.
(i) Let

$$\exp(-\beta(\hat{H} - \mu\hat{N})) \equiv (\exp(-\beta(\hat{H}_0 - \mu\hat{N})))S(\beta).$$

Find $S(\beta)$.
(ii) Calculate

$$Z(\beta) = \operatorname{tr}(\exp(-\beta(\hat{H}_0 - \mu\hat{N}))S(\beta)). \tag{1}$$

(iii) From (i) we find

$$S(\beta) = 1 + \sum_{n=1}^{\infty} (-1)^n \int_0^\beta d\tau_1 \int_0^{\tau_1} d\tau_2 \cdots \int_0^{\tau_{n-1}} d\tau_n \hat{H}_1(\tau_1)\hat{H}_1(\tau_2)\cdots\hat{H}_1(\tau_n) \tag{2}$$

where

$$\hat{H}_1(\beta) := \exp(\beta(\hat{H}_0 - \mu\hat{N}))\hat{H}_1 \exp(-\beta(\hat{H}_0 - \mu\hat{N}))$$

(the so-called *interaction picture*). It can be shown that $S(\beta)$ can be written as

$$S(\beta) = \sum_{n=0}^{\infty} \frac{(-1)^n}{n!} \int_0^\beta d\tau_1 \int_0^\beta d\tau_2 \cdots \int_0^\beta d\tau_n T_\tau[\hat{H}_1(\tau_1)\hat{H}_1(\tau_2)\cdots\hat{H}_1(\tau_n)]$$

where T_τ is the *time-ordering operator*. We have

$$T_\tau[A(\tau_1)B(\tau_2)] := \begin{cases} A(\tau_1)B(\tau_2), & \text{if } \tau_2 < \tau_1 \\ B(\tau_2)A(\tau_1), & \text{if } \tau_1 < \tau_2 \end{cases}.$$

We set

$$\langle T_\tau \exp(-\int_0^\beta \hat{H}_1(\tau)d\tau)\rangle_0 \equiv \exp(\langle T_\tau \exp(-\int_0^\beta \hat{H}_1(\tau)d\tau) - 1\rangle_c). \qquad (3)$$

Express $\langle \ldots \rangle_c$ in terms of $\langle \ldots \rangle_0$. This is the so-called *cumulant expansion*. The expansions given by (1) and (2) cannot be used for calculating Ω. The cumulant expansion must be used.

Solution 14. (i) We set

$$\exp(-\beta(\hat{H} - \mu\hat{N})) = (\exp(-\beta(\hat{H}_0 - \mu\hat{N})))S(\beta) =: \phi(\beta).$$

Taking the derivative of this equation with respect to β gives

$$\frac{\partial\phi}{\partial\beta} = -(\hat{H} - \mu\hat{N})\phi.$$

Therefore

$$\frac{\partial S}{\partial\beta} = -\hat{H}_1(\beta)S.$$

Obviously we have the "initial condition" $S(\beta = 0) \equiv S(0) = 1$. The integration of this equation with the initial condition gives

$$S(\beta) = 1 - \int_0^\beta \hat{H}_1(\tau)S(\tau)d\tau.$$

By iterating this equation we arrive at

$$S(\beta) = 1 - \int_0^\beta \hat{H}_1(\tau)d\tau + \int_0^\beta d\tau_1 \int_0^{\tau_1} d\tau_2 \hat{H}_1(\tau_1)\hat{H}_1(\tau_2) + \cdots$$

$$+ (-1)^n \int_0^\beta d\tau_1 \int_0^{\tau_1} d\tau_2 \cdots \int_0^{\tau_{n-1}} d\tau_n \hat{H}_1(\tau_1)\hat{H}_1(\tau_2) \cdots \hat{H}_1(\tau_n) + \cdots .$$

(ii) For the grand thermodynamical partition function Z we obtain

$$Z(\beta) = \text{tr}(\exp(-\beta(\hat{H}_0 - \mu\hat{N}))S(\beta))$$

$$\equiv \frac{[\text{tr}\exp(-\beta(\hat{H}_0 - \mu\hat{N}))][\text{tr}(\exp(-\beta(\hat{H}_0 - \mu\hat{N}))S(\beta))]}{\text{tr}\exp(-\beta(\hat{H}_0 - \mu\hat{N}))}.$$

Therefore we find

$$Z(\beta) = e^{-\beta\Omega_0}\left(1 - \int_0^\beta \langle\hat{H}_1(\tau)\rangle_0 d\tau + \int_0^\beta d\tau_1 \int_0^{\tau_1} d\tau_2 \langle\hat{H}_1(\tau_1)\hat{H}_1(\tau_2)\rangle_0 + \cdots\right)$$

where

$$\langle\cdots\rangle_0 := \frac{\text{tr}\cdots e^{-\beta(\hat{H}_0 - \mu\hat{N})}}{\text{tr}e^{-\beta(\hat{H}_0 - \mu\hat{N})}}.$$

Taking the logarithm of both sides of this equation we find

$$\Omega = \Omega_0 - \frac{1}{\beta} \ln \left(1 - \int_0^\beta \langle \hat{H}_1(\tau) \rangle_0 d\tau + \int_0^\beta d\tau_1 \int_0^{\tau_1} d\tau_2 \langle \hat{H}_1(\tau_1) \hat{H}(\tau_2) \rangle_0 - \cdots \right).$$

(iii) Taking the logarithm of both sides of (3) with $\hat{H}_1 \rightarrow \lambda \hat{H}_1$ it follows that we can set

$$\phi(\lambda) = \sum_{n=1}^\infty \frac{(-1)^n \lambda^n}{n!} \int_0^\beta d\tau_1 \cdots \int_0^\beta d\tau_n \langle T_\tau \hat{H}_1(\tau_1) \cdots \hat{H}_1(\tau_n) \rangle_c$$

and

$$\phi(\lambda) = \ln \langle T_\tau \exp(-\lambda \int_0^\beta \hat{H}_1(\tau) d\tau) \rangle_0$$

where λ is a real parameter. This equation can be written as

$$\phi(\lambda) = \lim_{\alpha \to 0} \ln \langle T_\tau \exp((\lambda + \alpha)(- \int_0^\beta \hat{H}_1(\tau) d\tau)) \rangle_0.$$

Since

$$\exp \left(\lambda \frac{\partial}{\partial \alpha} \right) f(\alpha) \equiv f(\alpha + \lambda)$$

we find

$$\phi(\lambda) = \lim_{\alpha \to 0} \exp \left(\lambda \frac{\partial}{\partial \alpha} \right) \ln \langle T_\tau \exp(\alpha(- \int_0^\beta \hat{H}_1(\tau) d\tau)) \rangle_0.$$

It follows that

$$\phi(\lambda) = \lim_{\alpha \to 0} \sum_{n=1}^\infty \frac{\lambda^n}{n!} \left(\frac{\partial}{\partial \alpha} \right)^n \ln \langle T_\tau \exp(\alpha(- \int_0^\beta \hat{H}_1(\tau) d\tau)) \rangle_0.$$

Comparing powers of λ^n we can express $\langle \cdots \rangle_c$ in terms of $\langle \cdots \rangle_0$. For the first two terms we find

$$\int_0^\beta d\tau \langle \hat{H}_1(\tau) \rangle_c = \int_0^\beta d\tau \langle \hat{H}_1(\tau) \rangle_0$$

$$\int_0^\beta d\tau_1 \int_0^\beta d\tau_2 \langle T_\tau \hat{H}_1(\tau_1) \hat{H}_1(\tau_2) \rangle_c = \int_0^\beta d\tau_1 \int_0^\beta d\tau_2 \langle T_\tau \hat{H}_1(\tau_1) \hat{H}_1(\tau_2) \rangle_0$$
$$- \left(\int_0^\beta d\tau \langle \hat{H}_1(\tau) \rangle_0 \right)^2.$$

Chapter 31

Bose Operators

Consider a family of linear operators b_j, $b_j^\dagger$, $j = 1, 2, \ldots, m$ on an inner product space V, satisfying the commutation relations (Heisenberg algebra)

$$[b_j, b_k] = [b_j^\dagger, b_k^\dagger] = 0, \qquad [b_j, b_k^\dagger] = \delta_{jk} I$$

where I is the identity operator. The operator $b_j^\dagger$ is called a Bose creation operator with mode j and the operator b_j is called a Bose annihilation operator with mode j. The inner product space must be infinite-dimensional for these equations to hold. For, if A and B are $n \times n$ matrices such that $[A, B] = \lambda I$, then $\mathrm{tr}([A, B]) = 0$ implies $\lambda = 0$. Let $|0\rangle = |00\ldots0\rangle$ be the *vacuum state*, i.e. $b_j|0\rangle = 0$, $\langle 0|0\rangle = 1$, $j = 1, 2, \ldots, m$. Now a *number state* is given by

$$(b_{j_1}^\dagger)^{n_1} (b_{j_2}^\dagger)^{n_2} \cdots (b_{j_k}^\dagger)^{n_k} |0\rangle$$

where $j_1, j_2, \ldots, j_k \in \{1, 2, \ldots, m\}$ and $n_1, n_2, \ldots, n_k \in \{0, 1, 2, \ldots\}$. For b_j we also use the notation

$$I \otimes I \otimes \cdots \otimes I \otimes b \otimes I \otimes \cdots \otimes I$$

where b is in the jth position.

Problem 1. Let $b^\dagger$, b be Bose creation and annihilation operators.
(i) In a *normal ordered product* all annihilation operators are placed to the right of all creation operators. Find the normal ordering of

$$((b^\dagger)^2 b^2)^2 .$$

(ii) Find the normal ordering of

$$(b^2(b^\dagger)^2)^2 .$$

Solution 1.   (i) Using the commutation relation $[b, b^\dagger] = I$ we find

$$((b^\dagger)^2 b^2)^2 = (b^\dagger)^4 b^4 + 4(b^\dagger)^3 b^3 + 2(b^\dagger)^2 b^2 .$$

(ii) Using the commutation relation $[b, b^\dagger] = I$ we find

$$(b^2(b^\dagger)^2)^2 = (b^\dagger)^4 b^4 + 12(b^\dagger)^3 b^3 + 38(b^\dagger)^2 b^2 + 32 b^\dagger b + 4I .$$

Problem 2.   (i) Calculate

$$\text{tr}(b^\dagger b \exp(-\epsilon b^\dagger b))$$

where $\epsilon \in \mathbb{R}$ ($\epsilon > 0$) and $b^\dagger b$ is the infinite-dimensional diagonal matrix

$$b^\dagger b = \text{diag}\ (0, 1, 2, \dots).$$

(ii) Calculate

$$\exp(-\epsilon b^\dagger) b \exp(\epsilon b^\dagger).$$

(iii) Calculate

$$\exp(-\epsilon b^\dagger b) b \exp(\epsilon b^\dagger b).$$

Solution 2.   (i) Since

$$\exp(-\epsilon b^\dagger b) = \text{diag}(1, e^{-\epsilon}, e^{-2\epsilon}, \dots)$$

and

$$b^\dagger b \exp(-\epsilon b^\dagger b) = \text{diag}(0, e^{-\epsilon}, 2e^{-2\epsilon}, 3e^{-3\epsilon}, \dots)$$

we obtain

$$\text{tr}(b^\dagger b \exp(-\epsilon b^\dagger b)) = \sum_{n=1}^{\infty} n e^{-\epsilon n}.$$

Now

$$f(\epsilon) := \sum_{n=1}^{\infty} e^{-\epsilon n} \equiv \frac{1}{e^\epsilon - 1} \quad \text{(geometric series)}.$$

The derivative of f yields

$$\frac{df}{d\epsilon} = -\sum_{n=1}^{\infty} n e^{-\epsilon n} = \frac{d}{d\epsilon} \frac{1}{e^\epsilon - 1} = \frac{-e^\epsilon}{(e^\epsilon - 1)^2}.$$

Consequently,

$$\mathrm{tr}(b^\dagger b \exp(-\epsilon b^\dagger b)) = \frac{e^\epsilon}{(e^\epsilon - 1)^2}.$$

(vi) We set

$$f(\epsilon) := \exp(-\epsilon b^\dagger) b \exp(\epsilon b^\dagger).$$

We seek the ordinary differential equation for $f(\epsilon)$, where $f(0) = b$. Taking the derivative with respect to ϵ yields

$$\frac{df}{d\epsilon} = e^{-\epsilon b^\dagger}(-b^\dagger)be^{\epsilon b^\dagger} + e^{-\epsilon b^\dagger}bb^\dagger e^{\epsilon b^\dagger} = e^{-\epsilon b^\dagger}(-b^\dagger b + bb^\dagger)e^{\epsilon b^\dagger} = I$$

where I is the identity operator. From this linear differential equation and the initial condition $f(0) = b$ we find

$$f(\epsilon) = \exp(-\epsilon b^\dagger) b \exp(\epsilon b^\dagger) = b + \epsilon I.$$

(vii) We set

$$g(\epsilon) := \exp(-\epsilon b^\dagger b) b \exp(\epsilon b^\dagger b).$$

We seek the ordinary differential equation for $g(\epsilon)$, where $g(0) = b$. Taking the derivative with respect to ϵ yields

$$\frac{dg}{d\epsilon} = e^{-\epsilon b^\dagger b}(-b^\dagger b)be^{\epsilon b^\dagger b} + e^{-\epsilon b^\dagger b}b(b^\dagger b)e^{\epsilon b^\dagger b} = e^{-\epsilon b^\dagger b}(-b^\dagger bb + bb^\dagger b)e^{\epsilon b^\dagger b}$$

$$= e^{-\epsilon b^\dagger b}be^{\epsilon b^\dagger b} = g(\epsilon).$$

From this linear differential equation and the initial condition $g(0) = b$ we obtain

$$g(\epsilon) = \exp(-\epsilon b^\dagger b) b \exp(\epsilon b^\dagger b) = b \exp(\epsilon).$$

Problem 3. (i) Let $b^\dagger$, b be Bose creation and annihilation operators, respectively. Calculate the commutators

$$[b^2, (b^\dagger)^2], \qquad [b^3, (b^\dagger)^3].$$

Extend the result to $[b^n, (b^\dagger)^n]$, where $n \in \mathbb{N}$.
(ii) Let $|0\rangle$ be the vacuum state with $b|0\rangle = 0$. Calculate $[b^n, (b^\dagger)^n]|0\rangle$.

Solution 3. (i) Using $bb^\dagger = I + b^\dagger b$ we obtain

$$[b^2, (b^\dagger)^2] = 2I + 4b^\dagger b, \qquad [b^3, (b^\dagger)^3] = 6I + 18b^\dagger b + 9b^\dagger b^\dagger bb.$$

To calculate $[b^n, (b^\dagger)^n]$ we can apply the formula

$$[f(b), g(b^\dagger)] = \sum_{j=1}^{\infty} \frac{\partial^j}{\partial b^{\dagger j}} g(b^\dagger) \frac{\partial^j}{\partial b^j} f(b)$$

where f and g are analytic functions. Thus we obtain

$$[b^n, (b^\dagger)^n] = \sum_{j=1}^{n} \frac{(n!)^2}{j!((n-j)!)^2} (b^\dagger)^{n-j} b^{n-j}.$$

(ii) Since $b|0\rangle = 0$ we obtain $[b^n, (b^\dagger)^n]|0\rangle = (n!)|0\rangle$.

Problem 4. Let b, $b^\dagger$ be Bose annihilation and creation operators. Let $m, n \geq 1$. We have the *ordering formula*

$$b^m (b^\dagger)^n = \sum_{j=0}^{\min\{m,n\}} \binom{m}{j} \frac{n!}{(n-j)!} (b^\dagger)^{n-j} b^{m-j}.$$

(i) Apply it to $b^2 b^\dagger$ and $b(b^\dagger)^2$.
(ii) Calculate the state $b^m (b^\dagger)^n |0\rangle$.

Solution 4. (i) Since $m = 2$, $n = 1$ we have $\min\{m, n\} = 1$. Thus we find

$$b^2 b^\dagger = \sum_{j=0}^{1} \binom{2}{j} \frac{1}{(1-j)!} (b^\dagger)^{1-j} b^{2-j} = b^\dagger b^2 + 2b.$$

For $m = 1$, $n = 2$ we have $\min\{m, n\} = 1$. Thus we find

$$b(b^\dagger)^2 = \sum_{j=0}^{1} \binom{1}{j} \frac{2}{(2-j)!} (b^\dagger)^{2-j} b^{1-j} = (b^\dagger)^2 b + 2b^\dagger.$$

(ii) If $m > n$ we find 0. If $m = n$ we find the state

$$b^m (b^\dagger)^m |0\rangle = m!|0\rangle.$$

If $n > m$ we obtain

$$\frac{n!}{(n-m)!} (b^\dagger)^{n-m} |0\rangle.$$

Problem 5. Let b be a Bose annihilation operator. Solve the eigenvalue problem

$$b|\beta\rangle = \beta|\beta\rangle. \tag{1}$$

Hint. Use the *number representation* $|n\rangle$, where

$$|n\rangle := \frac{(b^\dagger)^n}{\sqrt{n!}} |0\rangle \tag{2}$$

and $n = 0, 1, 2, \ldots$. Therefore

$$b|n\rangle = \sqrt{n}|n-1\rangle, \qquad b^\dagger|n\rangle = \sqrt{n+1}|n+1\rangle. \tag{3}$$

The states $|\beta\rangle$ ($\beta \in \mathbb{C}$) are called a *Bose coherent states*. The *commutation relation* for the Bose operators are given by

$$[b, b^\dagger] = I, \qquad [b, b] = [b^\dagger, b^\dagger] = 0 \tag{4}$$

where I is the identity operator.

Solution 5. To solve the eigenvalue problem we apply the completeness relation of the number representation. This means we expand $|\beta\rangle$ with respect to $|n\rangle$. We obtain

$$|\beta\rangle = \sum_{n=0}^{\infty} |n\rangle\langle n|\beta\rangle \equiv \sum_{n=0}^{\infty} c_n(\beta)|n\rangle \tag{5}$$

where $c_n(\beta) := \langle n|\beta\rangle$. If we insert (5) into (1) and use (3) we obtain

$$\sum_{n=1}^{\infty} c_n(\beta)\sqrt{n}|n-1\rangle = \sum_{n=0}^{\infty} \beta c_n(\beta)|n\rangle.$$

The first sum goes from 1 to ∞ since the term $n = 0$ vanishes. We can therefore shift indices and put $n \to n+1$. It follows that

$$\sum_{n=0}^{\infty} c_{n+1}(\beta)\sqrt{n+1}|n\rangle = \sum_{n=0}^{\infty} \beta c_n(\beta)|n\rangle.$$

If we apply the dual state $\langle m|$ and use $\langle m|n\rangle = \delta_{nm}$ we obtain the linear difference equation

$$c_{n+1}(\beta)\sqrt{n+1} = \beta c_n(\beta).$$

On inspection we see that

$$c_1 = \frac{\beta}{\sqrt{1}}c_0, \quad c_2 = \frac{\beta}{\sqrt{2}}c_1 = \frac{\beta^2}{\sqrt{2!}}c_0, \quad c_3 = \frac{\beta^3}{\sqrt{3!}}c_0, \quad \ldots, \quad c_n(\beta) = \frac{\beta^n}{\sqrt{n!}}c_0.$$

Therefore

$$|\beta\rangle = c_0 \sum_{n=0}^{\infty} \frac{\beta^n}{\sqrt{n!}}|n\rangle.$$

We normalize $|\beta\rangle$ to determine c_0. From the condition $\langle\beta|\beta\rangle = 1$ we obtain

$$1 = |c_0|^2 e^{|\beta|^2}$$

where from (1) we have $\langle\beta|b^\dagger = \langle\beta|\bar{\beta}$. It follows that

$$|\beta\rangle = e^{-|\beta|^2/2} \sum_{n=0}^{\infty} \frac{\beta^n}{\sqrt{n!}}|n\rangle \equiv e^{-|\beta|^2/2} e^{\beta b^\dagger}|0\rangle$$

where $\beta \in \mathbb{C}$.

Problem 6.   Let $|n\rangle$ with $n = 0, 1, 2, \dots$ be the number states (Fock states). Let $|\beta\rangle$, $\beta \in \mathbb{C}$ be coherent states.
(i) Calculate $\langle n|\beta\rangle$ and $\langle\beta|n\rangle \equiv \overline{\langle n|\beta\rangle}$.
(ii) Calculate the square of the distance

$$\| \, |n\rangle - |\beta\rangle \, \|^2 = ((\langle n| - \langle\beta|)(|n\rangle - |\beta\rangle)) \,.$$

Discuss.

Solution 6.   (i) Since

$$|\beta\rangle = \exp\left(-\frac{1}{2}|\beta|^2\right) \sum_{m=0}^{\infty} \frac{\beta^m}{\sqrt{m!}}|m\rangle$$

and $\langle n|m\rangle = \delta_{nm}$ we have

$$\langle n|\beta\rangle = \exp\left(-\frac{1}{2}|\beta|^2\right) \frac{\beta^n}{\sqrt{n!}} \,.$$

Thus

$$\overline{\langle n|\beta\rangle} = \exp\left(-\frac{1}{2}|\beta|^2\right) \frac{\bar{\beta}^n}{\sqrt{n!}}$$

since $\overline{\beta^n} = \bar{\beta}^n$.
(ii) We have

$$\begin{aligned}
\| \, |n\rangle - |\beta\rangle \, \|^2 &= ((\langle n| - \langle\beta|)(|n\rangle - |\beta\rangle)) \\
&= 2 - \langle n|\beta\rangle - \langle\beta|n\rangle \\
&= 2 - \langle n|\beta\rangle - \overline{\langle n|\beta\rangle} \,.
\end{aligned}$$

Using (i) we obtain

$$\| \, |n\rangle - |\beta\rangle \, \|^2 = 2 - \frac{e^{-|\beta|^2/2}}{\sqrt{n!}}\left(\beta^n + \bar{\beta}^n\right) \,.$$

If we set $\beta = re^{i\phi}$ with $r \geq 0$, $\phi \in \mathbb{R}$ and apply the identity $2\cos(n\phi) \equiv e^{in\phi} + e^{-in\phi}$ we obtain

$$\| \, |n\rangle - |\beta\rangle \, \|^2 = 2\left(1 - \frac{e^{-r^2/2}}{\sqrt{n!}}r^n \cos(n\phi)\right) \,.$$

Problem 7. Let $|\beta\rangle$ be a coherent state. Let $\hat{n} := b^\dagger b$. Calculate

$$\langle \beta | \hat{n} | \beta \rangle .$$

Solution 7. Since $b|\beta\rangle = \beta|\beta\rangle$, $\langle\beta|b^\dagger = \langle\beta|\beta^*$ we obtain

$$\langle \beta | \hat{n} | \beta \rangle = \beta\beta^* = |\beta|^2 .$$

Problem 8. Let $b^\dagger$, b be Bose creation and annihilation operators, respectively.
(i) Show that

$$e^{\epsilon b^\dagger} b = (b - \epsilon I) e^{\epsilon b^\dagger} .$$

(ii) Extend the identity to $e^{\epsilon b^\dagger} b^k$, where $k \in \mathbb{N}$.

Solution 8. Using the *Baker-Campbell-Hausdorff formula* we have

$$e^{\epsilon b^\dagger} b e^{-\epsilon b^\dagger} = b - \epsilon I .$$

It follows that

$$e^{\epsilon b^\dagger} b = e^{\epsilon b^\dagger} b e^{-\epsilon b^\dagger} e^{\epsilon b^\dagger} = (b - \epsilon I) e^{\epsilon b^\dagger} .$$

(ii) Using the result form (i) we obtain

$$e^{\epsilon b^\dagger} b^k = (b - \epsilon I)^k e^{\epsilon b^\dagger} .$$

Problem 9. (i) Let $b_j^\dagger$, b_j be Bose creation and annihilation operators, respectively and $j = 1, 2, \ldots, n$. We define

$$B(x) := \sum_{j=1}^{n} \frac{b_j}{1 - \epsilon_j x}$$

and

$$f(x) := \sum_{j=1}^{n} \frac{1}{1 - \epsilon_j x} .$$

Show that

$$[B(x), B^\dagger(y)] = \frac{I}{x - y}(xf(x) - yf(y)), \qquad [B(x), B(y)] = 0$$

where $[,]$ denotes the commutator, I is the identity operator and 0 is the zero operator.

(ii) Let

$$N(x) := \sum_{j=1}^{n} \frac{b_j^\dagger b_j}{1 - \epsilon_j x} \, .$$

Show that

$$[N(x), B^\dagger(y)] = \frac{1}{x - y}(xB^\dagger - yB^\dagger(y))$$

$$[N(x), B(y)] = -\frac{1}{x - y}(xB(x) - yB(y)) \, .$$

Solution 9.   (i) We have

$$B^\dagger(x) = \sum_{j=1}^{n} \frac{b_j^\dagger}{1 - \epsilon_j x} \, .$$

Thus we find

$$[B(x), B^\dagger(y)] = \sum_{j=1}^{n} \frac{b_j}{1 - \epsilon_j x} \sum_{k=1}^{n} \frac{b_k^\dagger}{1 - \epsilon_k y} - \sum_{k=1}^{n} \frac{b_k^\dagger}{1 - \epsilon_k y} \sum_{j=1}^{n} \frac{b_j}{1 - \epsilon_j x}$$

$$= \sum_{j=1}^{n} \sum_{k=1}^{n} \frac{b_j b_k^\dagger}{(1 - \epsilon_j x)(1 - \epsilon_k y)} - \sum_{k=1}^{n} \sum_{j=1}^{n} \frac{b_k^\dagger b_j}{(1 - \epsilon_k y)(1 - \epsilon_j x)}$$

$$= \sum_{j=1}^{n} \sum_{k=1}^{n} \frac{b_j b_k^\dagger - b_k^\dagger b_j}{(1 - \epsilon_j x)(1 - \epsilon y)}$$

$$= \sum_{j=1}^{n} \sum_{k=1}^{n} \frac{\delta_{jk} I}{(1 - \epsilon_j x)(1 - \epsilon_k y)}$$

$$= I \sum_{j=1}^{n} \frac{1}{(1 - \epsilon_j x)(1 - \epsilon_j y)}$$

$$= \frac{I}{x - y} \sum_{j=1}^{n} \frac{x - y}{(1 - \epsilon_j x)(1 - \epsilon_j y)}$$

$$= \frac{I}{x - y} \sum_{j=1}^{n} \frac{x(1 - \epsilon_j y) - y(1 - \epsilon_j x)}{(1 - \epsilon_j x)(1 - \epsilon_j y)}$$

$$= \frac{I}{x - y} \sum_{j=1}^{n} \left(\frac{x}{1 - \epsilon_j x} - \frac{y}{1 - \epsilon_j y} \right)$$

$$= \frac{I}{x - y} \left(x \sum_{j=1}^{n} \frac{1}{1 - \epsilon_j x} - y \sum_{j=1}^{n} \frac{1}{1 - \epsilon_j y} \right)$$

$$= \frac{I}{x - y}(xf(x) - yf(y))$$

where we used that $[b_j, b_k^\dagger] = \delta_{jk} I$. Since $[b_j, b_k] = 0$ the second result is obvious.

(ii) Using

$$[b_j^\dagger b_j, b_k] = -b_j \delta_{jk}, \qquad [b_j^\dagger b_j, b_k^\dagger] = b_j^\dagger \delta_{jk}$$

and a similar calculation as in (i) we obtain the results.

Problem 10. Let $b^\dagger$, b be Bose creation and annihilation operators. Let $z, w \in \mathbb{C}$.

(i) Calculate the commutator

$$[zb^\dagger - \bar{z}b, wb^\dagger - \bar{w}b].$$

(ii) Let $D(z) := e^{zb^\dagger - \bar{z}b}$ be the *displacement operator*. Show that

$$D(z)D(w) \equiv e^{z\bar{w} - \bar{z}w} D(w)D(z)$$

$$D(z + w) \equiv e^{-\frac{1}{2}(z\bar{w} - \bar{z}w)} D(z)D(w).$$

Solution 10. (i) We have

$$
\begin{aligned}
[zb^\dagger - \bar{z}b, wb^\dagger - \bar{w}b] &= -[\bar{z}b, wb^\dagger] - [zb^\dagger, \bar{w}b] \\
&= -\bar{z}w[b, b^\dagger] - z\bar{w}[b^\dagger, b] \\
&= (z\bar{w} - \bar{z}w)I.
\end{aligned}
$$

(ii) Using the result from (i) and that

$$e^{A+B} = e^{-\frac{1}{2}[A,B]} e^A e^B$$

we obtain the two identities.

Problem 11. Let $z \in \mathbb{C}$ and

$$D(z) := e^{zb^\dagger - \bar{z}b}$$

be the *displacement operator*. Let $|n\rangle$ be the number states (Fock states). Calculate the matrix elements

$$\langle n|D(z)|m\rangle.$$

Solution 11. The matrix elements of the displacement operator $D(z)$ are

$$n \le m \qquad \langle n|D(z)|m\rangle = e^{-|z|^2/2} \sqrt{\frac{n!}{m!}} (-\bar{z})^{m-n} L_n^{(m-n)}(|z|^2)$$

$$n \geq m \quad \langle n|D(z)|m \rangle = e^{-|z|^2/2} \sqrt{\frac{m!}{n!}} z^{n-m} L_m^{(n-m)}(|z|^2)$$

where $L_n^{(\alpha)}$ are the *associated Laguerre polynomials* defined by

$$L_n^{(\alpha)}(x) = \sum_{j=0}^{n} (-1)^j \binom{n+\alpha}{n-j} \frac{x^j}{j!}.$$

In particular $L_n^{(0)} = L_n$ are the usual Laguerre polynomials.

Problem 12.   Let $|\beta\rangle$ be a coherent state. Show that

$$\frac{1}{\pi} \int_{\mathbb{C}} |\beta\rangle\langle\beta| d^2\beta = I$$

where I is the identity operator and the integration is over the entire complex plane. Set $\beta = r\exp(i\phi)$ with $0 \leq r < \infty$ and $0 \leq \phi < 2\pi$.

Solution 12.   We have

$$\frac{1}{\pi} \int_{\mathbb{C}} |\beta\rangle\langle\beta| d^2\beta = \frac{1}{\pi} \sum_{n=0}^{\infty} \sum_{m=0}^{\infty} \frac{|n\rangle\langle m|}{\sqrt{n!\,m!}} \int_{\mathbb{C}} e^{-|\beta|^2} \beta^{*m} \beta^n d^2\beta.$$

Using $\beta = r\exp(i\phi)$ we arrive at

$$\frac{1}{\pi} \int_{\mathbb{C}} |\beta\rangle\langle\beta| d^2\beta = \frac{1}{\pi} \sum_{n=0}^{\infty} \sum_{m=0}^{\infty} \frac{|n\rangle\langle m|}{\sqrt{n!\,m!}} \int_0^{\infty} re^{-r^2} r^{n+m} dr \int_0^{2\pi} e^{i(n-m)\phi} d\phi.$$

Since

$$\int_0^{2\pi} e^{i(n-m)\phi} d\phi = 2\pi\delta_{nm}$$

we have

$$\frac{1}{\pi} \int_{\mathbb{C}} |\beta\rangle\langle\beta| d^2\beta = \sum_{n=0}^{\infty} \frac{|n\rangle\langle n|}{n!} \int_0^{\infty} e^{-s} s^n ds$$

where we set $s = r^2$ and therefore $ds = 2rdr$. Thus

$$\frac{1}{\pi} \int_{\mathbb{C}} |\beta\rangle\langle\beta| d^2\beta = \sum_{n=0}^{\infty} |n\rangle\langle n| = I$$

where we used the completeness relation for the number states.

Problem 13.   Let $|\beta\rangle$ be a coherent state and $\rho = |\beta\rangle\langle\beta|$. Calculate the *characteristic function*

$$\chi(\beta) := \text{tr}(\rho e^{\beta b^\dagger - \beta^* b}) \equiv \text{tr}(\rho D(\beta))$$

where $D(\beta)$ is the displacement operator.

Solution 13. We apply coherent states to calculate the trace. Then

$$\chi(\beta) = \text{tr}(|\beta\rangle\langle\beta|D(\beta))$$

$$= \frac{1}{\pi}\int_{\mathbb{C}} d\gamma\langle\gamma|\beta\rangle\langle\beta|D(\beta)|\gamma\rangle$$

$$= \frac{1}{\pi}\int_{\mathbb{C}} d\gamma\langle\gamma|\beta\rangle\langle\beta|D(\beta)D(\gamma)|0\rangle$$

$$= \frac{1}{\pi}\int_{\mathbb{C}} d\gamma\langle\gamma|\beta\rangle\langle\beta|e^{\frac{1}{2}(\beta^*\gamma-\beta\gamma^*)}D(\beta+\gamma)|0\rangle$$

$$= \frac{1}{\pi}\int_{\mathbb{C}} d\gamma\langle\gamma|\beta\rangle e^{\frac{1}{2}(\beta^*\gamma-\beta\gamma^*)}\langle\beta|\beta+\gamma\rangle$$

$$= \frac{1}{\pi}\int_{\mathbb{C}} d\gamma e^{-\frac{1}{2}(|\beta|^2+|\gamma|^2)+\beta\gamma^*}e^{\frac{1}{2}(\beta^*\gamma-\beta\gamma^*)}e^{-\frac{1}{2}(|\beta+\gamma|^2+|\beta|^2)+(\beta+\gamma)\beta^*}$$

$$= \frac{1}{\pi}\int_{\mathbb{C}} d\gamma e^{-\gamma\gamma^*-\frac{1}{2}\beta\beta^*+\beta^*\gamma}$$

$$= \frac{1}{\pi}e^{-\frac{1}{2}\beta\beta^*}\int_{\mathbb{C}} d\gamma e^{-\gamma\gamma^*+\beta^*\gamma}$$

$$= \frac{1}{\pi}e^{-\frac{1}{2}\beta\beta^*}\int_{r=0}^{\infty} dr e^{-r^2}\int_{\phi=0}^{2\pi} d\phi e^{\beta^*re^{i\phi}}$$

$$= 2e^{-\frac{1}{2}\beta\beta^*}\int_{r=0}^{\infty} dr e^{-r^2}$$

$$= \sqrt{\pi}e^{-\frac{1}{2}\beta\beta^*}.$$

Problem 14. The *thermal mixture of states* with the mean number of photon equal to $\bar{n}$ is represented by the density operator

$$\rho_{\bar{n}} = \sum_{n=0}^{\infty} \frac{\bar{n}^n}{(\bar{n}+1)^{n+1}}|n\rangle\langle n|$$

where $|n\rangle$ are the number states. Calculate the Husimi distribution $\langle\beta|\rho_{\bar{n}}|\beta\rangle$.

Solution 14. Since

$$|\beta\rangle = e^{-|\beta|^2/2}\sum_{k=0}^{\infty}\frac{\beta^k}{\sqrt{k!}}|k\rangle, \qquad \langle\beta| = e^{-|\beta|^2/2}\sum_{j=0}^{\infty}\frac{(\beta^*)^j}{\sqrt{j!}}\langle j|$$

we have

$$\langle\beta|\rho_{\bar{n}}|\beta\rangle = e^{-|\beta|^2}\sum_{j=0}^{\infty}\frac{(\beta^*)^j}{\sqrt{j!}}\langle j|\sum_{n=0}^{\infty}\frac{\bar{n}^n}{(\bar{n}+1)^{n+1}}|n\rangle\langle n|\sum_{k=0}^{\infty}\frac{\beta^k}{\sqrt{k!}}|k\rangle$$

$$= e^{-|\beta|^2} \sum_{j=0}^{\infty} \frac{\beta^*}{\sqrt{j!}} \langle j| \sum_{n=0}^{\infty} \frac{\overline{n}^n \beta^n}{\sqrt{n!}(\overline{n}+1)^{n+1}} |n\rangle$$

$$= e^{-|\beta|^2} \sum_{n=0}^{\infty} \frac{(\beta^*)^n \beta^n \overline{n}^n}{\sqrt{n!}\sqrt{n!}(\overline{n}+1)^{n+1}}$$

$$= \frac{e^{-|\beta|^2}}{(\overline{n}+1)} \sum_{n=0}^{\infty} \frac{1}{n!} \left(\frac{\beta^* \beta \overline{n}}{\overline{n}+1} \right)^n$$

$$= \frac{e^{-|\beta|^2}}{(\overline{n}+1)} e^{\overline{n}|\beta|^2/(\overline{n}+1)}$$

$$= \frac{1}{\overline{n}+1} e^{-|\beta|^2/(\overline{n}+1)}$$

where we used that $\langle n|k \rangle = \delta_{nk}$ and $\langle j|n \rangle = \delta_{jn}$.

Problem 15. Consider the *two-mode state*

$$|\psi\rangle = e^{r(b_1^\dagger b_2^\dagger - b_1 b_2)} |00\rangle$$

where $|00\rangle \equiv |0\rangle \otimes |0\rangle$ and r is the squeezing parameter. This state can also be written as

$$|\psi\rangle = \frac{1}{\cosh(r)} \sum_{n=0}^{\infty} (\tanh(r))^n |n\rangle \otimes |n\rangle .$$

This is the *Schmidt basis* for this state. The *density operator* ρ is given by $\rho = |\psi\rangle\langle\psi|$. Calculate the partial traces using the number states.

Solution 15. We have

$$\rho = \frac{1}{(\cosh(r))^2} \sum_{n=0}^{\infty} (\tanh(r))^n |n\rangle \otimes |n\rangle \sum_{m=0}^{\infty} (\tanh(r))^m \langle m| \otimes \langle m|$$

$$= \frac{1}{(\cosh(r))^2} \sum_{m=0}^{\infty} \sum_{n=0}^{\infty} (\tanh(r))^n (\tanh(r))^m |n\rangle\langle m| \otimes |n\rangle\langle m| .$$

Let I be the identity operator. Using that $\langle k|n \rangle = \delta_{kn}$ and $\langle m|k \rangle = \delta_{mk}$ we find

$$\rho_1 = \sum_{k=0}^{\infty} (I \otimes \langle k|)\rho(I \otimes |k\rangle)$$

$$= \frac{1}{(\cosh(r))^2} \sum_{k=0}^{\infty} (I \otimes \langle k|) \sum_{m,n=0}^{\infty} (\tanh(r))^n (\tanh(r))^m |n\rangle\langle m| \otimes |n\rangle\langle m|(I \otimes |k\rangle)$$

$$= \frac{1}{(\cosh(r))^2} \sum_{k=0}^{\infty} \sum_{m,n=0}^{\infty} (\tanh(r))^n (\tanh(r))^m |n\rangle\langle m|\delta_{kn}\delta_{mk}$$

$$= \frac{1}{(\cosh(r))^2} \sum_{k=0}^{\infty} (\tanh(r))^{2k} |k\rangle\langle k| .$$

We obtain the same result for ρ_2.

Problem 16. Let $b^\dagger$, b be Bose creation and Bose annihilation operators. Define

$$K_0 := \frac{1}{4}(b^\dagger b + b b^\dagger), \qquad K_+ := \frac{1}{2}\left(b^\dagger\right)^2, \qquad K_- := \frac{1}{2}b^2. \qquad (1)$$

(i) Show that K_0, K_+, K_- form a basis of a Lie algebra.
(ii) Define

$$L_0 := b_1^\dagger b_1 + b_2 b_2^\dagger, \qquad L_+ := b_1^\dagger b_2^\dagger, \qquad L_- := b_1 b_2. \qquad (2)$$

Show that L_0, L_+, L_- form a basis of a Lie algebra.
(iii) Are the two Lie algebras isomorphic ?
(iv) Do the operators

$$\{I, b, b^\dagger, b^\dagger b\} \qquad (3)$$

form a Lie algebra under the commutator? Here I denotes the identity operator.

Solution 16. (i) Using the commutation relations for Bose operators we obtain

$$[K_0, K_+] = K_+ \qquad [K_0, K_-] = -K_- \qquad [K_-, K_+] = 2K_0.$$

Thus the set $\{K_0, K_+, K_-\}$ forms a basis of a Lie algebra.
(ii) Using the commutation relations for Bose operators we find

$$[L_0, L_+] = 2L_+, \qquad [L_0, L_-] = -2L_-, \qquad [L_-, L_+] = L_0.$$

(iii) Two Lie algebras, L_1 and L_2, are called *isomorphic* if there is a map $\phi : L_1 \rightarrow L_2$ such that $\phi([a, b]) = [\phi(a), \phi(b)]$, where $a, b \in L_1$. Let

$$\phi(K_+) = L_+ \qquad \phi(K_-) = L_- \qquad \phi(K_0) = \frac{1}{2}L_0.$$

Then we find that the Lie algebras are isomorphic.
(iv) Since $[I, b] = 0$, $[I, b^\dagger] = 0$, $[I, b^\dagger b] = 0$ and

$$[b, b^\dagger] = I, \qquad [b, b^\dagger b] = b, \qquad [b^\dagger, b^\dagger b] = -b^\dagger$$

we find that the set of operators $\{I, b, b^\dagger, b^\dagger b\}$ forms a basis of a Lie algebra.

Problem 17. Let $b_1^\dagger$, $b_2^\dagger$ be the single-particle Bose creation operators for two bosonic modes. Let $\hat{N}_1 := b_1^\dagger b_1$, $\hat{N}_2 := b_2^\dagger b_2$ be the corresponding number operators. Consider the *Bose-Hubbard dimer Hamilton operator*

$$\hat{H} = \frac{k}{8}(\hat{N}_1 - \hat{N}_2)^2 - \frac{\mu}{2}(\hat{N}_1 - \hat{N}_2) - \frac{\mathcal{E}}{2}(b_1^\dagger b_2 + b_2^\dagger b_1). \tag{1}$$

The coupling k provides the strength of the scattering interaction between bosons, μ is the external potential and $\mathcal{E}$ is the coupling for tunneling. The change $\mathcal{E} \to -\mathcal{E}$ corresponds to the unitary transformation $b_1 \to b_1$, $b_2 \to -b_2$. The change $\mu \to -\mu$ corresponds to $b_1 \leftrightarrow b_2$.
(i) Show that the Hamilton operator $\hat{H}$ commutes with the total number operator

$$\hat{N} = \hat{N}_1 + \hat{N}_2 = b_1^\dagger b_1 + b_2^\dagger b_2.$$

Thus $\hat{N}$ is a conserved quantity.
(ii) The *Jordan-Schwinger realization* in the $su(2)$ Lie algebra is given by

$$S^+ := b_1^\dagger b_2, \qquad S^- := b_2^\dagger b_1, \qquad S^z := \frac{1}{2}(\hat{N}_1 - \hat{N}_2)$$

which is $(N+1)$-dimensional when the constraint of fixed particle number $N = N_1 + N_2$ is imposed. Express the Hamilton operator using S^+, S^-, S^z.
(iii) The same $(N+1)$-dimensional representation of $su(2)$ is given by the mapping to differential operators

$$S^+ = Nu - u^2 \frac{d}{du}, \qquad S^- = \frac{d}{du}, \qquad S^z = u\frac{d}{du} - \frac{N}{2}$$

acting on the $(N+1)$-dimensional vector space of polynomials with basis $\{1, u, u^2, \ldots, u^N\}$. Express the Hamilton operator in this representation. Then solving for the spectrum of the Hamilton operator (1) is equivalent for solving the eigenvalue equation

$$Hp(u) = Ep(u)$$

where p is a polynomial of u of order N. Find the energy eigenvalues $E(k, \mu, \mathcal{E})$.

Solution 17. (i) Since

$$[b_1^\dagger b_2, b_1^\dagger b_1] = -b_1^\dagger b_2, \qquad [b_2^\dagger b_1, b_1^\dagger b_1] = b_2^\dagger b_1$$

$$[b_1^\dagger b_2, b_2^\dagger b_2] = b_1^\dagger b_2, \qquad [b_2^\dagger b_1, b_2^\dagger b_2] = -b_2^\dagger b_1$$

we obtain $[\hat{H}, \hat{N}] = 0$.

(ii) We have

$$\hat{H} = \frac{k}{2}(S^z)^2 - \mu S^z - \frac{\mathcal{E}}{2}(S^+ + S^-).\qquad(2)$$

(iii) We obtain the second order differential operator for the Hamilton operator

$$\hat{H} = \frac{ku^2}{2}\frac{d^2}{du^2} + \frac{1}{2}((k(1-N)-2\mu)u+\mathcal{E}(u^2-1))\frac{d}{du} + \frac{kN^2}{8} + \frac{\mu N}{2} - \frac{\mathcal{E}Nu}{2}.\quad(3)$$

Solving for the spectrum of the Hamilton operator (1) is now equivalent for solving the eigenvalue equation

$$\hat{H}p(u) = Ep(u)$$

where $\hat{H}$ is given by (3) and p is a polynomial of u of order N. We first express p in terms of its roots $\{v_j\}$

$$p(u) = \prod_{j=1}^{N}(u - v_j).$$

Now we have

$$\frac{d}{du}p(u) = p(u)\sum_{j=1}^{N}\frac{1}{u - v_j}.$$

For the second order derivative we have

$$\frac{d^2}{du^2}p(u) = p(u)\left(\left(\sum_{j=1}^{N}\frac{1}{u - v_j}\right)^2 - \sum_{j=1}^{N}\frac{1}{(u - v_j)^2}\right).$$

Furthermore we need the identity

$$\left(\left(\sum_{j=1}^{N}\frac{1}{u - v_j}\right)^2 - \sum_{j=1}^{N}\frac{1}{(u - v_j)^2}\right) \equiv 2\sum_{\substack{i<j\\j=2\\i=1}}^{N}\frac{1}{v_i - v_j}\left(\frac{1}{u - v_i} - \frac{1}{u - v_j}\right).$$

Using these equations and evaluating the eigenvalue equation $\hat{H}p(u) = Ep(u)$ at $u = v_\ell$ for each ℓ we obtain the set of *Bethe ansatz equations*

$$\frac{(k(1 - N) - 2\mu)v_\ell + \mathcal{E}v_\ell^2 - \mathcal{E}}{kv_\ell^2} = 2\sum_{\substack{j\neq\ell\\j}}^{N}\frac{1}{v_j - v_\ell},\qquad \ell = 1,\ldots,N.\quad(4)$$

Writing the asymptotic expansion $p(u) \sim u^N - u^{N-1}\sum_{j=1}^{N} v_j$ and by considering the terms of order N in the eigenvalue equation, the energy eigenvalues are given as

$$E(k,\mu,\mathcal{E}) = \frac{kN^2}{8} - \frac{\mu N}{2} + \frac{\mathcal{E}}{2}\sum_{j=1}^{N} v_j.$$

For given N, k, μ and $\mathcal{E}$ we solve the Bethe ansatz equations for $v_1,\ldots,v_N$ and insert them into this equation to find $E(k,\mu,\mathcal{E})$.

Problem 18.   Consider the difference equation of first order

$$x_{n+1} = f(x_n), \qquad n = 0,1,2,\ldots \tag{1}$$

where $f : \mathbb{R} \to \mathbb{R}$ is an analytic function. Consider the state

$$|x,n\rangle := \exp\left(\frac{1}{2}(x_n^2 - x_0^2)\right)|x_n\rangle \tag{2}$$

where x_n satisfies (1) and $|x_n\rangle$ is a normalized coherent state. Suppose we are given a Boson operator

$$M := \sum_{k=0}^{\infty} \frac{1}{k!}b^{\dagger k}(f(b) - b)^k \tag{3}$$

where $b^\dagger$ and b are Bose operators. Find $M|x,n\rangle$.

Solution 18.   The *coherent states* $|z\rangle$, where $z \in \mathbb{C}$, are defined as the eigenvectors of the annihilation operator b, i.e. $b|z\rangle = z|z\rangle$. The normalized coherent states are given by

$$|z\rangle = \exp(-\frac{1}{2}|z|^2)\exp(zb^\dagger)|0\rangle. \tag{4}$$

Using (1), (2) and (4) we find that

$$|x,n+1\rangle = M|x,n\rangle. \tag{5}$$

Now let $x_n(x_0)$ designate the solution of (1) and let $|x_0,n\rangle$ be the solution of (5). Taking (2) into account we find that the following eigenvalue equation holds true

$$b|x_0,n\rangle = x_n(x_0)|x_0,n\rangle.$$

Thus the solution of (1) is equivalent to the solution of the linear abstract difference equation (5).

Chapter 32

Bose-Fermi Systems

Problem 1. Consider the Hamilton operator $\hat{H}$ for a single two-level atom coupled to a single mode of an electromagnetic field

$$\hat{H} = b^\dagger \otimes \sigma_- + b \otimes \sigma_+$$

where

$$\sigma_+ = \begin{pmatrix} 0 & 1 \\ 0 & 0 \end{pmatrix}, \qquad \sigma_- = \begin{pmatrix} 0 & 0 \\ 1 & 0 \end{pmatrix}.$$

Let $\theta \in \mathbb{R}$ and $U = e^{i\theta \hat{H}}$. Calculate

$$A(n, \beta) = (\langle n| \otimes I_2)U(|\beta\rangle \otimes I_2)$$

where $|\beta\rangle$ and $|n\rangle$ are *coherent states* and *number states* of the electromagnetic field, respectively. This means $b|\beta\rangle = \beta|\beta\rangle$ and $b|n\rangle = \sqrt{n}|n-1\rangle$ with $\beta \in \mathbb{C}$ and $n = 0, 1, \ldots$.

Solution 1. Since $\sigma_+\sigma_+ = 0_2$, $\sigma_-\sigma_- = 0_2$ and

$$\sigma_+\sigma_- = \begin{pmatrix} 1 & 0 \\ 0 & 0 \end{pmatrix}, \qquad \sigma_-\sigma_+ = \begin{pmatrix} 0 & 0 \\ 0 & 1 \end{pmatrix}$$

we find

$$(b^\dagger \otimes \sigma_- + b \otimes \sigma_+)(b^\dagger \otimes \sigma_- + b \otimes \sigma_+) = (I + b^\dagger b) \otimes \sigma_+\sigma_- + b^\dagger b \otimes \sigma_-\sigma_+$$

$$= I \otimes \sigma_+\sigma_- + b^\dagger b \otimes I_2.$$

347

Using this result we obtain

$$A(\beta, n) = e^{-|\beta^2|} \frac{|\beta|^2}{n!} \begin{pmatrix} \cos(\theta\sqrt{n}) & \frac{i\sqrt{n}}{\beta} \sin(\theta\sqrt{n}) \\ \frac{i\beta}{\sqrt{n+1}} \sin(\theta\sqrt{n+1}) & \cos(\theta\sqrt{n+1}) \end{pmatrix}$$

where $n = 0, 1, 2, \ldots$.

Problem 2. Consider the following form of the *Jaynes-Cummings Hamilton operator*

$$\hat{H} = \hat{H}_1 + \hat{H}_2 + \frac{1}{2}\hbar\omega(I_B \otimes I_2)$$

where

$$\hat{H}_1 = \hbar\omega(b^\dagger b \otimes I_2 + \frac{1}{2}I_B \otimes \sigma_z)$$

and

$$\hat{H}_2 = \hbar\kappa(b^\dagger \otimes \sigma_- + b \otimes \sigma_+) - \frac{\hbar}{2}(\omega - \omega_0)I_B \otimes \sigma_z$$

with I_B the identity operator, I_2 the 2×2 identity matrix and

$$\sigma_+ := \sigma_x + i\sigma_y = \begin{pmatrix} 0 & 2 \\ 0 & 0 \end{pmatrix}, \quad \sigma_- := \sigma_x - i\sigma_y = \begin{pmatrix} 0 & 0 \\ 2 & 0 \end{pmatrix}.$$

Calculate the commutator $[\hat{H}_1, \hat{H}_2]$.

Solution 2. Since we have the commutators

$$[b^\dagger b \otimes I_2, b^\dagger \otimes \sigma_-] = b^\dagger \otimes \sigma_-$$
$$[b^\dagger b \otimes I_2, b \otimes \sigma_+] = -b \otimes \sigma_+$$
$$[I_B \otimes \sigma_z, b^\dagger \otimes \sigma_-] = -2b^\dagger \otimes \sigma_-$$
$$[I_B \otimes \sigma_z, b \otimes \sigma_+] = 2b \otimes \sigma_+$$
$$[I_B \otimes \sigma_z, I_B \otimes \sigma_z] = 0$$

we obtain that $[\hat{H}_1, \hat{H}_2] = 0$, i.e. $\hat{H}_1$ and $\hat{H}_2$ commute.

Problem 3. Let $b_1, b_2, b_1^\dagger, b_2^\dagger$ be Bose annihilation and creation operators. Consider the three 2×2 matrices

$$\sigma_+ = \begin{pmatrix} 0 & 1 \\ 0 & 0 \end{pmatrix}, \quad \sigma_- = \begin{pmatrix} 0 & 0 \\ 1 & 0 \end{pmatrix}, \quad \sigma_z = \begin{pmatrix} 1 & 0 \\ 0 & -1 \end{pmatrix}.$$

Consider now the operators in the product space

$$J_+ = b_1^\dagger b_2 \otimes I_2 + I \otimes \sigma_+$$
$$J_- = b_2^\dagger b_1 \otimes I_2 + I \otimes \sigma_-$$
$$J_z = (b_1^\dagger b_1 - b_2^\dagger b_2) \otimes I_2 + I \otimes \sigma_z.$$

Find the commutators

$$[J_+, J_-], \qquad [J_+, J_z], \qquad [J_-, J_z].$$

Solution 3. Note that

$$[b_1^\dagger b_2, b_2^\dagger b_1] = b_1^\dagger b_1 - b_2^\dagger b_2, \quad [\sigma_+, \sigma_-] = \sigma_z.$$

Thus

$$[J_+, J_-] = [b_1^\dagger b_2, b_2^\dagger b_1] \otimes I_2 + I \otimes [\sigma_+, \sigma_-] = (b_1^\dagger b_1 - b_2^\dagger b_2) \otimes I_2 + I \otimes \sigma_z = J_z.$$

Analogously we find

$$[J_+, J_z] = -2J_+, \qquad [J_-, J_z] = 2J_-.$$

Problem 4. Let

$$Q := (b - \epsilon(b + b^\dagger)^2) \otimes c^\dagger \qquad (1)$$

be a linear operator, where b is a Bose annihilation operator, $c^\dagger$ is a Fermi creation operator and ϵ is a real parameter.
(i) Show that

$$Q^2 = 0. \qquad (2)$$

(ii) We define the Hamilton operator as

$$\hat{H}_\epsilon := [Q, Q^\dagger]_+ \equiv QQ^\dagger + Q^\dagger Q. \qquad (3)$$

Find $\hat{H}_\epsilon$. Calculate the commutators $[\hat{H}_\epsilon, Q]$ and $[\hat{H}_\epsilon, Q^\dagger Q]$.
(iii) Give a basis of the underlying Hilbert space.
(iv) Can the Hilbert space be decomposed into subspaces?
The so constructed $\hat{H}_\epsilon$ is called a *supersymmetric Hamilton operator*.

Solution 4. (i) Since $c^\dagger c^\dagger = 0$ we find that

$$Q^2 = 0.$$

(ii) From (1) we obtain

$$Q^\dagger = (b^\dagger - \epsilon(b^\dagger + b)^2) \otimes c.$$

Applying $[b, b^\dagger] = I_B$ and $[c, c^\dagger]_+ = I_F$ we arrive at

$$\hat{H}_\epsilon = [Q, Q^\dagger]_+ = QQ^\dagger + Q^\dagger Q$$
$$= I_B \otimes c^\dagger c + b^\dagger b \otimes I_F - 4\epsilon(b + b^\dagger) \otimes c^\dagger c$$
$$+ \epsilon^2(b^4 + b^{\dagger 4} + 6b^2 + 6b^{\dagger 2} + 4b^{\dagger 3}b + 4b^\dagger b^3 + 6b^{\dagger 2}b^2 + 12b^\dagger b + 3) \otimes I_F$$
$$- \epsilon(b^3 + b^{\dagger 3} + b + b^\dagger + 3b^{\dagger 2}b + 3b^\dagger b^2) \otimes I_F$$

where I_B is the identity operator in the Hilbert space $\mathcal{H}_B$ of the Bose operators and I_F is the identity operator in the Hilbert space $\mathcal{H}_F$ of the Fermi operators. Straightforward calculation yields for the commutators

$$[\hat{H}_\epsilon, Q] = 0, \qquad [\hat{H}_\epsilon, Q^\dagger Q] = 0.$$

Thus the three operators $\hat{H}_\epsilon$, Q, $Q^\dagger Q$ may be diagonalized simultaneously.
(iii) Let

$$|n\rangle := \frac{(b^\dagger)^n}{\sqrt{n!}} |0\rangle$$

be the number states, where $n = 0, 1, 2, \ldots$. Then a basis in the Hilbert space $\mathcal{H}_B \otimes \mathcal{H}_F$ is given by

$$\{ \, |n\rangle \otimes |0\rangle_F, \quad |n\rangle \otimes c^\dagger |0\rangle_F \, \}$$

where $n = 0, 1, 2, \ldots$.
(iv) Since

$$c^\dagger c |0\rangle_F = 0, \qquad c^\dagger c c^\dagger |0\rangle_F = c^\dagger |0\rangle_F$$

we find that the Hilbert space $\mathcal{H}_B \otimes \mathcal{H}_F$ decomposes, under the Hamilton operator, into two subspaces with the bases

$$\{ \, |n\rangle \otimes |0\rangle \, \}$$

and

$$\{ \, |n\rangle \otimes c^\dagger |0\rangle \, \}$$

where $n = 0, 1, 2, \ldots$.

Problem 5. Let

$$\hat{H} = -\Delta I_B \otimes \sigma_z + \frac{k}{2}(b^\dagger \otimes \sigma_- + b \otimes \sigma_+) + \Omega b^\dagger b \otimes I_2 \tag{1}$$

where

$$\sigma_- := \begin{pmatrix} 0 & 0 \\ 2 & 0 \end{pmatrix}, \quad \sigma_+ := \begin{pmatrix} 0 & 2 \\ 0 & 0 \end{pmatrix}, \quad \sigma_z := \begin{pmatrix} 1 & 0 \\ 0 & -1 \end{pmatrix}, \quad I_2 := \begin{pmatrix} 1 & 0 \\ 0 & 1 \end{pmatrix}$$

and Δ, k and Ω are constants. Let

$$\hat{U} := \exp[i\pi(b^\dagger b \otimes I_2 + \frac{1}{2}I_B \otimes \sigma_z + \frac{1}{2}I_B \otimes I_2)]$$

$$\hat{N} := b^\dagger b \otimes I_2 + \frac{1}{2}I_B \otimes \sigma_z.$$

(i) Calculate the commutators and discuss

$$[\hat{H}, \hat{U}], \quad [\hat{H}, \hat{N}], \quad [\hat{U}, \hat{N}], \quad [\hat{N}, I_B \otimes \sigma_z].$$

(ii) Find the eigenvalues of $\hat{U}$. Discuss how $\hat{U}$ can be used to simplify the eigenvalue problem.

(iii) Calculate the eigenvalues of $\hat{H}$.

Solution 5. (i) We find for the commutators

$$[\hat{H}, \hat{U}] = 0, \quad [\hat{H}, \hat{N}] = 0, \quad [\hat{U}, \hat{N}] = 0, \quad [\hat{N}, I_B \otimes \sigma_z] = 0.$$

Thus the operators $\hat{H}, \hat{U}, \hat{N}, I_B \otimes \sigma_z$ form a basis of an abelian Lie algebra.

(ii) A basis in the Hilbert space $\mathcal{H}_B \otimes \mathbb{C}^2$ is given by

$$\left\{ |n\rangle \otimes \begin{pmatrix} 1 \\ 0 \end{pmatrix}, \quad |n\rangle \otimes \begin{pmatrix} 0 \\ 1 \end{pmatrix} \right\}$$

where $|n\rangle$ are the Bose number states. Since the commutators of the operators $b^\dagger b \otimes I_2$, $I_B \otimes \sigma_z$ and $I_B \otimes I_2$ vanish we can write

$$\hat{U} = \exp(i\pi b^\dagger b \otimes I_2) \exp\left(\frac{1}{2}i\pi(I_B \otimes \sigma_z)\right) \exp\left(\frac{1}{2}i\pi(I_B \otimes I_2)\right).$$

We find

$$\hat{U}|n\rangle \otimes \begin{pmatrix} 1 \\ 0 \end{pmatrix} = e^{i\pi(n+1)}|n\rangle \otimes \begin{pmatrix} 1 \\ 0 \end{pmatrix}$$

$$\hat{U}|n\rangle \otimes \begin{pmatrix} 0 \\ 1 \end{pmatrix} = e^{i\pi n}|n\rangle \otimes \begin{pmatrix} 0 \\ 1 \end{pmatrix}.$$

Thus the basis elements are eigenfunctions of $\hat{U}$ and the eigenvalues of $\hat{U}$ are given by $1, -1$, because

$$e^{i\pi m} = 1 \quad \text{for } m \text{ even}, \qquad e^{i\pi m} = -1 \quad \text{for } m \text{ odd}.$$

The eigenvalues are infinitely degenerate. Since $\hat{U}$ and $I_B \otimes I_2$ form a finite group (isomorphic to C_2) we can decompose the Hilbert space $\mathcal{H}_B \otimes \mathbb{C}^2$ into two invariant subspaces. We obtain the two subspaces

$$S_1 = \left\{ |0\rangle \otimes \begin{pmatrix} 0 \\ 1 \end{pmatrix}, \quad |1\rangle \otimes \begin{pmatrix} 1 \\ 0 \end{pmatrix}, \quad |2\rangle \otimes \begin{pmatrix} 0 \\ 1 \end{pmatrix}, \quad |3\rangle \otimes \begin{pmatrix} 1 \\ 0 \end{pmatrix}, \ldots \right\}$$

$$S_2 = \left\{ |0\rangle \otimes \begin{pmatrix} 1 \\ 0 \end{pmatrix}, \quad |1\rangle \otimes \begin{pmatrix} 0 \\ 1 \end{pmatrix}, \quad |2\rangle \otimes \begin{pmatrix} 1 \\ 0 \end{pmatrix}, \quad |3\rangle \otimes \begin{pmatrix} 0 \\ 1 \end{pmatrix}, \ldots \right\}.$$

The operator $\hat{U}$ is the so-called *parity operator*.

(iii) Since the Hamilton operator $\hat{H}$ commutes with $\hat{N}$, we can write the infinite matrix representation of $\hat{H}$ as the direct sum of 2×2 matrices.

This can also be seen when we apply the Hamilton operator to the basis elements. We find

$$\hat{H}|n\rangle \otimes \begin{pmatrix} 1 \\ 0 \end{pmatrix} = (-\Delta + \Omega n)|n\rangle \otimes \begin{pmatrix} 1 \\ 0 \end{pmatrix} + k\sqrt{n+1}|n+1\rangle \otimes \begin{pmatrix} 0 \\ 1 \end{pmatrix}$$

$$\hat{H}|n\rangle \otimes \begin{pmatrix} 0 \\ 1 \end{pmatrix} = (\Delta + \Omega n)|n\rangle \otimes \begin{pmatrix} 0 \\ 1 \end{pmatrix} + k\sqrt{n}|n-1\rangle \otimes \begin{pmatrix} 1 \\ 0 \end{pmatrix}.$$

Thus for the subspace S_1 we have the direct sums

$$\begin{pmatrix} -\Delta & k \\ k & \Delta + \Omega \end{pmatrix} \oplus \begin{pmatrix} -\Delta + 2\Omega & \sqrt{3}k \\ \sqrt{3}k & \Delta + 3\Omega \end{pmatrix} \oplus \cdots$$

$$\oplus \begin{pmatrix} -\Delta + 2n\Omega & \sqrt{2n+1}k \\ \sqrt{2n+1}k & \Delta + (2n+1)\Omega \end{pmatrix} \oplus \cdots$$

For the subspace S_2 we find

$$\Delta \oplus \begin{pmatrix} -\Delta + \Omega & \sqrt{2}k \\ \sqrt{2}k & \Delta + 2\Omega \end{pmatrix} \oplus \cdots \oplus \begin{pmatrix} -\Delta + (2n+1)\Omega & \sqrt{2n+2}k \\ \sqrt{2n+2}k & \Delta + (2n+2)\Omega \end{pmatrix} \oplus \cdots$$

The eigenvalues of $\hat{H}$ are the eigenvalues of these 2×2 matrices.

Problem 6.   Calculate

$$f(\epsilon) = \exp(\epsilon\sigma_z \otimes (b - b^\dagger))(\sigma_z \otimes (b^\dagger + b)) \exp(-\epsilon\sigma_z \otimes (b - b^\dagger)) \qquad (1)$$

where ϵ is a real parameter, σ_z is the Pauli matrix and b and $b^\dagger$ are Bose annihilation and creation operators, i.e. $[b, b^\dagger] = I$.

Solution 6.   We find a differential equation for f and then solve the initial value problem of the linear differential equation. From (1) we find the initial condition

$$f(0) = \sigma_z \otimes (b^\dagger + b).$$

Differentiating (1) with respect to ϵ yields

$$\frac{df(\epsilon)}{d\epsilon} = \exp(\epsilon\sigma_z \otimes (b - b^\dagger))(\sigma_z \otimes (b^\dagger - b))(\sigma_z \otimes (b^\dagger + b)) \exp(-\epsilon\sigma_z \otimes (b - b^\dagger))$$

$$- \exp(\epsilon\sigma_z \otimes (b - b^\dagger))(\sigma_z \otimes (b^\dagger + b))(\sigma_z \otimes (b - b^\dagger)) \exp(-\epsilon\sigma_z \otimes (b - b^\dagger)).$$

We have $\sigma_z^2 = I_2$, where I_2 is the 2×2 unit matrix and

$$(b - b^\dagger)(b^\dagger + b) - (b^\dagger + b)(b - b^\dagger) = 2I_b$$

where I_b is the unit operator. Thus we find the linear first order differential equation

$$\frac{df(\epsilon)}{d\epsilon} = 2I_2 \otimes I_b.$$

Integrating this differential equation and inserting the initial condition we obtain

$$f(\epsilon) = 2\epsilon I_2 \otimes I_b + \sigma_z \otimes (b^\dagger + b).$$

Problem 7. Consider the Hamilton operator

$$\hat{H} = \hat{H}_0 + g(b^\dagger \otimes S^- + b \otimes S^+) + \gamma b^\dagger b^\dagger bb \otimes I_S + \gamma(I_B \otimes S_z)^2 \quad (1)$$

with

$$\hat{H}_0 = \omega b^\dagger b \otimes I_S + \omega_0 I_B \otimes S_z.$$

Here g and γ are real coupling constants, and

$$S^\pm := \sum_{n=1}^{N} \sigma_n^\pm, \qquad S_z := \frac{1}{2}\sum_{n=1}^{N} \sigma_n^z$$

are collective N-atom Dicke operators, i.e. spin operators for which total spin $S \le N/2$. The operators satisfy the $su(2)$ Lie algebra, i.e.,

$$[S^+, S^-] = 2S_z, \qquad [S_z, S^\pm] = \pm S^\pm.$$

The Bose operators $b^\dagger$ and b satisfy the commutation relation $[b, b^\dagger] = I$. The number operator $\hat{M}$ is given by

$$\hat{M} = I_B \otimes S_z + b^\dagger b \otimes I_S.$$

(i) Find the commutator $[\hat{H}, \hat{M}]$.

(ii) Find

$$\hat{K} := g^{-1}(\hat{H} + (\gamma - \omega)\hat{M} - \gamma \hat{M}^2).$$

(iii) The total spin operator of the model is given by the Casimir operator

$$\hat{S}^2 = S^+ S^- + S_z(S_z - I_S).$$

Find the commutator $[I_B \otimes \hat{S}^2, \hat{K}]$.

Solution 7. (i) From the commutation relations for the Bose operators we find

$$[b^\dagger b^\dagger bb, b^\dagger b] = 0$$

and

$$[b^\dagger, b^\dagger b] = -b^\dagger, \qquad [b, b^\dagger b] = b.$$

Using these commutation relations and

$$[\hat{A} \otimes I_S, I_B \otimes \hat{B}] = 0$$

we find that

$$[\hat{H}, \hat{M}] = 0.$$

(ii) We find

$$\hat{K} = \Delta I_B \otimes S_z + (b^\dagger \otimes S^- + b \otimes S^+) + cb^\dagger b \otimes S_z$$

where

$$\Delta := g^{-1}(\omega_0 - \omega + \gamma), \qquad c := -2g^{-1}\gamma.$$

The last term on the right-hand side of of $\hat{K}$ causes photon number dependent changes in the atomic transitions and therefore describes a *Stark shift*. Since $[\hat{H}, \hat{M}] = 0$ we obtain

$$[\hat{K}, \hat{M}] = 0.$$

(iii) We obtain

$$[I_B \otimes \hat{S}^2, \hat{K}] = 0.$$

Since $[\hat{M}, \hat{K}] = 0$ and $[\hat{S}^2, \hat{K}] = 0$, it is convenient to decompose the Hilbert space

$$\mathcal{H} = \mathcal{H}_B \otimes \mathbb{C}_1^2 \otimes \cdots \otimes \mathbb{C}_N^2$$

in terms of the irreducible representations of the Lie algebra $su(2)$ with spin S and excitation numbers M.

Chapter 33

Lax Representations and Bethe Ansatz

Problem 1. Let

$$\psi_{m+1} = L_m \psi_m \tag{1}$$

$$\frac{d\psi_m}{dt} = M_m \psi_m \tag{2}$$

where the entries of the square matrices L_m and M_m depend on time t, λ is the so-called spectral parameter which does not depend on time and $m \in \mathbb{Z}$ or $m \in \mathbb{Z}_N$. Show that

$$\frac{dL_m}{dt} = M_{m+1} L_m - L_m M_m. \tag{3}$$

L_m and M_m are called a *Lax pair*. We can find conservation laws from L_m.

For a number of solid state physics-models, the *Heisenberg equation of motion*

$$i\hbar \frac{d\hat{A}}{dt} = [\hat{A}, \hat{H}](t)$$

can be written as (3). An example is the XY-Hamilton operator

$$\hat{H} = -\frac{1}{2} \sum_{j=1}^{N} (J_x S_j^x S_{j+1}^x + J_y S_j^y S_{j+1}^y + h S_j^z)$$

with periodic boundary condition $S_{N+1}^\alpha \equiv S_N^\alpha$, where $\alpha = x, y, z$.

355

Solution 1.   From (1) we obtain

$$\frac{d\psi_{m+1}}{dt} = \frac{dL_m}{dt}\psi_m + L_m\frac{d\psi_m}{dt}. \tag{4}$$

Equation (2) yields

$$\frac{d\psi_{m+1}}{dt} = M_{m+1}\psi_{m+1}. \tag{5}$$

Inserting (5) and (2) into (4) gives

$$M_{m+1}\psi_{m+1} = \frac{dL_m}{dt}\psi_m + L_mM_m\psi_m.$$

Inserting (1) into this equation leads to

$$M_{m+1}L_m\psi_m = \frac{dL_m}{dt}\psi_m + L_mM_m\psi_m.$$

Consequently,

$$\frac{dL_m}{dt}\psi_m = (M_{m+1}L_m - L_mM_m)\psi_m.$$

Since ψ_m is arbitrary we obtain (3).

Problem 2.   The one-dimensional field-quantized *nonlinear Schrödinger model* is defined by the Hamilton operator

$$\hat{H} = \int_0^L \left(\frac{\partial\psi^\dagger}{\partial x}\frac{\partial\psi}{\partial x} + c\psi^\dagger\psi^\dagger\psi\psi \right) dx \tag{1}$$

where ψ is a nonrelativistic Bose field with canonical equal-time commutation relations

$$[\psi(x,t), \psi^\dagger(x',t)] = \delta(x - x') \tag{2a}$$

$$[\psi(x,t), \psi(x',t)] = 0 \tag{2b}$$

where $\delta(x-x')$ is the delta function and c is a real constant. The Hamilton operator (1) is in the standard form of a many body problem with the second term corresponding to a two-body delta function potential. The Hamilton operator (1) commutes with the particle number operator

$$\hat{N} := \int_0^L \psi^\dagger\psi dx. \tag{2c}$$

We impose cyclic boundary conditions, i.e. $\psi(0) = \psi(L)$. Let

$$|\phi(k_1,k_2)\rangle := \int_0^L \int_0^L dx_1 dx_2 e^{i(k_1x_1+k_2x_2)}(\theta(x_1 - x_2)$$
$$+S(k_{21})\theta(x_2 - x_1))\psi^\dagger(x_1)\psi^\dagger(x_2)|0\rangle$$

be a two-particle state, where $k_{21} := k_2 - k_1$ and

$$S(k_{21}) := \frac{k_2 - k_1 - ic}{k_2 - k_1 + ic}.$$

Moreover $\langle 0|0 \rangle = 1$ and θ denotes the step function with $\theta(0) = 1$. The right-hand side of this state is called a *Bethe ansatz*. Calculate the state

$$\hat{H}|\phi(k_1, k_2)\rangle$$

and discuss. Use $\psi|0\rangle = 0$, where $|0\rangle$ is the *vacuum state*, and integration by parts. Furthermore apply

$$\frac{\partial}{\partial x_1}\theta(x_1 - x_2) = \delta(x_1 - x_2).$$

Solution 2. The Hamilton operator can be written as $\hat{H} = \hat{H}_k + \hat{H}_I$, where $\hat{H}_k$ denotes the kinetic part and $\hat{H}_I$ denotes the interacting part. First we consider the interacting part, i.e.

$$\hat{H}_I = c \int_0^L \psi^\dagger \psi^\dagger \psi\psi dx.$$

We set

$$f(x_1, x_2) := e^{i(k_1 x_1 + k_2 x_2)}(\theta(x_1 - x_2) + S(k_{21})\theta(x_2 - x_1)).$$

It follows that

$$f(x_1, x_1) = e^{i(k_1 x_1 + k_2 x_1)}(\theta(0) + S(k_{21})\theta(0)).$$

Since $\theta(0) = 1$, the parenthesis of the right-hand side of this expression can be written as

$$1 + S(k_{21}) = 1 + \frac{k_{21} - ic}{k_{21} + ic} = \frac{2k_{21}}{k_{21} + ic}.$$

Using the commutation relations we find

$$H_I|\phi(k_1, k_2)\rangle = c \int dx dx_1 dx_2 \psi^\dagger(x)\psi^\dagger(x)\psi(x)\psi(x)f(x_1, x_2)\psi^\dagger(x_1)\psi^\dagger(x_2)|0\rangle$$

$$= c \int dx dx_1 dx_2 f(x_1, x_2)\psi^\dagger(x)\psi^\dagger(x)\psi(x)\delta(x - x_1)\psi^\dagger(x_2)|0\rangle$$

$$+ c \int dx dx_1 dx_2 f(x_1, x_2)\psi^\dagger(x)\psi^\dagger(x)\psi(x)\psi^\dagger(x_1)\psi(x)\psi^\dagger(x_2)|0\rangle$$

where the integration is from 0 to L. It follows that

$$\hat{H}_I|\phi(k_1,k_2)\rangle = c\int dx_1 dx_2 f(x_1,x_2)\psi^\dagger(x_1)\psi^\dagger(x_1)\psi(x_1)\psi^\dagger(x_2)|0\rangle$$

$$+ c\int dx dx_1 dx_2 f(x_1,x_2)\psi^\dagger(x)\psi^\dagger(x)\psi(x)\psi^\dagger(x_1)\psi(x)\psi^\dagger(x_2)|0\rangle$$

$$= c\int dx_1 dx_2 f(x_1,x_2)\psi^\dagger(x_1)\psi^\dagger(x_1)\psi(x_1)\psi^\dagger(x_2)|0\rangle$$

$$+ c\int dx dx_1 dx_2 f(x_1,x_2)\psi^\dagger(x)\psi^\dagger(x)\psi(x)\psi^\dagger(x_1)\delta(x-x_2)|0\rangle$$

$$= c\int dx_1 dx_2 f(x_1,x_2)\psi^\dagger(x_1)\psi^\dagger(x_1)\delta(x_1-x_2)|0\rangle$$

$$+ c\int dx_1 dx_2 f(x_1,x_2)\psi^\dagger(x_2)\psi^\dagger(x_2)\psi(x_2)\psi^\dagger(x_1)|0\rangle$$

$$= 2c\int dx_1 f(x_1,x_1)\psi^\dagger(x_1)\psi^\dagger(x_1)|0\rangle$$

$$= 2c\int dx_1 e^{i(k_1 x_1 + k_2 x_1)}(\theta(0) + S(k_{21})\theta(0))\psi^\dagger(x_1)\psi^\dagger(x_1)|0\rangle$$

$$= 2c\int dx_1 e^{i(k_1 x_1 + k_2 x_1)}\left(\frac{2(k_2-k_1)}{k_2-k_1+ic}\right)\psi^\dagger(x_1)\psi^\dagger(x_1)|0\rangle$$

$$= 4c\int_0^L dx_1 e^{i(k_1 x_1 + k_2 x_1)}\frac{k_2-k_1}{k_2-k_1+ic}\psi^\dagger(x_1)\psi^\dagger(x_1)|0\rangle.$$

Next we consider the kinetic part. Applying integration by parts we find

$$\hat{H}_k|\phi(k_1,k_2)\rangle = \int dx dx_1 dx_2 \frac{\partial \psi^\dagger}{\partial x}\frac{\partial \psi}{\partial x}f(x_1,x_2)\psi^\dagger(x_1)\psi^\dagger(x_2)|0\rangle.$$

Thus

$$\hat{H}_k|\phi(k_1,k_2)\rangle = -\int dx dx_1 dx_2 \frac{\partial^2 \psi^\dagger}{\partial x^2}f(x_1,x_2)\psi(x)\psi^\dagger(x_1)\psi^\dagger(x_2)|0\rangle$$

where we have used the cyclic boundary condition $\psi(0) = \psi(L)$. It follows that

$$\hat{H}_k|\phi(k_1,k_2)\rangle = -\int dx dx_1 dx_2 \frac{\partial^2 \psi^\dagger}{\partial x^2}f(x_1,x_2)\delta(x-x_1)\psi^\dagger(x_2)|0\rangle$$

$$- \int dx dx_1 dx_2 \frac{\partial^2 \psi^\dagger}{\partial x^2}f(x_1,x_2)\psi^\dagger(x_1)\psi(x)\psi^\dagger(x_2)|0\rangle$$

$$= -\int dx_1 dx_2 \frac{\partial^2 \psi^\dagger}{\partial x_1^2}f(x_1,x_2)\psi^\dagger(x_2)|0\rangle$$

$$- \int dx dx_1 dx_2 \frac{\partial^2 \psi^\dagger}{\partial x^2}f(x_1,x_2)\psi^\dagger(x_1)\delta(x-x_2)|0\rangle$$

$$= -\int dx_1 dx_2 \frac{\partial^2 \psi^\dagger}{\partial x_1^2} f(x_1, x_2) \psi^\dagger(x_2)|0\rangle$$

$$- \int dx_1 dx_2 \frac{\partial^2 \psi^\dagger}{\partial x_2^2} f(x_1, x_2) \psi^\dagger(x_1)|0\rangle$$

$$= -\int dx_1 dx_2 \frac{\partial^2 f}{\partial x_1^2} \psi^\dagger(x_1) \psi^\dagger(x_2)|0\rangle$$

$$- \int dx_1 dx_2 \frac{\partial^2 f}{\partial x_2^2} \psi^\dagger(x_2) \psi^\dagger(x_1)|0\rangle.$$

From the function f we obtain the derivatives in the sense of generalized functions

$$\frac{\partial f}{\partial x_1} = ik_1 e^{i(k_1 x_1 + k_2 x_2)}(\theta(x_1 - x_2) + S(k_{21})\theta(x_2 - x_1))$$

$$+ e^{i(k_1 x_1 + k_2 x_2)}(\delta(x_1 - x_2) - S(k_{21})\delta(x_2 - x_1))$$

$$\frac{\partial^2 f}{\partial x_1^2} = -k_1^2 e^{i(k_1 x_1 + k_2 x_2)}(\theta(x_1 - x_2) + S(k_{21})\theta(x_2 - x_1))$$

$$+ 2ik_1 e^{i(k_1 x_1 + k_2 x_2)}(\delta(x_1 - x_2) - S(k_{21})\delta(x_2 - x_1))$$

$$+ e^{i(k_1 x_1 + k_2 x_2)}(\delta'(x_1 - x_2) - S(k_{21})\delta'(x_2 - x_1))$$

$$\frac{\partial f}{\partial x_2} = ik_2 e^{i(k_1 x_1 + k_2 x_2)}(\theta(x_1 - x_2) + S(k_{21})\theta(x_2 - x_1))$$

$$+ e^{i(k_1 x_1 + k_2 x_2)}(-\delta(x_1 - x_2) + S(k_{21})\delta(x_2 - x_1))$$

$$\frac{\partial^2 f}{\partial x_2^2} = -k_2^2 e^{i(k_1 x_1 + k_2 x_2)}(\theta(x_1 - x_2) + S(k_{21})\theta(x_2 - x_1))$$

$$+ 2ik_2 e^{i(k_1 x_1 + k_2 x_2)}(-\delta(x_1 - x_2) + S(k_{21})\delta(x_2 - x_1))$$

$$+ e^{i(k_1 x_1 + k_2 x_2)}(-\delta'(x_1 - x_2) + S(k_{21})\delta'(x_2 - x_1)).$$

We arrive at

$$\hat{H}_k|\phi(k_1, k_2)\rangle = (k_1^2 + k_2^2)|\phi(k_1, k_2)\rangle$$

$$- \int dx_1 2ik_1 e^{i(k_1 x_1 + k_2 x_1)}(1 - S(k_{21}))\psi^\dagger(x_1)\psi^\dagger(x_1)|0\rangle$$

$$- \int dx_1 2ik_2 e^{i(k_1 x_1 + k_2 x_1)}(-1 + S(k_{21}))\psi^\dagger(x_1)\psi^\dagger(x_1)|0\rangle$$

$$= (k_1^2 + k_2^2)|\phi(k_1, k_2)\rangle$$

$$- \int_0^L dx_1 e^{i(k_1 x_1 + k_2 x_1)} 2i(k_1 - k_1 S(k_{21}) - k_2 + k_2 S(k_{21}))$$

$$\times \psi^\dagger(x_1)\psi^\dagger(x_1)|0\rangle.$$

Since

$$2i(k_1 - k_2 - k_1 S(k_{21}) + k_2 S(k_{21})) = 4c \frac{k_2 - k_1}{k_2 - k_1 + ic}$$

the second term of the right-hand side of this equation cancels with the term $\hat{H}_I|\phi(k_1, k_2)\rangle$. Consequently we find

$$\hat{H}|\phi(k_1, k_2)\rangle = \hat{H}_k|\phi(k_1, k_2)\rangle + \hat{H}_I|\phi(k_1, k_2)\rangle = (k_1^2 + k_2^2)|\phi(k_1, k_2)\rangle.$$

Thus the state $|\phi(k_1, k_2)\rangle$ is an eigenstate of $\hat{H}$ and the eigenvalue is given by $k_1^2 + k_2^2$. The Bethe ansatz can be extended to N particles.

Problem 3.   Consider the 2^N dimensional Hilbert space

$$\mathcal{H}_N = \prod_{n=1}^{N} \otimes \mathbb{C}^2$$

and the Hamilton operator

$$\hat{H}_N = \frac{J}{4} \sum_{n=1}^{N} (\sigma_{1,n}\sigma_{1,n+1} + \sigma_{2,n}\sigma_{2,n+1} + \sigma_{3,n}\sigma_{3,n+1} - I_N) , \qquad (1)$$

in this Hilbert space. Here J is a real constant and I_N is the identity operator ($2^N \times 2^N$ unit matrix) in the space $\mathcal{H}_N$ ($\dim \mathcal{H}_N = 2^N$). The operators $\sigma_{j,n}$ with ($j = 1, 2, 3$) have the following form

$$\sigma_{j,n} := I_2 \otimes \cdots \otimes I_2 \otimes \sigma_j \otimes I_2 \otimes \cdots \otimes I_2 \quad (N - \text{factors})$$
$$\downarrow$$
$$n - th \text{ place}$$

where I_2 is the 2×2 identity matrix, $n = 1, \ldots, N$ and σ_1, σ_2, σ_3 are the Pauli matrices. Here $\otimes$ denotes the Kronecker product. Let $\sigma_{j,N+1} \equiv \sigma_{j,1}$ (periodic boundary condition). Let

$$L_n(\lambda) := \begin{pmatrix} \lambda I_n + \frac{i}{2}\sigma_{3,n} & \frac{i}{2}\sigma_{-,n} \\ \frac{i}{2}\sigma_{+,n} & \lambda I_n - \frac{i}{2}\sigma_{3,n} \end{pmatrix}$$

be an operator valued matrix of order 2×2 with

$$\sigma_{+,n} := \sigma_{1,n} + i\sigma_{2,n}, \qquad \sigma_{-,n} := \sigma_{1,n} - i\sigma_{2,n} .$$

Here λ is a complex parameter. Thus $L_n(\lambda)$ is a $2^{N+1} \times 2^{N+1}$ matrix. L_n is called a *Lax operator*. The matrix $L_n(\lambda)$ can also be represented in the form

$$L_n(\lambda) = \lambda I_2 \otimes I_N + \frac{i}{2} \sum_{j=1}^{3} \sigma_j \otimes \sigma_{j,n} .$$

Show that (*Yang-Baxter relation*)

$$R(\lambda - \mu)(L_n(\lambda) \otimes L_n(\mu)) \equiv (L_n(\mu) \otimes L_n(\lambda))R(\lambda - \mu)$$

where

$$R(\lambda) = \frac{1}{\lambda + i}\left(\left(\frac{\lambda}{2} + i\right) I_2 \otimes I_2 + \frac{\lambda}{2}\sum_{a=1}^{3}\sigma^a \otimes \sigma^a\right).$$

We define

$$(\sigma_1 \otimes \sigma_1)(L_n(\lambda) \otimes L_n(\mu)) := (\sigma_1 L_n(\lambda)) \otimes (\sigma_1 L_n(\mu))$$

and

$$\begin{pmatrix} 0 & 1 \\ 1 & 0 \end{pmatrix}\begin{pmatrix} \lambda I_N + \frac{i}{2}\sigma_n^3 & \frac{i}{2}\sigma_n^- \\ \frac{i}{2}\sigma_n^+ & \lambda I_N - \frac{i}{2}\sigma_n^3 \end{pmatrix} = \begin{pmatrix} \frac{i}{2}\sigma_n^+ & \lambda I_N - \frac{i}{2}\sigma_n^3 \\ \lambda I_N + \frac{i}{2}\sigma_n^3 & \frac{i}{2}\sigma_n^- \end{pmatrix}$$

and so on. Owing to this definition, one says that the 2×2 identity matrix I_2 and the Pauli matrices σ_j act in the auxiliary space $\mathbb{C}^2$.

Solution 3. We put

$$L_n(\lambda) = \begin{pmatrix} a & b \\ c & d \end{pmatrix}, \qquad L_n(\mu) = \begin{pmatrix} e & f \\ g & h \end{pmatrix}$$

and $D := 1/(\lambda - \mu + i)$. Obviously $b = f$ and $c = g$ with

$$b = f = \frac{i}{2}\sigma_n^-, \qquad c = g = \frac{i}{2}\sigma_n^+.$$

Then we find

$$R(\lambda - \mu)(L_n(\lambda) \otimes L_n(\mu))$$

$$= D\begin{pmatrix} \lambda - \mu + i & 0 & 0 & 0 \\ 0 & i & \lambda - \mu & 0 \\ 0 & \lambda - \mu & i & 0 \\ 0 & 0 & 0 & \lambda - \mu + i \end{pmatrix}\begin{pmatrix} ae & af & be & bf \\ ag & ah & bg & bh \\ ce & cf & de & df \\ cg & ch & dg & dh \end{pmatrix}$$

$$= D\begin{pmatrix} (\lambda - \mu + i)ae & (\lambda - \mu + i)af & (\lambda - \mu + i)be & (\lambda - \mu + i)bf \\ iag + (\lambda - \mu)ce & iah + (\lambda - \mu)cf & ibg + (\lambda - \mu)de & ibh + (\lambda - \mu)df \\ (\lambda - \mu)ag + ice & (\lambda - \mu)ah + icf & (\lambda - \mu)bg + ide & (\lambda - \mu)bh + idf \\ (\lambda - \mu + i)cg & (\lambda - \mu + i)ch & (\lambda - \mu + i)dg & (\lambda - \mu + i)dh \end{pmatrix}$$

and

$$(L_n(\mu) \otimes L_n(\lambda))R(\lambda - \mu)$$

$$= D\begin{pmatrix} ea & eb & fa & fb \\ ec & ed & fc & fd \\ ga & gb & ha & hb \\ gc & gd & hc & hd \end{pmatrix}\begin{pmatrix} \lambda - \mu + i & 0 & 0 & 0 \\ 0 & i & \lambda - \mu & 0 \\ 0 & \lambda - \mu & i & 0 \\ 0 & 0 & 0 & \lambda - \mu + i \end{pmatrix}$$

$$= D\begin{pmatrix} (\lambda - \mu + i)ea & ieb + (\lambda - \mu)fa & (\lambda - \mu)eb + ifa & (\lambda - \mu + i)fb \\ (\lambda - \mu + i)ec & ied + (\lambda - \mu)fc & (\lambda - \mu)ed + ifc & (\lambda - \mu + i)fd \\ (\lambda - \mu + i)ga & igb + (\lambda - \mu)ha & (\lambda - \mu)gb + iha & (\lambda - \mu + i)hb \\ (\lambda - \mu + i)gc & igd + (\lambda - \mu)hc & (\lambda - \mu)gd + ihc & (\lambda - \mu + i)hd \end{pmatrix}$$

For identity Yang-Baxter relation to be true, the matrices on the right-hand sides of these equations must be equal. We number the entries as follows

$$\begin{pmatrix} 1 & 2 & 5 & 6 \\ 3 & 4 & 7 & 8 \\ 9 & 10 & 13 & 14 \\ 11 & 12 & 15 & 16 \end{pmatrix}.$$

Therefore we have to prove 16 identities. The first identity to prove is

$$ea = ae.$$

From $ea - ae$ we find

$$(\mu I_N + \frac{i}{2}\sigma_n^3)(\lambda I_N + \frac{i}{2}\sigma_n^3) - (\lambda I_N + \frac{i}{2}\sigma_n^3)(\mu I_N + \frac{i}{2}\sigma_n^3) = 0$$

since λ and μ are complex parameters. Analogously, we can prove identity 16, i.e. $dh = hd$. Identity 6 is given by $fb = bf$. This is obviously true, because $f = b$. Analogously, identity 11 is satisfied. Identity 2 is given by

$$(\lambda - \mu + i)af = ieb + (\lambda - \mu)fa$$

or

$$(\lambda - \mu)(af - fa) + i(af - eb) = 0.$$

Using the commutation relation $[\sigma_{-,n}, \sigma_{3,n}] = 2\sigma_{-,n}$ we find that the identity is satisfied. By using the same identity, we can similarly prove 5, 8 and 14. Applying the commutation relation

$$[\sigma_{+,n}, \sigma_{3,n}] = -2\sigma_{+,n}$$

we can prove 3, 9, 12 and 15. The technique is the same as that used for identity 2. Identity 4 is given by

$$ied + (\lambda - \mu)fc = iah + (\lambda - \mu)cf.$$

Using the commutation relation

$$[\sigma_{+,n}, \sigma_{-,n}] = 4\sigma_{3,n}$$

we find that identity 4 holds. By using the same identity, we can prove identity 13 in the same way. Identity 7 is given by

$$(\lambda - \mu)ed + ifc = ibg + (\lambda - \mu)de.$$

Thus we have to prove that

$$(\lambda - \mu)(ed - de) + i(fc - bg) = 0.$$

For $ed - de$ we find

$$ed - de = (\mu I_N - \frac{i}{2}\sigma_{3,n})(\lambda I_N - \frac{i}{2}\sigma_{3,n}) - (\lambda I_N - \frac{i}{2}\sigma_{3,n})(\mu I_N - \frac{i}{2}\sigma_{3,n}) = 0.$$

For $fc - bg$ we obtain

$$fc - bg = \frac{i}{2}\sigma_n^- \frac{i}{2}\sigma_n^+ - \frac{i}{2}\sigma_n^- \frac{i}{2}\sigma_n^+ = 0.$$

Identity 11 is proved in the same way. Consequently, we have shown that the corresponding entries of the two matrices are equal. Thus the Yang-Baxter relation follows.

Problem 4. The classical massless *Thirring model* in (1+1) dimensions is given by

$$\frac{\partial\phi}{\partial t} + v\frac{\partial\phi}{\partial x} = -2iJ\chi^*\chi\phi \tag{1a}$$

$$\frac{\partial\chi}{\partial t} - v\frac{\partial\chi}{\partial x} = -2iJ\phi^*\phi\chi \tag{1b}$$

where v is the constant velocity and J is the coupling constant.
(i) Find the general solution of the initial-value problem

$$\phi(x,0) = a(x)\exp(ib(x)), \qquad \chi(x,0) = c(x)\exp(id(x)) \tag{2}$$

where a, b, c and d are real-valued functions.
(ii) Find the Lax pair for (1). To find the Lax pair we consider

$$\frac{\partial\Psi}{\partial t} = U\Psi, \qquad \frac{\partial\Psi}{\partial x} = V\Psi.$$

Here U and V are 4×4 matrices. The consistency condition $\partial^2\Psi/\partial x\partial t = \partial^2\Psi/\partial t\partial x$ for this system yields

$$\frac{\partial U}{\partial t} - \frac{\partial V}{\partial x} + [U, V] = 0$$

where $[U, V]$ denotes the commutator.

Solution 4. (i) Using the new independent variables $\xi := x - vt$, $\eta := x + vt$ we can write (1) in the form

$$v\frac{\partial\phi}{\partial\eta} = -iJ\chi^*\chi\phi, \qquad v\frac{\partial\chi}{\partial\xi} = iJ\phi^*\phi\chi. \tag{3}$$

Since

$$\frac{\partial(\phi^*\phi)}{\partial\eta} = 0, \qquad \frac{\partial(\chi^*\chi)}{\partial\xi} = 0$$

we set

$$\phi(\xi, \eta) = f(\xi) \exp(ik(\xi, \eta)), \qquad \chi(\xi, \eta) = g(\eta) \exp(i\ell(\xi, \eta)) \qquad (4)$$

where f and g are real arbitrary functions. The real functions k and ℓ are determined as follows. Substituting the expressions (4) into (3) yields

$$v\frac{\partial k}{\partial \eta} = -Jg^2(\eta), \qquad v\frac{\partial \ell}{\partial \xi} = Jf^2(\xi). \qquad (5)$$

Integrating this system of partial differential equations yields

$$k(\xi, \eta) = -\frac{J}{v} \int^\eta g^2(\eta)d\eta + \theta(\xi), \qquad \ell(\xi, \eta) = \frac{J}{v} \int^\xi f^2(\xi)d\xi + \epsilon(\eta) \quad (6)$$

where θ and ϵ are real arbitrary functions. With (4), (6) and (3) we arrive at the general solutions for (1)

$$\phi(x, t) = f(x - vt) \exp\left(-i\frac{J}{v} \int^{x+vt} g^2(\eta)d\eta + i\theta(x - vt)\right)$$

$$\chi(x, t) = g(x + vt) \exp\left(i\frac{J}{v} \int^{x-vt} f^2(\xi)d\xi + i\epsilon(x + vt)\right).$$

Suppose that the initial conditions are given by (2) we obtain

$$f(x) = a(x), \qquad \theta(x) = \frac{J}{v} \int^x c^2(\eta)d\eta + b(x)$$

$$g(x) = c(x), \qquad \epsilon(x) = -\frac{J}{v} \int^x a^2(\xi)d\xi + d(x).$$

Thus we find

$$\phi(x, t) = a(x - vt) \exp\left(-i\frac{J}{v} \int_{x-vt}^{x+vt} c^2(\eta)d\eta + ib(x - vt)\right)$$

$$\chi(x, t) = c(x + vt) \exp\left(-i\frac{J}{v} \int_{x-vt}^{x+vt} a^2(\xi)d\xi + id(x + vt)\right).$$

(ii) We assume that the 4×4 matrices U and V take the following form

$$U = \lambda U^{(1)} + U^{(0)}, \qquad V = \lambda V^{(1)} + V^{(0)}.$$

Substituting the this expression into the consistency condition and equating the terms with the same powers of λ yields

$$[U^{(1)}, V^{(1)}] = 0$$

$$\frac{\partial U^{(1)}}{\partial t} - \frac{\partial V^{(1)}}{\partial x} + [U^{(1)}, V^{(0)}] + [U^{(0)}, V^{(1)}] = 0$$

$$\frac{\partial U^{(0)}}{\partial t} - \frac{\partial V^{(0)}}{\partial x} + [U^{(0)}, V^{(0)}] = 0.$$

We set

$$U^{(1)} := \begin{pmatrix} A & 0 \\ 0 & B \end{pmatrix}, \qquad U^{(0)} := \begin{pmatrix} C & 0 \\ 0 & D \end{pmatrix},$$

$$V^{(1)} := \begin{pmatrix} -vA & 0 \\ 0 & vB \end{pmatrix}, \qquad V^{(0)} := \begin{pmatrix} vC & 0 \\ 0 & -vD \end{pmatrix}$$

where A, B, C and D are 2×2 matrices. With this choice, $[U^{(1)}, V^{(1)}] = 0$ is satisfied identically. For the other two equations we obtain

$$\frac{\partial A}{\partial t} + v\frac{\partial A}{\partial x} + 2v[A, C] = 0, \qquad \frac{\partial B}{\partial t} - v\frac{\partial B}{\partial x} - 2v[B, D] = 0$$

and

$$\frac{\partial C}{\partial t} - v\frac{\partial C}{\partial x} = 0, \qquad \frac{\partial D}{\partial t} + v\frac{\partial D}{\partial x} = 0.$$

Thus we can set

$$A = \begin{pmatrix} 0 & \phi \\ \phi^* & 0 \end{pmatrix}, \qquad B = \begin{pmatrix} 0 & \chi \\ \chi^* & 0 \end{pmatrix},$$

$$C = -\frac{iJ}{2v}\begin{pmatrix} \chi^*\chi & 0 \\ 0 & -\chi^*\chi \end{pmatrix}, \qquad D = \frac{iJ}{2v}\begin{pmatrix} \phi^*\phi & 0 \\ 0 & -\phi^*\phi \end{pmatrix}.$$

These partial differential equations are the massless Thirring model (1).

Problem 5. The quantum three wave interaction equation in one-space dimension is described by the Hamilton operator

$$\hat{H} = \int dx \left(\sum_{j=1}^{3} c_j b_j^* \left(\frac{1}{i}\frac{\partial}{\partial x} \right) b_j + g(b_2^* b_1 b_3 + b_1^* b_3^* b_2) \right) \qquad (1)$$

where c_j's ($j = 1, 2, 3$) are constant velocities and g is the coupling constant. We assume that all c_j's are distinct. The three fields b_j's are bosons which satisfy the equal time commutation relations

$$[b_j(x, t), b_k^*(y, t)] = \delta_{jk}\delta(x - y) \qquad (2a)$$

$$[b_j(x, t), b_k(y, t)] = [b_j^*(x, t), b_k^*(y, t)] = 0. \qquad (2b)$$

We define the *vacuum state* by

$$b_j(x, t)|0\rangle = 0, \qquad j = 1, 2, 3. \qquad (3)$$

(i) Find the equation of motion for $b_j(x, t)$.
(ii) Find the commutator of $\hat{H}$ with the operators

$$\hat{N}_a := \int dx(b_1^* b_1 + b_2^* b_2), \qquad \hat{N}_b := \int dx(b_2^* b_2 + b_3^* b_3).$$

We define the states

$$\|0,0\rangle\rangle := |0\rangle$$

$$\|M,0\rangle\rangle := \int \cdots \int dx_1 \cdots dx_M$$
$$\times \exp(i(p_1 x_1 + \cdots + p_M x_M)) b_1^*(x_1,t) \cdots b_1^*(x_M,t)|0\rangle$$

$$\|0,N\rangle\rangle := \int \cdots \int dx_1 \cdots dx_N$$
$$\times \exp(i(q_1 x_1 + \cdots + q_N x_N)) b_3^*(x_1,t) \cdots b_3^*(x_N,t)|0\rangle$$

where $M = 1, 2, \ldots$ and $N = 1, 2, \ldots$.
(iii) Find $\hat{H}\|M,0\rangle\rangle$. Find $\hat{H}\|0,N\rangle\rangle$.

Solution 5.   (i) The *Heisenberg equation of motion* is given by

$$\frac{\partial}{\partial t} b_j = i[\hat{H}, b_j(x,t)], \qquad j = 1, 2, 3.$$

Using the commutation relations it follows that

$$\frac{\partial b_1}{\partial t} + c_1 \frac{\partial b_1}{\partial x} = -igb_3^* b_2$$

$$\frac{\partial b_2}{\partial t} + c_2 \frac{\partial b_2}{\partial x} = -igb_1 b_3$$

$$\frac{\partial b_3}{\partial t} + c_3 \frac{\partial b_3}{\partial x} = -igb_1^* b_2.$$

(ii) Using the commutation relation we find that the operators $\hat{N}_A$ and $\hat{N}_B$ commute with $\hat{H}$.
(iii) Using

$$\left(\frac{\partial}{\partial x} b_1(x,t) \right) b_1^*(x_j,t) = \frac{\partial}{\partial x} (b_1(x,t) b_1^*(x_j,t))$$

$$= \frac{\partial}{\partial x} \delta(x - x_j) + b_1^*(x_j,t) \frac{\partial}{\partial x} b_1(x,t)$$

and the derivative of the delta function we obtain

$$\hat{H}\|M,0\rangle\rangle = \int dx \int \cdots \int dx_1 \ldots dx_M \exp(i(p_1 x_1 + \cdots + p_M x_M))$$

$$\times \left(b_1^*(x,t) c_1 \frac{1}{i} \frac{\partial}{\partial x} b_1(x,t) \right) b_1^*(x_1,t) \cdots b_1^*(x_M,t)|0\rangle$$

$$= E\|M,0\rangle\rangle$$

where the energy E is given by $E = c_1(p_1 + \cdots + p_M)$. Thus $\|M,0\rangle\rangle$ is an eigenstate. Similarly we find that $\|0,N\rangle\rangle$ is an eigenstate with the energy E given by

$$E = c_3(q_1 + \cdots + q_N).$$

Chapter 34

Gauge Transformations

Problem 1. Let H be a Hamilton function and

$$\omega := \sum_{j=1}^{n} p_j dq_j - H(\mathbf{p}, \mathbf{q}, t)dt \qquad (1)$$

be a *Cartan form* of a Hamilton system. Consider the transformation

$$q_j \to q_j, \quad p_j \to p_j + \frac{\partial \Omega(\mathbf{q}, t)}{\partial q_j}, \quad H(\mathbf{p}, \mathbf{q}, t) \to H(\mathbf{p}, \mathbf{q}, t) - \frac{\partial \Omega(\mathbf{q}, t)}{\partial t} \qquad (2)$$

where Ω is a smooth function of $\mathbf{q}$ and t. This transformation is called a *gauge transformation*.
(i) Show that under this transformation, $d\omega \to d\omega$.
(ii) Calculate the exterior derivative of ω, i.e. $d\omega$.
(iii) Calculate

$$Z \rfloor d\omega = 0$$

where $\rfloor$ denotes the *contraction* $(\partial/\partial x_j \rfloor dx_k = \delta_{jk})$ and Z is the vector field

$$Z := \sum_{j=1}^{n} \left(V_j(\mathbf{p}, \mathbf{q}, t)\frac{\partial}{\partial p_j} + W_j(\mathbf{p}, \mathbf{q}, t)\frac{\partial}{\partial q_j} \right) + \frac{\partial}{\partial t}.$$

Solution 1. (i) From (1) and transformation (2) we obtain

$$\omega \to \sum_{j=1}^{n} \left(p_j + \frac{\partial \Omega}{\partial q_j} \right) dq_j - \left(H - \frac{\partial \Omega}{\partial t} \right) dt = \omega + \sum_{j=1}^{n} \frac{\partial \Omega}{\partial q_j}dq_j + \frac{\partial \Omega}{\partial t}dt.$$

Now

$$d\left(\sum_{j=1}^{n}\frac{\partial\Omega}{\partial q_j}dq_j + \frac{\partial\Omega}{\partial t}dt\right) = \sum_{j=1}^{n}\sum_{k=1}^{n}\frac{\partial^2\Omega}{\partial q_j\partial q_k}dq_k \wedge dq_j + \sum_{j=1}^{n}\frac{\partial^2\Omega}{\partial t\partial q_j}dt \wedge dq_j$$

$$+ \sum_{j=1}^{n}\frac{\partial^2\Omega}{\partial t\partial q_j}dq_j \wedge dt.$$

Since $dq_j \wedge dq_k = -dq_k \wedge dq_j$ and $dq_j \wedge dt = -dt \wedge dq_j$ we find that the right-hand side of this equation vanishes.
(ii) The exterior derivative of ω yields

$$d\omega = \sum_{j=1}^{n}\left(dp_j \wedge dq_j - \frac{\partial H}{\partial q_j}dq_j \wedge dt - \frac{\partial H}{\partial p_j}dp_j \wedge dt\right).$$

(iii) The corresponding equations of motion of the vector field Z are given by

$$\frac{dp_j}{d\epsilon} = V_j(\mathbf{p},\mathbf{q},t), \qquad \frac{dq_j}{d\epsilon} = W_j(\mathbf{p},\mathbf{q},t), \qquad \frac{dt}{d\epsilon} = 1$$

where $t(\epsilon = 0) = 0$. Therefore, we find

$$Z\rfloor d\omega = \sum_{j=1}^{n}\left(V_j dq_j - W_j dp_j - W_j\frac{\partial H}{\partial q_j}dt - V_j\frac{\partial H}{\partial p_j}dt + \frac{\partial H}{\partial q_j}dq_j + \frac{\partial H}{\partial p_j}dp_j\right).$$

Thus the condition $Z\rfloor d\omega = 0$ yields the *Hamilton equations of motion*

$$V_j(\mathbf{p},\mathbf{q},t) = -\frac{\partial H}{\partial q_j}, \qquad W_j(\mathbf{p},\mathbf{q},t) = \frac{\partial H}{\partial p_j}.$$

Problem 2.   The equation of motion of a charged particle with charge q in an electromagnetic field is given by

$$m\frac{d\mathbf{v}}{dt} = q(\mathbf{E} + \mathbf{v} \times \mathbf{B}). \tag{1}$$

(i) Show that the equation of motion is invariant under

$$\mathbf{A}(\mathbf{r},t) \to \mathbf{A}(\mathbf{r},t) + \text{grad}\phi(\mathbf{r},t), \qquad U(\mathbf{r},t) \to U(\mathbf{r},t) - \frac{\partial\phi(\mathbf{r},t)}{\partial t} \tag{2}$$

where ϕ is a smooth function. Transformation (2) is called a *gauge transformation*.
(ii) Show that the equation of motion (1) can be derived from the *Lagrange function*

$$L(\mathbf{r},\mathbf{v},t) = \frac{m\mathbf{v}^2}{2} + q(-U(\mathbf{r},t) + \mathbf{A}(\mathbf{r},t) \cdot \mathbf{v}) \tag{3}$$

where
$$\mathbf{A} \cdot \mathbf{v} = A_1 v_1 + A_2 v_2 + A_3 v_3, \qquad \mathbf{v}^2 = v_1^2 + v_2^2 + v_3^2.$$
Is the Lagrange function (3) invariant under the gauge transformation?

Solution 2.   (i) We write
$$\mathbf{A}' = \mathbf{A} + \mathrm{grad}\phi, \qquad U' = U - \frac{\partial \phi}{\partial t}.$$
Since
$$\mathbf{B} = \mathrm{curl}\mathbf{A}, \qquad \mathbf{E} = -\mathrm{grad}U - \frac{\partial \mathbf{A}}{\partial t}$$
we obtain
$$\mathbf{B}' = \mathrm{curl}\mathbf{A}' = \mathrm{curl}(\mathbf{A} + \mathrm{grad}\phi) = \mathrm{curl}\mathbf{A} = \mathbf{B}$$
where we used $\mathrm{curl}(\mathrm{grad}\phi) = 0$ for any smooth ϕ. Analogously $\mathbf{E}' = \mathbf{E}$. Thus $\mathbf{E}$ and $\mathbf{B}$ are invariant under the gauge transformation as well as the equation of motion (1) which only depends on $\mathbf{E}$ and $\mathbf{B}$.
(ii) The *Euler-Lagrange equations* are given by
$$\frac{d}{dt}\frac{\partial L}{\partial v_1} - \frac{\partial L}{\partial x} = 0, \qquad \frac{d}{dt}\frac{\partial L}{\partial v_2} - \frac{\partial L}{\partial y} = 0, \qquad \frac{d}{dt}\frac{\partial L}{\partial v_3} - \frac{\partial L}{\partial z} = 0.$$
Since
$$\frac{\partial L}{\partial v_1} = mv_1 + qA_1$$
we obtain
$$\frac{d}{dt}\frac{\partial L}{\partial v_1} = m\frac{dv_1}{dt} + q\frac{\partial A_1}{\partial t} + q\frac{\partial A_1}{\partial x}v_1 + q\frac{\partial A_1}{\partial y}v_2 + q\frac{\partial A_1}{\partial z}v_3.$$
Now
$$\frac{\partial L}{\partial x} = -q\frac{\partial U}{\partial x} + qv_1\frac{\partial A_1}{\partial x} + qv_2\frac{\partial A_2}{\partial x} + qv_3\frac{\partial A_3}{\partial x}.$$
Therefore, from the Euler-Lagrange equations we obtain
$$m\frac{dv_1}{dt} - qE_1 - q(v_2 B_3 - v_3 B_2) = 0.$$
Analogously we obtain the two other components of (1). Under the gauge transformation the Lagrange function L becomes $L' = L + qd\phi/dt$, where
$$\frac{d\phi}{dt} = \frac{\partial \phi}{\partial t} + \frac{\partial \phi}{\partial x}v_1 + \frac{\partial \phi}{\partial y}v_2 + \frac{\partial \phi}{\partial z}v_3.$$

Problem 3.   Consider the linear partial differential equations
$$\frac{\partial \Phi}{\partial x} = U\Phi, \qquad \frac{\partial \Phi}{\partial t} = V\Phi \tag{1}$$

and

$$\frac{\partial Q}{\partial x} = WQ, \qquad \frac{\partial Q}{\partial t} = ZQ \qquad (2)$$

where $\boldsymbol{\Phi} = (\Phi_1, \Phi_2)^T$ and U, V, W, Z are smooth x, t dependent 2×2 matrices. Let

$$Q(x,t) = g(x,t)\boldsymbol{\Phi}(x,t) \qquad (3)$$

where the 2×2 matrix g is invertible. Show that

$$U = -g^{-1}\frac{\partial g}{\partial x} + g^{-1}Wg, \qquad V = -g^{-1}\frac{\partial g}{\partial t} + g^{-1}Zg. \qquad (4)$$

Solution 3.   From $g^{-1}g = I_2$ we obtain

$$\frac{\partial g^{-1}}{\partial x}g + g^{-1}\frac{\partial g}{\partial x} = 0, \qquad \frac{\partial g^{-1}}{\partial t}g + g^{-1}\frac{\partial g}{\partial t} = 0.$$

From $Q = g\boldsymbol{\Phi}$ we obtain

$$\frac{\partial Q}{\partial x} = \frac{\partial g}{\partial x}\boldsymbol{\Phi} + g\frac{\partial \boldsymbol{\Phi}}{\partial x}, \qquad \frac{\partial Q}{\partial t} = \frac{\partial g}{\partial t}\boldsymbol{\Phi} + g\frac{\partial \boldsymbol{\Phi}}{\partial t}$$

and

$$\frac{\partial \boldsymbol{\Phi}}{\partial x} = \frac{\partial g^{-1}}{\partial x}Q + g^{-1}\frac{\partial Q}{\partial x}, \qquad \frac{\partial \boldsymbol{\Phi}}{\partial t} = \frac{\partial g^{-1}}{\partial t}Q + g^{-1}\frac{\partial Q}{\partial t}.$$

Inserting these equations into $\partial\boldsymbol{\Phi}/\partial x = U\boldsymbol{\Phi}$ and using $\boldsymbol{\Phi} = g^{-1}Q$ yields

$$\frac{\partial g^{-1}}{\partial x}Q + g^{-1}WQ = U\boldsymbol{\Phi} = Ug^{-1}Q.$$

Thus

$$\left(-g^{-1}\frac{\partial g}{\partial x}g^{-1} + g^{-1}W - Ug^{-1}\right)Q = 0.$$

It follows that the expression in the parenthesis of the left-hand side of this equation must vanish. Multiplying this expression with g on the right-hand side, we obtain the first equation of (4). Analogously we can prove the second one.

Problem 4.   Consider the *derivative nonlinear Schrödinger equation* in one-space dimension

$$\frac{\partial \phi}{\partial t} - i\frac{\partial^2 \phi}{\partial x^2} + \frac{\partial}{\partial x}(|\phi|^2\phi) = 0 \qquad (1)$$

and another derivative nonlinear Schrödinger equation in one-space dimension

$$\frac{\partial q}{\partial t} - i\frac{\partial^2 q}{\partial x^2} + |q|^2\frac{\partial q}{\partial x} = 0 \qquad (2)$$

where ϕ and q are complex-valued functions. Both partial differential equations are integrable by the inverse scattering method. Both arise as consistency conditions of a system of linear differential equations

$$\frac{\partial \Phi}{\partial x} = U(x, t, \lambda)\Phi \tag{3a}$$

$$\frac{\partial \Phi}{\partial t} = V(x, t, \lambda)\Phi \tag{3b}$$

$$\frac{\partial Q}{\partial x} = W(x, t, \lambda)Q \tag{4a}$$

$$\frac{\partial Q}{\partial t} = Z(x, t, \lambda)Q \tag{4b}$$

where λ is a complex parameter and U, V, W, Z are x, t dependent 2×2 matrices. The consistency conditions have the form

$$\frac{\partial U}{\partial t} - \frac{\partial V}{\partial x} + [U, V] = 0, \qquad \frac{\partial W}{\partial t} - \frac{\partial Z}{\partial x} + [W, Z] = 0.$$

For (1) we have

$$U = -i\lambda^2 \sigma_3 + \lambda \begin{pmatrix} 0 & \phi \\ -\phi^* & 0 \end{pmatrix}$$

$$V = -(2i\lambda^4 - i|\phi|^2\lambda^2)\sigma_3 + 2\lambda^3 \begin{pmatrix} 0 & \phi \\ -\phi^* & 0 \end{pmatrix} + \lambda \begin{pmatrix} 0 & i\phi_x - |\phi|^2\phi \\ i\phi_x^* + |\phi|^2\phi^* & 0 \end{pmatrix}$$

where σ_3 is the Pauli spin matrix. For (2) we have

$$W = \left(-i\lambda^2 + \frac{i}{4}|q|^2\right)\sigma_3 + \lambda \begin{pmatrix} 0 & q \\ -q^* & 0 \end{pmatrix}$$

$$Z = \left(-2i\lambda^4 - i|q|^2\lambda^2 + \frac{1}{4}(qq_x^* - q^*q_x) - \frac{i}{8}|q|^4\right)\sigma_3 + 2\lambda^3 \begin{pmatrix} 0 & q \\ -q^* & 0 \end{pmatrix}$$

$$+ \lambda \begin{pmatrix} 0 & iq_x - |q|^2q/2 \\ iq_x^* + |q|^2q^*/2 & 0 \end{pmatrix}.$$

Show that (1) and (2) are gauge-equivalent.

Solution 4. We look for a transformation which converts (4a) with W into the form of (3a) with U. For this purpose the transformation which eliminates the term

$$\frac{i|q|^2}{4}\sigma_3$$

in W is introduced, i.e., $Q = g\Phi$, where

$$g = \begin{pmatrix} f & 0 \\ 0 & f^{-1} \end{pmatrix}, \qquad f(x, t) := \exp\left(\frac{i}{4}\int^x |q(s, t)|^2 ds\right).$$

Note that $g(x,t) = Q(x,t,\lambda = 0)$. Thus we obtain

$$U = -g^{-1}\frac{\partial g}{\partial x} + g^{-1}Wg, \qquad V = -g^{-1}\frac{\partial g}{\partial t} + g^{-1}Zg.$$

It follows that

$$-i\lambda^2\sigma_3 + \lambda \begin{pmatrix} 0 & \phi \\ -\phi^* & 0 \end{pmatrix} = -i\lambda^2\sigma_3 + \lambda \begin{pmatrix} 0 & qf^{-2} \\ -q^*f^2 & 0 \end{pmatrix}.$$

We find that the transformation of eigenfunction and the transformation of the field variables $\phi = qf^{-2}$ keep the inverse scattering method invariant. We call the transformation which changes an eigenvalue problem into another eigenvalue problem the *gauge transformation*. Here g is the gauge transformation.

Problem 5.   (i) Consider the general linear group $GL(n,\mathbb{R})$ with the usual coordinates. Let $X \in GL(n,\mathbb{R})$. We write

$$X = (x_{ij}), \qquad dX = (dx_{ij})$$

where $\det X \neq 0$. Consider the $n \times n$ matrix

$$\Omega := X^{-1}dX \tag{1}$$

whose entries are differential one-forms on G. Show that

$$d\Omega + \Omega \wedge \Omega = 0 \tag{2}$$

where $\wedge$ denotes the wedge product of matrices.
(ii) Show that

$$d\Omega + \Omega \wedge \Omega = 0 \tag{3}$$

is invariant under the transformation

$$\Omega \rightarrow R^{-1}dR + R^{-1}\Omega R \tag{4}$$

where R is an $n \times n$ invertible matrix. This is called a *gauge transformation*.

Solution 5.   (i) From (1) we obtain

$$dX = X\Omega. \tag{5}$$

Taking the exterior derivative of this equation we obtain

$$0 = (dX) \wedge \Omega + X \wedge d\Omega \tag{6}$$

since $ddX = 0$. Inserting (5) into (6) we yields

$$X(\Omega \wedge \Omega + d\Omega) = 0.$$

Since X is invertible ($X \in GL(n, \mathbb{R})$), we find (3).
(ii) From $R^{-1}R = I_n$, where I_n is the $n \times n$ unit matrix, we obtain

$$(dR^{-1})R + R^{-1}dR = 0$$

since $dI_n = 0$. Thus $dR^{-1} = -R^{-1}(dR)R^{-1}$. Inserting (4) into (3) we obtain

$$d(R^{-1}dR + R^{-1}\Omega R) + (R^{-1}dR + R^{-1}\Omega R) \wedge (R^{-1}dR + R^{-1}\Omega R) = 0.$$

From $dR = -R^{-1}(dR)R^{-1}$, $ddR = 0$ and

$$d(R^{-1}dR) = dR^{-1} \wedge dR$$
$$d(R^{-1}\Omega R) = dR^{-1} \wedge \Omega R + R^{-1}d\Omega R - R^{-1}\Omega \wedge dR$$

we find $R^{-1}(d\Omega + \Omega \wedge \Omega)R = 0$. Thus (2) follows.

Problem 6. Let

$$i\hbar \frac{\partial \psi(\mathbf{x}, t)}{\partial t} = \hat{H}_0 \psi(\mathbf{x}, t) \tag{1}$$

be the *Schrödinger equation* in three space dimensions, where

$$\hat{H}_0 = -\frac{\hbar^2}{2m}\triangle \tag{2}$$

with $j = 1, 2, 3$, $\mathbf{x} = (x_1, x_2, x_3)$. ϵ is a real dimensionless parameter and

$$\triangle := \sum_{j=1}^{3} \frac{\partial^2}{\partial x_j^2}.$$

(i) Show that (1) is invariant under the *global gauge transformation*

$$x_j'(\mathbf{x}, t) = x_j, \qquad t'(\mathbf{x}, t) = t \tag{2a}$$

$$\psi'(\mathbf{x}'(\mathbf{x}, t), t'(\mathbf{x}, t)) = \exp(i\epsilon)\, \psi(\mathbf{x}, t). \tag{2b}$$

(ii) Let

$$i\hbar \frac{\partial \psi}{\partial t} = -\frac{\hbar^2}{2m} \sum_{j=1}^{3} \left(\frac{\partial}{\partial x_j} - i\frac{q}{\hbar}A_j \right)^2 \psi + qU\psi \tag{3}$$

where $\mathbf{A}$ is the vector potential and U is the scalar potential. Show that this partial differential equation is invariant under the transformation

$$x_j'(\mathbf{x}, t) = x_j, \qquad t'(\mathbf{x}, t) = t$$

$$\psi'(\mathbf{x}'(\mathbf{x},t),t'(\mathbf{x},t)) = \exp\left(i\epsilon(\mathbf{x},t)\right)\psi(\mathbf{x},t)$$

$$U'(\mathbf{x}'(\mathbf{x},t),t'(\mathbf{x},t)) = U(\mathbf{x},t) - \frac{\hbar}{q}\frac{\partial\epsilon}{\partial t}$$

$$A'_j(\mathbf{x}'(\mathbf{x},t),t'(\mathbf{x},t)) = A_j(\mathbf{x},t) + \frac{\hbar}{q}\frac{\partial\epsilon}{\partial x_j}$$

where ϵ is a smooth function of $\mathbf{x} = (x_1,x_2,x_3)$ and t, and $j = 1,2,3$.

Solution 6.   (i) Since ϵ is a space and time independent parameter we find

$$\frac{\partial^2\psi'}{\partial x_j'^2} = e^{i\epsilon}\frac{\partial\psi}{\partial x_j^2}, \qquad \frac{\partial\psi'}{\partial t'} = e^{i\epsilon}\frac{\partial\psi}{\partial t}$$

where $j = 1,2,3$. Thus (1) is invariant under transformation(2).
(ii) Since ϵ is a smooth function of $\mathbf{x} = (x_1,x_2,x_3)$ and t we find

$$\frac{\partial\psi'}{\partial t'} = ie^{i\epsilon}\frac{\partial\epsilon}{\partial t}\psi + e^{i\epsilon}\frac{\partial\psi}{\partial t}, \qquad \frac{\partial\psi'}{\partial x_j'} = ie^{i\epsilon}\frac{\partial\epsilon}{\partial x_j}\psi + e^{i\epsilon}\frac{\partial\psi}{\partial x_j}$$

and

$$\frac{\partial^2\psi'}{\partial x_j'^2} = -e^{i\epsilon}\left(\frac{\partial\epsilon}{\partial x_j}\right)^2\psi + ie^{i\epsilon}\frac{\partial^2\epsilon}{\partial x_j^2}\psi + 2ie^{i\epsilon}\frac{\partial\epsilon}{\partial x_j}\frac{\partial\psi}{\partial x_j} + e^{i\epsilon}\frac{\partial^2\psi}{\partial x_j^2}.$$

Therefore

$$\frac{\partial\psi}{\partial x_j} = e^{-i\epsilon}\left(\frac{\partial\psi'}{\partial x_j'} - i\frac{\partial\epsilon}{\partial x_j}\psi'\right)$$

$$\frac{\partial\psi}{\partial t} = e^{-i\epsilon}\left(\frac{\partial\psi'}{\partial t'} - i\frac{\partial\epsilon}{\partial t}\psi'\right)$$

$$\frac{\partial^2\psi}{\partial x_j^2} = e^{-i\epsilon}\left(\frac{\partial^2\psi'}{\partial x_j'^2} - \left(\frac{\partial\epsilon}{\partial x_j}\right)^2\psi' - i\frac{\partial^2\epsilon}{\partial x_j^2}\psi' - 2i\frac{\partial\epsilon}{\partial x_j}\frac{\partial\psi'}{\partial x_j'}\right).$$

Since

$$\left(\frac{\partial}{\partial x_j} - i\frac{q}{\hbar}A_j\right)^2\psi \equiv \frac{\partial^2\psi}{\partial x_j^2} - \frac{2iq}{\hbar}A_j\frac{\partial\psi}{\partial x_j} - \frac{iq}{\hbar}\psi\frac{\partial A_j}{\partial x_j} - \frac{q^2}{\hbar^2}A_j^2\psi$$

we obtain using

$$\frac{\partial A_j'}{\partial x_j'} = \frac{\partial A_j}{\partial x_j} + \frac{\hbar}{q}\frac{\partial^2\epsilon}{\partial x_j^2}$$

that

$$\left(\frac{\partial}{\partial x_j} - i\frac{q}{\hbar}A_j\right)^2\psi = e^{-i\epsilon}\left(\frac{\partial}{\partial x_j'} - i\frac{q}{\hbar}A_j'\right)^2\psi'.$$

Moreover

$$i\hbar\frac{\partial\psi}{\partial t} - qU\psi = i\hbar e^{-i\epsilon}\left(\frac{\partial\psi'}{\partial t'} - i\frac{\partial\epsilon}{\partial t'}\psi'\right) - qe^{-i\epsilon}\left(U' + \frac{\hbar}{q}\frac{\partial\epsilon}{\partial t}\right)\psi'$$

$$= e^{-i\epsilon}\left(i\hbar\frac{\partial\psi'}{\partial t'} - qU'\psi'\right).$$

Since $e^{-i\epsilon} \neq 0$, we obtain

$$i\hbar\frac{\partial\psi'}{\partial t'} = -\frac{\hbar^2}{2m}\sum_{j=1}^{3}\left(\frac{\partial}{\partial x'_j} - i\frac{q}{\hbar}A'_j\right)^2\psi' + qU'\psi'.$$

Consequently, (3) is invariant under the gauge transformation.

Problem 7. Apply the approach described in the previous problem to the linear Dirac equation. The linear *Dirac equation* with nonvanishing rest mass m_0 is given by

$$\left(i\hbar\left(\gamma_0\frac{\partial}{\partial x_0} + \gamma_1\frac{\partial}{\partial x_1} + \gamma_2\frac{\partial}{\partial x_2} + \gamma_3\frac{\partial}{\partial x_3}\right) - m_0c\right)\psi(\mathbf{x}) = 0$$

where $\mathbf{x} = (x_0, x_1, x_2, x_3)$, $x_0 = ct$ and the complex valued spinor ψ is

$$\psi(\mathbf{x}) := \begin{pmatrix} \psi_1(\mathbf{x}) \\ \psi_2(\mathbf{x}) \\ \psi_3(\mathbf{x}) \\ \psi_4(\mathbf{x}) \end{pmatrix}.$$

The *gamma matrices* are defined by

$$\gamma_0 \equiv \beta := \begin{pmatrix} 1 & 0 & 0 & 0 \\ 0 & 1 & 0 & 0 \\ 0 & 0 & -1 & 0 \\ 0 & 0 & 0 & -1 \end{pmatrix}, \quad \gamma_1 := \begin{pmatrix} 0 & 0 & 0 & 1 \\ 0 & 0 & 1 & 0 \\ 0 & -1 & 0 & 0 \\ -1 & 0 & 0 & 0 \end{pmatrix}$$

$$\gamma_2 := \begin{pmatrix} 0 & 0 & 0 & -i \\ 0 & 0 & i & 0 \\ 0 & i & 0 & 0 \\ -i & 0 & 0 & 0 \end{pmatrix}, \quad \gamma_3 := \begin{pmatrix} 0 & 0 & 1 & 0 \\ 0 & 0 & 0 & -1 \\ -1 & 0 & 0 & 0 \\ 0 & 1 & 0 & 0 \end{pmatrix}.$$

The linear Dirac equation and the *free wave equation*

$$\left(\frac{\partial^2}{\partial x_0^2} - \frac{\partial^2}{\partial x_1^2} - \frac{\partial^2}{\partial x_2^2} - \frac{\partial^2}{\partial x_3^2}\right)A_\mu = 0$$

can be derived from the Lagrange densities

$$\mathcal{L}_D = c\bar{\psi}\left(i\hbar\gamma_\mu\frac{\partial}{\partial x_\mu} - m_0c\right)\psi, \quad \mathcal{L}_E = -\frac{1}{2\mu_0}\left(\frac{\partial A_\mu}{\partial x^\nu} - \frac{\partial A_\nu}{\partial x^\mu}\right)\frac{\partial A^\mu}{\partial x_\nu},$$

respectively. We used sum convention and $\bar{\psi} = \psi^\dagger \gamma_0$, $A_\mu = (A_0, A_1, A_2, A_3)$, $A^\mu = (A_0, -A_1, -A_2, -A_3)$, $x_\mu = (x_0, -x_1, -x_2, -x_3)$, $x^\mu = (x_0, x_1, x_2, x_3)$.

Solution 7.   From gauge theory described above we obtain the Lagrangian density

$$\mathcal{L} = -\frac{1}{2\mu_0}\left(\frac{\partial A_\mu}{\partial x^\nu} - \frac{\partial A_\nu}{\partial x^\mu}\right)\frac{\partial A^\mu}{\partial x_\nu} + c\bar{\psi}\left(i\hbar\gamma_\mu\frac{\partial}{\partial x_\mu} - e\gamma_\mu A^\mu - m_0 c\right)\psi$$

where $\bar{\psi} = \psi^\dagger \gamma_0 \equiv (\psi_1^*, \psi_2^*, -\psi_3^*, -\psi_4^*)$ and

$$\gamma_\mu A^\mu = \gamma_0 A^0 + \gamma_1 A^1 + \gamma_2 A^2 + \gamma_3 A^3 = \gamma_0 A_0 - \gamma_1 A_1 - \gamma_2 A_2 - \gamma_3 A_3.$$

Here we make use of the sum convention. From this Lagrangian density and the *Euler-Lagrange equation*

$$\frac{\partial \mathcal{L}}{\partial \phi_r} - \frac{\partial}{\partial x^\mu}\frac{\partial \mathcal{L}}{\partial(\partial \phi_r/\partial x^\mu)} = 0$$

we obtain the *Maxwell-Dirac equation*

$$\left(\frac{\partial^2}{\partial x_0^2} - \frac{\partial^2}{\partial x_1^2} - \frac{\partial^2}{\partial x_2^2} - \frac{\partial^2}{\partial x_3^2}\right)A_\mu = \mu_0 ec\bar{\psi}\gamma_\mu\psi$$

$$\left(i\hbar\left(\gamma_0\frac{\partial}{\partial x_0} + \sum_{j=1}^{3}\gamma_j\frac{\partial}{\partial x_j}\right) - m_0 c\right)\psi = e\left(\sum_{j=1}^{3}\gamma_j A^j + \gamma_0 A_0\right)\psi$$

$$\frac{\partial A_0}{\partial x_0} + \frac{\partial A_1}{\partial x_1} + \frac{\partial A_2}{\partial x_2} + \frac{\partial A_3}{\partial x_3} = 0 \qquad (\textit{Lorentz gauge condition})$$

is the Dirac equation which is coupled with the vector potential A_μ where $A_0 = U/c$ and U denotes the scalar potential. Here $\mu = 0, 1, 2, 3$ and

$$\bar{\psi} = (\psi_1^*, \psi_2^*, -\psi_3^*, -\psi_4^*).$$

Here c is the speed of light, e the charge, μ_0 the permeability of free space, $\hbar = h/(2\pi)$ where h is Planck's constant and m_0 the particle rest mass. We set $\lambda = \hbar/(m_0 c)$. The Maxwell-Dirac equation can be written as the following coupled system of thirteen partial differential equations

$$\left(\frac{\partial^2}{\partial x_0^2} - \frac{\partial^2}{\partial x_1^2} - \frac{\partial^2}{\partial x_2^2} - \frac{\partial^2}{\partial x_3^2}\right)A_0 = (\mu_0 ec)\sum_{j=1}^{4}(u_j^2 + v_j^2)$$

$$\left(\frac{\partial^2}{\partial x_0^2} - \frac{\partial^2}{\partial x_1^2} - \frac{\partial^2}{\partial x_2^2} - \frac{\partial^2}{\partial x_3^2}\right)A_1 = (2\mu_0 ec)(u_1 u_4 + v_1 v_4 + u_2 u_3 + v_2 v_3)$$

$$\left(\frac{\partial^2}{\partial x_0^2} - \frac{\partial^2}{\partial x_1^2} - \frac{\partial^2}{\partial x_2^2} - \frac{\partial^2}{\partial x_3^2}\right)A_2 = (2\mu_0 ec)(u_1 v_4 - u_4 v_1 + u_3 v_2 - u_2 v_3)$$

$$\left(\frac{\partial^2}{\partial x_0^2} - \frac{\partial^2}{\partial x_1^2} - \frac{\partial^2}{\partial x_2^2} - \frac{\partial^2}{\partial x_3^2}\right)A_3 = (2\mu_0 ec)(u_1 u_3 + v_3 v_1 - u_2 u_4 - v_2 v_4)$$

$$\lambda \frac{\partial u_1}{\partial x_0} + \lambda \frac{\partial u_3}{\partial x_3} + \lambda \frac{\partial u_4}{\partial x_1} + \lambda \frac{\partial v_4}{\partial x_2} - v_1 = \frac{e}{m_0 c}(A_0 v_1 - A_3 v_3 - A_1 v_4 + A_2 u_4)$$

$$-\lambda \frac{\partial v_1}{\partial x_0} - \lambda \frac{\partial v_3}{\partial x_3} - \lambda \frac{\partial v_4}{\partial x_1} + \lambda \frac{\partial u_4}{\partial x_2} - u_1 = \frac{e}{m_0 c}(A_0 u_1 - A_3 u_3 - A_1 u_4 - A_2 v_4)$$

$$\lambda \frac{\partial u_2}{\partial x_0} + \lambda \frac{\partial u_3}{\partial x_1} - \lambda \frac{\partial v_3}{\partial x_2} - \lambda \frac{\partial u_4}{\partial x_3} - v_2 = \frac{e}{m_0 c}(A_0 v_2 - A_1 v_3 - A_2 u_3 + A_3 v_4)$$

$$-\lambda \frac{\partial v_2}{\partial x_0} - \lambda \frac{\partial v_3}{\partial x_1} - \lambda \frac{\partial u_3}{\partial x_2} + \lambda \frac{\partial v_4}{\partial x_3} - u_2 = \frac{e}{m_0 c}(A_0 u_2 - A_1 u_3 + A_2 v_3 + A_3 u_4)$$

$$-\lambda \frac{\partial u_1}{\partial x_3} - \lambda \frac{\partial u_2}{\partial x_1} - \lambda \frac{\partial v_2}{\partial x_2} - \lambda \frac{\partial u_3}{\partial x_0} - v_3 = \frac{e}{m_0 c}(A_3 v_1 + A_1 v_2 - A_2 u_2 - A_0 v_3)$$

$$\lambda \frac{\partial v_1}{\partial x_3} + \lambda \frac{\partial v_2}{\partial x_1} - \lambda \frac{\partial u_2}{\partial x_2} + \lambda \frac{\partial v_3}{\partial x_0} - u_3 = \frac{e}{m_0 c}(A_3 u_1 + A_1 u_2 + A_2 v_2 - A_0 u_3)$$

$$-\lambda \frac{\partial u_1}{\partial x_1} + \lambda \frac{\partial v_1}{\partial x_2} + \lambda \frac{\partial u_2}{\partial x_3} - \lambda \frac{\partial u_4}{\partial x_0} - v_4 = \frac{e}{m_0 c}(A_1 v_1 + A_2 u_1 - A_3 v_2 - A_0 v_4)$$

$$\lambda \frac{\partial v_1}{\partial x_1} + \lambda \frac{\partial u_1}{\partial x_2} - \lambda \frac{\partial v_2}{\partial x_3} + \lambda \frac{\partial v_4}{\partial x_0} - u_4 = \frac{e}{m_0 c}(A_1 u_1 - A_2 v_1 - A_3 u_2 - A_0 u_4)$$

$$\frac{\partial A_0}{\partial x_0} + \frac{\partial A_1}{\partial x_1} + \frac{\partial A_2}{\partial x_2} + \frac{\partial A_3}{\partial x_3} = 0.$$

Problem 8. In *string theory* one considers 2-dimensional surfaces in d-dimensional Minkowski space called the world sheet of strings. As a string propagates through space-time, it describes a surface in space-time which is called its world sheet. The action S is proportional to an area of the world sheet. This is the simplest generalization of the mechanics of relativistic particles with action, proportional to the length of the world line. One considers a fixed background pseudo-Riemannian space-time manifold M of dimension D, with coordinates $X = (X^\mu)$ and $\mu = 0, 1, \dots, D - 1$. The metric is $G_{\mu\nu}$ and we take the signature $(-, (+)^{D-1})$. In Minkowski space we have $D = 4$ and $(-, +, +, +)$. The motion of a relativistic string in M is described by its generalized world line, a two-dimensional surface Σ, which is called the *world-sheet*. For a single non-interacting string the world-sheet has the form of an infinite strip. One introduces coordinates $\sigma = (\sigma^0, \sigma^1)$ on the world-sheet. Sometimes one also writes $\sigma^0 = \tau$ and $\sigma^1 = \sigma$. The embedding of the world-sheet into space-time is given by the smooth maps

$$X : \Sigma \to M : \sigma \to X(\sigma).$$

The background metric induces a metric on the world-sheet

$$G_{\alpha\beta} := \frac{\partial X^\mu}{\partial \sigma^\alpha} \frac{\partial X^\nu}{\partial \sigma^\beta} G_{\mu\nu} \qquad \text{summation convention}$$

where $\alpha, \beta = 0, 1$ are the world-sheet indices. The induced metric $G_{\alpha\beta}$ is to be distinguished from the intrinsic metric $h_{\alpha\beta}$ on Σ. An intrinsic metric is used as an auxiliary field in the Polyakov formulation of the bosonic string. This setting can be viewed from two perspectives. In the space-time perspective we interpret the system as a relativistic string moving in the space-time M. On the other hand it can be considered as a two-dimensional field theory living on the world sheet, with fields X which take values in the target-space M. This is the world-sheet perspective. Thus we can use methods of the two-dimensional field theory. The natural action for a relativistic string is its area, measured with the induced metric

$$ S_{NG} = \frac{1}{2\pi\alpha'} \int_\Sigma d\sigma_0 d\sigma_1 |\det G_{\alpha\beta}|^{1/2} . $$

This is the *Nambu-Goto action*. The prefactor $(2\pi\alpha')^{-1}$ is the energy per length or tension of the string, which is the fundamental parameter of the theory. The *Polyakov action*, is equivalent to the Nambu-Goto action. It is a standard two-dimensional field theory action. For this action one introduces an intrinsic metric on the world-sheet, $h_{\alpha\beta}(\sigma)$. The action takes the form of a *nonlinear sigma model* on the world-sheet

$$ S_P = \frac{1}{4\pi\alpha'} \int_\Sigma d\sigma^0 d\sigma^1 \sqrt{h} h^{\alpha\beta} \frac{\partial X^\mu}{\partial \sigma^\alpha} \frac{\partial X^\nu}{\partial \sigma^\beta} G_{\mu\nu}(X) $$

where $h := |\det h_{\alpha\beta}|$ and $h_{\alpha\beta}$ is the inverse of $h^{\alpha\beta}$.
(i) Let

$$ G_{\mu\nu} = \eta_{\mu\nu} = \mathrm{diag}(-1,+1,\ldots,+1) $$

be a flat Minkowski metric and let $h_{00} = -1$, $h_{11} = 1$, $h_{01} = h_{10} = 0$. Find the Polyakov action for this special case. Give the Lagrangian density and the equations of motion.
(ii) Let $D = 3$ and use the results from (i). Consider

$$ X^0(\tau,\sigma) = A\tau $$
$$ X^1(\tau,\sigma) = A\cos(2\tau)\cos(2\sigma) $$
$$ X^2(\tau,\sigma) = A\sin(c\tau)\cos(c\sigma) $$

where A is a nonzero constant and c is a positive constant. Find the condition on c such that the equations of motion are satisfied. Are the boundary conditions $X^\mu(\tau,\sigma+\pi) = X^\mu(\tau,\sigma)$ satisfied? Find the condition on c such that

$$ \frac{\partial X}{\partial \tau} \cdot \frac{\partial X}{\partial \sigma} := -\frac{\partial X^0}{\partial \tau}\frac{\partial X^0}{\partial \sigma} + \frac{\partial X^1}{\partial \tau}\frac{\partial X^1}{\partial \sigma} + \frac{\partial X^2}{\partial \tau}\frac{\partial X^2}{\partial \sigma} = 0 . $$

Solution 8. (i) We have $h = 1$ and

$$ \sum_{\mu,\nu=0}^{D-1}\sum_{\alpha,\beta=0}^{1} h^{\alpha\beta}\frac{\partial X^\mu}{\partial \sigma^\alpha}\frac{\partial X^\nu}{\partial \sigma^\beta}\eta_{\mu\nu} = \sum_{\mu=0,\nu=0}^{D-1}\left(h^{00}\frac{\partial X^\mu}{\partial \sigma^0}\frac{\partial X^\nu}{\partial \sigma^0}\eta_{\mu\nu} + h^{11}\frac{\partial X^\mu}{\partial \sigma^1}\frac{\partial X^\nu}{\partial \sigma^1}\eta_{\mu\nu} \right) $$

$$= \sum_{\mu,\nu=0}^{D-1} \left(-\frac{\partial X^\mu}{\partial \sigma^0} \frac{\partial X^\nu}{\partial \sigma^0} \eta_{\mu\nu} + \frac{\partial X^\mu}{\partial \sigma^1} \frac{\partial X^\nu}{\partial \sigma^1} \eta_{\mu\nu} \right)$$

$$= \frac{\partial X^0}{\partial \sigma^0} \frac{\partial X^0}{\partial \sigma^0} - \sum_{\mu=1}^{D-1} \frac{\partial X^\mu}{\partial \sigma^0} \frac{\partial X^\mu}{\partial \sigma^0}$$

$$- \frac{\partial X^0}{\partial \sigma^1} \frac{\partial X^0}{\partial \sigma^1} + \sum_{\mu=1}^{D-1} \frac{\partial X^\mu}{\partial \sigma^1} \frac{\partial X^\mu}{\partial \sigma^1}.$$

The *Euler-Lagrange equations* are given by ($\sigma^0 = \tau, \sigma^1 = \sigma$)

$$\frac{\partial \mathcal{L}}{\partial X^\mu} - \frac{\partial}{\partial \tau} \left(\frac{\partial \mathcal{L}}{\partial(\partial X^\mu/\partial \tau)} \right) - \frac{\partial}{\partial \sigma} \left(\frac{\partial \mathcal{L}}{\partial(\partial X^\mu/\partial \sigma)} \right) = 0$$

with the *Lagrange density*

$$\mathcal{L} = -\frac{\partial X^0}{\partial \tau} \frac{\partial X^0}{\partial \tau} + \sum_{\mu=1}^{D-1} \frac{\partial X^\mu}{\partial \tau} \frac{\partial X^\mu}{\partial \tau} + \frac{\partial X^0}{\partial \sigma} \frac{\partial X^0}{\partial \sigma} - \sum_{\mu=1}^{D-1} \frac{\partial X^\mu}{\partial \sigma} \frac{\partial X^\mu}{\partial \sigma}.$$

Since $\partial \mathcal{L}/\partial X^\mu = 0$ for $\mu = 0, 1, \ldots, D-1$ we obtain a system of one-dimensional linear wave equations

$$\frac{\partial^2 X^\mu}{\partial \tau^2} - \frac{\partial^2 X^\mu}{\partial \sigma^2} = 0$$

where $\mu = 0, 1, \ldots, D-1$.
(ii) Inserting the ansatz for X^0, X^1, X^2 into these linear wave equations we find that the equations are satisfied with c arbitrary. We obtain

$$\frac{\partial X}{\partial \tau} \cdot \frac{\partial X}{\partial \sigma} := -\frac{\partial X^0}{\partial \tau} \frac{\partial X^0}{\partial \sigma} + \frac{\partial X^1}{\partial \tau} \frac{\partial X^1}{\partial \sigma} + \frac{\partial X^2}{\partial \tau} \frac{\partial X^2}{\partial \sigma}$$

$$= 4A^2 \sin(2\tau) \cos(2\tau) \sin(2\sigma) \cos(2\sigma)$$

$$- A^2 c^2 \sin(c\tau) \cos(c\tau) \sin(c\sigma) \cos(c\sigma).$$

Thus c must be 2. Since $\cos(2\pi) = 1$ the boundary conditions $X^\mu(\tau, \sigma + \pi) = X^\mu(\tau, \sigma)$ are satisfied.

Problem 9. The AdS/CFT correspondence is the equivalence between a string theory or supergravity defined on some sort of anti de Sitter space (AdS) and a conformal field theory (CFT) defined on its conformal boundary whose dimension is lower by one. Therefore it refers to the existence of dualities between theories with gravity and theories without gravity. An example of such a correspondence is the exact equivalence between type IIB string theory compactified on $AdS^5 \times S^5$, and four-dimensional $N = 4$

supersymmetric Yang-Mills theory. Here AdS^5 refers to an Anti-de Sitter space in five dimensions, S^5 refers to a five-dimensional sphere

$$S^5 := \{\, (X_1, X_2, \dots, X_6) \,:\, X_1^2 + X_2^2 + \cdots + X_6^2 = 1 \,\}.$$

Anti-de Sitter spaces are maximally symmetric solutions of the Einstein equations with a negative cosmological constant. The large symmetry group of 5d anti-de Sitter space matches with the group of conformal symmetries of the $N = 4$ super Yang-Mills theory. Integrable nonlinear dynamical systems can be derived from this correspondence. Consider the bosonic part of the classical closed string propagating in the $AdS_5 \times S^5$ space time. The two metrics have the standard form in terms of the $5 + 5$ angular coordinates

$$(ds^2)_{AdS_5} = -\cosh^2 \rho\, dt^2 + d\rho^2 + \sinh^2 \rho (d\theta^2 + \sin^2 \theta d\phi^2 + \cos^2 \theta d\varphi^2)$$

$$(ds^2)_{S^5} = d\gamma^2 + \cos^2 \gamma d\varphi_3^2 + \sin^2 \gamma (d\psi^2 + \cos^2 \psi d\varphi_1^2 + \sin^2 \psi d\varphi_2^2).$$

Consider the compact Lie group $O(6)$ and the non-compact Lie group $SO(4,2)$ The bosonic part of the sigma model action as an action S for the $O(6) \times SO(4,2)$ sigma-model is given by

$$S = \frac{\sqrt{\lambda}}{2\pi} \int d\tau d\sigma (\mathcal{L}_S + \mathcal{L}_{AdS}), \qquad \sqrt{\lambda} \equiv \frac{R^2}{\alpha'}$$

with the Lagrange densities

$$\mathcal{L}_S = -\frac{1}{2} \sum_{M=1}^{6} \sum_{\alpha=0}^{1} \partial_\alpha X_M \partial^\alpha X_M + \frac{1}{2}\Lambda \sum_{M=1}^{6} (X_M X_M - 1)$$

$$\mathcal{L}_{AdS} = -\frac{1}{2} \sum_{M=0}^{5} \sum_{N=0}^{5} \sum_{\alpha=0}^{1} \eta_{MN} \partial_\alpha Y_M \partial^\alpha Y_N + \frac{1}{2}\tilde{\Lambda} \sum_{M=0}^{5} \sum_{N=0}^{5} (\eta_{MN} Y_M Y_N + 1).$$

Here X_M, $M = 1, 2, \dots, 6$ and Y_M, $M = 0, 1, \dots, 5$ are the embedding coordinates of $\mathbb{R}^6$ with the Euclidean metric in $\mathcal{L}_S$ and with

$$\eta_{MN} = (-1, +1, +1, +1, +1, -1)$$

in $\mathcal{L}_{AdS}$, respectively. Λ and $\tilde{\Lambda}$ are the Lagrange multipliers. The worldsheet coordinates are σ^α, $\alpha = 0, 1$ with $(\sigma^0, \sigma^1) = (\tau, \sigma)$ and the worldsheet metric is $\mathrm{diag}(-1, 1)$. The action is to be supplemented with the usual conformal gauge constraints. The embedding coordinates are related to the angular ones as follows

$$X_1 + iX_2 = \sin \gamma \cos \psi e^{i\varphi_1}, \qquad X_3 + iX_4 = \sin \gamma \sin \psi e^{i\varphi_2}$$
$$X_5 + iX_6 = \cos \gamma e^{i\varphi_3},$$
$$Y_1 + iY_2 = \sinh \rho \sin \theta e^{i\phi}, \qquad Y_3 + iY_4 = \sinh \rho \cos \theta e^{i\varphi}$$
$$Y_5 + iY_0 = \cosh \rho e^{it}.$$

Consider the case that the string is located at the centre of AdS_5 and rotating in S^5. This means it is embedded in AdS_5 as $Y_5 + iY_0 = e^{i\kappa\tau}$ with $Y_1 = Y_2 = Y_3 = Y_4 = 0$. The S^5 metric has three translational isometries in φ_i which give rise to three global commuting integrals of motion (spins) J_i. When one considers period motion with three J_i non-zero, one chooses the following ansatz for X_M

$$X_1 + iX_2 = x_1(\sigma)e^{i\omega_1\tau}, \quad X_3 + iX_4 = x_2(\sigma)e^{i\omega_2\tau}, \quad X_5 + iX_6 = x_3(\sigma)e^{i\omega_3\tau}$$

where the real radial functions x_i are independent of time and as a consequence of the condition $X_M^2 = 1$ lie on a 2-sphere S^2, $x_1^2 + x_2^2 + x_3^2 = 1$. Then the spins $J_1 = J_{12}$, $J_2 = J_{34}$, $J_3 = J_{56}$ forming a Cartan subalgebra of the compact Lie algebra $so(6)$ are

$$J_i = \sqrt{\lambda}\omega_i \int_0^{2\pi} \frac{d\sigma}{2\pi} x_i^2(\sigma).$$

Substitute the ansatz for X_1+iX_2, X_3+iX_4 and X_5+iX_6 into the Lagrange density $\mathcal{L}_S$ and find the Lagrange function and the corresponding equations of motion.

Solution 9. Inserting the ansatz for X_1+iX_2, X_3+iX_4, X_5+iX_6 into the Lagrange density $\mathcal{L}_S$ we find the Lagrange function of a one-dimensional nonlinear system

$$L(\mathbf{x'}, \mathbf{x}) = \frac{1}{2}\sum_{i=1}^{3}(x_i'^2 - \omega_i^2 x_i^2) + \frac{1}{2}\Lambda\sum_{i=1}^{3}(x_i^2 - 1).$$

This Lagrange function describes an $n = 3$ harmonic oscillator constrained to remain on a unit $n-1 = 2$ sphere. It is a special case of the n-dimensional Neumann dynamical system which is know to be integrable. Solving the equation of motion for the Lagrange multiplier Λ we obtain the autonomous nonlinear sytem of ordinary differential equations

$$\frac{d^2 x_j}{d\sigma^2} + \omega_j^2 x_j = -x_j \sum_{k=1}^{3}\left(\left(\frac{dx_k}{d\sigma}\right)^2 - \omega_k^2 x_k^2\right), \quad j = 1, 2, 3.$$

The canonical momenta conjugate to x_j are defined by

$$\pi_j := \frac{dx_j}{d\sigma}, \quad \sum_{j=1}^{3}\pi_j x_j = 0.$$

The n-dimensional Neumann system has the following n first integrals

$$F_j(\mathbf{x}, \boldsymbol{\pi}) = x_j^2 + \sum_{k\neq j}\frac{(x_j\pi_k - x_k\pi_j)^2}{\omega_j^2 - \omega_k^2}, \quad \sum_{j=1}^{n}F_j(\mathbf{x}, \boldsymbol{\pi}) = 1$$

where $n = 3$ for the present case.

Problem 10.   The *Nambu-Goto action* in a flat target space of signature $(2, 2)$ is given by

$$S_{NG} = -\frac{T}{2} \int d\tau d\sigma \sqrt{-\det(\gamma)}$$

where (summation convention, $\tau = \sigma^0, \sigma = \sigma^1$)

$$\gamma_{A''B''} := \frac{\partial X^\mu(\tau, \sigma)}{\partial \sigma^{A''}} \frac{\partial X^\nu(\tau, \sigma)}{\partial \sigma^{B''}} \eta_{\mu\nu}$$

with $\eta = \mathrm{diag}(1, 1, -1, -1)$, $A'', B'' = 0, 1$ and $\mu, \nu = 0, 1, 2, 3$. Consider the 2×2 matrix

$$X^{AA'} := \frac{1}{\sqrt{2}} \begin{pmatrix} -X^0 + X^2 & X^1 - X^3 \\ -X^1 - X^3 & -X^0 - X^2 \end{pmatrix}$$

where $A, A' = 0, 1$. We define the $2 \times 2 \times 2$ hypermatrix a by

$$a_{AA'0} := \frac{\partial X^{AA'}}{\partial \tau}, \qquad a_{AA'1} := \frac{\partial X^{AA'}}{\partial \sigma}$$

where $A, A' = 0, 1$. The *hyperdeterminant* $\mathrm{Det}(a)$ of a $2 \times 2 \times 2$ hypermatrix a defined by

$$\mathrm{Det}(a) := -\frac{1}{2} \epsilon^{AB} \epsilon^{A'B'} \epsilon^{CD} \epsilon^{C'D'} \epsilon^{A''D''} \epsilon^{B''C''} a_{AA'A''} a_{BB'B''} a_{CC'C''} a_{DD'D''}$$

where $\epsilon^{00} = \epsilon^{11} = 0$ and $\epsilon^{01} = -\epsilon^{10} = 1$. Find the connection between the 2×2 matrix γ and the $2 \times 2 \times 2$ hypermatrix a. Express $\det(\gamma)$ using the hypermatrix and hyperdeterminant.

Solution 10.   Obviously we have (summation convention)

$$\gamma_{A''B''} = \epsilon^{AB} \epsilon^{A'B'} a_{AA'A''} a_{BB'B''}$$

We obtain $\det(\gamma) = -\mathrm{Det}(a)$. The non-compact Lie group $SL(2, \mathbb{R})$ acting on the index A and the $SL(2, \mathbb{R})$ acting on the index A' is the $O(2, 2) \sim SL(2, \mathbb{R}) \times SL(2, \mathbb{R})$ spacetime symmetry. The Lie goup $SL(2, \mathbb{R})$ acting on the index A'' is the worldsheet symmetry

$$\begin{pmatrix} \partial X^\mu / \partial \tau \\ \partial X^\mu / \partial \sigma \end{pmatrix} \rightarrow \begin{pmatrix} c_{11} & c_{12} \\ c_{21} & c_{22} \end{pmatrix} \begin{pmatrix} \partial X^\mu / \partial \tau \\ \partial X^\mu / \partial \sigma \end{pmatrix}$$

where for the Lie group $SL(2, \mathbb{R})$ the c_{jk} are real constants with $c_{11}c_{22} - c_{12}c_{21} = 1$.

Chapter 35

Chaos, Fractals and Complexity

Problem 1. Let

$$x_{t+1} = 4x_t(1 - x_t) \tag{1}$$

where $t = 0, 1, 2, \ldots$ and $x_0 \in [0, 1]$. This nonlinear difference equation (map) is the so-called *logistic equation* or *logistic map*. It can also be considered as a map $f : [0, 1] \to [0, 1]$, $f(x) = 4x(1 - x)$ and $x \in [0, 1]$.

(i) Show that $x_t \in [0, 1]$ for $t = 0, 1, 2, \ldots$

(ii) Find the *fixed points*. The fixed points are the solutions of the equation $f(x^*) = x^*$.

(iii) Assume that f is differentiable (which is the case for the logistic map). Then

$$y_{t+1} = \frac{df}{dx}(x_t)y_t$$

is called the (difference) *variational equation* or *linearized equation*. Give the variational equation for (1).

(iv) Study the stability of the fixed points.

(v) Show that the general solution is given by

$$x_t = \frac{1}{2} - \frac{1}{2}\cos(2^t \cos^{-1}(1 - 2x_0)) \tag{2}$$

where x_0 is the initial value.

(vi) Find the periodic orbits.

Solution 1.   (i) Let $x \in [0,1]$. Then $x(1 - x) \leq 1/4$. Consequently, $4x(1 - x) \leq 1$. If $x \neq 1/2$, then $x(1 - x) < 1/4$. In other words, the function $g(x) = x(1 - x)$ has one maximum at $x = 1/2$ with $g(1/2) = 1/4$.
(ii) The fixed points x^* are solutions of the algebraic equation

$$4x^*(1 - x^*) = x^*.$$

We obtain as fixed points

$$x_1^* = 0, \qquad x_2^* = \frac{3}{4}.$$

The fixed points are time-independent solutions.
(iii) Let

$$x_{t+1} = f(x_t)$$

be a one-dimensional difference equation. Since $f(x) = 4x(1 - x)$ we have

$$\frac{df}{dx} = 4 - 8x$$

we obtain the variational equation

$$y_{t+1} = (4 - 8x_t)y_t$$

where $t = 0, 1, 2, \dots$.
(iv) Inserting the fixed point $x_1^* = 0$ into the variational equation yields the linear difference equation $y_{t+1} = 4y_t$. The solution is given by $y_t = 4^t y_0$. Therefore the fixed point $x_1^* = 0$ is unstable, since $y_t \to \infty$ as $t \to \infty$. Inserting the fixed point $x_2^* = 3/4$ into the variational equation yields

$$y_{t+1} = -2y_t.$$

Therefore this fixed point is also unstable, since $|y_t| \to \infty$ as $t \to \infty$.
(v) Let

$$\alpha = \cos^{-1}(1 - 2x_0).$$

Then

$$x_t = \frac{1}{2} - \frac{1}{2}\cos(2^t\alpha).$$

It follows that

$$x_{t+1} = \frac{1}{2} - \frac{1}{2}\cos(2^{t+1}\alpha) = \frac{1}{2} - \frac{1}{2}(2\cos^2(2^t\alpha) - 1) = 1 - \cos^2(2^t\alpha).$$

The left hand side of (1) is given by

$$4x_t(1 - x_t) = 4x_t - 4x_t^2 = 1 - \cos^2(2^t\alpha).$$

This proves that (2) is the general solution of (1), where x_0 is the initial value.

(vi) The periodic orbits are given by the initial values

$$x_0 = \frac{1}{2} - \frac{1}{2}\cos\left(\frac{r\pi}{2^s}\right)$$

where r and s are positive integers. We have

$$\arccos(1 - 2x_0) = \arccos\left(\cos\left(\frac{r\pi}{2^s}\right)\right) = \frac{r\pi}{2^s}.$$

It follows that

$$x_t = \frac{1}{2} - \frac{1}{2}\cos\left(\frac{2^t r\pi}{2^s}\right).$$

Problem 2. Consider the logistic map $f : [0, 1] \rightarrow [0, 1]$

$$f(x) = 4x(1 - x).$$

(i) The *Ljapunov exponent* λ is given by

$$\lambda(x_0) := \lim_{T\to\infty} \frac{1}{T} \sum_{t=0}^{T-1} \ln\left|\frac{df(x)}{dx}\right|_{x=x_t}$$

where x_0 is the initial value, i.e., the Ljapunov exponent depends on the initial value. The Ljapunov exponent measures the exponential rate at which the derivative grows. Find the maximal Ljapunov exponent.

(ii) The *time average* is defined by

$$\langle x_t \rangle := \lim_{T\to\infty} \frac{1}{T} \sum_{t=0}^{T} x_t.$$

For almost all initial values we find $\langle x_t \rangle = 1/2$. The *autocorrelation function* is defined by

$$C_{xx}(\tau) := \lim_{T\to\infty} \frac{1}{T} \sum_{t=0}^{T} (x_t - \langle x_t \rangle)(x_{t+\tau} - \langle x_t \rangle)$$

where $\tau = 0, 1, 2, \ldots$. Find the autocorrelation function.

Solution 2. (i) Since

$$\frac{df}{dx} = 4 - 8x$$

we find that the Ljapunov exponent is given by $\lambda(x_0) = \ln 2$ for almost all initial values, i.e., the periodic orbits given in problem 1 are excluded.

(ii) Using the exact solution

$$x_t = \frac{1}{2} - \frac{1}{2}\cos(2^t \cos^{-1}(1 - 2x_0))$$

we find that the autocorrelation function is given by

$$C_{xx}(\tau) = \begin{cases} \frac{1}{8} & \text{for} \quad \tau = 0 \\ 0 \text{ otherwise} \end{cases}$$

Problem 3.   (i) Let $f : [-1, 1] \mapsto [-1, 1]$ be defined by

$$f(x) := 1 - 2x^2. \tag{1}$$

Let $-1 \le a \le b \le 1$ and

$$\mu([a, b]) := \frac{1}{\pi}\int_a^b \frac{dx}{\sqrt{1 - x^2}}. \tag{2}$$

Calculate $\mu([-1, 1])$.
(ii) Show that

$$\mu(f^{-1}([a, b])) = \mu([a, b]) \tag{3}$$

where $f^{-1}([a, b])$ denotes the set S which is mapped under f to $[a, b]$, i.e., $f(S) = [a, b]$. The quantity μ is called the *invariant measure* of the map f.

Solution 3.   (i) Since

$$\int_a^b \frac{dx}{\sqrt{1 - x^2}} = \arcsin(b) - \arcsin(a)$$

and $\arcsin(1) = \pi/2$, $\arcsin(-1) = -\pi/2$ we obtain $\mu([-1, 1]) = 1$.
(ii) The inverse function f^{-1} is not globally defined. We set $f_1 : [-1, 0] \mapsto [-1, 1]$ with $f_1(x) = 1 - 2x^2$. Then $f_1^{-1} : [-1, 1] \mapsto [-1, 0]$,

$$f_1^{-1}(x) = -\sqrt{\frac{1 - x}{2}}$$

with $f_1^{-1}(-1) = -1$ and $f_1^{-1}(1) = 0$. Analogously, we set $f_2 : [0, 1] \mapsto [-1, 1]$ with $f_2(x) = 1 - 2x^2$. Then $f_2^{-1} : [-1, 1] \mapsto [0, 1]$,

$$f_2^{-1}(x) = \sqrt{\frac{1 - x}{2}}$$

with $f_2^{-1}(-1) = 1$ and $f_2^{-1}(1) = 0$. It follows that

$$f_1^{-1}(a) = -\sqrt{\frac{1 - a}{2}}, \qquad f_1^{-1}(b) = -\sqrt{\frac{1 - b}{2}}$$

$$f_2^{-1}(a) = \sqrt{\frac{1-a}{2}}, \qquad f_2^{-1}(b) = \sqrt{\frac{1-b}{2}}.$$

Thus f maps the intervals

$$\left[-\sqrt{\frac{1-a}{2}}, -\sqrt{\frac{1-b}{2}}\right], \qquad \left[\sqrt{\frac{1-b}{2}}, \sqrt{\frac{1-a}{2}}\right]$$

into the interval $[a, b]$. We notice that

$$\pi\mu([a, b]) = \arcsin(b) - \arcsin(a).$$

Furthermore $\arcsin(-x) \equiv -\arcsin(x)$. Now

$$\int_{-\sqrt{(1-a)/2}}^{-\sqrt{(1-b)/2}} \frac{dx}{\sqrt{1-x^2}} + \int_{\sqrt{(1-b)/2}}^{\sqrt{(1-a)/2}} \frac{dx}{\sqrt{1-x^2}} = \arcsin(b) - \arcsin(a).$$

Condition (3) can also be written as

$$\frac{1}{\pi}\frac{1}{\sqrt{1-x^2}} = \frac{d}{dx}\int_{f^{-1}([-1,x])} \frac{ds}{\pi\sqrt{1-s^2}}.$$

Problem 4. Let $g : [0, 1] \mapsto [0, 1]$ be defined by

$$g(x) := \begin{cases} 2x & 0 \leq x \leq 1/2 \\ 2(1-x) & 1/2 < x \leq 1 \end{cases}.$$

This map is called the *tent map*. Let $0 \leq a \leq b \leq 1$ and

$$\nu([a, b]) := \int_a^b dx.$$

(i) Show that

$$\nu(g^{-1}([a, b])) = \nu([a, b])$$

where $g^{-1}([a, b])$ is the set S which is mapped under g to $[a, b]$, i.e., $g(S) = [a, b]$.

(ii) Find the Ljapunov exponent of the tent map.

Solution 4. (i) Obviously

$$\nu([a, b]) = \int_a^b dx = b - a.$$

The set $g^{-1}([a, b])$ is given by

$$g^{-1}([a, b]) = \left[\frac{a}{2}, \frac{b}{2}\right] \cup \left[1 - \frac{b}{2}, 1 - \frac{a}{2}\right]$$

where $0 \le a \le b \le 1$. Therefore

$$\nu(g^{-1}([a,b])) = \int_{a/2}^{b/2} dx + \int_{1-b/2}^{1-a/2} dx = b - a = \nu([a,b]).$$

(ii) Since g is not differentiable at $x = 1/2$ we define the *Ljapunov exponent* as

$$\lambda(x_0) := \lim_{T \to \infty} \lim_{\epsilon \to 0} \frac{1}{T} \ln \left| \frac{g^{(T)}(x_0 + \epsilon) - g^{(T)}(x_0)}{\epsilon} \right|$$

where x_0 is the initial value and $g^{(T)}$ is the T-th iterate of the function g. Inserting the tent map into this expression we obtain

$$\lambda(x_0) = \ln 2$$

for almost all initial values, i.e., the initial conditions

$$x_0 \in \mathbb{Q} \cap [0,1]$$

are excluded, where $\mathbb{Q}$ denotes the rational numbers. Thus the Ljapunov exponent is equal to $\ln(2)$ if the initial value $x_0 \in [0,1]$ is an irrational number. If $x_0 \in \mathbb{Q} \cap [0,1]$, then we obtain a periodic orbit. For example, if $x_0 = \frac{1}{7}$, then we find

$$x_1 = \frac{2}{7}, \quad x_2 = \frac{4}{7}, \quad x_3 = \frac{6}{7}, \quad x_4 = \frac{2}{7}, \cdots .$$

Problem 5. Show that the logistic map $f(x) = 4x(1-x)$ can be transformed into the tent map given in problem 4.

Solution 5. The transformation is

$$g = \phi \circ f \circ \phi^{-1}$$

where $\phi : [0,1] \to [0,1]$ is given by

$$\phi(x) = \frac{2}{\pi} \arcsin(\sqrt{x}).$$

Here $\circ$ denotes the composition of maps. Both maps have the same Ljapunov exponent.

Problem 6. The *Cantor set* is constructed as follows: We set

$$E_0 = [0,1]$$

$$E_1 = \left[0, \frac{1}{3}\right] \cup \left[\frac{2}{3}, 1\right]$$

$$E_2 = \left[0, \frac{1}{9}\right] \cup \left[\frac{2}{9}, \frac{1}{3}\right] \cup \left[\frac{2}{3}, \frac{7}{9}\right] \cup \left[\frac{8}{9}, 1\right]$$

$$\cdots = \cdots$$

(1)

In other words: Delete the middle third of the line segment $[0,1]$ and then the middle third from all the resulting segments and so on, ad infinitum. The set defined by

$$C := \bigcap_{k=0}^{\infty} E_k \tag{2}$$

is called the *standard Cantor set* (or *Cantor ternary set*).

(i) Show that the Cantor set is of *Lebesgue measure* zero.

(ii) Show that there is a bijective mapping $f : C \mapsto [0,1]$.

(iii) Let X be a subset of $\mathbb{R}^n$. Let $N(\epsilon)$ be the number of n-dimensional cubes (boxes) of side ϵ to cover the set X. The *capacity* D of X is defined as

$$D := \lim_{\epsilon \to 0} \frac{\ln N(\epsilon)}{\ln \left(\frac{1}{\epsilon}\right)}. \tag{3}$$

Find the capacity D of the interval $I = [0,2]$. Find the capacity of the standard Cantor set C. The capacity is a so-called *fractal dimension*.

(iv) Let X be a subset of $\mathbb{R}^n$. A *cover* of the set X is a (possibly infinite) collection of balls, the union of which contains X. The diameter of a cover $\mathcal{A}$ is the maximum diameter of the balls in $\mathcal{A}$. For $d, \epsilon > 0$, we define

$$\alpha(d, \epsilon) := \inf_{\substack{\mathcal{A} = \text{cover of } X \\ \text{diam} \mathcal{A} \le \epsilon}} \sum_{A \in \mathcal{A}} (\text{diam} A)^d \tag{4}$$

and

$$\alpha(d) := \lim_{\epsilon \to 0} \alpha(d, \epsilon). \tag{5}$$

There is a unique d_0 such that

$$d < d_0 \Rightarrow \alpha(d) = \infty$$

$$d > d_0 \Rightarrow \alpha(d) = 0.$$

This d_0 is defined to be the *Hausdorff dimension* of the set X, written $HD(X)$. Show that for the standard Cantor set $HD(C) \le (\ln 2)/(\ln 3)$.

Solution 6. (i) We define

$$\chi_{E_k} := \begin{cases} 1 & x \in E_k \\ 0 & x \notin E_k. \end{cases}$$

Owing to the construction of the Cantor set the sets E_k are the union of 2^k pairwise disjoint intervals with length $1/3^k$. Thus

$$\int_0^1 \chi_{E_k} d\mu = 2^k \frac{1}{3^k} = \left(\frac{2}{3}\right)^k$$

and

$$\lim_{k \to \infty} \int_0^1 \chi_{E_k} d\mu = 0.$$

On the other hand $\chi_{E_{k+1}} \leq \chi_{E_k}$ and

$$\lim_{k \to \infty} \chi_{E_k} = \chi_C.$$

Therefore χ_C is integrable and

$$\mu(C) = \int_0^1 \chi_C d\mu = 0.$$

(ii) Every number $x \in [0, 1]$ can be represented as (*binary representation*)

$$x = \sum_{j=1}^{\infty} a_j 2^{-j}, \qquad a_j \in \{0, 1\}.$$

Every $y \in C$ can be written in the form (*ternary representation*)

$$y = \sum_{j=1}^{\infty} c_j 3^{-j}, \qquad c_j \in \{0, 2\}.$$

Let $f : C \mapsto [0, 1]$ be defined by

$$f\left(\sum_{j=1}^{\infty} c_j 3^{-j}\right) = \sum_{j=1}^{\infty} a_j 2^{-j}$$

with

$$a_j = \begin{cases} 0 \text{ for } c_j = 0 \\ 1 \text{ for } c_j = 2 \end{cases}$$

A mapping $g : A \mapsto B$ is called *surjective* if $g(A) = B$. Obviously f, given by (15), is surjective. A mapping g is called *injective* (one-to-one) when

$$\text{for all} \quad a, a' \in A \quad g(a) = g(a') \Rightarrow a = a'.$$

Since different numbers have different binary and ternary representations, the mapping f is injective. Since the mapping f is surjective and injective   -
we find that the mapping f is bijective.

(iii) Let $I = [0, 2]$. Let $\epsilon > 0$ and $N \cdot \epsilon = 2$. Then $N(\epsilon) = 2/\epsilon$. Inserting this into definition (3) we find

$$D = \lim_{\epsilon \to 0} \frac{\ln\left(\frac{2}{\epsilon}\right)}{\ln\left(\frac{1}{\epsilon}\right)} = 1.$$

To calculate the capacity of the Cantor set we note the following: In the first step of the construction we have $\epsilon = 1/3$ and $N = 2$. In the second step of the construction we have $\epsilon = 1/9$ and $N = 4$ and at the p-th step of the construction we find

$$\epsilon = \left(\frac{1}{3}\right)^p, \qquad N = 2^p.$$

Consequently, the capacity of the Cantor set is given by

$$D = \frac{\ln 2}{\ln 3}.$$

(iv) Owing to the construction of the standard Cantor set the following cover is obvious. For $n = 1, 2, \ldots$, let $\mathcal{A}_n$ be the cover of C consisting of 2^n intervals of length $1/3^n$ each. We set

$$\sum_{A \in \mathcal{A}_n} (\text{diam} A)^d = 1.$$

It follows that

$$2^n \left(\frac{1}{3^n}\right)^d = 1.$$

This leads to

$$d = \frac{\ln 2}{\ln 3}.$$

We conclude that with d given by $\ln 2 / \ln 3$ we have

$$\alpha\left(d, \frac{1}{3^n}\right) \leq \sum_{A \in \mathcal{A}_n} (\text{diam} A)^d = 1$$

for every n. Therefore

$$\alpha\left(\frac{\ln 2}{\ln 3}\right) \leq 1.$$

This implies that

$$HD(C) \leq \frac{\ln 2}{\ln 3}.$$

One can show that $HD(C) = \ln 2 / \ln 3$. Thus we have $D(C) = HD(C)$ for the ternary Cantor set. In general we have the inequality

$$HD(X) \leq D(X).$$

The capacity of a set does not distinguish between a set and its closure, while the Hausdorff dimension of the two is different. For example, the set of rationals has $HD = 0$ because it is a countable union of points, and has capacity $= 1$ because its closure $= \mathbb{R}$. Both the capacity and the Hausdorff dimension are so-called *fractal dimensions*. Another fractal dimension is the similarity dimension. The topological dimension can only take integer values. Within the topological dimension a point has dimension 0, curves have dimension 1, surfaces have dimension 2 and so on.

If $X \subset Y \subset \mathbb{R}^n$, then

$$HD(X) \leq HD(Y).$$

The Hausdorff dimension of a set does not change under smooth reparametrizations. That is, if $f : \mathbb{R}^n \to \mathbb{R}^n$ is a C^1 diffeomorphism and $X \subset \mathbb{R}^n$, then

$$HD(X) = HD(f(X)).$$

Problem 7.   (i) Show that the two-dimensional map

$$\mathbf{f} : x_{k+1} = \lambda x_k + \cos \theta_k, \qquad \theta_{k+1} = 2\theta_k \bmod (2\pi) \tag{1}$$

when $2 > \lambda > 1$, has two attractors, at $x = \pm\infty$. Show that the map has no finite attractor.
(ii) Find the *boundary of basins of attractions*.

Solution 7.   (i) There are no finite attractors because the eigenvalues of the *Jacobian matrix* of the two-dimensional map (1) are 2 and λ, where $\lambda > 1$. Thus

$$\mathbf{f}^{(n)}(x_0, \theta_0) = (x_n, \theta_n \bmod (2\pi)) \tag{2}$$

and x_n either tends to $+\infty$ or $-\infty$ as $n \to \infty$, except for the (unstable) boundary set $x = g(\theta)$, for which x_n remains finite.
(ii) To find this boundary set, we note that

$$\theta_k = 2^k \theta_0 \bmod (2\pi). \tag{3}$$

The map $\mathbf{f}$ is noninvertible, but we can select any x_n and find one orbit that ends at (x_n, θ_n), by using the above θ_k and taking

$$x_{k-1} = \lambda^{-1} x_k - \lambda^{-1} \cos(2^{k-1}\theta_0). \tag{4}$$

For the given (x_n, θ_0) we find that this orbit started at

$$x_0 = \lambda^{-n} x_n - \sum_{l=0}^{n-1} \lambda^{-l-1} \cos(2^l \theta_0).$$

The boundary between the two basins are those (x_0, θ_0) such that x_n is finite as $n \to \infty$. Thus the x and θ are related by

$$x = -\sum_{l=0}^{\infty} \lambda^{-l-1} \cos(2^l \theta) \equiv g(\theta).$$

Since $\lambda > 1$, this sum converges absolutely and uniformly. On the other hand

$$\frac{dg(\theta)}{d\theta} = \frac{1}{2} \sum_{l=0}^{\infty} (2/\lambda)^{l+1} \sin(2^l \theta)$$

and the sum diverges, since $\lambda < 2$. Hence $g(\theta)$ is nondifferentiable. The curve has a fractal dimension

$$d_c = 2 - (\ln \lambda)(\ln 2)^{-1}.$$

Problem 8. Consider the *tent map* $f : \mathbb{R} \to \mathbb{R}$ defined by

$$f(x) := \begin{cases} 3x & \text{if } x \leq \frac{1}{2} \\ 3 - 3x & \text{if } x > \frac{1}{2} \end{cases}. \tag{1}$$

Discuss the dynamics of the map f on the set Σ of all points x whose iterates stay in the unit interval $[0, 1]$. For any other point the iterates approach $-\infty$.

Solution 8. First we note that $f(x) \leq 1$ for $x \in [1/3, 2/3]$ and $f(0) = 0$, $f(1) = 0$. The fixed points are $x^* = 0$ and $x^* = 3/4$. If $x \in (1/3, 2/3)$ then the orbit escapes to $-\infty$. We show that Σ is the ternary Cantor set. To prove that Σ is the ternary Cantor set we use the representation of the Cantor set in ternary numbers. We adopt the convention that, where two alternative representations of a given number exist, we choose the form that ends with an infinite string of 2's, rather than the form that ends with a 1 followed by an infinite string of 0's. The Cantor set then consists of the numbers in $[0, 1]$ whose ternary expansions contain only 0's and 2's. Let

$$x = 0.x_1 x_2 x_3 \ldots \tag{2}$$

in ternary form and introduce the notation

$$\overline{x}_i := 2 - x_i.$$

Multiplication by 3 shifts x one digit to the left, and $3 = 2.222\ldots$ in ternary form. Hence

$$f(x) = \begin{cases} x_1.x_2 x_3 x_4 \ldots & \text{if } 0 \leq x \leq \frac{1}{2} \\ \overline{x}_1.\overline{x}_2 \overline{x}_3 \overline{x}_4 \ldots & \text{if } \frac{1}{2} < x \leq 1. \end{cases}$$

Note that $\overline{1} = 1$. Hence $f(x) \notin [0, 1]$ if $x_1 = 1$. Otherwise, since $\overline{0} = 2$ and $\overline{2} = 0$,

$$f(x) = \begin{cases} 0.x_2 x_3 x_4 \ldots & \text{if } x_1 = 0 \\ 0.\overline{x}_2 \overline{x}_3 \overline{x}_4 \ldots & \text{if } x_1 = 2. \end{cases}$$

Hence $f(x) \in [0,1]$ if and only if $x_1 = 0$ or 2. An easy inductive argument shows that for all $i \in \mathbb{N}$,

$$f^{(i)}(x) \in [0,1] \quad \text{if and only if} \quad x_i = 0 \text{ or } x_i = 2.$$

Thus $x \in \Sigma$ if and only if x is in the ternary Cantor set and so Σ is the ternary Cantor set. Hence inspection of the formula (1) shows that f maps Σ into itself.

Problem 9.   Let M be a manifold or simply $\mathbb{R}^n$. Assume the map $f : M \to M$ is continuous. μ is called a *Borel probability measure* if μ assigns to every Borel subset $E \subset M$ a certain probability or measure that we denote by $\mu(E)$. A property is said to hold a.e. (almost everywhere) if it is enjoyed by every $x \in M$ except possibly on a set E with $\mu(E) = 0$. Given a probabilistic dynamical system

$$f : (M, \mu) \to (M, \mu).$$

Such a system is said to be *ergodic* if

$$f^{-1}(E) = E \implies \mu(E) = 0 \quad \text{or} \quad \mu(E) = 1.$$

That is, an ergodic system cannot be decomposed into two nontrivial sub-systems that do not interact with each other.

Birkhoff Ergodic Theorem. For any integrable function $\phi : (M, \mu) \to \mathbb{R}$, the *time average*

$$\frac{1}{n} \sum_{i=0}^{n-1} \phi(f^{(i)}(x)) \tag{1}$$

converges for a.e. x. Denote this limit by $\phi^*(x)$. If the system is ergodic, then $\phi^*(x)$ equals a.e. the *space average*

$$\int_M \phi(x) d\mu(x). \tag{2}$$

Apply the theorem to the logistic map $f : [0,1] \to [0,1]$,

$$f(x) = 4x(1-x)$$

where

$$\phi(x) = \ln \left| \frac{df}{dx} \right|.$$

Solution 9.   It can be shown that

$$d\mu(x) = \frac{1}{\pi \sqrt{x(1-x)}} dx$$

is an ergodic invariant measure and that

$$\ln \left| \frac{df}{dx} \right|$$

is integrable. Substituting $\phi(x) = \ln |df/dx|$ in the ergodic theorem, we have for a.e. x

$$\frac{1}{n} \ln \left| \frac{f^{(n)\prime}(x)}{dx} \right| = \frac{1}{n} \ln \prod_{i=0}^{n-1} \left| f'(f^{(i)}(x)) \right| = \frac{1}{n} \sum_{i=0}^{n-1} \phi(f^{(i)}(x)).$$

For $n \to \infty$ we obtain

$$\int_0^1 \ln |4 - 8x| \frac{1}{\pi \sqrt{x(1-x)}} dx = \ln 2$$

where we used that

$$\frac{df}{dx} = 4 - 8x.$$

This says that for a.e. x, $|(f^{(n)})'(x)|$ is in the order of 2^n for large enough n. The number $\ln 2$ is called the *Ljapunov exponent*.

Problem 10. Given a compact set X embedded in $\mathbb{R}^n$. The *capacity* of X can be calculated by counting the number of grid boxes of side $\epsilon = B^{-\ell}$ that cover X. $B > 1$ and $\ell > 0$ are integers. Let $N(\ell)$ be the number of such boxes. The capacity dimension is assumed to satisfy

$$\lim_{\ell \to \infty} \frac{\ln N(\ell)}{\ell \cdot \ln B} = D_c. \tag{1}$$

The definition of the capacity dimension is

$$D_c := \overline{\lim_{\epsilon \to 0}} \frac{\ln(\bar{N}(\epsilon))}{\ln(1/\epsilon)} \tag{2}$$

where $\bar{N}(\epsilon)$ is the minimum number of boxes of side ϵ of any type that covers the set X. If the limit in the definition exists the limit in (1) will also exist. Discuss how D_c could be calculated numerically.

Solution 10. Denote by G_ℓ the set of all grid boxes of side $B^{-\ell}$. Numerical calculations of the capacity are based on the heuristic idea that the set X has finite D_c-dimensional measure V. The quantity V satisfies

$$V \approx (\bar{N}(B^{-\ell})) \cdot (B^{-\ell})^{D_c}$$

for sufficiently large ℓ. Since

$$N(B^{-\ell}) \leq N(\ell) \leq N(B^{-\ell}) \cdot c$$

for $c > 1$, a constant independent of X or ℓ, implies that

$$V \approx O(1) \cdot N(\ell) \cdot [B^{-\ell}]^{D_c}$$

where $O(1)$ holds as $\ell \to \infty$. Thus if V is constant and $O(1)$ is also assumed to be a constant then

$$\ln N(\ell) \approx \ln(\text{const} \cdot V) + \ell \cdot \ln B \cdot D_c$$

that is, for large ℓ a plot of $\ln N$ versus $\ell \cdot \ln B$ is approximately a straight line with slope D_c. The replacement of $O(1)$ by a constant introduces a possible error in the computation that results from the use of boxes to cover X.

Problem 11.   Consider the *Rössler model*

$$\frac{dx}{dt} = -y - \epsilon z$$

$$\frac{dy}{dt} = x + \epsilon r y$$

$$\frac{dz}{dt} = 1 + (x - C)z$$

where ϵ, r and C are positive constants. There is numerical evidence that the system shows chaotic behaviour for $\epsilon = 0.2$, $r = 1$ and $C = 4$. The system can be considered as a harmonic oscillator in x and y nonlinearly coupled to a third variable. Find a *Poincaré map* for the system with $\epsilon \ll 1$.

Solution 11.   For $\epsilon = 0$ we have a harmonic oscillator

$$\frac{dx}{dt} = -y, \qquad \frac{dy}{dt} = x$$

with the solution

$$x(t) = a_0 \cos(\omega_0 t), \qquad y(t) = a_0 \sin(\omega_0 t)$$

where $\omega_0 = 1$. For $\epsilon \ll 1$ we consider the expansion

$$x(t) = x_0(t) + \epsilon x_1(t) + \cdots$$
$$y(t) = y_0(t) + \epsilon y_1(t) + \cdots$$
$$z(t) = z_0(t) + \epsilon z_1(t) + \ldots$$

where the zeroth solution of x and y is a *harmonic oscillator*

$$x_0(t) = a_0 \cos(\omega t), \qquad y_0(t) = a_0 \sin(\omega t)$$

with

$$\omega = \omega_0 + \epsilon\omega_1 + \cdots .$$

We still have to find ω_1. The terms of first order are given by

$$x_1(t) = a_1(t)\cos(\omega t) + b_1(t)\sin(\omega t)$$

$$y_1(t) = a_1(t)\sin(\omega t) - b_1(t)\cos(\omega t).$$

Inserting this ansatz into (1) we find up to the zeroth order that

$$(\omega_0 - 1)x_0 = 0, \qquad (\omega_0 - 1)y_0 = 0, \qquad \frac{dz_0}{dt} = 1 + (x_0 - C)z_0 .$$

The first two equations are satisfied for $\omega_0 = 1$. The third equation can be integrated. We obtain

$$z_0(t) = \frac{z_0(0) + \int_0^t f(s)ds}{f(t)}$$

where $z_0(0)$ is the initial value of z_0 and the function f is given by

$$f(t) = \exp\left(-\int_0^t (x_0(s) - C)ds\right) = \exp\left(-\frac{a_0}{\omega}\sin(\omega t) + Ct\right).$$

Since we are only interested in the attractor, we have to find $z_0(t \to \infty)$. At time $t + n\tau$ with $0 \le t \le \tau$ the quantity z_0 is given by

$$z_0(t + n\tau) = \frac{z_0 + \int_0^{t+n\tau} f(s)ds}{f(t + n\tau)} .$$

The function f has the property

$$f(t + n\tau) = f(t)\exp(nC\tau).$$

Using this property we find

$$z_0(t + n\tau) = \frac{(z_0 - z_C(\tau))\exp(-nC\tau) + z_C(\tau) + \int_0^t f(s)ds}{f(t)}$$

where

$$z_C(\tau) = \frac{\int_0^\tau f(s)ds}{\exp(C\tau) - 1} .$$

For $n \to \infty$ the function $z_0(t + n\tau)$ approaches the asymptotic function $\tilde{z}_0(t)$

$$z_0(t + n\tau) \to \tilde{z}_0(t) = \frac{z_C(\tau) + \int_0^t f(s)ds}{f(t)} .$$

We also find $\tilde{z}_0(t + n\tau) = \tilde{z}_0(t)$. In first-order we find

$$\frac{da_1}{dt} = (\omega_1 y_0 - z_0)\cos(\omega t) + (-\omega_1 x_0 + r y_0)\sin(\omega t)$$

$$\frac{db_1}{dt} = (\omega_1 y_0 - z_0)\sin(\omega t) - (-\omega_1 x_0 + r y_0)\cos(\omega t)$$

where x_0, y_0, z_0 are given above. This system of differential equations can be integrated to give

$$a_1(\tau) = \frac{r}{2} a_0 \tau - \int_0^\tau \tilde{z}_0(t)\cos(\omega t)dt$$

$$b_1(\tau) = \omega_1 a_0 \tau - \int_0^\tau \tilde{z}_0(t)\sin(\omega t)dt$$

with the initial values $a_1(0) = b_1(0) = 0$ and

$$x(0) = a_0, \quad y(0) = 0, \quad z(0) = z_C(\tau)$$

After one period τ, we have

$$x(\tau) = a_0 + \epsilon a_1(\tau) + O(\epsilon^2) = a(\tau)$$
$$y(\tau) = -\epsilon b_1(\tau) + O(\epsilon^2) = b(\tau)$$
$$z(\tau) = z_C(\tau) + O(\epsilon).$$

The initial values a_0 and $b_0 = 0$ are mapped, after one period, into

$$a(\tau) = a_0 + \epsilon \tau \left(\frac{r}{2} a_0 - \frac{1}{\tau}\int_0^\tau \tilde{z}_0(t)\cos(\omega t)dt\right)$$

$$b(\tau) = \epsilon \tau \left(\omega_1 a_0 - \frac{1}{\tau}\int_0^\tau \tilde{z}_0(t)\sin(\omega t)dt\right).$$

This is the discrete Poincaré map.

Problem 12. Consider the one-dimensional map $f : [0, 1] \to [0, 1]$

$$x_{t+1} = f(x_t) \tag{1}$$

where $t = 0, 1, 2, \ldots$ and $x_0 \in [0, 1]$. The n-th moment of the time evolution of x_t is defined as

$$\langle x_t^n \rangle := \lim_{T \to \infty} \frac{1}{T}\sum_{t=0}^{T} x_t^n \tag{2}$$

where $n = 1, 2, \ldots$. The moments depend on the initial conditions. In the case of ergodic systems, the moments and the probability density ρ are related by

$$\langle x_t^n \rangle = \int_0^1 y^n \rho(y)dy \tag{3}$$

where $\rho > 0$ for $x \in [0,1]$ and

$$\int_0^1 \rho(x)dx = 1.\qquad(4)$$

We can calculate the Ljapunov exponent as follows

$$\lambda = \int_0^1 \rho(x)\ln\left|\frac{df}{dx}\right|dx.\qquad(5)$$

The missing information function (entropy) of a probability density ρ is defined as

$$I = -\int_0^1 \rho(x)\ln\rho(x)\,dx.\qquad(6)$$

Apply the *maximum entropy formalism* to obtain the probability density approximately using as information, N moments. Apply it to the logistic map $f(x) = 4x(1-x)$ with $N = 2$.

Solution 12. In the maximum entropy formalism, one maximizes the missing information subject to the constraints of the available information and to the normalization of the probability density. We assume that we have the N lowest moments. The constraints are introduced via the method of the *Lagrange multipliers* $\lambda_1, \lambda_2, \ldots, \lambda_N$. Our aim is to find the approximate probability density ρ_{app} which minimizes

$$I' = -\int_0^1 \rho_{app}\ln\rho_{app}\,dx + \lambda_0\left(1 - \int_0^1 \rho_{app}dx\right) + \sum_{n=1}^N \lambda_n\left(\langle x^n\rangle - \int_0^1 x^n\rho_{app}dx\right)$$

where λ_n, $n = 0,1,2,\ldots,N$ are the Lagrange multipliers. Performing the minimization we obtain

$$\rho_{app}(x) = \exp\left(-1 - \sum_{n=0}^N \lambda_n x^n\right) \equiv \frac{1}{Z}\exp\left(-\sum_{n=1}^N \lambda_n x^n\right)$$

where $Z = \exp(1 + \lambda_0)$. From the normalization condition, we obtain

$$1 = \frac{1}{Z}\int_0^1 \exp\left(-\sum_{n=1}^N \lambda_n x^n\right)dx.$$

The remaining Lagrange multipliers are obtained by solving the following set of N coupled nonlinear equations for λ_m, $m = 1,2,\ldots,N$,

$$\langle x^m\rangle = \frac{1}{Z}\int_0^1 x^m \exp\left(-\sum_{n=1}^N \lambda_n x^n\right)dx, \qquad m = 1,2,\ldots,N.$$

For the logistic map $f(x) = 4x(1-x)$, the moments are given by

$$\langle x^n \rangle = \frac{1}{2^{2n}} \binom{2n}{n}.$$

Thus $\langle x \rangle = 1/2$ and $\langle x^2 \rangle = 3/8$. Thus we have to solve

$$1 = \int_0^1 \exp(-1 - \lambda_0 - \lambda_1 x - \lambda_2 x^2) dx$$

$$\frac{1}{2} = \int_0^1 \exp(-1 - \lambda_0 - \lambda_1 x - \lambda_2 x^2) dx$$

$$\frac{3}{8} = \int_0^1 \exp(-1 - \lambda_0 - \lambda_1 x - \lambda_2 x^2) dx.$$

We solve this system numerically and find

$$\lambda_0 = 2.69242, \qquad \lambda_1 = -6.76825, \qquad \lambda_2 = -\lambda_1.$$

Obviously $\lambda_1 = -\lambda_2$.

Problem 13.   The *Anosov map* is defined as follows: $\Omega = [0,1)^2$,

$$\phi(x, y) = (x + y, x + 2y).$$

In matrix form we have

$$\begin{pmatrix} x \\ y \end{pmatrix} \mapsto \begin{pmatrix} 1 & 1 \\ 1 & 2 \end{pmatrix} \begin{pmatrix} x \\ y \end{pmatrix} \quad \mod 1.$$

(i) Show that the map preserves Lebesgue measure.
(ii) Show that ϕ is invertible. Show that the entire sequence can be recovered from one term.
(iii) Show that ϕ is mixing.

Solution 13.   (i) Since the *Jacobian determinant* of the matrix of the map is equal to one, i.e.

$$\det \begin{pmatrix} 1 & 1 \\ 1 & 2 \end{pmatrix} = 1$$

the map ϕ preserves the Lebesgue measure.
(ii) The inverse of the matrix which belongs to the Lie group $SL(2, \mathbb{R})$

$$\begin{pmatrix} 1 & 1 \\ 1 & 2 \end{pmatrix}$$

is given by

$$\begin{pmatrix} 2 & -1 \\ -1 & 1 \end{pmatrix}.$$

Since ϕ is invertible, the entire sequence can be recovered from one term. Thus ϕ maps Ω $1-1$ onto itself.

(iii) We observe that the n-th iterate is given by

$$\phi^{(n)}(x, y) = (a_{2n-2}x + a_{2n-1}y, a_{2n-1}x + a_{2n}y)$$

where the a_n are the *Fibonacci numbers* given by $a_0 = a_1 = 1$ and $a_{n+1} = a_n + a_{n-1}$ for $n \geq 1$. To check this notice that

$$a_{2n-2} + 2a_{2n-1} + a_{2n} = a_{2n} + a_{2n+1}.$$

Consider the Hilbert space $L_2([0, 1) \times [0, 1))$. Let f and g be two elements of this Hilbert space defined by

$$f(x, y) := \exp(2\pi i(px + qy)) \equiv \exp(2\pi ipx)\exp(2\pi iqy)$$
$$g(x, y) := \exp(2\pi i(rx + sy)) \equiv \exp(2\pi irx)\exp(2\pi isy)$$

where $p, q, r, s \in \mathbb{Z}$. Since $(k \in \mathbb{Z})$

$$\int_0^1 \exp(2\pi ikx)dx = 0$$

unless $k = 0$. It follows that

$$\int_0^1 \int_0^1 f(x, y)g(\phi^{(n)}(x, y))dxdy = 0$$

unless

$$ra_{2n-2} + sa_{2n-1} + p = 0$$
$$ra_{2n-1} + sa_{2n} + q = 0.$$

Now the difference equation $b_{n+1} - b_n - b_{n-1} = 0$ has a two-parameter family of solutions given by

$$b_n = C_1\left(\frac{1 + \sqrt{5}}{2}\right)^n + C_2\left(\frac{1 - \sqrt{5}}{2}\right)^n$$

so the *Fibonacci numbers* are given by

$$a_n = \frac{1}{2}\left(\frac{1 + \sqrt{5}}{2}\right)^n + \frac{1}{2}\left(\frac{1 - \sqrt{5}}{2}\right)^n \quad \text{for } n \geq 0.$$

From the last formula we see that

$$\lim_{n \to \infty} a_{2n}/a_{2n-1} = \lim_{n \to \infty} a_{2n-1}/a_{2n-2} = \frac{1 + \sqrt{5}}{2}.$$

One sees that the two equations for a_n cannot hold for infinitely many n unless $p = q = r = s = 0$. The last result implies that if

$$f(x, y) = \sum_{j=1}^{k} a_j \exp(2\pi i(p_j x + q_j y))$$

$$g(x, y) = \sum_{j=1}^{l} b_j \exp(2\pi i(r_j x + s_j y)).$$

Then as $n \to \infty$,

$$\int_0^1 \int_0^1 f(x, y) g(\phi^{(n)}(x, y)) dx dy \to 0.$$

Since the f and g for which the last equation holds are dense in

$$L_0^2([0, 1]^2) := \left\{ f : \int_0^1 \int_0^1 f(x, y) dx dy = 0, \quad \int_0^1 \int_0^1 f^2(x, y) dx dy < \infty \right\}$$

it follows that the map ϕ is mixing.

Problem 14.   Let $\mathbb{C}$ be the complex plane. Let $c \in \mathbb{C}$. The Mandelbrot set M is defined as follows

$$M := \{ c \in \mathbb{C} : c, c^2 + c, (c^2 + c)^2 + c, \ldots \nrightarrow \infty \}. \tag{1}$$

(i) Show that to find the Mandelbrot set one has to study the recursion relation

$$z_{t+1} = z_t^2 + c \tag{2}$$

where $t = 0, 1, 2, \ldots$ and $z_0 = 0$.
(ii) Write the recursion relation in terms of real and imaginary parts. For a given $c \in \mathbb{C}$ (or $(c_1, c_2) \in \mathbb{R}^2$) we can now study whether or not c belongs to M.
(iii) Show that $(c_1, c_2) = (0, 0)$ belongs to M.

Solution 14.   (i) This is obvious since

$$z_1 = c, \quad z_2 = c^2 + c, \quad z_3 = (c^2 + c)^2 + c \tag{3}$$

etc..
(ii) Using $z = x + iy$ and $c = c_1 + ic_2$ with $x, y, c_1, c_2 \in \mathbb{R}$ we can write (2) as

$$x_{t+1} = x_t^2 - y_t^2 + c_1, \qquad y_{t+1} = 2x_t y_t + c_2$$

(iii) With the initial value $(x_0, y_0) = (0, 0)$ and $(c_1, c_2) = (0, 0)$ we obtain $(x_1, y_1) = (0, 0)$. Therefore $z_t = 0$ for all t. The Mandelbrot set lies within $|c| < 2$. For $|c| > 2$ the sequence diverges.

Problem 15. Describe the filled Julia set for the map $f : \mathbb{C} \to \mathbb{C}$, $f(z) = z^3$.

Solution 15. The filled Julia set for $f(z) = z^3$ is the closed unit disk. We have

$$f(z) = z^3, \quad f^{(2)} = z^9, \quad \ldots, \quad f^{(n)} = z^{3^n}.$$

Now we have

$$z^k = r^k(\cos(k\theta) + i\sin(k\theta)) = r^k \exp(ik\theta)$$

where $r = |z|$ and $\tan(\theta) = \Im z / \Re z$. Thus we have

$$f^{(n)}(z) = z^{3^n} = r^{3^n} \exp(i 3^n \theta).$$

It follows that $|z| = 1$ implies that $|f(z)| = 1$, that is the unit circle is invariant, since

$$|f(z)| = |z^3| = |z|^3.$$

Thus if $|z| = 1$ we have $|f(z)| = 1$.

Problem 16. Given a binary string

$$\text{"100001010110010101110} \cdots \text{10111"}$$

of finite length n. Let A^* denote the set of all finite-length sequences (strings) over a finite alphabet A (in our case $\{0, 1\}$). The quantity $S(i, j)$ denotes the substring $S(i, j) := s_i s_{i+1} \cdots s_j$. A *vocabulary* of a string S, denoted by $v(S)$, is the subset of A^* formed by all the substrings, or words, $S(i, j)$ of S. The *complexity*, in the sense of Lempel and Ziv, of a finite string is evaluated from the point of view of a simple self-delimiting learning machine which, as it scans a given n digit string $S = s_1 s_2 \cdots s_n$ from left to right, adds a new word to its memory every time it discovers a substring of consecutive digits not previously encountered. Thus the calculation of the complexity $c(n)$ proceeds as follows. Let us assume that a given string $s_1 s_2 \cdots s_n$ has been reconstructed by the program up to the digit s_r and that s_r has been newly inserted, i.e., it was not obtained by simply copying it from $s_1 s_2 \cdots s_{r-1}$. The string up to s_r will be denoted by $R := s_1 s_2 \cdots s_r \circ$, where the $\circ$ indicates that s_r is newly inserted. In order to check whether the rest of R, i.e., $s_{r+1} s_{r+2} \cdots s_n$ can be reconstructed by simple copying or whether one has to insert new digits, we proceed as follows: First, one takes

$Q \equiv s_{r+1}$ and asks whether this term is contained in the vocabulary of the string R so that Q can simply be obtained by copying a word from R. This is equivalent to the question of whether Q is contained in the vocabulary $v(RQ\pi)$ of $RQ\pi$ where $RQ\pi$ denotes the string which is composed of R and Q (concatenation) and π means that the last digit has to be deleted. This can be generalized to situations where Q also contains two (i.e., $Q = s_{r+1}s_{r+2}$) or more elements. Let us assume that s_{r+1} can be copied from the vocabulary of R. Then we next ask whether $Q = s_{r+1}s_{r+2}$ is contained in the vocabulary of $RQ\pi$ and so on, until Q becomes so large that it can no longer be obtained by copying a word from $v(RQ\pi)$ and one has to insert a new digit. The number c of production steps to create the string S, i.e., the number of newly inserted digits (plus one if the last copy step is not followed by inserting a digit), is used as a measure of the complexity of a given string.

(i) Find the complexity of a string which contains only zeros.

(ii) Find the complexity of a string which is only composed of units of 01, i.e.,

$$01010101\cdots 01$$

(iii) Find the complexity of the string 0010.

(iv) Write a C++ program that finds the complexity $c(n)$ for a given binary string of length n.

Solution 16.   (i) If we have a sequence which contains only zeros we could say that it should have the smallest possible complexity of all strings (equivalent to a string consisting only of 1's). One has only to insert the first zero and can then reconstruct the whole string by copying this digit, i.e.,

$$00000\cdots \rightarrow 0 \circ 000\cdots .$$

Thus the complexity of this string is $c = 2$.

(ii) Similarly one finds for a sequence which is only composed of units 01, i.e.,

$$010101\cdots 01 \rightarrow 0 \circ 1 \circ 0101\cdots 01$$

the value $c = 3$.

(iii) The complexity c of the string $S = 0010$ can be determined as follows:

(1) The first digit has always to be inserted $\rightarrow 0\circ$

(2) $R = 0$, $Q = 0$, $RQ = 00$, $RQ\pi = 0$, $Q \in v(RQ\pi) \rightarrow 0 \circ 0$

(3) $R = 0$, $Q = 01$, $RQ = 001$, $RQ\pi = 00$, $Q \notin v(RQ\pi) \rightarrow 0 \circ 01\circ$

(4) $R = 001$, $Q = 0$, $RQ = 0010$, $RQ\pi = 001$, $Q \in v(RQ\pi) \rightarrow 0 \circ 01 \circ 0$.

Now c is equal to the number of parts of the string that are separated by $\circ$, i.e., $c = 3$.

The set of the rational numbers in the interval $[0, 1]$ are of Lebesgue measure zero. This implies that for almost all numbers in $[0, 1]$ (i.e., for all irrationals) the string of zeros and ones which represents their binary representation is not periodic. Therefore we expect that almost all strings which correspond to a binary representation of a number $x \in [0, 1]$ should be random and have maximal complexity. For almost all $x \in [0, 1]$ the complexity $c(n)$ (of the string which represents the binary representation) tends to the same value, namely

$$\lim_{n \to \infty} c(n) = b(n) \equiv \frac{n}{\log_2 n}.$$

$b(n)$ gives, therefore, the asymptotic behaviour of $c(n)$ for a random string. One often normalizes $c(n)$ via this limit.

(iv) A computer program in C++ for finding the complexity uses a do-while loop, a for loop and a while loop as follows:

```cpp
// ziv.cpp

#include <iostream>    // for cout, cin
#include <string>
using namespace std;

long complexity(const string& s)   // s: string
{
    long c=1, l=1, n=s.length(), i;
    do
    {
        long k=0, kmax=1;
        for(i=0;i<l;i++)
        {
        while(s[i+k]==s[l+k])
        {
        ++k;
        if(l+k >= n-1) return (++c);
        }
        if(k >= kmax) kmax=k+1; k=0;
        }
        ++c; l += kmax;
    } while(l < n);
    return c;
}

int main(void)
```

```
{
    string s;
    cout << "enter string: ";
    cin >> s;
    cout << "complexity of the string: " << s
         << " is " << complexity(s) << endl;
    return 0;
}
```

Bibliography

Becker K., Becker M. and Schwarz J. H., *String Theory and M-Theory: A Modern Introduction*, Cambridge University Press (2007)

Constantinescu F. and Magyari E., *Problems in Quantum Mechanics*, Pergamon Press, Oxford (1971)

Flügge S., *Practical Quantum Mechanics*, Springer, Berlin (1974)

Hall B. C., *Lie Groups, Lie Algebras and Representations*, Springer, New York (2004)

Hardy Y., Kiat Shi Tan and Steeb W.-H., *Computer Algebra with SymbolicC++*, World Scientific, Singapore (2008)

Helgason S., *Differential Geometry, Lie Groups and Symmetric Spaces*, American Mathematical Society (2001)

Hirota R., *The Direct Method in Soliton Theory*, Cambridge University Press (2004)

Knapp A. W., *Lie Groups: Beyond an Introduction*, 2nd edition, Birkhäuser, Boston 2002

Kriele M., *Spacetime: Foundations of General Relativity and Differential Geometry*, Springer (2001)

Louisell W. H., *Quantum Statistical Properties of Radiation*, Wiley, New York (1973)

Manteanu L. and Donescu S., *Introduction to Soliton Theory: Applications to Mechanics*, Springer (2004)

Polyanin A. D. and Zaitsev V. F., *Handbook of Nonlinear Partial Differential Equations*, Chapman and Hall/CRC, (2003)

Procesi C., *Lie Groups: An Approach through Invariants and Representations*, Springer (2009)

Rassias J. M., *Counter Examples in Differential Equations and Related Topics*, World Scientific, Singapore (1991)

Spiegel M. R., *Advanced Calculus*, Schaum's Outline Series, McGraw Hill, New York (1974)

Spiegel M. R., *Finite Differences and Difference Equations*, Schaum's Outline Series, McGraw Hill, New York (1971)

Spiegel M. R., *Complex Variables*, Schaum's Outline Series, McGraw Hill, New York (1971)

Steeb W.-H., *Matrix Calculus and Kronecker Product with Applications and C++ Programs*, World Scientific, Singapore (1997)

Steeb W.-H., *Problems and Solutions in Introductory and Advanced Matrix Calculus*, World Scientific, Singapore (2006)

Steeb W.-H. and Hardy Y. *Problems and Solutions in Quantum Computing and Quantum Information*, World Scientific, Singapore (2006)

Steeb W.-H., *The Nonlinear Workbook*, 4th edition, World Scientific, Singapore (2008)

Steeb W.-H., *Continuous Symmetries, Lie Algebras Differential Equations and Computer Algebra*, 2nd edition, World Scientific, Singapore (2007)

Tai-Pei Cheng and Ling-Fong Li, *Gauge Theory of Elementary Particle Physics: Problems and Solutions*, Oxford University Press, USA (2000)

Tomescu I., *Problems in Combinatorics and Graph Theory*, Wiley, New York (1985)

Zeidler E., *Quantum Field Theory II: Quantum Electrodynamics: A Bridge between Mathematicians and Physicists*, Springer, Heidelberg (2008)

Index